physica status solidi (b)

ISSN 0370-1972 * VOL. 131 · NO. 2 · OCTOBER 1985

AKADEMIE-VERLAG · BERLIN

Classification Scheme

1. Structure of Crystalline Solids
   1.1 Perfectly Periodic Structure
   1.2 Solid-State Phase Transformations
   1.3 Alloys. Metallurgy
   1.4 Microstructure (Magnetic Domains See 18; Ferroelectric Domains See 14.4.1)
   1.5 Films
   1.6 Surfaces
2. Non-Crystalline State
3. Crystal Growth
4. Bonding Properties
5. Mössbauer Spectroscopy
6. Lattice Dynamics. Phonons
7. Acoustic Properties
8. Thermal Properties
9. Diffusion
10. Defect Properties (Irradiation Defects See 11)
    10.1 Metals
    10.2 Non-Metals
11. Irradiation Effects (X-Ray Diffraction Investigations See 1 and 10)
12. Mechanical Properties (Plastic Deformations See 10)
    12.1 Metals
    12.2 Non-Metals
13. Electron States
    13.1 Band Structure
    13.2 Fermi Surfaces
    13.3 Surface and Interface States
    13.4 Impurity and Defect States
    13.5 Elementary Excitations (Phonons See 6)
        13.5.1 Excitons
        13.5.2 Plasmons
        13.5.3 Polarons
        13.5.4 Magnons
14. Electrical Properties. Transport Phenomena
    14.1 Metals. Semi-Metals
    14.2 Superconductivity. Superconducting Materials and Devices
    14.3 Semiconductors
        14.3.1 Films
        14.3.2 Surfaces and Interfaces
        14.3.3 Devices. Junctions (Contact Problems See 14.3.4)
        14.3.4 High-Field Phenomena, Space-Charge Effects, Inhomogeneities, Injected Carriers (Electroluminescence See 20.3; Junctions See 14.3.3)
    14.4 Dielectrics
        14.4.1 Ferroelectrics
15. Thermoelectric and Thermomagnetic Properties
16. Photoconductivity. Photovoltaic Effects
17. Emission of Electrons and Ions
    17.1 Field Emission Microscope Investigations
18. Magnetic Properties
    18.1 Paramagnetic Properties
    18.2 Ferromagnetic Properties
        18.2.1 Ferromagnetic Films
    18.3 Ferrimagnetic Properties
    18.4 Antiferromagnetic Properties

(Continued on cover three)

physica status solidi (b)

basic research

Volume 131 · Number 2 · Pages 401 to 784, K97 to K184, and A9 to A16

October 1, 1985

AKADEMIE-VERLAG · BERLIN

Subscriptions and orders for single copies should be directed
in the GDR:

to the Postzeitungsvertrieb or
to the Akademie-Verlag Berlin, DDR-1086 Berlin, Leipziger Str. 3–4;

in the other socialist countries:

to a book-shop for foreign language literature or
to the competent news-distributing agency;

in the FRG and BERLIN (WEST):

to a book-shop or
to the wholesale distributing agency Kunst und Wissen, Erich Bieber, OHG, D-7000 Stuttgart 1, Wilhelmstr. 4–6;

in the other Western European countries:

to Kunst und Wissen, Erich Bieber GmbH, CH-8008 Zürich, Dufourstr. 51;

in USA and CANADA:

to VCH Publishers, Inc., 303 NW 12th Avenue, Deerfield Beach, FL 33442–1705, USA;

in other countries:

to the international book- and journal-selling trade,
to Buchexport, Volkseigener Außenhandelsbetrieb der Deutschen Demokratischen Republik, DDR-7010 Leipzig, Postfach 160, or
to the Akademie-Verlag Berlin, DDR-1086 Berlin, Leipziger Str. 3–4.

Editorial Note:

"physica status solidi (b)" undertakes that an original paper *accepted for publication* before the 8th of any month will be published within 50 days of this date unless the author requests a postponement. In special cases there may be some delay between *receipt* and *acceptance* of a paper due to the review and, if necessary, revision of the paper.

Schriftleiter und verantwortlich für den Inhalt:
Professor Dr. Dr. h. c. P. Görlich, DDR-1086 Berlin, Leipziger Str. 3–4, Postfach 1233 bzw. DDR-6900 Jena, Schillbachstr. 24.

Verlag:
Akademie-Verlag Berlin, DDR-1086 Berlin, Leipziger Str. 3–4; Fernruf: 2236221 und 2236229; Telex-Nr.: 114420; Bank: Staatsbank der DDR, Berlin, Kto.-Nr.: 6836-26-20712.

Chefredakteur:
Dr. H.-J. Hänsch.

Redaktionskollegium:
Prof. Dr. E. Gutsche, Dr. H.-J. Hänsch, Dr. H. Lange, Dr. S. Oberländer.

Anschrift der Redaktion:
DDR-1086 Berlin, Leipziger Str. 3–4, Postfach 1233; Fernruf: 2236279.

Veröffentlicht unter der Lizenznummer 1310 des Presseamtes beim Vorsitzenden des Ministerrates der Deutschen Demokatischen Republik.

Gesamtherstellung:
VEB Druckerei „Thomas Müntzer", DDR-5820 Bad Langensalza.

Erscheinungsweise:
Die Zeitschrift „physica status solidi (b)" erscheint jeweils am 1. eines jeden Monats. Jährlich erscheinen 6 Bände zu je 2 Heften.

Bezugspreis:
Je Band 200,– M zuzüglich Versandspesen (Preis für die DDR: 130,– M).

Bestellnummer dieses Bandes: 1068/131.

Printed in the German Democratic Republic.

AN (EDV) 20831

THIS ISSUE

IS DEDICATED TO

PROFESSOR Dr. Dr. h. c. Dr. E. h. PAUL GÖRLICH

IN HONOUR OF

HIS 80th BIRTHDAY

On the 7th October 1985 the Editor-in-Chief of this journal, Professor Paul Görlich, celebrates his 80th birthday.

For more than 5 decades of his life he has successfully worked as research scientist and scientific manager in industry, as university teacher and member of national and international committees, as well as editor of numerous scientific journals. Professor Görlich has been associated with the journal "physica status solidi" for 25 years. He is one of the co-founders of this journal in 1960 and was appointed as Editor-in-Chief in 1962. The successful progress of "physica status solidi" and its international reputation are largely due to his personal involvement and his untiring activity. Inspite of health problems he is still fully involved in the continuing progress of the journal. In addition, he is editor or coeditor of the journals "Infrared Physics", "Experimentelle Technik der Physik", "Astronomische Nachrichten", "Crystal Research and Technology", and "Feingerätetechnik" and the series "Technical-Physical Monographs". He continues to be active even now as chairman of the Subcommittee on Photon Detectors of the IMEKO, as member of the Board of the Physical Society of the GDR, of the German Society for Measuring Techniques and Automatization, and of the Society for Photogrammetry as well as of the National Committee for Physics and of the Research Council of the GDR. He is a member of the Academy of Sciences of the GDR, of the German Academy of Scientists Leopoldina in Halle, of the Saxonian Academy of Sciences in Leipzig, and of the Royal Microscopical Society (London and Oxford). Professor Görlich also performed numerous additional duties at national and international level, and many honours and high decorations have been bestowed on him. All his admirers find it astonishing how he manages to combine so effectively his many-sided and diverse activities.

He acquired an international reputation early in his scientific career, following his dissertation "The External Photoelectric Effect in Liquids", as director of the laboratory of the firm Zeiss Ikon, Dresden, and through his invention of the cesium anti-

mony alloy cathode (1936) and his thesis on photoemission and secondary emission (1942).

During his activities in the Soviet Union from 1946 to 1952, where he contributed to the reconstruction of the optical industry, he edited a two-volume monograph on photo effects. This monograph was then extended to form a three-volume edition (1962 to 1966) and has since become an international standard work in this field.

The majority of his more than 200 publications on problems of radiation detectors, solid state and laser physics, nonlinear optics, spectroscopy and the construction of scientific instruments appeared when he was scientific chief and director of the Zeiss Research Centre. At the same time he was appointed Professor of Solid State Physics at the Friedrich-Schiller-University of Jena (1954) and was a member for many years of the Senate Council and Faculty of this University. These appointments were fully in line with the original intentions of Ernst Abbe regarding the development of good relations between industrial research on the one hand and University research and teaching on the other.

While a member of the Academy of Sciences of the GDR, and especially as Director at the Institute of Optics and Spectroscopy in Berlin, he recognized the importance of fundamental research at the Academy Institutes for the rapid development of the field of construction of scientific instruments. He has also been responsible for the long-term cooperation between the Academy and industry.

All this demonstrates Professor Görlich's strength and creative personality. However, his human greatness and cordiality, which has also decisively influenced his pupils and coworkers, surpasses even his extraordinary creativity, which it helped to establish. Thus he has become a genuine friend of many people all over the world. May he be blessed with health and may this good friend continue to be with us for a long time.

phys. stat. sol. (b) **131**, No. 2 (1985)

Contents

Review Article

Original Papers and Short Notes

Structure

Lattice properties

Extended electronic states and transitions

Localized electronic states and transitions

Electric transport

Pre-Printed Titles

Systematic List

Subject classification: Corresponding papers begin on the following pages (pages given in italics refer to the principle subject classification):

The Author Index of Volume 131 Begins on Page 785

(It will be delivered together with Volume 132, Number 1.)

Review Article

phys. stat. sol. (b) **131**, 415 (1985)

Subject classification: 13.1 and 20.1; 22; 22.2.1; 22.2.3

Upper Colwall, Malvern[1])

Relations between the Refractive Index and Energy Gap of Semiconductors

By

T. S. Moss

Dedicated to Prof. Dr. Dr. h. c. Dr. E. h. P. GÖRLICH on the occasion of his 80th birthday

Contents

1. Introduction

The two most interesting optical properties of a semiconductor are the absorption edge, or optical energy gap, and the refractive index. It is therefore natural that attempts should have been made to find a general relationship between these parameters, both from the point of view of fundamental interest and also as a technological aid in estimating the refractive index if only the energy gap is known.

---

[1]) Malvern, Worcestershire WR 13 6PX, Great Britain.

The first proposal was made by the author in 1950 [1, 2] on the very general grounds that all energy levels in a solid are scaled down by a factor $1/\varepsilon_{\mathrm{opt}}^2$, where $\varepsilon_{\mathrm{opt}} = n^2$ is the optical dielectric constant. The data on the then known materials were correlated and as a result the so-called "Moss relation"

$$n_0^4/\lambda_g = 77\ \mu\mathrm{m}^{-1} \tag{1a}$$

was proposed. In terms of the energy gap E_g this is equivalent to

$$n_0^4 E_g = 95\ \mathrm{eV}\,. \tag{1b}$$

Recently there has been renewed interest in this topic, especially by Ravindra and his co-workers [3 to 10]. From an analysis of modern data, Ravindra and Srivastava [4] proposed a revised value of the constant

$$n_0^4 E_g = 108\ \mathrm{eV}\,. \tag{1c}$$

From a study of a wide range of materials, Ravindra et al. [3, 5] have proposed an alternative empirical relationship of linear form, namely

$$n_0 = 4.084 - 0.62 E_g\,. \tag{2}$$

In the following sections, the fundamental interdependence of the optical constants will first be considered, and this will be seen to lead inevitably to the absorption behaviour in the UV region being the key to the refractive index, and to its dependence upon environmental parameters. Qualitative and quantitative comparison of the two relations will then be made, followed by a discussion of the variation of n and E_g with environmental parameters.

2. Fundamental Dependence of Index on Absorption

The optical constants of a material are not independent but are integrally related. The inter-relation between the real and imaginary parts of the dielectric constant was first given by Kramers [11] and Kronig [12] in 1929/1931 but the most useful inter-relation, namely that between n and K, was derived by Moss [13] viz:

$$n_a - 1 = \frac{1}{2\pi^2}\int_0^\infty \frac{K\,\mathrm{d}\lambda}{1 - \lambda^2/\lambda_a^2}, \tag{3}$$

which for the long wavelength, zero frequency, index becomes

$$n_0 - 1 = \frac{1}{2\pi^2}\int_0^\infty K\,\mathrm{d}\lambda\,. \tag{4}$$

In the present context, a significant feature of this function is the fact that it is integrated over *wavelength*, not energy. Thus, if for example a given level of absorption persists over a given *energy* interval — as occurs where absorption is due to transitions between two allowed energy bands — then the smaller the width of the forbidden zone the greater will be the spread of absorption in *wavelength*, resulting in a greater value of the integral and a larger n_0.

Taking a simple, hypothetical absorption spectrum which has constant K absorption at a level K_0 from essentially zero wavelength to the band-edge at λ_g, we would have

$$2\pi^2(n_0 - 1) = \int_0^{\lambda_g} K_0\,\mathrm{d}\lambda = K_0\lambda_g\,, \tag{5}$$

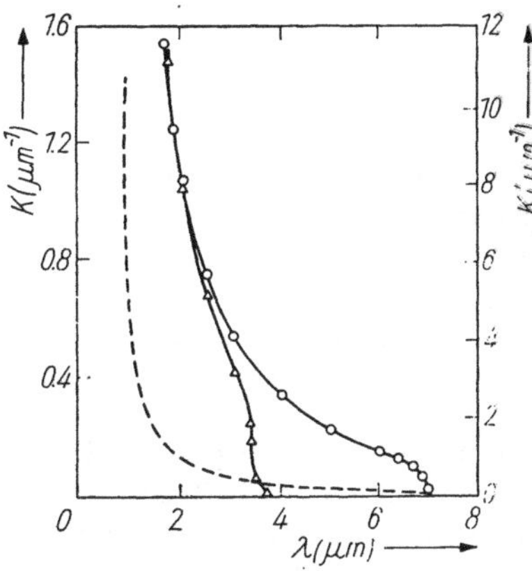

Fig. 1. Absorption constant of InSb (○) and InAs (△). The dashed (InSb) curve uses right-hand scale

i.e. the index would be proportional to the wavelength of the absorption edge. Thus this highly idealised example shows how a strong dependence of the index on the long wavelength edge could arise — in fact this example shows a much stronger dependence than the Moss relation (1) i.e. $n_0 \sim \lambda_g^{1/4}$.

We will now consider the contribution of real absorption spectra to the refractive index, taking as an example InSb, where the absorption has been measured accurately over a wide wavelength range, viz. from 1.5 μm to beyond the edge at 7 μm [14]. The results are plotted in Fig. 1, together with results from InAs, to the same scale [15]. From the difference in area under the two curves we can calculate the extra contribution to the refractive index arising from the absorption in the InSb tail-band beyond ≈ 3.4 μm. We find

$$\int_{\text{InAs}}^{\text{InSb}} K \, d\lambda = 1.1$$

so that the contribution to the refractive index is 0.055. Now the measured indices are: InSb, 3.96; InAs, 3.44 giving a measured $\Delta n = 0.52$.

Hence the contribution to the index from the extra, long wavelength, absorption in InSb is barely 10% of the actual increase in index. Noting that we are now dealing with precise theoretical inter-relations, the calculated Δn can be considered quite accurate — as are also the measured indices.

We are thus forced to conclude that the InSb tail-band absorption i.e. that lying beyond the long wavelength edge of InAs, is not a significant factor in determining the high index of this material.

This analysis of the InSb absorption spectrum can be extended to shorter wavelengths. For example, the contribution from the whole spectrum from 1 to 8 μm (shown by the dashed curve of Fig. 1) is

$$\int_1^8 K \, d\lambda = 4.45$$

giving a contribution to $n_0 - 1$ of only 0.22. (It may be noted that this is approximately the excess absorption tail of InSb relative to GaAs, in contrast to the actual difference in indices of $\Delta n = 0.66$.)

The clear implication of this analysis is that the majority of the refractive index arises from a relatively narrow band of intense absorption at wavelengths well below 1 μm, i.e. at very short wavelengths compared with the absorption edge.

3. Classical Oscillator Theory

As a model to provide such resonant absorption, we may consider the simple classical oscillator. For this, the optical constants are given by

$$n^2 - k^2 - 1 = \frac{(\omega_0^2 - \omega^2) \, Ne^2 f/m\varepsilon_0}{(\omega_0^2 - \omega^2)^2 + \gamma^2\omega^2}, \tag{6}$$

which for $\omega = 0$ gives

$$n_0^2 - 1 = Ne^2 f/m\varepsilon_0\omega_0^2 , \tag{7}$$

where N is the density of oscillators and f the oscillator strength, i.e. essentially the number of valence electrons per atom, ω_0 is the angular frequency at resonance and ε_0 is the dielectric constant of free space. Note that the index does not depend on the oscillator bandwidth (γ) but only on the resonant frequency.

The fact that this simple classical oscillator model could give a good quantitative explanation of the refractive index of Ge was pointed out by the author in 1952 [2] (p. 21).

From analysis of UV reflection spectra, Phillip and Ehrenreich [16] showed that, for Ge there is a strong peak in the absorption spectrum at 4.3 eV, or 0.29 μm. Hence, taking $f = 4$ electrons/atom, i.e. assuming any possible contribution from inner shell electrons is negligible, we calculate: $n_0^2 - 1 = 13.6$ or $n_0 = 3.84$. Considering that the UV spectrum has considerable structure — although the main peak is well-pronounced — this is very satisfactory agreement with the experimental value. Similarly, for the very high index material PbTe, the measured peak of reflectivity is at 2.4 eV [17]. The *absorption* peak will be at almost the same energy (actually a few per-cent less) so

$$\omega_0 = 3.5 \times 10^{15} \text{ and } f = 2 + 6 \text{ electrons/molecule,}$$

giving

$$n_0^2 - 1 = 30.6 \quad \text{or} \quad n_0 = 5.62 ,$$

again very close to the known value.

Now we may write (7) as

$$\varepsilon_{\text{opt}} - 1 = n_0^2 - 1 = (\hbar\omega_p/E_0)^2 , \tag{8}$$

where $\omega_p^2 = Nfe^2/m\varepsilon_0$ is the valence electron plasma frequency and $E_0 = \hbar\omega_0$ is the energy of the UV resonance.

In 1962, Penn [18] gave a modified expression for the refractive index viz.

$$n_0^2 - 1 = (\hbar\omega_p/E_0)^2 S_0 , \tag{9}$$

where the factor

$$S_0 = 1 - E_0/4E_f + \tfrac{1}{3} (E_0/4E_f)^2 . \tag{10}$$

For most materials there is little nett contribution from the second and third terms, and Penn proposed that S_0 could be taken as unity. Also Heine and Jones [19] confirmed that the energy E_0 in the above expression (which, following Penn's paper is often referred to as the Penn gap) should be identified with the UV absorption peak. For a discussion of the influence of inner shell electrons, see Ravindra et al. [7].

Thus we are forced to conclude that the intense, relatively narrow absorption band which semiconductors have in the UV region i.e. far removed from the long wavelength absorption edge, is the absorption feature which essentially determines the refractive index.

There are two important consequences from this conclusion, namely:

(i) Any relationship between the index and the long wavelength absorption edge must arise from some relation between the UV oscillator frequency and the energy gap, and —

(ii) Modifications of the absorption edge which do not at the same time vary the resonance frequency cannot be expected to produce change in the index.

4. Relation between the Resonance Energy and Energy Gap

The simplest assumption, in mathematical terms, which we can make to relate the UV resonant energy to the energy gap is to say that there is a constant difference between the two i.e.

$$E_0 = E_g + B\,. \tag{11a}$$

This possibility was first postulated by Moss [2] (p. 68 and 246) and such an assumption has been made recently by Ravindra [9] on the grounds that the valence and conduction bands are more or less parallel to each other. The consequences of this assumption will be considered. We have

$$E_0 = B(1 + E_g/B) \tag{11b}$$

and writing (7) as

$$n_0^2 - 1 = (\hbar\omega_p/E_0)^2\,, \tag{8a}$$

then, if we ignore unity compared with n_0^2 and also assume that $E_g \ll B$, both reasonable assumptions for many semiconductors, we obtain

$$n_0 = \omega_p/\omega_0 = \hbar\omega_p/E_0 = (1 - E_g/B)\,\hbar\omega_p/B = \hbar\omega_p/B - (\hbar\omega_p/B^2)E_g\,. \tag{12}$$

This form of relationship closely resembles the linear relation (2) put forward by Ravindra and co-workers in 1979 [3]. However, it has significant disadvantages in comparison with the Moss relation, as will be discussed in the next section.

5. Comparison of the Moss and Ravindra Relations

As stated in the introduction, the two important attempts to relate the refractive index to the energy gap are the Moss (1950) relation [1], with the revised value of the constant [4] (1c) and the Ravindra et al. linear relation (2).

The latter form of relationship has the slight disadvantage that it has two arbitrary constants compared with one in (1c), but it has two more serious defects, namely the behaviour at high and low values of n respectively.

(i) *High n*

The relation indicates that $n \geqq 4.1$ cannot occur, whereas it is well-known that materials with much higher indices exist e.g.

PbSe, $n = 4.7$ and $Pb_{0.76}Sn_{0.24}Te$, $n = 7.0$,

the latter being probably the highest reliably known index [32]. In contrast, the Moss relation, using the revised constant (1c) gives: PbSe ($E_g = 0.25$), $n = 4.6$; PbSnTe ($E_g = 0.082$), $n = 6.03$, which are very reasonable predictions.

(ii) *Low n*

Here again the linear formula gives unrealistic results, predicting that the index would become zero for $E_g = 6.6$ eV, whereas (1) would require an infinite gap. For the condition $n = 1$, the linear relation requires $E_g = 5$ eV only — a value exceeded by many materials — although unity index is of course impossible in a material. By contrast, (1c) would need $E_g > 100$ eV, much larger than any real material.

More realistically, if we consider the highest energy gaps likely to exist (perhaps 20 to 25 eV) and the smallest indices found in solids (approximately 1.4) we find these two parameters quite compatible with the Moss relation.

These points are well illustrated by Fig. 2 where a wide range of values of E_g and n is shown on a log–log plot. Most of the data is taken from Ravindra and his co-

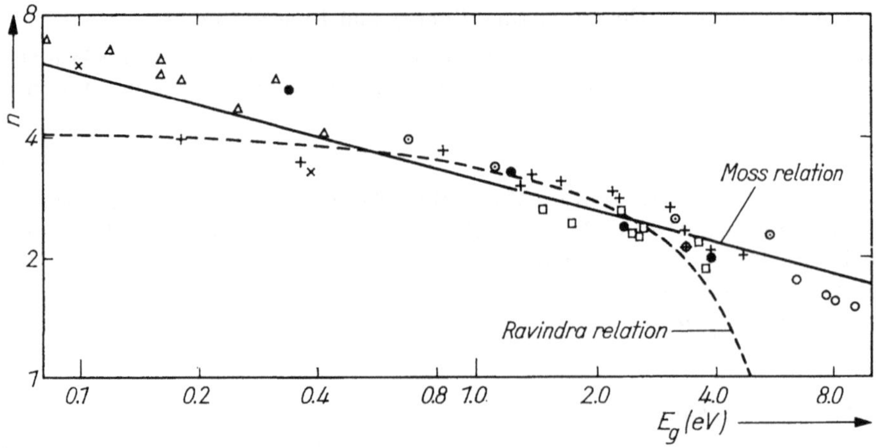

Fig. 2. Refractive index versus energy gap. Full line, Moss relation; dashed line, Ravindra relation; ⊙ group IV, ● group VI and As, + III–V, □ II–VI, × GeTe and CdHgTe, △ Pb and Pb/Sn chalcogenides, ○ halides

workers [4, 5], with Gopal's [20] data for the Pb/Sn chalcogenides. They are listed in Table 1, in order of increasing E_g. On this plot the Moss relation is shown by the full, straight line, and the Ravindra relation by the dashed curve.

It is clear from the figure that:

(i) There is no sign of the sharp fall in n at high energy gaps predicted by Ravindra, the points for the nitrides, halides, and diamond lying quite close to the straight line.

(ii) At low energy gaps the Ravindra equation predicts almost constant n for gaps below ≈ 0.4 eV, whereas the measured values continue to increase quite sharply as the gap decreases down to 0.082 eV. In fact, by careful analysis of experimental data, Oron et al. [31] have shown that (for Pb/SnTe) n continues to increase as $(1/E_g)^{1/4}$ down to $E_g = 0.095$ eV. (If anything, the indices rise more rapidly than predicted by (1) and a straight line corresponding to $1/E_g \propto n^{3.5}$ might be an improvement.²))

Plotting the data in this way thus shows very clearly that the Moss relation is superior, and gives a good general representation of the data over the whole range of measured energy gaps (more than 100:1 in E_g).

The shortcomings of the Ravindra relation at low and high values of n can mostly be explained by reference to (12) and its derivation from (11a).

(I) *Low n*

Clearly for materials with low index and correspondingly high energy gaps the approximations $n_0^2 \gg 1$ and $E_g \ll B$ are inadequate. Also, identification of (2) with (12) implies that $\hbar\omega_p/B$ is a constant. Although for many materials the variation in plasmon energy is small (it lies between 12.7 for InSb and 16.6 for Si for a wide range of semiconductors), it does depend directly on the lattice constant and has considerably larger values for low index solids e.g. AlN, 23; diamond, 31.

(II) *High n*

The assumption of an approximately constant difference (B) between the resonance energy and the band edge energy on the grounds that the valence and conduction bands are roughly parallel is not unreasonable, but to be justifiable in the present context it is necessary for any variation in B to be small compared with the change

²) For example, $E_g n^{3.5} = 63$.

in E_g between one material and another. For example, a 5% change in B would completely outweigh the difference in E_g between InSb and InAs. In contrast, for the ten materials listed by Gupta and Ravindra [5], the energy difference varies from 3.5 to 4.5, while for PbTe it is only 2.1 eV. Hence the assumption of a constant energy difference is not justified for low gap (high index) materials.

Table 1
Energy gap and refractive index data (plotted in Fig. 1)

material	gap (E_g)	index	ref.
Pb/SnTe	0.082	7.0	[32]
GeTe	0.10	6.0	[4]
Pb/SnTe	0.12	6.58	[20]
Pb/SnTe	0.16	6.23	[20]
Pb/SnSe	0.16	5.74	[20]
Pb/SnSe	0.18	5.52	[20]
InSb	0.18	3.95	[4]
PbSe	0.25	4.70	[25]
PbTe	0.32	5.64	[25]
Te	0.335	5.3 (mean)	[13]
InAs	0.36	3.5	[4]
Cd/HgTe	0.38	3.29	[20]
PbS	0.41	4.15	[5]
Ge	0.67	4.0	[5]
GaSb	0.81	3.79	[4]
Si	1.1	3.46	[5]
As	1.2	3.35	[2]
InP	1.27	3.1	[4]
GaAs	1.35	3.3	[4]
CdTe	1.44	2.7	[4]
AlSb	1.6	3.19	[4]
CdSe	1.7	2.49	[4]
AlAs	2.16	3.0	[4]
ZnTe	2.26	2.7	[4]
Se	2.3	2.45	[13]
CdS	2.4	2.38	[4]
CdO	2.5	2.32	[4]
ZnSe	2.58	2.43	[4]
AlP	3.0	2.75	[4]
SiC	3.1	2.59	[4]
GaN	3.25	2.4	[4]
CuCl	3.31	2.19	[4]
ZnS	3.54	2.27	[4]
ZnO	3.7	1.92	[4]
AlN	3.8	2.16	[4]
S	3.82	2.04 (mean)	[29]
BN	4.6	2.1	[4]
C (diamond)	5.4	2.35	[13]
CsI	6.3	1.82	[4]
CsBr	7.5 (mean)	1.67	[4]
CsCl	8.0	1.61	[4]
NiF_2	9.0	1.56	[30]

Finally, the two constants found empirically by Ravindra in (2) can be compared quantitatively with (12). The mean value of $\hbar\omega_p$ for ten well-known materials [5] is 14.9 eV and the mean energy difference is $B = 3.82$ eV. Hence, (a) the first Ravindra term should be: $14.9/3.82 = 3.9$, which is not far from Ravindra's value of 4.084, and (b) the second term should be: $(\hbar\omega_p/B^2) = 1.02$. This is much higher (65%) than Ravindra's value of $0.62E_g$. Alternatively, his value implies that B would be 6.6 eV, far above observed values.

6. Finkenrath Theory

A fundamental approach to the problem has been followed by Finkenrath [21]. This author carried out by far the most thorough theoretical derivation of the optical polarizability resulting from direct interband transitions corresponding to an assembly of classical oscillators similar to the single resonant oscillator considered above. In this treatment, the transitions from the valence to the conduction band are represented by differential oscillator densities at continuously distributed resonant frequencies, the summation of their combined polarizability being replaced by an integral over the relevant frequency range. The differential oscillators are specified in terms of wave number and occupation probability on the basis of spherical energy surfaces. The oscillator mass is taken to be the reduced electron/hole mass.

The outcome of the analysis, for direct interband transitions, for nondegenerate semiconductors, is a relation which shows that $(\varepsilon_{opt} - 1)^2 E_g$ equals a complex expression which is essentially constant. Hence, if we again ignore unity with respect to ε_{opt}, we obtain the Moss relation, $\varepsilon_{opt}E_g =$ constant.

Finkenrath makes a quantitative estimate that for a variety of materials, the mean value of the constant should be approximately 95 eV, with some scatter for different materials and different band structures.

There is no doubt that this analysis is the most fundamental and most soundly-based theoretical treatment of the problem, and thus, within the limitations specified, it must be considered to give a valid relationship. Finkenrath points out that deviations from the relation could well arise if transitions are not direct, and from neglecting the transition probabilities between other bands and the differing densities of states and matrix elements. Also, the assumption of parabolic and spherical energy bands is not, of course, generally true, and could thus affect the results.

7. Variations of Energy Gap and Refractive Index

The main variables concerned are pressure, temperature, and doping. Of these, doping and temperature effects are the most marked and are of technological significance.

Doping is most pronounced in very low gap, low mass materials e.g. InSb, where filling the lower levels in the conduction band, thus effectively preventing optical transitions at energies below the Fermi level, produces a marked Burstein-Moss shift of the absorption edge. This effect of course always produces a positive $\mathrm{d}E/\mathrm{d}N$ effect and there is a corresponding negative $\mathrm{d}n/\mathrm{d}N$ change.

By contrast, the temperature change in E_g can be either negative or positive — the lead salts in particular showing a marked increase in the long wavelength absorption edge on cooling, in contra-distinction to most other materials. However, in all cases the signs of $\mathrm{d}n/\mathrm{d}T$ and $\mathrm{d}E/\mathrm{d}T$ are consistent with (1) and (2).

It is useful to compare the derivatives of (1) and (2) in terms of predicting the variation in n for a known variation in E_g. From (1),

$$(\mathrm{d}n/\mathrm{d}E_g)_{\text{Moss}} = -n/4E_g \tag{13}$$

and from (2),

$$(\mathrm{d}n/\mathrm{d}E_g)_{\mathrm{Rav}} = -0.62\,. \tag{14}$$

These derivatives are thus equal when $E_g \approx 1.23$ eV and $n \approx 3.1$, and the difference between the two predictions will be most marked for very small or very large values of E_g.

7.1 Pressure effects

As stated earlier, it would be expected that high pressure would — by compressing the whole lattice — affect the UV resonance and so produce a significant change in refractive index. Unfortunately, good experimental data for the refractive index under pressure are scarce, especially for materials of high or low E_g. The best data are probably that for Si and Ge, the best for Si being that for amorphous material [22], namely

$$\mathrm{d}E_g/\mathrm{d}P = 0.25 \times 10^{-6}\ \mathrm{eV/bar}\,, \qquad \mathrm{d}n_0/\mathrm{d}P = -0.175 \times 10^{-6}\ \mathrm{bar}^{-1}\,.$$

The Moss relation in its derivative form — which it may be noted contains no arbitrary constants — gives

$$\mathrm{d}n_0/\mathrm{d}P = -\,(n_0/4E_g)\,\mathrm{d}E_g/\mathrm{d}P = -0.193 \times 10^{-6}\ \mathrm{bar}^{-1}\,,$$

while the Ravindra formula gives

$$\mathrm{d}n_0/\mathrm{d}P = -0.62\,\mathrm{d}E_g/\mathrm{d}P = -0.155 \times 10^{-6}\ \mathrm{bar}^{-1}\,.$$

Hence for the pressure change in Si, both predictions are only 10% from the observed value — which is probably as good as the experimental data itself.

Earlier results for Ge [6] give values of

$$\mathrm{d}E/\mathrm{d}P = +5.0 \text{ or } +7.3 \times 10^{-6}\ \mathrm{eV/kg\ cm^{-2}}\,,$$

so the expected $\mathrm{d}n/\mathrm{d}P$ is

(i) from Moss relation, $\quad -10.9$ or $-7.6 \times 10^{-6}/\mathrm{kg\ cm^{-2}}$,

(ii) from Ravindra, $\quad -4.53$ or $-3.1 \times 10^{-6}/\mathrm{kg\ cm^{-2}}$.

The observed value is $-9.8 \times 10^{-6}/\mathrm{kg\ cm^{-2}}$, which is well within the experimental uncertainty for (1) but well outside the values calculated from (2).

Thus, for pressure-dependent effects, where we would expect reasonable correlation, the predictions from the Moss relation are as good as are justified by the experimental results and rather better than the Ravindra estimates.

7.2 Doping dependence

The most dramatic effect of doping is shown by n-type InSb, where the Burstein-Moss shift of the absorption edge results from the filling of the lower energy levels of the conduction band so that optical transitions can only take place to the Fermi level or above it.

The reflectivity measurements of Spitzer and Fan [23] were analysed by Finkenrath [21] to obtain optical dielectric constants. The corresponding refractive indices (n_x) and energy gaps are given in Table 2, together with the changes in index predicted by the Moss and Ravindra relations. Over the full range of doping, the absorption edge wavelength is seen to decrease by the large factor of 2.8:1 but n_x^4 decreases by only 1.36:1 and the changes (Δn_x) are only $\approx \frac{1}{3}$ of those predicted $(\Delta n)_{\mathrm{Moss}}$. Thus, as expected from the earlier discussion, changing the energy gap by doping — where

Table 2

Energy gap and refractive index of doped InSb. Data from Spitzer and Fan [23] analysed by Finkenrath [21]

doping ($\times 10^{12}$ cm^{-3})	0	3.5	6.2	12	28	40
Fermi level ΔE (eV)	0	0.83	0.114	0.168	0.268	0.324
energy gap E_g^* (eV)	0.18	0.263	0.294	0.348	0.448	0.504
index n_x	3.96	3.852	3.808	3.748	3.692	3.655
Δn_x	0	0.108	0.152	0.212	0.268	0.305
$(\Delta n)_{Moss}$	0	0.360	0.456	0.602	0.808	0.900
$(\Delta n)_{Rav}$	0	0.051	0.071	0.104	0.166	0.201

little effect on UV resonance would be expected — does not greatly affect the refractive index.

The Ravindra relation, by contrast, predicts a smaller change in n than that observed, the predicted values of $(\Delta n)_{Rav} = 0.62\,\Delta E_g$ being $\approx \frac{2}{3}$ of the observed.

For a very different system, we may consider the results on amorphous Si, heavily doped with hydrogen during preparation by glow discharge decomposition of silane, which have been analysed extensively by Ravindra et al. [8a]. The samples thus prepared contain many per-cent of hydrogen. Some of the results are shown in Table 3a. For the sample with the smallest amount of hydrogen, the energy gap is 50% above that for pure crystalline Si, but n_x is also *above* the normal Si value. Thereafter n_x falls rapidly — much faster than predicted by either (1) or (2) — as the H_2 content increases.

In the second of their papers [8b], the results in Table 3b for samples made by chemical vapour deposition show that for five samples with virtually identical energy gaps (1.54 to 1.57 eV) the indices vary from 3.195 to 3.632 so that again there is no coherent dependence of n on E_g.

Table 3

Energy gap and refractive index of hydrogenerated amorphous Si

a) Glow discharge in silane [8a]

% H_2	E_g (eV)	n_x
8	1.67	3.561
12	1.70	3.327
17	1.78	3.074
35	2.05	2.485

b) Chemical vapour deposition [8b], in order of decreasing n_x

sample	QA 342	QAH 342	QB 232	QB 224	QBH 224
E_g (eV)	1.57	1.64	1.56	1.55	1.655
n_x	3.632	3.536	3.356	3.319	3.311
sample	**QBH 232**	**CVD(S)**	**QA 222**	**QAH 222**	
E_g (eV)	1.655	1.54	1.54	1.63	
n_x	3.306	3.195	3.148	3.123	

Hence for this system, where both heavy doping and non-crystalline structures are involved, neither relationship is relevant.

A new paper by Ravindra et al. (1985) [10] shows that heavy doping of Si does indeed greatly modify the UV reflectivity. Both reflectivity peak and refractive index decrease steadily with increasing doping.

7.3 Temperature variation

As regards dependence on temperature, two separate influences can be expected, viz:

7.3.1 Lattice dilation

This can be expected to vary the UV resonance in the same sort of way as varying pressure or changing lattice constants by varying the composition. This effect can be of either sign. The lead salts for example all show a marked decrease in energy gap on cooling.

7.3.2 Broadening of the energy bands by lattice vibrations

The sign of this effect should always be to decrease the gap, by broadening the band edges, on heating. This part of the energy gap change would thus be expected to have only a small effect on the refractive index. Thus, overall, it would be expected that for a given change in refractive index, when dn/dT is positive, the net shift of the absorption edge will be greater than it would be in the absence of thermal broadening. Hence the dn/dT predicted from the measured edge shift will be overestimated.

Ravindra [6] has compared measured values of dn/dT with predictions from (1) and (2) for Si, Ge and the III–V compounds. The agreement is moderate for Ge, Si, and the Ga compounds, both predictions being in error by $\approx 1.5:1$, but for the In compounds the errors are $\approx 3:1$. However, it may be noted that all the predictions from (1) are *overestimates*, as expected from the argument above, since all these materials have positive dn/dT values. Thus, if some allowance were made for the thermal broadening of the absorption edge, it seems likely that the predictions of (1) would be somewhat more accurate than those from (2).

Regarding the lead salts, Avery [24] found for PbS, $dn/dT = -6 \times 10^{-4}$. As $dE/dT = 4 \times 10^{-4}$ eV/K, the predictions for dn/dT are:

$$\text{Moss: } -9.8 \times 10^{-4}\,, \qquad \text{Ravindra: } -2.3 \times 10^{-4}\,.$$

Thus the Moss relation is the better estimate, though it again overestimates. However, although conventional theory says that the electron–phonon contribution should always be negative, pressure measurements on the lead salts indicate that it is actually positive, so that this could well explain the overestimate.

Ravindra et al. [8] have analysed extensive data on samples of amorphous Si containing several per-cent of hydrogen, over a wide temperature range (100 to 700 K). The value of $dn/dE_g = 0.44$ to $0.84\ \text{eV}^{-1}$, so that deviation from the Ravindra prediction is approximately $\pm 30\%$. It was also shown that all samples followed accurately a law of $E_g \propto 1/n^4$, although there was a sizeable intercept on the axis as opposed to the zero intercept required by (1).

8. Other Relations

Two other relations which have been proposed recently are worthy of mention, namely those of Dionne and Woolley [26] and of Ghosh et al. [27].

The former relation is of logarithmic form, i.e.

$$n^2 = 19.2 - 11.6 \ln E_g\,. \tag{15}$$

It has been applied only to Pb/SnTe alloys.

The latter is essentially based on (8) with the assumption of (11) of a constant difference between resonance energy and energy gap but with $\hbar\omega_p$ replaced by a function of E_g instead being treated as a constant. The resulting relation is

$$n_0^2 - 1 = (mE_g + c)/(E_g + B)^2 . \tag{16}$$

By introducing two more arbitrary constants, m and c, and furthermore by redetermining their values for each class of material, more accurate agreement with experiment is achieved and the expression is useful for interpolating data on mixed crystal systems. See also a relationship by Singh et al. [33[.

9. Conclusions

It is shown from the fundamental inter-relations between the refractive index and the absorption, that the main factor determining the index is the frequency of the UV absorption peak, and a good quantitative explanation of index values can be obtained by postulating a single classical oscillator at this frequency.

Thus any relation between the index and the band edge energy must result from a close relation between the band edge and the resonance energy. Making the simplest possible assumption i.e. a constant energy difference, one can — with some approximations — derive a relation resembling the Ravindra one.

However, this relation is found to have specific disadvantages and to be inferior to the Moss relation, which is shown to give a good representation of data on materials over more than 100:1 range of energy gaps (from 0.082 to 9 eV). Fundamental support for the Moss relation is given by the formal theory of Finkenrath.

In spite of this general agreement, bearing in mind the fact that three of the best known semiconductors, InSb, PbS and Ge, all have refractive indices of $n = 4$ but have energy gaps varying by almost 4:1, it is clear a simple relationship can only be of a general and approximate nature. Nevertheless, the Moss relation has been found useful in estimating the refractive indices of technologically interesting materials [28].

References

[1] T. S. Moss, Proc. Phys. Soc. **B63**, 167 (1950).
[2] T. S. Moss, Photoconductivity in the Elements, Butterworths, London 1952.
[3] N. M. Ravindra, Suhil Auluck, and V. K. Srivastava, phys. stat. sol. (b) **93**, K155 (1979).
[4] N. M. Ravindra and V. K. Srivastava, Infrared Phys. **19**, 603 (1979).
[5] V. P. Gupta and N. M. Ravindra, phys. stat. sol. (b) **100**, 715 (1980).
[6] N. M. Ravindra, Infrared Phys. **21**, 283 (1981).
[7] N. M. Ravindra, R. P. Bhardwaj, K. S. Kuman, and V. K. Srivastava, Infrared Phys. **21**, 369 (1981).
[8a] N. M. Ravindra, C. Ance, S. P. Coulibaly, F. de Chelle, J. M. Berger, J. P. Ferraton, and A. Donnadieu, Infrared Phys. **23**, 99 (1983).
[8b] N. M. Ravindra, C. Ance, J. P. Ferraton, J. M. Berger, F. de Chelle, and S. Robin, Infrared Phys. **23**, 223 (1983).
[9] N. M. Ravindra, Phil. Mag. (1985), in the press.
[10] N. M. Ravindra, J. F. Mhoronge, and M. Jouanne, Infrared Phys. **25**, 707 (1985).
[11] H. A. Kramers, Phys. Z. **30**, 522 (1929).
[12] R. de L. Kronig, Proc. Roy. Soc. **124**, 409 (1929); **133**, 245 (1931).
[13] T. S. Moss, Optical Properties of Semiconductors, Butterworths, London 1959.
[14] T. S. Moss, S. D. Smith, and T. D. Hawkins, Proc. Phys. Soc. **B70**, 776 (1957).

[15] J. R. Dixon and J. M. Ellis, Phys. Rev. **123**, 1560 (1961).
[16] H. R. Phillip and H. Ehrenreich, in: Semiconductors and Semimetals, Vol. 3, Ed A. C. Beer and R. K. Willardson, Academie Press, New York 1967 (Chap. 4).
[17] M. Cardona, Phys. Rev. **129**, 69 (1963).
[18] D. R. Penn, Phys. Rev. **128**, 2093 (1962).
[19] V. Heine and R. O. Jones, J. Phys. C **2**, 719 (1969).
[20] V. Gopal, Infrared Phys. **22**, 255 (1982).
[21] H. Z. Finkenrath, Naturf. **19a**, 794 (1964); Z. angew. Phys. **16**, 503, (1964).
[22] J. Tauc, Amorphous and Liquid Semiconductors, Plenum Press, London 1976 Chap 6.
[23] W. G. Spitzer and H. Y. Fan, Phys. Rev. **106**, 882 (1957).
[24] D. G. Avery, Proc. Phys. Soc. **B67**, 2 (1954).
[25] T. S. Moss, G. J. Burrell, and B. Ellis, Semiconductor Optoelectronics, Butterworths, London 1973.
[26] G. Dionne and J. C. Woolley, Phys. Rev. **B6**, 3898 (1972).
[27] D. K. Ghosh, L. K. Samanta, and G. C. Bhar, Infrared Phys. **24**, 43 (1984).
[28] B. Seraphin and A. B. Meinel, Optical Properties of Solids, North-Holland Publ. Co., Amsterdam 1976 (p. 950).
[29] C. A. Mead, Phys. Letters **3**, 212 (1964).
[30] G. J. Gevers, J. Richon, and A. S. Barriere, phys. stat. sol. (b) **104**, 641 (1981).
[31] M. Oron, A. Zussman, and A. Katzir, Infrared Phys. **22**, 317 (1982).
[32] W. W. Anderson, Infrared Phys. **20**, 353 (1980); quoting thesis by W. M. Walz.
[33] Sing et al. Infrared Phys. **26**, No. 1, in the press.

(Received July 17, 1985)

Original Papers

phys. stat. sol. (b) **131**, 429 (1985)
Subject classification: 1.6

A. N. Shevchenko Scientific Research Institute for Applied Physics Problems[1]) *(a) and Department of Physics*[2]) *(b) V. I. Lenin Belorussian State University, Minsk*

The Tensor Reflection Coefficient for the Surface Diffraction

By

S. A. Maksimenko (a) and V. V. Skadorov (b)

Dedicated to Prof. Dr. Dr. h. c. Dr. E. h. P. Görlich on his 80th birthday

On the basis of the covariant method the matrix equation describing the dynamical diffraction in the Laue-to-Bragg geometries is obtained. The tensor reflection and transmission coefficients are found for the case of symmetric surface diffraction. It is shown that unenriched Mössbauer crystals are more suitable for observing the surface diffraction.

Ковариантными методами получено матричное уравнение динамической дифракции для резкоассиметричной и поверхностной дифракционных геометрий. В случае симметричной поверхностной дифракции найдены тензорные коэффициенты прохождения и отражения. Показано, что необогащенные мёссбауэровские кристаллы являются более подходящими для наблюдения поверхностной дифракции.

1. Introduction

To develop the investigations of physical properties of surfaces new procedures which should be adequate to solve the complicated problems are required. In this connection recently considerable attention has been attracted to the Laue-to-Bragg diffraction geometries (extremely asymmetric diffraction (EAD), surface diffraction (SD)) which are extremely sensitive to the surface properties. Taking into account that the EAD geometry is studied well enough (see, for example, [1, 2]), in the present paper we will pay mainly attention to the second case which has been predicted by Baryshevskii in [3] and which has been observed experimentally in [4, 5]. Thereby one should have in mind that the diffraction equations which will be obtained below describe both geometries. The disposition of wave vectors and reflected planes relative to the crystal surface for the SD is represented in Fig. 1.

The peculiarities of these diffraction schemes are conditioned first of all by the fact that either an incident or a diffracted wave (in EAD case) or both of them (in SD-case) are propagated at a grazing angle to the crystal surface ($\approx 10^{-3}$ rad). The extinction length becomes so small under these conditions that it becomes possible to obtain information about thin surface layers. The Laue-to-Bragg diffraction pattern analysis becomes much complicated because the small quadratic term ξ^2 must be considered in the expression for the square of the wave vector inside the crystal $\boldsymbol{k}_{\text{ins}} = \boldsymbol{k} + \xi \boldsymbol{N}$ (vector $\boldsymbol{N}$ is normal to the crystal surface). This follows from the fact that for the single wave (EAD) or for both waves (SD) inside the crystal $(\boldsymbol{kN}) \sim \xi$. This fact, in turn, means that the second-order derivative terms in the Maxwell equations cannot be neglected when writing the diffraction equations in the Takagi form [6], i.e. for the Bloch wave amplitudes.

---

[1]) 220106 Minsk 106, USSR.
[2]) 220080 Minsk 80, USSR.

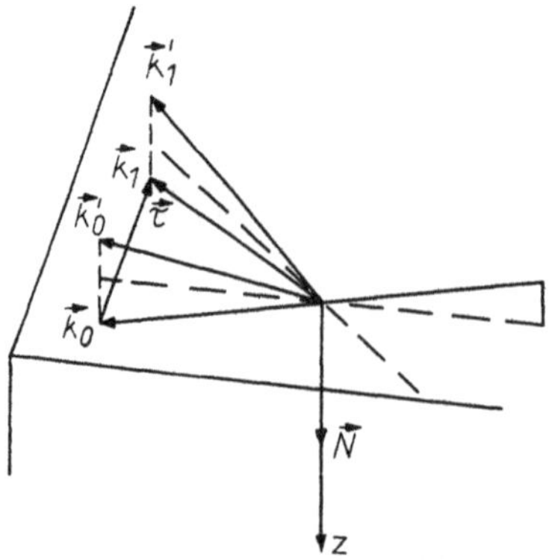

Fig. 1. Surface diffraction scheme

The covariant method developed by Fedorov [7] for the anisotropic medium optics and successfully applied in the γ-region [8 to 10] is used to discuss the EAD-SD diffraction schemes in the present paper. The matrix form of the two-wave diffraction Takagi equations in the required approximation is obtained. For the Buryshevskii geometry in the symmetric case the boundary problem is solved and the tensor reflection coefficient is found. The generally accepted approximation that the wave vector of the incident wave and the ones of the waves inside the crystal are coplanar, is used.

2. Diffraction Equations

In accordance with [11] if we introduce the electromagnetic induction by means of the relation

$$\boldsymbol{D}(\boldsymbol{r}, t) = \boldsymbol{E}(\boldsymbol{r}, t) + 4\pi \int_{-\infty}^{t} \mathrm{d}t' \, \boldsymbol{J}(\boldsymbol{r}, t') \,. \tag{1}$$

The Maxwell equations in the medium without external sources have the following form:

$$\nabla \times \boldsymbol{H} - \frac{1}{c} \frac{\partial \boldsymbol{D}}{\partial t} = 0 \,, \qquad \nabla \times \boldsymbol{E} + \frac{1}{c} \frac{\partial \boldsymbol{H}}{\partial t} = 0 \,, \tag{2}$$

$$\nabla \boldsymbol{D} = 0 \,.$$

The electromagnetic field inside the crystal which is formed by γ-quanta is a superposition of Bloch waves, and for monochromatic incident radiation with the frequency ω in the stationary case, it has the form

$$\begin{aligned} \boldsymbol{E}(\boldsymbol{r}, t) &= \sum_{\boldsymbol{\tau}} \boldsymbol{E}_{\boldsymbol{\tau}}(\boldsymbol{r}) \, \mathrm{e}^{i(\boldsymbol{k}+\boldsymbol{\tau})\boldsymbol{r} + i\omega t} \,, \\ \boldsymbol{H}(\boldsymbol{r}, t) &= \sum_{\boldsymbol{\tau}} \boldsymbol{H}_{\boldsymbol{\tau}}(\boldsymbol{r}) \, \mathrm{e}^{i(\boldsymbol{k}+\boldsymbol{\tau})\boldsymbol{r} + i\omega t} \,. \end{aligned} \tag{3}$$

The current density induced in the crystal by these γ-quanta is defined by the conductivity tensor $\boldsymbol{\sigma}$ of the crystal,

$$\boldsymbol{J}(\boldsymbol{r}, t) = \mathrm{e}^{i\omega t} \sum_{\boldsymbol{\tau}} \mathrm{e}^{i(\boldsymbol{k}+\boldsymbol{\tau})\boldsymbol{r}} \sum_{\boldsymbol{\tau}'} \boldsymbol{\sigma}(\boldsymbol{k} + \boldsymbol{\tau}, \boldsymbol{k} + \boldsymbol{\tau} + \boldsymbol{\tau}') \, \boldsymbol{E}_{\boldsymbol{\tau}+\boldsymbol{\tau}'}(\boldsymbol{r}) \,. \tag{4}$$

In (3) and (4) the summation is carried out over all reciprocal vectors $\boldsymbol{\tau}$ of the crystal (the coefficient 2π is included to define $\boldsymbol{\tau}$). Using (1), (3), and (4) and the transverse character of the conductivity tensor $((\boldsymbol{k} + \boldsymbol{\tau}) \, \boldsymbol{\sigma} = 0)$ the vector differential equation system for the Bloch wave amplitude corresponding to a given $\boldsymbol{\tau}$ can be found,

$$[\nabla^{\mathrm{x}} + i(\boldsymbol{k} + \boldsymbol{\tau})^{\mathrm{x}}]\,\boldsymbol{H}_{\boldsymbol{\tau}}(\boldsymbol{r}) + \frac{i\omega}{c} \times$$

$$\times\,[\boldsymbol{E}_{\boldsymbol{\tau}}(\boldsymbol{r}) + \sum_{\boldsymbol{\tau}'} \chi(\boldsymbol{k} + \boldsymbol{\tau}, \boldsymbol{k} + \boldsymbol{\tau} + \boldsymbol{\tau}')\,\boldsymbol{E}_{\boldsymbol{\tau}+\boldsymbol{\tau}'}(\boldsymbol{r})] = 0\,,$$

$$[\nabla^{\mathrm{x}} + i(\boldsymbol{k} + \boldsymbol{\tau})^{\mathrm{x}}]\,\boldsymbol{E}_{\boldsymbol{\tau}}(\boldsymbol{r}) + \frac{i\omega}{c}\,\boldsymbol{H}_{\boldsymbol{\tau}}(\boldsymbol{r}) = 0\,, \tag{5}$$

$$i(\boldsymbol{k} + \boldsymbol{\tau}) \cdot \boldsymbol{E}_{\boldsymbol{\tau}}(\boldsymbol{r}) + \nabla\,[\boldsymbol{E}_{\boldsymbol{\tau}}(\boldsymbol{r}) + \sum_{\boldsymbol{\tau}'} \chi(\boldsymbol{k} + \boldsymbol{\tau}, \boldsymbol{k} + \boldsymbol{\tau} + \boldsymbol{\tau}')\,\boldsymbol{E}_{\boldsymbol{\tau}+\boldsymbol{\tau}'}(\boldsymbol{r})] = 0\,.$$

The polarizability tensor of the crystal is defined by the relation $\boldsymbol{\chi} = (4\pi i/c)\,\boldsymbol{\sigma}$. The superscript x marks the dual tensor of a given vector ($\boldsymbol{a}^{\mathrm{x}}\boldsymbol{b} = [\boldsymbol{a} \times \boldsymbol{b}]$). Hence, if we limit ourselves to the two-wave approximation, the Laue-to-Bragg diffraction will be described by six vector equations for incident (marked by subscript „0" below) and diffracted (subscript "1", respectively) waves. Let us introduce the six-dimensional diffraction space [8] in accordance with covariant methods. It can be obtained after some transformations

$$\left\{\left[\nabla^{\mathrm{x}} + i\begin{pmatrix}\boldsymbol{k}_0^{\mathrm{x}} & 0\\ 0 & \boldsymbol{k}_1^{\mathrm{x}}\end{pmatrix}\right]^2 + \frac{\omega^2}{c^2}\,\boldsymbol{\epsilon}\right\}|\boldsymbol{E}((\boldsymbol{r})\rangle = 0\,,$$

$$\nabla\,\boldsymbol{\epsilon}\,|\boldsymbol{E}(\boldsymbol{r})\rangle + i\begin{pmatrix}\boldsymbol{k}_0 & 0\\ 0 & \boldsymbol{k}_1\end{pmatrix}|\boldsymbol{E}(\boldsymbol{r})\rangle = 0\,, \tag{6}$$

where

$$|\boldsymbol{E}(\boldsymbol{r})\rangle = \begin{pmatrix}\boldsymbol{E}_0(\boldsymbol{r})\\ \boldsymbol{E}_1(\boldsymbol{r})\end{pmatrix}, \qquad \boldsymbol{\epsilon} = \mathbf{I} + \boldsymbol{\chi} = \mathbf{I} + \begin{pmatrix}\chi_{00} & \chi_{01}\\ \chi_{10} & \chi_{11} - \alpha\end{pmatrix},$$

$\alpha = (2\boldsymbol{k}_0 + \boldsymbol{\tau})\,\boldsymbol{\tau}/k_0^2$ is the deviation parameter from the exact Bragg condition, $\boldsymbol{k}_1 = \boldsymbol{k}_0 + \boldsymbol{\tau}$.

In the general case the electromagnetic wave inside the crystal is nontransverse. In order to divide (6) into transverse and nontransverse parts we make use of the projectors $\mathbf{P}_{\parallel}$ and $\mathbf{P}_{\perp}$ ($|\boldsymbol{E}(\boldsymbol{r})\rangle = \mathbf{P}_{\parallel}\,|\boldsymbol{E}_{\parallel}(\boldsymbol{r})\rangle + \mathbf{P}_{\perp}\,|\boldsymbol{E}_{\perp}(\boldsymbol{r})\rangle$) which are given by [7]

$$\mathbf{P}_{\perp} = -\frac{1}{k_0^2}\begin{pmatrix}\boldsymbol{k}_0^{\mathrm{x}}\boldsymbol{k}_0^{\mathrm{x}} & 0\\ 0 & \boldsymbol{k}_1^{\mathrm{x}}\boldsymbol{k}_1^{\mathrm{x}}\end{pmatrix}, \qquad \mathbf{P}_{\parallel} = \mathbf{I} - \mathbf{P}_{\perp} = \frac{1}{k_0^2}\begin{pmatrix}\boldsymbol{k}_0 \otimes \boldsymbol{k}_0 & 0\\ 0 & \boldsymbol{k}_1 \otimes \boldsymbol{k}_1\end{pmatrix},$$

where the symbol $\otimes$ signifies the tensor (diadic) product, $|\boldsymbol{E}_{\perp,\parallel}(\boldsymbol{r})\rangle$ are column vectors consisting of the projections of $\boldsymbol{E}_0$ and $\boldsymbol{E}_1$ on the directions normal and parallel to $\boldsymbol{k}_0$ and $\boldsymbol{k}_1$, respectively. Using the well-known properties of the orthogonal projectors and the relations $\mathbf{P}_{\parallel}\boldsymbol{\epsilon} = \mathbf{P}_{\parallel}$, $\mathbf{P}_{\perp}\boldsymbol{\epsilon}\mathbf{P}_{\parallel} = 0$ following from the transverse character of $\boldsymbol{\chi}$ we have from (6)

$$\left[\nabla^2 - i\begin{pmatrix}\boldsymbol{k}_0 & 0\\ 0 & \boldsymbol{k}_1\end{pmatrix} \otimes \nabla + 2i\begin{pmatrix}\boldsymbol{k}_0\nabla & 0\\ 0 & \boldsymbol{k}_1\nabla\end{pmatrix} + k_0^2\right]|\boldsymbol{E}_{\parallel}(\boldsymbol{r})\rangle + \mathbf{P}_{\parallel}\nabla \otimes \nabla\boldsymbol{\chi}|\boldsymbol{E}_{\perp}(\boldsymbol{r})\rangle +$$

$$+\left[\nabla^2 + 2i\begin{pmatrix}\boldsymbol{k}_0\nabla & 0\\ 0 & \boldsymbol{k}_1\nabla\end{pmatrix} + k_0^2\boldsymbol{\chi}\right]|\boldsymbol{E}_{\perp}(\boldsymbol{r})\rangle = 0\,, \tag{7a}$$

$$\mathbf{P}_{\perp}\nabla \otimes \nabla\boldsymbol{\chi}\,|\boldsymbol{E}_{\perp}(\boldsymbol{r})\rangle + \left[\nabla^2 + 2i\begin{pmatrix}\boldsymbol{k}_0\nabla & 0\\ 0 & \boldsymbol{k}_1\nabla\end{pmatrix} + k_0^2\boldsymbol{\chi}\right]|\boldsymbol{E}_{\parallel}(\boldsymbol{r})\rangle = 0\,. \tag{7b}$$

The term $\nabla \oplus \nabla\boldsymbol{\chi}\,|\boldsymbol{E}_\perp\rangle \sim \chi\,\partial^2/\partial r^2|\boldsymbol{E}_\perp\rangle \sim \chi^2$, hence, its contribution is negligibly small. As a result the first term in (7a) gives the same-order contribution, too. Thus, neglecting the term $\sim \chi^2$, which is quite possible for the X-ray region ($\chi \approx 10^{-4}$ to 10^{-6}), we obtain the matrix equation for the transverse component of the electric field vector inside the crystal,

$$\left[\nabla^2 + 2i\begin{pmatrix} \boldsymbol{k}_0\nabla & 0 \\ 0 & \boldsymbol{k}_1\nabla \end{pmatrix} + k_0^2\boldsymbol{\chi}\right]|\boldsymbol{E}(\boldsymbol{r})\rangle = 0\,, \tag{8}$$

here and below the symbol $\perp$ is omitted.

Equation (8) is the matrix form of the Takagi equations in the case of Laue-to-Bragg two-wave diffraction. It should be noted that the substitution

$$|\boldsymbol{E}(\boldsymbol{r})\rangle = \mathbf{U}(\boldsymbol{r})\,|\boldsymbol{E}'(\boldsymbol{r})\rangle\,, \qquad \mathbf{U}(\boldsymbol{r}) = \exp\left\{-\,i\begin{pmatrix} \boldsymbol{k}_0 & 0 \\ 0 & \boldsymbol{k}_1 \end{pmatrix}\boldsymbol{r}\right\}$$

reduces (8) to the Helmholtz equation

$$[\Delta^2 + k_0^2\boldsymbol{\epsilon}'(\boldsymbol{r})]\,|\boldsymbol{E}'(\boldsymbol{r})\rangle = 0 \tag{9}$$

with dielectric matrix function to be defined as follows:

$$\boldsymbol{\epsilon}'(\boldsymbol{r}) = \mathbf{U}^+(\boldsymbol{r})\,\boldsymbol{\epsilon}\mathbf{U}(\boldsymbol{r})\,.$$

If we choose the z-axis directed along the vector $\boldsymbol{N}$, then, due to the continuity of the tangential components of the wave vectors at the boundary, the solution (8) can be represented in the form

$$|\boldsymbol{E}(\boldsymbol{r})\rangle = \mathbf{U}^+(\boldsymbol{r})\,|\boldsymbol{E}_z(\boldsymbol{r})\rangle\,, \tag{10}$$

where $|\boldsymbol{E}_z(\boldsymbol{r})\rangle$ is a solution of the one-dimensional matrix equation

$$\left[\frac{\mathrm{d}^2}{\mathrm{d}z^2} + 2i\begin{pmatrix} \boldsymbol{k}_0\boldsymbol{N} & 0 \\ 0 & \boldsymbol{k}_1\boldsymbol{N} \end{pmatrix}\frac{\mathrm{d}}{\mathrm{d}z} + k_0^2\boldsymbol{\chi}\right]|\boldsymbol{E}_z(\boldsymbol{r})\rangle = 0\,. \tag{11}$$

The case where one of the scalar products $(\boldsymbol{k}_0\boldsymbol{N})/k_0$ or $(\boldsymbol{k}_1\boldsymbol{N})/k_0$ is of the order unity and the second one is much less than unity, corresponds to EAD. But, when $(\boldsymbol{k}_0\boldsymbol{N})/k_0$, $(\boldsymbol{k}_1\boldsymbol{N})/k_0 \ll 1$, (11) describes the SD geometry. To find a solution of (11), in the general case, is a sufficiently complicated problem. In the present paper we are restricted by the consideration of the symmetric ($\boldsymbol{\tau} \perp \boldsymbol{N}$) SD when $\boldsymbol{k}_0\boldsymbol{N} = \boldsymbol{k}_1\boldsymbol{N}$ and this solution can be found easily,

$$|\boldsymbol{E}_z(\boldsymbol{r})\rangle = \mathrm{e}^{ik_{0z}\boldsymbol{n}_+\boldsymbol{N}\boldsymbol{r}}\,|\boldsymbol{E}_+\rangle + \mathrm{e}^{ik_{0z}\boldsymbol{n}_-\boldsymbol{N}\boldsymbol{r}}\,|\boldsymbol{E}_-\rangle\,. \tag{12}$$

Here $\mathbf{n}_\pm = -\,\mathbf{I} \pm (\mathbf{I} + \boldsymbol{\chi}/\cos^2\theta)^{1/2}$, $k_{0z} = k_0\cos\theta$, θ is the incident angle. To define the column vectors independent of the coordinates $|\boldsymbol{E}_\pm\rangle$, (11) should be added to the boundary conditions. However, we make some preliminary remarks before starting the discussion of the boundary problem.

3. Absorption Depth

Because of the exponential divergence, which occurs in the first term of (12) $\big(\mathrm{Im}\,(\mathbf{n}_+) < 0\big)$, the amplitude $|\boldsymbol{E}_+\rangle$ must contain the compensating factor of the form

$$\exp\big(-k_{0z}\,\mathrm{Im}\,(\mathbf{n}_+)\,L\big) \sim \exp\,(-k_{0z}\,\mathrm{Im}\sum_h \sqrt{1 + \lambda_h/\cos^2\theta}\,L)\,,$$

where λ_h are the eigenvalues of the operator $\boldsymbol{\chi}$. As a result, we have the expression of the absorption depth inside the crystal

$$L_{\mathrm{abs}} = \big(k_{0z}\,\mathrm{Im}\sqrt{1 + \lambda_h/\cos^2\theta}\big)^{-1}\,. \tag{13}$$

It can be determined from (13) that for $\cos^2\theta \sim \lambda_h$ the absorption depth is about 10^{-5} to 10^{-6} cm, but for the Mössbauer crystal ^{57}Fe enriched to 90%, it is about 10^{-6} to 10^{-8} cm. Thus, the term with $|\boldsymbol{E}_+\rangle$ in (12) can be omitted in the case of sufficiently thick crystals, i.e. the medium can be considered as half-infinite. We will make use of this fact in the analysis of the boundary problem. It should be noted that (13), just as the preceding one, can be obtained rigorously if we take into account the boundary conditions on the exit surface of the crystal.

The thickness of the layer over which the field of diffracted γ-quanta is formed is defined by L_{abs}. On the other hand, the surface of any real crystal has defects with a size of the same order. As a result, to observe SD the precision of the surface treatment for X-rays must be better than 10^{-6} cm, but for Mössbauer crystals of 90% enrichment it must be better than 10^{-7} cm. Because of so rigorous requirements concerning the surface quality the optical anisotropy of Mössbauer crystals should be studied preferably with unenriched targets for which the extinciton length is large.

4. The Reflection Coefficient

As the inequality $|n - 1| \ll 1$ is satisfied for the refraction index n in the X-ray region, the difference between $\boldsymbol{D}$ and $\boldsymbol{E}$ in the boundary conditions can be neglected (according to the analysis [10] taking into account exact boundary conditions leads to a correction of the order $\approx \chi/\cos\theta \ll 1$). Let us make use of the following form of the boundary conditions:

$$\boldsymbol{E}^{\mathrm{I}} = \boldsymbol{E}^{\mathrm{II}}\,, \qquad \nabla\times\boldsymbol{E}^{\mathrm{I}} = \nabla\times\boldsymbol{E}^{\mathrm{II}}\,, \tag{14}$$

where superscripts I and II correspond to the fields outside and inside the crystal, respectively. According to [3] the field $\boldsymbol{E}^{\mathrm{I}}$ is a superposition of three waves,

$$\boldsymbol{E}^{\mathrm{I}}(\boldsymbol{r}) = \boldsymbol{E}_0\,\mathrm{e}^{i\boldsymbol{k}_0\boldsymbol{r}} + \boldsymbol{E}_0^{\mathrm{I}}\,\mathrm{e}^{i\boldsymbol{k}_0'\boldsymbol{r}} + \boldsymbol{E}_1^{\mathrm{I}}\,\mathrm{e}^{i\boldsymbol{k}_1'\boldsymbol{r}}\,,$$

where $\boldsymbol{E}_0$, $\boldsymbol{k}_0$, $\boldsymbol{E}_0^{\mathrm{I}}$, $\boldsymbol{k}_0'$, $\boldsymbol{E}_1^{\mathrm{I}}$, $\boldsymbol{k}_1'$ are amplitudes and wave vectors of the incident, reflected, and diffracted-reflected waves. From simple considerations it follows that

$$\boldsymbol{k}_0' = \boldsymbol{k}_0 - 2k_{0z}\boldsymbol{N}\,, \qquad \boldsymbol{k}_1' = \boldsymbol{k}_1 - (k_{0z} + k_{0z}')\,\boldsymbol{N}\,. \tag{15}$$

The expression for the normal component of the wave vector of the diffracted-reflected wave follows from the energy conservation law for elastic scattering [3],

$$k_{0z}' = \sqrt{k_0^2 - \boldsymbol{k}_{1\tau}'^2}\,, \qquad \boldsymbol{k}_{1\tau}' = \boldsymbol{k}_{0\tau} + \boldsymbol{\tau}\,,$$

$\boldsymbol{k}_{0\tau}$, $\boldsymbol{k}_{1\tau}'$ are the tangential components of $\boldsymbol{k}_0$, $\boldsymbol{k}_1'$. The field inside the crystal is defined by (10) and (12). In accordance with the ideas mentioned above, the term in (12) containing $|\boldsymbol{E}_+\rangle$ is omitted. Having in mind that

$$\nabla\times\mathrm{e}^{i\boldsymbol{k}\boldsymbol{r}}\,\boldsymbol{E} = i\boldsymbol{k}^{\times}\,\mathrm{e}^{i\boldsymbol{k}\boldsymbol{r}}\,\boldsymbol{E}\,, \qquad \nabla\times\mathrm{e}^{ik_{0z}n_-\boldsymbol{N}\boldsymbol{r}}\,\boldsymbol{E} = ik_{0z}\boldsymbol{N}^{\times}\mathrm{n}_-\,\mathrm{e}^{ik_{0z}n_-\boldsymbol{N}\boldsymbol{r}}\,\boldsymbol{E}\,,$$

and that boundary conditions are fulfilled independently for incident and diffracted waves the system of matrix equations is obtained from (14),

$$\begin{pmatrix}\boldsymbol{k}_0'^{\times} & 0\\ 0 & \boldsymbol{k}_0'^{\times}\end{pmatrix}|\boldsymbol{E}^{\mathrm{I}}\rangle + \begin{pmatrix}\boldsymbol{k}_0^{\times} & 0\\ 0 & \boldsymbol{k}_1^{\times}\end{pmatrix}|\boldsymbol{E}_0\rangle = k_{0z}\boldsymbol{N}^{\times}\left\{\mathrm{n}_+ + \begin{pmatrix}\boldsymbol{k}_0^{\times} & 0\\ 0 & \boldsymbol{k}_1^{\times}\end{pmatrix}\right\}|\boldsymbol{E}_-\rangle\,, \tag{16}$$

$$|\boldsymbol{E}^{\mathrm{I}}\rangle + |\boldsymbol{E}_0\rangle = |\boldsymbol{E}_-\rangle\,,$$

where

$$|\boldsymbol{E}^{\mathrm{I}}\rangle = \begin{pmatrix}\boldsymbol{E}_0^{\mathrm{I}}\\ \boldsymbol{E}_1^{\mathrm{I}}\end{pmatrix}, \qquad |\boldsymbol{E}_0\rangle = \begin{pmatrix}\boldsymbol{E}_0\\ 0\end{pmatrix}.$$

The system (16) permits us to obtain the tensor coefficients of reflection **R** and transmission **T** by means of simple transformations,

$$\mathbf{R} = \mathbf{A}\mathbf{A}^+ , \qquad \mathbf{T} = \mathbf{B}\mathbf{B}^+ . \tag{17}$$

The matrices **A**, **B** define the field vectors outside and inside the crystal through the incident one and have the form

$$\mathbf{A} = -\frac{\mathbf{I} - \sqrt{\mathbf{I} + \boldsymbol{\chi}/\cos^2\theta}}{\mathbf{I} + \sqrt{\mathbf{I} + \boldsymbol{\chi}/\cos^2\theta} + \begin{pmatrix} 0 & 0 \\ 0 & \delta \end{pmatrix}}, \tag{18}$$

$$\mathbf{B} = \frac{2}{\mathbf{I} + \sqrt{\mathbf{I} + \boldsymbol{\chi}/\cos^2\theta} + \begin{pmatrix} 0 & 0 \\ 0 & \delta \end{pmatrix}}; \qquad \delta = \frac{k'_{0z} - k_{0z}}{k_{0z}}.$$

Expressions (18) are the tensor analogue of the well-known Fresnel formulae in optics. For our purposes it is important that (17) and (18) were obtained without any assumptions concerning the form of tensor $\boldsymbol{\chi}$ (only its obvious property to be transverse, was used). Hence, these formulae are correct for arbitrary $\boldsymbol{\chi}$ [12, 13],

$$\boldsymbol{\chi}_{\alpha\beta} = \chi_{\text{el}} - \frac{4\pi\eta}{\hbar\omega^2 V_0} \sum_{j,M,M_0} W(M_0, \boldsymbol{r}) \exp\left\{-\frac{1}{2} Z_j(\boldsymbol{k}_\alpha) - \frac{1}{2} Z_j(\boldsymbol{k}_\beta)\right\} \times$$

$$\times \exp(i\boldsymbol{\tau}\boldsymbol{r}_j) \frac{\langle M_0| J_n^j(\boldsymbol{k}_\alpha) |M\rangle \otimes \langle M| J_n^j(\boldsymbol{k}_\beta) |M_0\rangle}{\omega - \omega(M, M_0) + i\Gamma/2}. \tag{19}$$

Here $\alpha, \beta = 0, 1$, η is the Mössbauer isotope concentration, V_0 the unit cell volume, $|M_0\rangle$ and $|M\rangle$ are the spin wave functions of the Mössbauer nuclei in the ground and excited states, the symbol j denotes the j-th nucleus in the unit cell. The value $W(M_0, \boldsymbol{r})$ is the probability of M_0 level population. It should be noted that this value leads to the appearance of temperature effects for Mössbauer crystals at low-temperature investigations [13]. The term χ_{el} corresponds to the nonresonant Rayleigh scattering.

Equations (17) and (18) obtained above permit to simplify sufficiently the reflection analysis by reducing it to a set of operations on matrices, which can be performed numerically. The latter is especially important because the evident analytical expressions for the field inside the crystal required in the usual calculations can be obtained for the same cases when only the eigenvalues and eigenvectors of the operator $\mathbf{I} + \boldsymbol{\chi}/\cos^2\theta$ are found analytically [9, 14]. We do not intend to discuss these particular cases in detail. We consider one of them only, as a demonstration of the possibility of the technique developed in the present paper. If we take into account the evident form of the transition current matrix elements [15, 9], (19) becomes diadic provided the nonresonant term χ_{el} is neglected,

$$\boldsymbol{\chi}_{\alpha\beta} = \chi_{\alpha\beta} [\boldsymbol{k}_\alpha \times \boldsymbol{e}_q] \otimes [\boldsymbol{k}_\beta \times \boldsymbol{e}_q]^* , \tag{20}$$

where constants $\chi_{\alpha\beta}$ include the scalar values from (19) as well as the resonant factor, $q = 0, \pm 1$. The vectors $\boldsymbol{e}_q$ define a cyclic basis connected with the Cartesian one by means of relations

$$\boldsymbol{e}_0 = \boldsymbol{e}_z , \qquad \boldsymbol{e}_{\pm 1} = \mp \frac{1}{\sqrt{2}} (\boldsymbol{e}_x \pm i\boldsymbol{e}_y) .$$

The vector $\boldsymbol{e}_z$ is parallel to the hyperfine field.

Due to the transverse character of the six-vector $|E_-\rangle$ the unit tensor $\mathbf{I}$ in (18) is transverse too: $\mathbf{I} = \mathbf{P}_\perp$. Using this it can be shown by expanding $(\mathbf{I} + \boldsymbol{\chi}/\cos^2\theta)^{1/2}$ into a series in $\boldsymbol{\chi}$ that the following relation for $\boldsymbol{\chi}$ of the form (20) takes place:

$$\sqrt{\mathbf{I} + \boldsymbol{\chi}/\cos^2\theta} = \mathbf{I} + \begin{pmatrix} k_0^x & 0 \\ 0 & k_1^x \end{pmatrix} (-1 + \sqrt{1 + \boldsymbol{\mu}\boldsymbol{\nu}})\, \frac{1}{\boldsymbol{\mu}}\, e_q \otimes e_q^* \begin{pmatrix} k_0^x & 0 \\ 0 & k_1^x \end{pmatrix}^*,$$

where

$$\boldsymbol{\mu} = \begin{pmatrix} \chi_{00} & \chi_{01} \\ \chi_{10} & \chi_{11} - \alpha \end{pmatrix}, \qquad \boldsymbol{\nu} = -\frac{1}{k_0^2} \begin{pmatrix} k_0^2 - |\boldsymbol{k}_0 e_q|^2 & 0 \\ 0 & k_0^2 - |\boldsymbol{k}_1 e_q|^2 \end{pmatrix}.$$

As a result, (18) reduces to

$$\mathbf{A} = \begin{pmatrix} k_0^x & 0 \\ 0 & k_1^x \end{pmatrix} \frac{1}{-1 + \sqrt{1 + \boldsymbol{\mu}\boldsymbol{\nu}} + \begin{pmatrix} 0 & 0 \\ 0 & \delta \end{pmatrix}}\, \frac{1}{\boldsymbol{\mu}}\, e_q \otimes e_q^* \begin{pmatrix} k_0^x & 0 \\ 0 & k_1^x \end{pmatrix}^*.$$

A corresponding expression for the matrix $\mathbf{B}$ can be obtained analogously. In this manner we get rid of the necessity to operate on 6×6 matrices (of fourth rank) in the particular case (20) and the operations of taking a root and inversion are not difficult for 2×2 matrices. Carrying out these operations we shall obtain the results found in [10] in the usual way.

5. Conclusion

The explicit expressions for the reflection and transmission coefficients have been obtained for the symmetric case of Baryshevskii diffraction geometry. It has been shown that it is preferable to use an unenriched Mössbauer analysing target in comparison with an enriched one.

Acknowledgement

The authors are particularly indebted to Prof. Dr. V. G. Baryshevskii for the formulation of the problem and numerous helpful discussions.

References

[1] S. K. Kishino and K. Kohra, Japan J. appl. Phys. **10**, 551 (1971); J. Phys. Soc. Japan **31**, 1168 (1971).
[2] V. M. Kaganer, V. L. Indenbom, M. Vrána, and B. Chalupa, phys. stat. sol. (a) **71**, 371 (1982).
[3] V. G. Baryshevskii, Zh. tekh. Fiz., Pisma **2**, 112 (1976); Nuclear Optics of Polarized Media, Belorussian State University, Minsk 1976 (in Russian).
[4] W. C. Marra, P. Eizenberger, and A. Y. Cho, J. appl. Phys. **50**, 6927 (1979).
[5] A. L. Golovin and R. M. Imamov, phys. stat. sol. (a) **77**, K91 (1983).
[6] S. Takagi, Acta cryst. **15**, 1131 (1962); J. Phys. Soc. Japan **26**, 1239 (1969).
[7] F. I. Fedorov, Theory of Gyrotropic Media, Izd. Nauka, Minsk 1976 (in Russian).
[8] I. D. Feranchuk, Izv. Akad. Nauk BSSR, Ser. fiz.-mat. Nauk, No. 4, 109 (1981).
[9] M. A. Andreeva and R. N. Kuzmin, Mössbauer Gamma-Optics, Moscow State University, Moscow 1982 (in Russian).
[10] M. A. Andreeva and R. N. Kuzmin, Solid. State Commun. **49**, 743 (1984).
[11] V. P. Silin and A. A. Ruchadse, Electromagnetic Properties of Plasma, Izd. Nauka, Moscow 1961 (in Russian).
[12] V. G. Baryshevskii and V. V. Skadorov, phys. stat. sol. (b) **108**, 631 (1981).
[13] V. G. Baryshevskii, J. Phys. C **17**, 3255 (1984).
[14] A. V. Kolpakov, V. A. Bushuev, and R. N. Kuzmin, Uspekhi fiz. Nauk **126**, 479 (1978).
[15] P. A. Aleksandrov, Zh. eksper. teor. Fiz. **67**, 728 (1974).

(Received February 12, 1985)

phys. stat. sol. (b) **131**, 437 (1985)

Subject classification: 1.3; 21.1

Institute of Nuclear Research and Nuclear Energy, Bulgarian Academy of Sciences, Sofia[1])

Atomic Ordering in Multicomponent Alloys

By

L. DĄBROWSKI

Dedicated to Prof. Dr. Dr. h. c. Dr. E. h. P. GÖRLICH on the occasion of his 80th birthday

The pair interaction model of binary alloys is generalized to cover the case of an arbitrary number of components. Both, the interaction radius and the number of short- and long-range ordering parameters are unlimited. The problem is solved by applying the high-temperature series expansion method, using a new diagram technique, to calculate the fluctuation potential.

Модель парных взаимодействий бинарного сплава обобщается на случай произвольного числа компонентов. Радиус взаимодействия, а также количество параметров ближнего и дальнего порядка неограничены. Проблема решается методом высокотемпературного разложения флюктуационного потенциала с применением новой диаграмной техники.

1. Introduction

The study of multicomponent solid solutions in general and of binary alloys in particular not only occupies a large space in experimental and theoretical physics but constitutes a great interest for metallography as well [1, 2]. There are numerous and diverse theories of the atomic ordering in binary alloys [3]. Several attempts have been made to solving the problem of ternary alloys [4 to 6], the basic idea being to reduce it to the Ising model with pseudo-spin one. Introducing the pseudo-spin complicates a great deal the starting Hamiltonian. This renders difficult generalizing the model to a higher number of components. An average-field method has been developed [7, 8] for solving the ternary-alloy problem, account being taken, however, of the long-range ordering only.

This paper presents a general solution of the problem for an arbitrary number of components. It makes use of the ideas of a long-range entropy, as found in a general form in [9], and of the short-range entropy derived in [10]. The results obtained here are a generalization of those reported in [10] for the case of a unlimited number of components.

2. Formulation of the Problem

Assuming the existence of pair interactions only, the Hamiltonian of an alloy composed of n components can be written in the following general form:

$$H = \frac{1}{2} \sum_{\mu,\nu=1}^{n} \sum_{ij} V_{ij}^{\mu\nu} \hat{c}_i^{\mu} \hat{c}_j^{\nu} , \tag{1}$$

[1]) Boul. Lenin 72, 1184 Sofia, Bulgaria.

where $V_{ij}^{\mu\nu}$ is the interaction potential between an atom of sort μ located in the point $\boldsymbol{r}_i$ of the crystalline lattice, and an atom of sort ν located at $\boldsymbol{r}_j$; $\hat{c}_i^{\mu}(\hat{c}_j^{\nu})$ is a random function, equal to unity when a $\mu(\nu)$ atom is located at $\boldsymbol{r}_i(\boldsymbol{r}_j)$, and equal to zero otherwise.

The partition function corresponding to Hamiltonian (1) is

$$Z = \sum_{\hat{c}_i^{\mu}, \hat{c}_j^{\nu}=0}^{1} \exp\left(-\frac{\beta}{2} \sum_{\mu,\nu} \sum_{i,j} V_{ij}^{\mu\nu} \hat{c}_i^{\mu} \hat{c}_j^{\nu}\right), \tag{2}$$

where $\beta = 1/kT$, k and T are the Boltzmann constant and the absolute temperature.

Starting from [2] and making the familiar manipulations [9, 10] we obtain an expression for the free energy of the form

$$F = \langle H \rangle - kTS + \Delta F\,, \tag{3}$$

where $\langle H \rangle$ is the Hamiltonian (1) averaged thermodynamically, S the entropy,

$$\Delta F = -\,kT \ln \langle \exp(-\beta\,\Delta H) \rangle$$

the fluctuation potential of the alloy,

$$\Delta H = H - \langle H \rangle\,.$$

$\langle H \rangle$ can be presented in the form

$$\langle H \rangle = \tfrac{1}{2} \sum_{\mu,\nu} \sum_{ij} V_{ij}^{\mu\nu} \langle \hat{c}_i^{\mu} \hat{c}_j^{\nu} \rangle = \tfrac{1}{2} \sum_{\mu\nu} \sum_{ij} V_{ij}^{\mu\nu} P_{ij}^{\mu\nu}$$
$$= \tfrac{1}{2} \sum_{\mu\nu} \sum_{ij} V_{ij}^{\mu\nu} (c_i^{\mu} c_j^{\nu} + G_{ij}^{\mu\nu})\,, \tag{4}$$

where $P_{ij}^{\mu\nu}$ is the probability that μ atom at $\boldsymbol{r}_i$ is adjoining a ν atom at $\boldsymbol{r}_j$; $c_i^{\mu}(c_j^{\nu})$ is the concentration of $\mu(\nu)$ atoms at $\boldsymbol{r}_i(\boldsymbol{r}_j)$; $G_{ij}^{\mu\nu} = \langle (\hat{c}_i^{\mu} - c_i^{\mu})\,(\hat{c}_j^{\nu} - c_j^{\nu}) \rangle$ is the static function of pair correlations between μ and ν atoms.

The entropy S consists of the entropy $S_{\rm L}$ related to the long-range ordering and $S_{\rm sh}$ arising from the short-range ordering [10]. To find the entropy of long-range ordering, one has to compute all possible combinations for distributing n sorts of atoms over m sublattices of the system at given average concentration of the sublattices. In its ultimately simplified form the problem reduces to that of the distribution of n different balls over N different cells under the condition that the total number of balls is equal to the number of the cells. The ball distribution follows the polynomial distribution so that the number of the respective combinations W is a coefficient of the latter, and is given by [11]

$$W = \frac{N!}{X_1!\, X_2! \ldots X_n!}\,. \tag{5}$$

Based on (5), the entropy $S_{\rm L}$ is finally obtained as

$$S_{\rm L} = -\sum_{\mu} \sum_{i} c_i^{\mu} \ln c_i^{\mu}\,. \tag{6}$$

To derive the entropy $S_{\rm sh}$, one has first to find the effective concentration of a μ atom, $c_{i\,\rm eff}^{\mu}$, which as localized at $\boldsymbol{r}_i$ due to its interaction with another atom of sort ν located at $\boldsymbol{r}_j$. This concentration is equal to

$$c_{i\,\rm eff}^{\mu} = \frac{P_{ij}^{\mu\nu}}{c_j^{\nu}} = c_i^{\mu}(1 + D_{ij}^{\mu\nu})\,, \tag{7}$$

where

$$D_{ij}^{\mu\nu} = \frac{G_{ij}^{\mu\nu}}{c_i^{\mu} c_j^{\nu}}\,.$$

The entropy associated with all such atoms is

$$S^{\nu}_{j\,\mathrm{eff}} = \sum_{\mu} \sum_{i} c^{\mu}_{i}(1 + D^{\mu\nu}_{ij}) \ln c^{\mu}_{i}(1 + D^{\mu\nu}_{ij}) \,. \tag{8}$$

The entropy obtained (cf. [10]) contains the entropy of the noninteracting atoms, as well as the long-range ordering entropy. Both have to be subtracted from (8). This can easily be carried out using, as in [10], the condition that $D^{\mu\nu}_{ij} \to 0$ as $V^{\mu\nu}_{ij}/kT \to \to 0$. One obtains then

$$S^{\nu}_{j} = \sum_{\mu} \sum_{i} c^{\mu}_{i}(1 + D^{\mu\nu}_{ij}) \ln (1 + D^{\mu\nu}_{ij}) \,. \tag{9}$$

Taking into account the statistical weight of different sorts of atoms, the short-range entropy is finally found to be

$$S_{\mathrm{sh}} = -\tfrac{1}{2} \sum_{\mu\nu} \sum_{ij} c^{\mu}_{i} c^{\nu}_{j}(1 + D^{\mu\nu}_{ij}) \ln (1 + D^{\mu\nu}_{ij}) \,. \tag{10}$$

For a binary alloy, when the number of components is 2, (6) coincides with the one derived in [9], while (10) for S_{sh} reduces to the one obtained in [10].

3. Ordering Parameters

For a binary alloy, the number of long-range ordering parameters is by 1 smaller than the number of sublattices. For a multicomponent alloy that number grows, as one has to separately account for each component. It should be kept in mind that the equality $\sum_{\mu} c^{\mu}_{k} = 1$ holds good for each sublattice "k". Consequently, the number of independent long-range ordering parameters M is

$$M = (m - 1)\,(n - 1)\,, \tag{11}$$

where m is the number of sublattices, while n is the number of components.

The short-range ordering parameters are given in [10] in general form for a binary alloy as follows:

$$\delta_{ij} = \frac{G_{ij}}{c_i c_j (1 - c_i)\,(1 - c_j)}\,, \tag{12}$$

where G_{ij} are the pair correlation functions between atoms of one sort, while $c_i(c_j)$ are the respective concentrations of these atoms.

For the multicomponent solid solution under consideration it is more convenient to use instead the correlation functions $G^{\mu\nu}_{ij}$ without normalizing factors as short-range ordering parameters. Inasmuch as $\sum_{\mu} \hat{c}^{\mu}_{i} = 1$ at each point, the fluctuations obey $\sum_{\mu} \Delta c^{\mu}_{i} = 0$. This implies that at any given subscripts i and j the number of independent short-range parameters is $(n - 1)^2$ where n is the number of components.

4. Fluctuation Potential

Using the high-temperature expansion method, we obtain the following perturbation series for the fluctuation potential (3):

$$-\frac{\Delta F}{kT} = \ln \langle \exp(-\beta\,\Delta H)\rangle = \sum_{n=1}^{\infty} \frac{(-\beta)^n}{n!} M_n(\Delta H)\,, \tag{13}$$

where M_n is the n-th cumulant. We then get by elementary manipulations

$$\Delta H = \tfrac{1}{2} \sum_{\mu\nu} \sum_{ij} V^{\mu\nu}_{ij}(\Delta c^{\mu}_{i} \Delta c^{\nu}_{j} - G^{\mu\nu}_{ij} + c^{\mu}_{i}\,\Delta c^{\nu}_{j} + \Delta c^{\mu}_{i} c^{\nu}_{j}) = \tfrac{1}{2} \sum_{\mu\nu} \sum_{ij} B^{\mu\nu}_{ij}\,. \tag{14}$$

To calculate M_n one can adapt the diagram expansion technique, as introduced in [12] and developed in [13], with ΔH given by (14). Cleary, the vertexes of the diagrams, appearing graphically as before, now symbolize the conjugated summation subscripts (μ, i). Single bonds connecting two different vertexes symbolize the expressions $B_{ij}^{\mu\nu}$, while n-fold bonds symbolize n-th powers of these expressions. For instance, diagrams

 (15)

signify

$$\tfrac{1}{2} \sum_{\mu\nu} \sum_{ij} B_{ij}^{\nu\mu} , \qquad 3 \sum_{\mu\nu\lambda} \sum_{ijk} (B_{ij}^{\mu\nu})^2 B_{jk}^{\nu\lambda} , \qquad \sum_{\mu\nu\lambda} \sum_{ijk} B_{ij}^{\mu\nu} B_{jk}^{\nu\lambda} B_{ki}^{\lambda\mu} , \tag{16}$$

respectively. The numerical factors to the sums in (16) are calculated in the same way as for a binary alloy [12, 13], namely, they are equal to $n!/g$ where n is the number of bonds, while g is the number of identity transforms of the diagram symmetry. These are old-type diagrams.

5. Summing of Two- and Three-Vertex Diagrams

The contribution to the fluctuation potential of all two-vertex diagrams can be written in the following way:

$\frac{(-\beta)}{1!}$  $+ \frac{(-\beta)^2}{2!}$  $+ \frac{(-\beta)^3}{3!}$  $+ \cdots =$

$$= \frac{1}{2} \sum_{n=1}^{\infty} \sum_{\mu\nu} \sum_{ij} \frac{(-\beta)^n (V_{ij}^{\mu\nu})^n}{n!} \langle (\Delta c_i^\mu \, \Delta c_j^\nu - G_{ij}^{\mu\nu} + \Delta c_i^\mu c_j^\nu + c_i^\mu \, \Delta c_j^\nu)^n \rangle . \tag{17}$$

Having in mind the definition of $\hat{c}_i^\mu$ (equation (1)) we see that

$$(\hat{c}_i^\mu)^n = \hat{c}_i^\mu , \tag{18}$$

wherefrom

$$(\Delta c_i^\mu)^n = (\hat{c}_i^\mu - c_i^\mu)^n = \hat{c}_i^\mu (1 - c_i^\mu)^n + (1 - \hat{c}_i^\mu) (-c_i^\mu)^n \tag{19}$$

follows. For the sake of further calculations it is convenient to present (19) into another form,

$$(\Delta c_i^\mu)^n = \sum_{p=0}^{1} (-1)^{p+1} \hat{\gamma}_{i,p}^\mu (\gamma_{i,1-p}^\mu)^n , \tag{20}$$

where

$$\hat{\gamma}_{i,0}^\mu = - \hat{c}_i^\mu , \qquad \hat{\gamma}_{i,1}^\mu = 1 - \hat{c}_i^\mu , \qquad \gamma_{i,p}^\mu = \langle \hat{\gamma}_{i,p}^\mu \rangle .$$

Using (20) we finally obtain the contribution of the two-vertex diagrams as

$\frac{(-\beta)}{1!}$  $+ \cdots = \frac{1}{2} \sum_{\mu\nu} \sum_{(ip)(jq)} (-1)^{p+q} \langle \hat{\gamma}_{i,p}^\mu \hat{\gamma}_{j,q}^\nu \rangle \, \alpha_{(ip)(jq)}^{\mu\nu}$, (21)

where

$$\alpha_{(ip)(jq)}^{\mu\nu} = \exp \left[-\beta V_{ij}^{\mu\nu} (\gamma_{i,1-p}^\mu \gamma_{j,1-q}^\nu - G_{ij}^{\mu\nu} + \gamma_{i,1-p}^\mu c_j^\nu + c_i^\mu \gamma_{j,1-q}^\nu) \right] - 1 .$$

Similary, we get an expression for the contribution to the fluctuation potential of the three-vertex diagrams,

$\frac{(-\beta)^2}{2!}$  $+ \frac{(-\beta)^3}{3!}$  $+ \frac{(-\beta)^4}{4!}$  $+ \cdots =$

$$= \frac{1}{2} \sum_{\mu\nu\lambda} \sum_{(ip)(jq)(kr)} (-1)^{p+q+r+1} \langle \hat{\gamma}_{i,p}^\mu \hat{\gamma}_{j,q}^\nu \hat{\gamma}_{k,r}^\lambda \rangle \, \alpha_{(ip)(jq)}^{\mu\nu} \alpha_{(jq)(kr)}^{\nu\lambda} \tag{22}$$

and, correspondingly,

$$\frac{(-\beta)^3}{3!}$$  $$+ \frac{(-\beta)^4}{4!}$$  $$+ \dots =$$

$$= \frac{1}{3!} \sum_{\mu\nu\lambda} \sum_{(ip)(jq)(kr)} (-1)^{p+q+r+1} \langle \hat{\gamma}^{\mu}_{i,p} \hat{\gamma}^{\nu}_{j,q} \hat{\gamma}^{\lambda}_{k,r} \rangle \, \alpha^{\mu\nu}_{(ip)(jq)} \alpha^{\nu\lambda}_{(jq)(kr)} \alpha^{\lambda\mu}_{(kr)(ip)} \,. \tag{23}$$

6. New-Type Diagrams

The above expressions for the contribution to the fluctuation potential of two- and three-vertex diagrams (21) to (23)) can be generalized so as to cover an arbitrary number of vertexes. These sums can be represented as new diagrams. Appearing graphically as the old ones, they, however, contain single bonds only. Vertexes now symbolize the expressions $(-1)^{p+1} \hat{\gamma}^{\mu}_{i,p}$ while bonds symbolize $\alpha^{\mu\nu}_{(ip)(jq)}$. Inasmuch as the $n!$ factor is now included into the diagrams, the new factor to the sums is $n!$ times smaller than the respective former terms of the same diagrams.

7. Discussion of the Results

The new-diagram method applied herein makes it possible to reveal the structure of the flutuation potential. As a matter of fact, terms of the $\langle \dots \hat{\gamma}^{\mu}_{i,p} \hat{\gamma}^{\nu}_{j,q} \hat{\gamma}^{\lambda}_{k,r} \dots \rangle$-type entering the diagrams ((21) to (23)) are given by

$$\langle \dots \hat{\gamma}^{\mu}_{i,p} \hat{\gamma}^{\nu}_{j,q} \hat{\gamma}^{\lambda}_{k,r} \dots \rangle = \dots \gamma^{\mu}_{i,p} \gamma^{\nu}_{j,q} \gamma^{\lambda}_{k,r} \dots + \dots G^{\mu\nu}_{ij} \gamma^{\lambda}_{k,r} \dots + \\ + \dots G^{\mu}_{\cdot i} \gamma^{\nu}_{j,q} \gamma^{\lambda}_{k,r} \dots + \dots + R(G^{\mu\nu\lambda}_{ijk}, G^{\mu\nu\lambda\theta}_{ijkl}, \dots) \,, \tag{24}$$

where R contains multiparticle correlation functions. The factors to the respective terms in the fluctuation potential play the role of effective multiparticle potentials. Each n-vertex diagram contains multiparticle correlation functions and multiparticle effective potentials up to and including the n-particle ones. Neglecting these multiparticle terms in the perturbation series should not be expected to lead to large errors for the following two reasons:

(i) if the starting Hamiltonian contains no multiparticle interactions, as it does in the present case, they will be generated by the fluctuation potential in higher orders of the perturbation series; consequently, these should be small compared to the basic two-particle potentials;

(ii) the multiparticle correlation functions must, in principle, be smaller or much smaller than the two-particle ones, independent of the multiparticle interaction strength. For example,

$$\langle \Delta c^{\mu}_i \, \Delta c^{\nu}_j \, \Delta c^{\lambda}_k \rangle \leqq \langle \Delta c^{\mu}_i \, (\Delta c^{\nu}_j)^2 \rangle = (1 - 2c^{\nu}_j) \, G^{\mu\nu}_{ij} \,. \tag{25}$$

Similar conditions are obtained for the correlation functions of more particles. Having in mind the considerations under (i) and (ii), as well as the fact that they enter the fluctuation potential as products, it follows that their contribution should be negligible.

References

[1] T. Zemčik, Proc. Internat. Conf. Mössbauer Spectroscopy, Kracov 1975, (p. 59).

[2] H. Sato, Physical Chemistry: An Advanced Treatise, Vol. 10, Ed. W. Jost, Academic Press, New York 1970 (p. 1)

[3] F. Reynaud, phys. stat. sol. (a) 72, 11 (1982).

[4] RAZA A. TAHIR-KHELI, Phys. Rev. **169**, 517 (1968).
[5] G. BRUCE TAGGART, Phys. Rev. B **19**, 2895 (1979).
[6] G. BRUCE TAGGAGT and RAZA A. TAHIR-KHELI, Phys. Rev. B **3**, 1059 (1971).
[7] L. P. GRANKINA, V. I. LOZOVOI, E. F. RYZHKOVA, and V. I. RYZHKOV, Izv. Vuz. Fiz. **8**, 47 (1982).
[8] L. I. SMIRNOV, Fiz. Metallov i Metallovedenie **54**, 629 (1982).
[9] A. G. KHACHATURYAN, Progr. Mater. Sci. **22**, 1 (1978).
[10] L. DĄBROWSKI, phys. stat. sol. (b) **128**, 371 (1985).
[11] G. A. KORN and T. M. KORN, Spravotchnik po Matematike, Nauka, Moskva 1973 (p. 567).
[12] G. HORVITZ and H. CALLEN, Phys. Rev. **124**, 1757 (1961).
[13] D. A. BADELYAN and A. G. KHACHATURYAN, Fiz. tverd. Tela **12**, 439 (1970).

(Received May 30, 1985)

phys. stat. sol. (b) 131, 443 (1985)

Subject classification: 1.6; 21.1

Laboratorium voor Kristallografie en Studie van de Vaste Stof, Rijksuniversiteit Gent[1])

Kinetics of Surface Segregation in Binary Alloys

By

L. T. WILLE and J. VENNIK

Dedicated to Prof. Dr. Dr. h. c. Dr. E. h. P. GÖRLICH on the occasion of his 80th birthday

A new theory for the kinetics of surface segregation in binary alloys is proposed, which takes into account the concentration in the subsurface layers. This more realistic segregation profile is derived within the regular solution pair-bonding model. The surface composition as a function of time is determined analytically and it is in qualitative agreement with the experimentally observed behaviour.

Eine neue Theorie für die Kinetik der Oberflächensegregation in binären Verbindungen wird vorgeschlagen, die die Konzentration in den Suboberflächenschichten berücksichtigt. Dieses realistischere Segregationsprofil wird im Rahmen des regulären Lösungs-Paarbindungsmodells abgeleitet. Die Oberflächenzusammensetzung als Funktion der Zeit wird analytisch bestimmt und ist in qualitativer Übereinstimmung mit dem experimentell beobachteten Verhalten.

1. Introduction

The segregation of one of the constituents to planar defects in binary alloys is a phenomenon with important consequences. In the case of grain boundaries [1] it can cause embrittlement, whereas at a free surface [2] it influences the catalytic behaviour. Experimental studies of surface segregation are often hindered by preferential sputtering or because the segregation kinetics are not known. The standard theory of these kinetics is due to McLean [3] in the case of grain boundaries. Obviously surface segregation can be treated completely similarly, the only difference being that the solute diffuses to the surface from one side only. McLean's original theory has been extended by several authors [4 to 6]. In this paper we propose an alternative theory in which the actual composition profile is approximated more closely. It will be our aim to investigate the consequences of this approach, rather than obtain exact agreement with experiments.

McLean [3] considered the grain boundary at temperature T_1 as a reservoir with a concentration C_{gb} in equilibrium with the crystal interior at a concentration C_1, the ratio being $\alpha_1 = C_{gb}/C_1$. After quenching to a temperature T_2, at which the equilibrium concentration ratio is α_2, the change in grain boundary concentration with time is described by a one-dimensional diffusion equation. It is assumed that the concentration in the grain boundary is equal to α_2 times the concentration in the adjacent crystal plane, during this process of diffusion. The interface condition expresses that the flux through the interface equals the rate of accumulation of the segregant in the grain boundary. In this way the problem is completely determined and it can be solved by the method of Laplace transforms. It is found that the grain boundary concentration builds up rather quickly but that it approaches equilibrium slowly.

---
[1]) Krijgslaan 281, B-9000 Gent, Belgium.

This theory was applied to the case of surface segregation by Lea and Seah [4]. These authors also took into account the possibility of evaporation of the segregant at the surface, assuming the rate of evaporation to be proportional to the solute content in the surface layer. The alloy system FeSn was studied with this method and it was found that the assumption of a constant α_2 during segregation was completely inadequate. Moreover in reality α_2 might vary over many orders of magnitude depending on the adsorption relation. Using an experimentally obtained relation between α_2 and the surface concentration, which gives a much larger initial value for α_2, satisfactory experimental agreement was reached.

A further analysis of the segregation kinetics was made by Rowlands and Woodruff [5], focussing on the driving forces behind surface segregation. Only parameters of direct physical interest such as diffusion coefficients and the heat of segregation entered this theory. Again the FeSn-system was investigated and the agreement with the data of Lea and Seah [4] was satisfactory. A similar, somewhat more phenomenological theory was proposed independently by Brailsford [6].

2. The Equilibrium Segregation Profile

In this paper we will investigate the consequences of using a more realistic omposition profile in the surface region. Most theories of segregation are based on the regular solution model and the pair-bonding treatment as introduced by Swalin [7] in the study of bulk thermodynamics in binary alloys. Assuming nearest neighbour pair interactions w_{ij} $(i, j = \text{A, B})$ between the atoms of an AB alloy, the equilibrium concentration in the different layers parallel to the surface is described by a set of transcendental equations [8],

$$c_0 = \left\{1 + \exp\left[\frac{1}{kT}\left(v_{\text{AB}} n_{\text{out}} + \frac{1}{2} w n_{\text{out}} (2c_{\text{B}} - 1) - \right.\right.\right.$$
$$\left.\left.\left. - w(n_{\text{in}} \,\Delta c_0 + n_{\text{out}} \,\Delta c_1)\right) + \log \frac{1 - c_{\text{B}}}{c_{\text{B}}}\right]\right\}^{-1}, \tag{1a}$$

$$c_n = \left\{1 + \exp\left[-\frac{w}{kT}\left(n_{\text{out}} \,\Delta c_{n-1} + n_{\text{in}} \,\Delta c_n + n_{\text{out}} \,\Delta c_{n+1}\right) + \right.\right.$$
$$\left.\left. + \log \frac{1 - c_{\text{B}}}{c_{\text{B}}}\right]\right\}^{-1}; \qquad n = 1, 2, \dots . \tag{1b}$$

Here c_{B} denotes the bulk concentration and Δc_n is the excess concentration in the n-th layer,

$$\Delta c_n = c_n - c_{\text{B}} , \tag{2}$$

where c_n is the concentration in the n-th layer and the surface is taken to be the zeroth layer. The interaction energies are

$$v_{\text{AB}} = w_{\text{AA}} - w_{\text{BB}} , \tag{3}$$

i.e. the difference in heats of sublimation of the pure metals A and B, and

$$w = w_{\text{AA}} + w_{\text{BB}} - 2w_{\text{AB}} , \tag{4}$$

which is the heat of mixing of the alloy. These are the only parameters in the model. Often these quantities are obtained from tabulated values of thermodynamic properties of alloys. However, it would be very interesting to derive them from an electronic structure calculation (e.g. tight binding or pseudopotential method). Finally n_{in} is

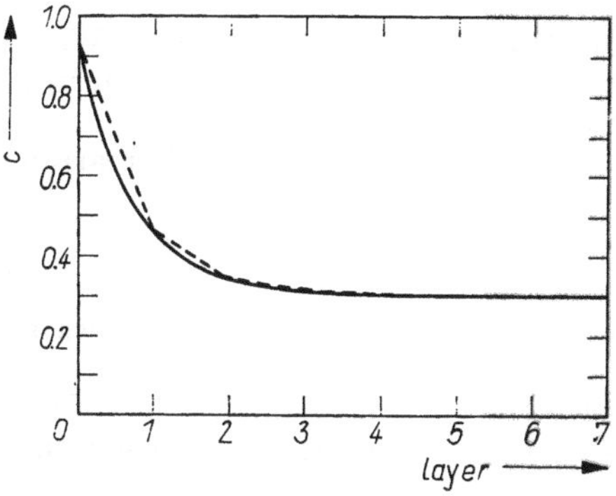

Fig. 1. Composition profile within the regular solution pair-bonding model (dashed line) and exponential fit (full line)

the number of nearest neighbours in a layer and n_{out} is the number of nearest neighbours in an adjacent layer.

The set of equations (1), which shows that the alloy can be considered as a system of (interacting) fermions, has been obtained [8] by using a mean-field approximation to the concentration functional theory proposed by Gyorffy and Stocks [9]. Similar equations have been obtained by a number of authors [10 to 12]. It is evident from (1a) that the surface composition is determined by v_{AB}, i.e. the component with the lowest heat of sublimation segregates to the surface. We will assume that c_n are the concentrations of this component in the different layers, so that $c_0 > 0$. When $w = 0$, the alloy forms an ideal solution and (1b) shows that only the surface layer is enriched. This is evident since it is the only layer with a deficiency of nearest neighbours. For non-ideal solutions two cases can be considered depending on the sign of w. When $w > 0$ the alloy has a miscibility gap and the bulk value is approached monotonically. This is the case for a clustering alloy, like CuNi. When $w < 0$ the alloy is an ordering one (e.g. CuZn) and the composition profile behaves in an oscillating manner. In both cases the bulk value is reached within a few layers [10 to 12]. Recently the model has been extended to treat the concentration profile of irradiatied alloys [13] and to determine the atomic distribution in bimetallic clusters [14].

In Fig. 1 we show the calculated composition profile at the (111) surface of an f.c.c. crystal, using (1), for an alloy with bulk concentration $c_B = 0.3$ and parameters $w = 0.0162$ eV, $v_{AB} = 0.1$ eV, $T = 770$ K. These values are taken from [13]. We note that the bulk concentration is reached monotonically after approximately five layers.

3. Segregation Kinetics

When the alloy forms an ideal solution the McLean model using a variable concentration ratio [5] might very well be valid since only the surface layer is enriched. It is our purpose to investigate the consequences of using a composition profile that approximates the above results for non-ideal clustering solutions. To this end we consider the concentration in the different layers, $c_n(t)$, as a continuous variable $c(x, t)$ (the crystal is located in $x > 0$) and we will approximate the solutions of (1) by an exponential profile,

$$c(x, 0) = c_B + c_1^s \exp(-\beta_1 x) . \tag{5}$$

The constants c_1^s and β_1, which are temperature dependent, should be obtained from a fit to the numerical solution of (1).

In Fig. 1 we have also plotted an exponential fit to the segregation profile. We note that the assumption of an exponential profile gives an adequate description in this case. The parameters have the values $c_1^s = 0.62$ and $\beta_1 = 1.354546$.

After quenching at $t = 0$ equilibrium is disturbed and there is an atomic rearrangement leading to an asymptotic profile

$$c(x, +\infty) = c_B + c_2^s \exp(-2\beta_2 x) . \tag{6}$$

The factor 2 is introduced for convenience later on. Inspection of (1) shows that $c_1^s < c_2^s$.

For the description of the kinetics we need a diffusion equation that has (6) as a quasi-stationary solution. This is provided by the equation describing diffusion in the presence of an external field [15],

$$D \frac{\partial^2 c}{\partial x^2} + 2\beta_2 D \frac{\partial c}{\partial x} = \frac{\partial c}{\partial t}, \tag{7}$$

where D is the diffusion coefficient (at temperature T_2) of the segregating species, which has been tabulated for a variety of elements [15]. The mechanism by which diffusion occurs (vacancy, interstitial, or interstitialcy) is left unspecified in our treatment. Note that it might prove necessary to define an "effective" diffusion coefficient, since not only the segregant is moving towards the surface, but also the other species is diffusing into the bulk. In fact one might even argue that it would be necessary to use a position-dependent diffusion coefficient, since the concentration in different layers is different. These extra complications will be ignored here.

In order to completely determine the problem we need an additional boundary condition. Evidently we would like to have some kind of interface condition, e.g. no net flux outside the surface. However, the matter is complicated because we can only conserve the total number of segregating particles, whereas we should conserve the total number of solute particles. This extra complication arises because in (1) the bulk is considered as an infinite reservoir, held at a fixed concentration c_B. Nevertheless we will assume conservation of the total number of segregating particles and we give further some evidence why we expect this to give meaningful results.

Conservation of the total number of segregating particles is expressed as

$$\int_0^{+\infty} \left(c(x, t) - c_B\right) \mathrm{d}x = N ; \qquad t > 0 , \tag{8}$$

which upon substitution of (5) and (6) gives

$$N = \frac{c_1^s}{\beta_1} = \frac{c_2^s}{2\beta_2}. \tag{9}$$

It is now evident that c_2^s might be obtained from a numerical solution of (1), but then β_2 is fixed by (9). The full solution of (1) will be different from (6) in that it approaches the bulk concentration more slowly. However, in view of the fact that we ultimately want to determine the surface composition as a function of time, i.e. $c(0, t)$, we believe that the use of (6) should give accurate results for this observable. The surface layer will be almost exclusively fed by the very first layers underneath it and this is expressed in (6) and (8). Although (6) is only a quasi-stationary state, in our opinion it should describe correctly the segregation kinetics in the surface layer. Nevertheless this is an approximation that has to be kept in mind when performing calculations for real systems. We find some support for this assumption in the experimental results of Ng et al. [16]. These authors found for the CuNi system a slight depletion in Cu in the subsurface layers. This nonmonotonic behaviour is not expected theoretically. We interpret it as an indication that the surface layer has reached its equilibrium concentration, drawing Cu atoms from the deeper layers, whereas these deeper layers are not yet in thermodynamic equilibrium. Our model of particle conservation should give a satisfactory description of this kind of behaviour, even though we cannot find depleted layers, since it is the excess number of particles that is conserved.

The problem is now completely determined by the diffusion equation (7), which automatically ensures the asymptotic solution (6), in combination with the initial condition (5) and the particle conservation condition (8). Referring all concentrations relative to c_B, we can solve (9) by Laplace transformation [17], putting

$$\bar{c}(x, p) = \int_0^{+\infty} e^{-pt}\, c(x, t)\, dt\,. \tag{10}$$

The Laplace transform of (7) is

$$\frac{\partial^2 \bar{c}}{\partial x^2} + 2\beta_2 \frac{\partial \bar{c}}{\partial x} - q^2 \bar{c} = -\frac{c_1^s}{D} \exp(-\beta_1 x)\,, \tag{11}$$

where we have used (5) and $q^2 = p/D$. The solution of (11) which is bounded at $+\infty$ is

$$\bar{c} = M \exp\{-[(q^2 + \beta_2^2)^{1/2} + \beta_2]\, x\} + \frac{c_1^s}{D} \frac{\exp(-\beta_1 x)}{q^2 - \beta_1(\beta_1 - 2\beta_2)}\,. \tag{12}$$

Next, M can be determined by substitution of (12) into the Laplace transform of (8). This yields

$$\bar{c} = \frac{c_1^s}{\beta_1 D} [(q^2 + \beta_2^2)^{1/2} + \beta_2] \left[\frac{1}{q^2} - \frac{1}{q^2 - \beta_1(\beta_1 - 2\beta_2)}\right] \times$$
$$\times \exp\{-[(q^2 + \beta_2^2)^{1/2} + \beta_2]\} + \frac{c_1^s}{D} \frac{\exp(-\beta_1 x)}{q^2 - \beta_1(\beta_1 - 2\beta_2)}\,. \tag{13}$$

Inverse Laplace transformation [15] yields

$$c(x, t) = \left(\frac{1}{2}\right) c_2^s \operatorname{erfc}\left(\frac{x}{2\sqrt{Dt}} - \beta_2 \sqrt{Dt}\right) \exp(-2\beta_2 x) -$$
$$- \left(\frac{1}{2}\right) c_1^s \operatorname{erfc}\left(\frac{x}{2\sqrt{Dt}} - (\beta_1 - \beta_2)\sqrt{Dt}\right) \times$$
$$\times \exp\left(-\beta_1 x + \beta_1(\beta_1 - 2\beta_2)\, Dt\right) + \left(\frac{1}{2}\right) (c_1^s - c_2^s) \times$$
$$\times \operatorname{erfc}\left(\frac{x}{2\sqrt{Dt}} + (\beta_1 - \beta_2)\sqrt{Dt}\right) \exp\left((\beta_1 - 2\beta_2)\, x +\right.$$
$$\left. + \beta_1(\beta_1 - 2\beta_2)\, Dt\right) + c_1^s \exp\left(-\beta_1 x + \beta_1(\beta_1 - 2\beta_2)\, Dt\right)\,. \tag{14}$$

Using units of time $1/\beta_1^2 D$, the surface concentration as a function of time is

$$c \equiv c(0, t) = \frac{1}{2} c_2^s \left\{\operatorname{erfc}\left(-\frac{c_2^s}{2c_1^s}\sqrt{t}\right) - \operatorname{erfc}\left(\left(1 - \frac{c_2^s}{2c_1^s}\right)\sqrt{t}\right) \times\right.$$
$$\left.\times \exp\left(\left(1 - \frac{c_2^s}{2c_1^s}\right) t\right)\right\} + c_1^s \exp\left(\left(1 - \frac{c_2^s}{c_1^s}\right) t\right) \operatorname{erfc}\left(\left(1 - \frac{c_2^s}{2c_1^s}\right)\sqrt{t}\right). \tag{15}$$

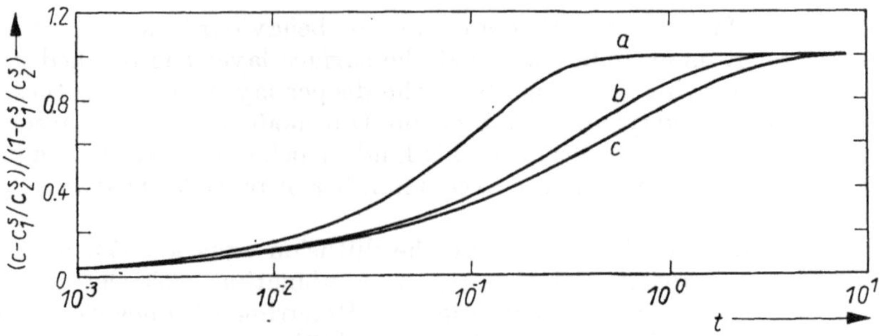

Fig. 2. Surface composition (scaled in order to vary from 0 to 1) as a function of time (in $1/\beta_1^2 D$). The equilibrium concentration ratios are (a) $c_1^s/c_2^s = 0.1$, (b) 0.4, (c) 0.7

In Fig. 2 we show the quantity $(c - c_1^s/c_2^s)/(1 - c_1^s/c_2^s)$ as a function of time, for different values of the ratio c_1^s/c_2^s. We note that the equilibrium surface composition is approached faster than in the McLean model with a constant α_1 [4]. This is a consequence of the fact that the concentration in the first layer below the surface is higher than the bulk concentration. It is possible to include evaporation from the surface layer in our framework, but since the algebra becomes quite involved we will not elaborate it here.

4. Conclusions

In summary, we have shown that we can reproduce qualitatively the behaviour of the kinetics of surface segregation in alloys, using a realistic composition profile. Our work has the character of a model calculation, since we have made several approximations in order to keep the calculations analytically tractable. As an improvement over other theories, we have assumed an exponential profile perpendicular to the surface. We expect this to be an adequate description of the actual situation in ordering alloys and it certainly is a satisfactory approximation to the solutions of the regular solution pair-bonding model, as described by (1). Furthermore, we have described the kinetics by a one-dimensional diffusion equation, as is customary. One might expect ordering phenomena, both within and perpendicular to the layers, to complicate the matter. Also, in order to obtain the correct quasi-stationary solution we have added an external field term in the diffusion equation, whereas the real profile is due to internal interactions. Our main approximation consists of conservation of the number of segregating particles. This certainly does not give the correct asymptotic behaviour, but we have argued that it should give a plausible description of the time evolution of the concentration in the surface layer.

This simple model illustrates that the segregation profile should be taken into account in any realistic theory of the segregation kinetics. We suggest that the discrepancy between the McLean model and the experiment is partly due to neglecting the concentration variation in the subsurface layers, and partly to the modified adsorption relation during the approach to equilibrium.

References

[1] M. P. SEAH, J. Phys. F **10**, 1043 (1980).
[2] W. M. H. SACHTLER and R. A. VAN SANTEN, Appl. Surface Sci. **3**, 121 (1979).
[3] D. MCLEAN, Grain Boundaries in Metals, Clarendon Press, Oxford 1957.
[4] C. LEA and M. P. SEAH, Phil. Mag. **35**, 213 (1977).
[5] G. ROWLANDS and D. P. WOODRUFF, Phil. Mag. **A40**, 459 (1979).

[6] A. D. Brailsford, Surface Sci. **94**, 387 (1980).
[7] R. A. Swalin, Thermodynamics of Solids, Wiley, 1962.
[8] L. T. Wille, Ph. D. Thesis, Rijksuniversiteit Gent, 1983, unpublished.
[9] B. L. Gyorffy and G. M. Stocks, Phys. Rev. Letters **50**, 374 (1983).
[10] F. L. Williams and D. Nason, Surface Sci. **45**, 377 (1974).
[11] J. J. Burton, E. Hyman, and D. G. Fedak, J. Catalysis **37**, 106 (1975).
[12] D. Kumar, A. Mookerjee, and V. Kumar, J. Phys. F **6**, 725 (1976).
[13] F. Mejia-Lira and J. L. Moran-Lopez, Surface Sci. **143**, L427 (1984).
[14] C. A. Balseiro and J. L. Moran-Lopez, Proc. 3rd Internat. Symp. Small Particles and Inorganic Clusters, Berlin (West) **1984**, to be published in Surface Sci.
[15] W. Jost, Diffusion, Academic Press, 1960.
[16] Y. S. Ng, T. T. Tsong, and S. B. McLane, Phys. Rev. Letters **42**, 588 (1979).
[17] P. M. Morse and H. Feshbach, Methods of Theoretical Physics, McGraw-Hill Publ. Co., New York 1953.

(Received July 17, 1985)

phys. stat. sol. (b) 131, 451 (1985)

Subject classification: 12

Department of Physics, Agra College[1])

Lattice Theory of Fourth-Order Elastic Constants of Primitive Lattices

By

S. C. GOYAL and K. KUMAR

Dedicated to Prof. Dr. Dr. h. c. Dr. E. h. P. GÖRLICH on the occasion of his 80th birthday

The theoretical expressions for the fourth order elastic (FOE) constants of the primitive lattices are derived. First the equation of wave propagation in a homogeneously strained medium is obtained and then it is compared with the equation derived from the macroscopic theory of elasticity which yields the required relations for (FOE) constants and the Kun-Huang relation among the fourth order coupling parameters.

Theoretische Ausdrücke für die elastischen Konstanten vierter Ordnung (FOE) der primitiven Gitter werden abgeleitet. Zuerst wird die Wellenausbreitungsgleichung in einem homogen verspannten Medium erhalten und dann mit der Gleichung verglichen, die aus der makroskopischen Elastizitätstheorie abgeleitet wird, was zu den geforderten Beziehungen für die FOE-Konstanten und der Kun-Huang-Beziehung zwischen den Kopplungsparametern vierter Ordnung führt.

1. Introduction

Interatomic potentials in real crystals are not harmonic and the anharmonicity of crystal lattices is noticeable when the atomic displacements are not small compared to the interatomic spacing. The crystal anharmonicity can be treated by higher-order elasticity theory when the continuum approximation is adopted. Thus the study of the higher-order elastic constants (HOEC) has become a subject of great interest from both theoretical [1 to 5] and experimental sides as these are used to accommodate higher-order nonlinearities appearing in phenomena such as pressure dependence of elastic constants at higher pressure [6], temperature dependence of the second-order elastic constants [7] and the shock waves in solids that sustain large compressions [8, 9].

Leibfried and Ludwig [10] and Coldwell-Horsefall [11] have derived the lattice theoretical expressions to study third-order elastic (TOE) constants of the primitive lattices. Later on in 1966 the above theory was extended by Srinivasan [12] to study (TOE) constants of the nonprimitive lattices.

Due to the availability of the experimental data of fourth-order elastic (FOE) constants and the second-order pressure derivatives of the second-order elastic (SOE) constants for some ionic crystals recently few attempts [2 to 5] have been made to derive the theoretical expressions of (FOE) constants of simple centro-symmetric cubic structure solids. These studies are generally based on the finite deformations method.

The present paper deals with the extension of the lattice theory [12] to obtain the general expressions of the (FOE) constants for the primitive lattices. These general expressions of elastic constants yield not only the eleven (FOE) constants of the cubic structure solids but also the Kun-Huang relations among the fourth-order coupling parameters.

[1]) Agra 282002, India.

In Section 2 we have given the basic steps of the theory and the conclusion is given in Section 3.

2. Theory

Consider a homogeneously deformed lattice in which the infinitesimal displacements $S_j(L\lambda)$ are produced from the deformed state due to the applied force.

The equation of motion for λ type atom where mass is $M(\lambda)$ can be expressed as

$$M(\lambda)\,\ddot{S}_j(L\lambda) = -[\sum_M \sum_\mu \sum_k \psi_{jk}(L\lambda, M\mu)\, S_k(M\mu) + \\ + \sum_{MN} \sum_{kr} \sum_{\mu\nu} \psi_{jkr}(L\lambda, M\mu, N\nu)\, S_k(M\mu)\, S_r(N\nu)]\,, \tag{1}$$

where

$$\psi_{jk}(L\lambda, M\mu) = \varphi_{jk}(L\lambda, M\mu) + \Delta_{jk}(L\lambda, M\mu)\,, \tag{2a}$$

$$\psi_{jkr}(L\lambda, M\mu, N\nu) = \varphi_{jkr}(L\lambda, M\mu, N\nu) + \Delta_{jkr}(L\lambda, M\mu, N\nu)\,, \tag{2b}$$

where

$$\Delta_{jk} = \sum_N \sum_\nu \sum_r \varphi_{jkr}(L\lambda, M\mu, N\nu)\, u_r(N\nu)\,, \tag{3a}$$

$$\Delta_{jkr} = \sum_P \sum_\gamma \sum_l \varphi_{jkrl}(L\lambda, M\mu, N\nu, P\gamma)\, u_l(P\gamma)\,, \tag{3b}$$

and

$$u_j(L\lambda) = \sum_p \varepsilon_{jp} X_p(L\lambda) + \omega_j(\lambda)\,. \tag{3c}$$

Here the other symbols have their same meaning as in [12]. We assume the following solution for (1):

$$S_j(L\lambda) = S_j(\lambda) \exp i[\omega t - YNX(L\lambda)]\,. \tag{4}$$

On using the above definitions equations (2a), (2b), (3a) to (3c), and (4) in (1) we get

$$M(\lambda)\,\omega^2 S_j(\lambda) = \sum_M \sum_\mu \sum_k \psi_{jk}(L\lambda, M\mu)\,[\exp[-\{iYNX(L\lambda, M\mu)\}]]\, S_k(\mu) + \\ + \sum_{MN} \sum_{\mu\nu} \sum_{rk} \psi_{jkr}(L\lambda, M\mu, N\nu) \times \\ \times [\exp[-\{iYNX(L\lambda, M\mu, N\nu)\}]]\, S_k(\mu)\, S_r(\nu)\,. \tag{5}$$

Here Y is the wave number of the wave.

Following method originally developed by Born and Huang [13], (5) is obtained as follows:

$$\varrho_0 W^2 S_j = \sum_{kpm} S_k N_p N_m \Big[-\frac{1}{2v_a} \sum_M \sum_{\lambda\mu} \psi_{jk}(o\lambda, M\mu)\, x_p(o\lambda, M\mu)\, x_m(o\lambda, M\mu) + \\ + \frac{1}{v_a} \sum_{MP} \sum_{\lambda\mu\sigma} \sum_{\pi q r} \psi_{jq}(o\lambda, M\mu)\, x_p(o\lambda, M\mu)\, \Gamma^{-1}(\mu q, \sigma r)\, \psi_{rk}(o\sigma, P\pi)\, x_m(o\sigma, P\pi) \Big] + \\ + \frac{1}{2} \sum_{rk} \sum_{pms} S_k S_r N_p N_m N_S \Big[-\frac{1}{2v_a} \sum_{MN} \sum_{\lambda\mu\nu} \psi_{jkr}(o\lambda, M\mu, N\nu)\, x_p(o\lambda, M\mu) \times \\ \times x_m(o\lambda, M\mu)\, x_S(o\lambda, N\mu) - \\ - \frac{2}{v_a} \sum_{MNP} \sum_{\lambda\mu\sigma\pi\nu} \sum_{qr} \psi_{jqn}(o\lambda, M\mu, N\nu)\, x_p(o\lambda, M\mu)\, x_S(o\lambda, N\nu) \times \\ \times \Gamma^{-1}(\mu q, \sigma r)\, \psi_{rk}(o\sigma, P\pi)\, x_m(o\sigma, P\pi) \Big]. \tag{6}$$

The matrix $\Gamma^{-1}(\mu q, \sigma r)$ can be obtained by following matrix $A(3n-3)\times(3n-3)$ matrix which is obtained from $\sum_M \psi_{jq}(o\lambda, M\mu)$ by omitting the three rows and three columns with $\lambda = 1$ or $\mu = 1$.

The macroscopic elasticity condition of the wave recently reported by the authors [15] is as follows:

$$\varrho_0 W^2 S_j = \sum_{kpm} A_{(jp),(km)} N_p N_m S_k + \tfrac{1}{2} \sum_{kr} \sum_{pms} A_{(jp),[rs],(km)} N_p N_m N_S S_k S_r\,, \tag{7}$$

where ϱ_0 is the density of the crystal in the undeformed state, W the natural velocity of propagation normal to natural normal N, S_j be the j-th component of the displacement in strained state.

On comparing (6) and (7), the lattice theoretical expressions of $A_{(jp),(km)}$ and $A_{(jp),[rs],(km)}$ are as follows:

$$A_{(jp),(km)} = -\frac{1}{2v_a}\sum_M \sum_{\lambda\mu} \psi_{jk}(o\lambda, M\mu)\, x_p(o\lambda, M\mu)\, x_m(o\lambda, M\mu) +$$

$$+\frac{1}{v_a}\sum_{Mp}\sum_{\lambda\mu\sigma\pi}\sum_{qr} \psi_{jq}(o\lambda, M\mu)\, x_p(o\lambda, M\mu)\, \Gamma^{-1}(\mu q, \sigma r)\, \psi_{rk}(o\sigma, P\pi)\, x_m(o\sigma, P\pi)\,, \tag{8}$$

$$A_{(jp),[rs],(km)} = -\frac{1}{2v_a}\sum_{MN}\sum_{\lambda\mu\nu} \psi_{jkr}(o\lambda, M\mu, N\nu)\, x_p(o\lambda, M\mu)\, x_m(o\lambda, M\mu)\,\times$$

$$\times\, x_S(o\lambda, N\nu) - \frac{2}{v_a}\sum_{MNP}\sum_{\lambda\mu\nu\sigma\pi}\sum_{qr} \psi_{jqn}(o\lambda, M\mu, N\nu)\, x_p(o\lambda, M\mu)\,\times$$

$$\times\, x_S(o\lambda, N\nu)\, \Gamma^{-1}(\mu q, \sigma r)\, \psi_{rk}(o\sigma, P\pi)\, x_m(o\sigma, P\pi)\,. \tag{9}$$

On substituting the value of ψ_{jk} from (2a) and $u_j(L\lambda)$ from (3c) in (8) and by assuming that the lattice has no interlattice displacement we get (8) as follows:

$$A_{(jp),(km)} = -\frac{1}{2v_a}\sum_M\sum_{\lambda\mu} \varphi_{jk}(o\lambda, M\mu)\, x_p(o\lambda, M\mu)\, x_m(o\lambda, M\mu) -$$

$$-\frac{1}{2v_a}\sum_{MN}\sum_{\lambda\mu\nu}\sum_{rs} \varphi_{jkr}(o\lambda, M\mu, N\nu)\, x_p(o\lambda, M\mu)\, x_m(o\lambda, M\mu)\,\times$$

$$\times\, x_S(o\lambda, N\nu)\,\varepsilon_{rs} \equiv \hat{C}_{[jk,pm]} + \sum_{rs} \hat{C}_{jk,[rs],pm}\varepsilon_{rs}\,, \tag{10}$$

where coupling coefficients $\hat{C}_{[jk,pm]}$ and $\hat{C}_{jk,[rs],pm}$ have the previous definitions [12].

Now $A_{(jp),(km)} + A_{(jm),(kp)}$ obtained using (10) must be equal to the same quantity obtained by using macroscopic elasticity theory [14]. Thus equating the two values we get

$$\hat{C}_{[jk,pm]} + \hat{C}_{[jk,mp]} + \sum_{rs}[\hat{C}_{jk,[rs],pm} + C_{jk,[rs],mp}]\,\varepsilon_{rs} =$$

$$= C_{pj,mk} + C_{mj,pk} + \sum_{rs}[C_{pj,mk,rs} + C_{mj,pk,rs} + 2C_{pm,rs}\delta_{jk} +$$

$$+ (C_{ps,mk} + C_{ms,pk})\,\delta_{rj} + (C_{pj,ms} + C_{mj,ps})\,\delta_{rk}]\,\varepsilon_{rs}\,. \tag{11}$$

Taking the term of zero order in ε on the both sides

$$C_{pj,mk} + C_{mj,pk} = 2\hat{C}_{[jk,pm]}\,. \tag{12}$$

This leads to the Kun-Huang relation on second-order coupling parameters, $\hat{C}_{[jk,pm]}$ is symmetric in $jk \leftrightarrow pm$.

Now we adding (12) with the equation obtained by interchanging $j \leftrightarrow p$ we get

$$2\hat{C}_{[jk,pm]} + 2\hat{C}_{[kp,jm]} = C_{jp,km} + C_{jm,kp} + C_{pj,km} + C_{pm,jk} ,$$

and using again equation (12) we can write the above equation as

$$C_{jp,km} = \hat{C}_{[jk,pm]} + \hat{C}_{[kp,jm]} - \hat{C}_{[jp,km]} . \tag{13}$$

This is the relation for second-order elastic constants.

Now comparing the coefficients of ε_{rs} in (11) and using (12) we can write

$$2[\hat{C}_{jk,[rs],pm} - \hat{C}_{[sk,pm]}\delta_{rj} - \hat{C}_{[js,pm]}\delta_{rk}] - 2C_{pm,rs}\delta_{jk} = C_{pj,mk,rs} + C_{mj,pk,rs} . \tag{14}$$

This is the Kun-Huang relation on the third-order coupling parameters which is identical to (16) of Srinivasan [12].

We derive an expression for $C_{jp,km,rs}$ (third-order elastic constants) from (14). Now adding (14) with the equation obtained by interchanging $j \leftrightarrow p$ and subtract the resulting equation from the interchanging $p \leftrightarrow k$,

$$\begin{aligned} C_{jp,km,rs} = {} & \hat{C}_{jk,[rs],pm} + \hat{C}_{pk,[rs],jm} - \hat{C}_{jp,[rs],km} - [\hat{C}_{[sk,pm]} - \hat{C}_{[sp,km]}]\,\delta_{jr} - \\ & - [\hat{C}_{[js,pm]} + \hat{C}_{[ps,jm]}]\,\delta_{kr} - [\hat{C}_{[sk,jm]} - \hat{C}_{[js,km]}]\,\delta_{pr} - \\ & - [C_{pm,rs}\,\delta_{jk} + C_{jm,rs}\delta_{kp} - C_{km,rs}\delta_{jp}] . \end{aligned} \tag{15}$$

This is the relation for third-order elastic constants. It is interesting to note that (13) and (15) are identical to the relations derived by Srinivasan [12].

Similarly substituting the value of ψ_{jkr} from (2b), $u_j(L\lambda)$ from (3c) in (9) and by assuming that the lattice has no interlattice displacement we find

$$\begin{aligned} A_{(jp),[rs],(km)} = {} & -\frac{1}{2v_a} \sum_{MN} \sum_{\lambda\mu\nu} \varphi_{jkr}(o\lambda, M\mu, N\nu)\, x_p(o\lambda, M\mu)\, x_m(o\lambda, M\mu) \times \\ & \times x_S(o\lambda, N\nu) - \frac{1}{2v_a} \sum_{MNP} \sum_{\lambda\mu\nu\gamma} \sum_{ld} \varphi_{jkrl}(o\lambda, M\mu, N\nu, P\gamma) \times \\ & \times x_p(o\lambda, M\mu)\, x_m(o\lambda, M\mu)\, x_S(o\lambda, N\nu)\, x_d(o\varepsilon, P\gamma)\, \varepsilon_{ld} \equiv \\ \equiv {} & \hat{C}_{jk,[rs],pm} + \sum_{ld} \hat{C}_{jk,[rs],[ld],pm}\varepsilon_{ld} , \end{aligned} \tag{16}$$

where coupling coefficients $\hat{C}_{jk,[rs],pm}$, $\hat{C}_{jk,[rs],[ld],pm}$ and relation between elastic constants and coefficients are defined as

$$\hat{C}_{jk,[rs],pm} = -\frac{1}{2v_a} \sum_{MN} \sum_{\lambda\mu\nu} \varphi_{jkr}(o\lambda, M\mu, N\nu)\, x_p(o\lambda, M\mu)\, x_m(o\lambda, M\mu)\, x_s(o\lambda, N\nu) , \tag{17a}$$

$$\begin{aligned} \hat{C}_{jk,[rs],[ld],pm} = {} & -\frac{1}{2v_a} \sum_{MNP} \sum_{\lambda\mu\nu\gamma} \varphi_{jkrl}(o\lambda, M\mu, N\nu, P\gamma)\, x_p(o\lambda, M\mu)\, x_m(o\lambda, M\mu) \times \\ & \times x_s(o\lambda, N\nu)\, x_d(o\lambda, P\gamma) , \end{aligned} \tag{17b}$$

$$C_{jp,km,rs} = \hat{C}_{jk,[rs],pm} - C_{pm,rs}\delta_{jk} - C_{km,ps}\delta_{jr} - C_{jp,ms}\delta_{kr} , \tag{17c}$$

$$\begin{aligned} C_{jp,km,rs,ld} = {} & \hat{C}_{jk,[rs],[ld],pm} - C_{jp,km,sd}\delta_{rl} - C_{jp,md,rs}\delta_{kl} - C_{jp,ms,ld}\delta_{kr} - \\ & - C_{km,pd,rs}\delta_{jl} - C_{pm,rs,ld}\delta_{jk} - C_{km,ps,ld}\delta_{jr} - C_{pm,sd}\delta_{jk}\delta_{rl} - \\ & - C_{ps,md}\delta_{jr}\delta_{kl} - C_{pd,ms}\delta_{jl}\delta_{kr} . \end{aligned} \tag{17d}$$

Now the values of $A_{(jp),[rs],(km)} + A_{(jm),[rs],(kp)}$ evaluated from (16) must be equal to the values recently obtained by macroscopic elasticity theory of the fourth-order elastic constants [15]. Thus equating the two values, we get

$$\hat{C}_{jk,[rs],pm} + \hat{C}_{jk,[rs],mp} + \sum_{ld} [\hat{C}_{jk,[rs],[ld],pm} + \hat{C}_{jk,[rs],[ld],mp}] \varepsilon_{ld} =$$
$$= C_{jp,rs,km} + C_{jm,rs,kp} + C_{pm,rs}\delta_{jk} + C_{mp,rs}\delta_{jk} + C_{ps,km}\delta_{jr} +$$
$$+ C_{ms,kp}\delta_{jr} + C_{jp,ms}\delta_{rk} + C_{jm,sp}\delta_{rk} + \sum_{ld} \varepsilon_{ld}[C_{jp,km,rs,ld} +$$
$$+ C_{jm,kp,rs,ld} + (C_{pm,rs,ld} + C_{mp,rs,ld})\,\delta_{jk} + (C_{ps,km,ld} + C_{ms,kp,ld})\,\delta_{jr} +$$
$$+ (C_{pd,km,rs} + C_{md,kp,rs})\,\delta_{jl} + (C_{jp,ms,ld} + C_{jm,'ps,ld})\,\delta_{kr} +$$
$$+ (C_{jp,md,rs} + C_{jm,pd,rs})\,\delta_{kl} + (C_{jm,kp,sd} + C_{jp,km,sd})\,\delta_{rl} +$$
$$+ (C_{pm,sd} + C_{mp,sd})\,\delta_{jk}\delta_{rl} + (C_{ps,md} + C_{ms,pd})\,\delta_{jr}\delta_{kl} +$$
$$+ (C_{pd,ms} + C_{md,ps})\,\delta_{jl}\delta_{kr}]\,. \tag{18}$$

On comparing the coefficients of zero order in ε on the both sides

$$\hat{C}_{jk,[rs],pm} + \hat{C}_{jk,[rs],mp} - (C_{ps,km} + C_{ms,kp})\,\delta_{jr} - (C_{jp,sm} + C_{jm,sp})\,\delta_{rk} =$$
$$= C_{jp,km,rs} + C_{jm,kp,rs} + (C_{pm,rs} + C_{mp,rs})\,\delta_{jk}\,. \tag{19}$$

On using (12) we can write this equation as

$$2[\hat{C}_{jk,[rs],pm} - \hat{C}_{[sk,pm]}\delta_{jr} - \hat{C}_{[js,pm]}\delta_{rk}] - 2C_{pm,rs}\delta_{jk} = C_{pj,mk,rs} + C_{mj,pk,rs} \tag{20}$$

which is identical to (14).

On comparing the coefficients of the ε_{ld} in (18) on both sides, we have

$$\hat{C}_{jk,[rs],[ld],pm} + \hat{C}_{jk,[rs],[ld],mp} = C_{jp,km,rs,ld} + C_{jm,kp,rs,ld} +$$
$$+ (C_{pm,rs,ld} + C_{mp,rs,ld})\,\delta_{jk} + (C_{ps,km,ld} + C_{ms,kp,ld})\,\delta_{jr} +$$
$$+ (C_{pd,km,rs} + C_{md,kp,rs})\,\delta_{jl} + (C_{jp,ms,ld} + C_{jm,ps,ld})\,\delta_{kr} +$$
$$+ (C_{jp,md,rs} + C_{jm,pd,rs})\,\delta_{kl} + (C_{jp,km,sd} + C_{jm,kp,sd})\,\delta_{rl} +$$
$$+ (C_{pm,sd} + C_{mp,sd})\,\delta_{jk}\delta_{rl} + (C_{ps,md} + C_{ms,pd})\,\delta_{jr}\delta_{kl} +$$
$$+ (C_{pd,ms} + C_{md,ps})\,\delta_{jl}\delta_{kr}\,. \tag{21}$$

Now using (12) and (20) and symmetrical conditions $r \leftrightarrow s$, $l \leftrightarrow d$ we get

$$2[\hat{C}_{jk,[rs],[ld],pm} - \hat{C}_{dk,[rs],pm}\delta_{jl} - \hat{C}_{sk,[ld],pm}\delta_{jr} - \hat{C}_{js,[ld],pm}\delta_{kr} -$$
$$- \hat{C}_{jd,[rs],pm}\delta_{kl} - \hat{C}_{jk,[sd],pm}\delta_{rl} + \hat{C}_{[ds,pm]}\delta_{jr}\delta_{kl} + \hat{C}_{[ds,pm]}\delta_{jl}\delta_{kr} +$$
$$+ \hat{C}_{[sk,pm]}\delta_{jl}\delta_{kr} + \hat{C}_{[dk,pm]}\delta_{js}\delta_{rl} + \hat{C}_{[dk,pm]}\delta_{jr}\delta_{ls} + \hat{C}_{[jd,pm]}\delta_{ks}\delta_{rl} +$$
$$+ \hat{C}_{[js,pm]}\delta_{rd}\delta_{kl} + \hat{C}_{[jd,pm]}\delta_{rl}\delta_{ks}] = C_{jp,km,rs,ld} + C_{jm,kp,rs,ld} +$$
$$+ (C_{pm,rs,ld} + C_{mp,rs,ld})\,\delta_{jk} + (C_{mp,sd} + C_{pm,sd})\,\delta_{jk}\delta_{rl} -$$
$$- (C_{pm,rs} + C_{mp,rs})\,\delta_{jl}\delta_{kd} - (C_{pm,ld} + C_{mp,ld})\,\delta_{jr}\delta_{sk} -$$
$$- (C_{ps,ld} + C_{sp,ld})\,\delta_{jm}\delta_{kr} - (C_{pd,rs} + C_{dp,rs})\,\delta_{jm}\delta_{kl} -$$
$$- (C_{pm,sd} + C_{mp,sd})\,\delta_{jk}\delta_{rl}. \tag{22}$$

Now further using the conditions $jp \leftrightarrow km \leftrightarrow rs \leftrightarrow ld$, we get

$$\begin{aligned}
&2[\hat{C}_{jk,[rs],[ld],pm} - \hat{C}_{dk,[rs],pm}\delta_{jl} - \hat{C}_{sk,[ld],pm}\delta_{jr} - \hat{C}_{js,[ld],pm}\delta_{kr} - \\
&- \hat{C}_{jd,[rs],pm}\delta_{kl} - \hat{C}_{jk,[sd],pm}\delta_{rl} + \hat{C}_{[ds,pm]}\delta_{jr}\delta_{kl} + \hat{C}_{[ds,pm]}\delta_{jl}\delta_{kr} + \\
&+ \hat{C}_{[sk,pm]}\delta_{jl}\delta_{kr} + \hat{C}_{[dk,pm]}\delta_{js}\delta_{rl} + \hat{C}_{[dk,pm]}\delta_{jr}\delta_{ls} + \hat{C}_{[jd,pm]}\delta_{ks}\delta_{rl} + \\
&+ \hat{C}_{[js,pm]}\delta_{rd}\delta_{kl} + \hat{C}_{[jd,pm]}\delta_{rl}\delta_{ks} - C_{pm,rs,ld}\delta_{jk} - C_{pm,sd}\delta_{jk}\delta_{rl} + \\
&+ C_{pm,rs}\delta_{jl}\delta_{dk} + C_{pm,ld}\delta_{jr}\delta_{sk} + C_{ps,ld}\delta_{jm}\delta_{kr} + C_{pd,rs}\delta_{jm}\delta_{kl} + C_{pm,rs}\delta_{jk}\delta_{rl}] = \\
&= C_{jp,km,rs,ld} + C_{jm,kp,rs,ld} \,. \qquad (23)
\end{aligned}$$

This is Kun-Huang relation on fourth-order coupling parameters.

We can derive an expression for FOE constants $C_{jp,km,rs,ld}$ from (21). On adding (21) with the equation obtained by interchanging $j \leftrightarrow p$ and subtract the resulting equation from the interchanging $p \leftrightarrow k$ and using (12) we find

$$\begin{aligned}
2C_{jp,km,rs,ld} = {} & 2[\hat{C}_{jk,[rs],[ld],pm} + \hat{C}_{pk,[rs],[ld],jm} - \hat{C}_{jp,[rs],[ld],km}] - \\
& - 2[\hat{C}_{[sd,pm]}\delta_{jr}\delta_{kl} + \hat{C}_{[ds,pm]}\delta_{jl}\delta_{kr} - \hat{C}_{[ds,km]}\delta_{jl}\delta_{pr} + \\
& + \hat{C}_{[sd,jm]}\delta_{pr}\delta_{kl} + \hat{C}_{[ds,jm]}\delta_{pl}\delta_{kr} - \hat{C}_{[sd,km]}\delta_{jr}\delta_{pl}] - \\
& - [C_{jp,km,sd} + C_{pm,kj,sd} + C_{pj,km,sd} - C_{jk,pm,sd}]\,\delta_{rl} - \\
& - [C_{js,km,ld} + C_{ms,kj,ld} - C_{jk,ms,ld} - C_{jm,ks,ld}]\,\delta_{pr} - \\
& - [C_{pd,km,rs} + C_{md,kp,rs} - C_{kd,pm,rs} - C_{md,pk,rs}]\,\delta_{jl} - \\
& - [C_{jp,ms,ld} + C_{jm,ps,ld} + C_{pj,ms,ld} + C_{pm,js,ld}]\,\delta_{kr} - \\
& - [C_{jp,md,rs} + C_{jm,pd,rs} + C_{pj,md,rs} + C_{pm,jd,rs}]\,\delta_{kl} - \\
& - [C_{ps,km,ld} + C_{ms,kp,ld} - C_{ks,pm,ld} - C_{ms,pk,ld}]\,\delta_{jr} - \\
& - [C_{jd,km,rs} + C_{md,kj,rs} - C_{jk,md,rs} - C_{jm,kd,rs}]\,\delta_{pl} - \\
& - 2[C_{jm,rs,ld}\,\delta_{pk} + C_{mp,rs,ld}\delta_{jk} - C_{km,rs,ld}\delta_{jp}] - \\
& - 2[C_{pm,sd}\delta_{jk}\delta_{rl} + C_{jm,sd}\delta_{pk}\delta_{rl} - C_{km,sd}\delta_{jp}\delta_{rl}] \,. \qquad (24)
\end{aligned}$$

This is the general expression for the fourth-order elastic constants for primitive lattice.

On using relations (12), (14), and (24) fourth-order elastic constants for cubic structure solids in terms of coupling coefficients are as follow:

$$C_{1111} = \hat{C}_{11,[11],[11],11} - 6\hat{C}_{11,[11],11} + 15\hat{C}_{[11,11]} \,, \qquad (25\text{a})$$

$$C_{1112} = \hat{C}_{11,[11],[22],11} - 3\hat{C}_{11,[22],11} + 6\hat{C}_{[12,12]} - 3\hat{C}_{[11,22]} \,, \qquad (25\text{b})$$

$$C_{1122} = \hat{C}_{11,[22],[22],11} - 2\hat{C}_{11,[22],11} + 2\hat{C}_{[12,12]} - \hat{C}_{[11,22]} \,, \qquad (25\text{c})$$

$$C_{1123} = \hat{C}_{11,[22],[33],11} - 2\hat{C}_{12,[33],12} + \hat{C}_{11,[33],22} - 2\hat{C}_{[12,12]} + \hat{C}_{[11,22]} \,, \qquad (25\text{d})$$

$$\begin{aligned}
C_{1144} = {} & \hat{C}_{11,[23],[23],11} - \hat{C}_{11,[22],11} - 2\hat{C}_{12,[23],13} + \hat{C}_{11,[23],23} + \\
& + 2\hat{C}_{[12,12]} - \hat{C}_{[11,22]} \,, \qquad (25\text{e})
\end{aligned}$$

$$C_{1166} = \hat{C}_{11,[12],[12],11} - \hat{C}_{11,[22],11} - 5\hat{C}_{11,[12],12} + 10\hat{C}_{[12,12]} + 3\hat{C}_{[11,22]} \,, \qquad (25\text{f})$$

$$C_{1244} = 2\hat{C}_{12,[23],[32],12} - \hat{C}_{11,[23],[32],22} - 2\hat{C}_{12,[23],13} + \hat{C}_{11,[23],23} - \\ - \hat{C}_{22,[31],13} + \hat{C}_{11,[12],12} + 2\hat{C}_{[13,13]} - 2\hat{C}_{[12,12]} - \hat{C}_{[11,22]}, \tag{25g}$$

$$C_{1266} = 2\hat{C}_{12,[12],[12],12} - \hat{C}_{11,[12],[12],22} - 2\hat{C}_{12,[22],12} + \hat{C}_{11,[22],22} + \\ + 2\hat{C}_{21,[21],22} - \hat{C}_{22,[21],12} + 4\hat{C}_{[12,12]} + 2\hat{C}_{[11,11]} - \hat{C}_{[11,22]}, \tag{25h}$$

$$C_{1456} = \hat{C}_{22,[31],[12],23} - \hat{C}_{22,[31],13} + \hat{C}_{[11,22]}, \tag{25i}$$

$$C_{4444} = \hat{C}_{22,[23],[23],33} - 6\hat{C}_{11,[12],12} + 12\hat{C}_{[12,12]} - 3\hat{C}_{[33,33]} + 6\hat{C}_{[11,22]}, \tag{25j}$$

$$C_{4466} = \hat{C}_{22,[12],[12],33} - \hat{C}_{11,[12],12} - 2\hat{C}_{12,[23],13} + \hat{C}_{11,[23],23} + \\ + 2\hat{C}_{[12,12]} + 2\hat{C}_{[13,13]} - 2\hat{C}_{[23,23]} + \hat{C}_{[22,33]}. \tag{25k}$$

3. Conclusion

Equation (24) is the required general expression of the FOE constants ($C_{jp,km,rs,ld}$) of the cubic structure solids in terms of the coupling coefficients. These coupling coefficients are related to the derivatives of the interatomic potentials of the crystals.

It is interesting to note that the relations for the SOE constants (13), TOE constants (15), Kun-Huang relation (12, 14) obtained during the present derivation are similar to those equations obtained by Srinivasan [12]. Equation (23) represents Kun-Huang relation on fourth-order coupling parameters.

The application of the above relations (25) to derive the equations of the FOE constants for the NaCl and CsCl structure solids is recently in hand and will be communicated soon.

References

[1] R. C. Hollinger and G. R. Barsch, J. Phys. Chem. Solids 37, 845 (1976).
[2] V. K. Garg, D. S. Puri, and M. P. Verma, phys. stat. sol. (b) 80, 63 (1977).
[3] S. Mori and Y. Hiki, J. Phys. Soc. Japan 45, 1449 (1978).
[4] Mcrkenscoff, colloqe C8, suppl. au n. o11, tome 40, C8–213 (1979).
[5] U. C. Sharma and M. P. Verma, phys. stat. sol. (b) 102, 487 (1980).
[6] Z. P. Chang and G. R. Barsch, Phys. Rev. Letters 19, 1381 (1967).
[7] J. A. Garber and I. V. Granato, Phys. Rev. B 11, 3998 (1975).
[8] R. A. Graham, J. Accoust. Soc. Amer. 51, 1576 (1972).
[9] G. Perrin and M. Delannoy, J. Mech. Phys. Solids 25, 41 (1977).
[10] G. Leibfried and W. Ludwig, Z. Physik 169, 80 (1960).
[11] R. Coldwell-Horsefall, Phys. Rev. 129, 22 (1963).
[12] R. Srinivasan, Phys. Rev. 144, 620 (1966).
[13] M. Born and K. Huang, Dynamical Theory of Crystal Lattices, Oxford University Press, London/New York 1954.
[14] R. N. Thurston and K. Brugger, Phys. Rev. 133, 1605 (1964).
[15] K. Kumar nad S. C. Goyal, Solid State Commun. 54, 433 (1985).

(Received May 6, 1985)

phys. stat. sol. (b) **131**, 459 (1985)

Subject classification: 6 and 20.1; 22.6

Laboratorium voor Kristallografie en Studie van de Vaste Stof, Rijksuniversiteit Gent[1])

Lattice Vibrations of V_2O_5

Calculation of Normal Vibrations in a Urey-Bradley Force Field

By

P. CLAUWS, J. BROECKX[2]), and J. VENNIK

Dedicated to Prof. Dr. Dr. h. c. Dr. E. h. P. GÖRLICH on the occasion of his 80th birthday

A normal coordinate analysis of the vibration spectrum of crystalline V_2O_5 is carried out in the assumption of a Urey-Bradley force field. The calculated frequencies are adjusted to 29 experimental infrared and Raman frequencies by an automatic force constant refinement program. Analysis of the potential energy distribution and the atomic displacements allows the classification of the modes into nine types of oxygen vibrations and three types of chain modes. A discussion is given of effective charges and infrared intensities. An assignment of the infrared spectrum of polycrystalline V_2O_5 is added.

Unter der Annahme eines Urey-Bradley-Kraftfeldes wird eine Normalkoordinatenanalyse des Schwingungsspektrums von kristallinem V_2O_5 durchgeführt. Die berechneten Frequenzen werden an 29 experimentelle Infrarot- und Ramanfrequenzen mittels eines automatischen Kraftkonstanten-Ausgleichsprogramms angepaßt. Die Analyse der Potentialenenergieverteilung und der atomaren Verzerrungen erlaubt die Einteilung der Moden in neun Arten von Sauerstoffschwingungen und drei Typen von Kettenmoden. Die effektiven Ladungen und Infrarotintensitäten werden diskutiert, zusätzlich erfolgt eine Zuordnung des Infrarotspektrums von polykristallinem V_2O_5.

1. Introduction

The infrared (IR) spectrum of V_2O_5, and to a lesser extent the Raman spectrum, have been the subject of several papers [1 to 11]. Due to the complicated lattice structure and the high number of fundamentals, the assignment of experimental frequencies without normal-mode calculations has necessarily been restricted to a few bands in the higher-frequency range. A thorough understanding of the vibration spectrum is however of both academic and practical importance. The latter is illustrated by the frequent use of the IR spectrum of V_2O_5 catalysts as a monitor for bond changes which occur at different stages of the reaction.

Some years ago, Gilson et al. [8] have published an extended list of IR and Raman frequencies of V_2O_5, together with calculated frequencies and bond assignments. As we have already shown in a previous paper [10], the experimental frequencies as determined by Gilson et al. deviate substantially from the correct values. We have therefore carried out new normal-mode calculations starting from our own IR and Raman data. The calculated frequencies are fitted to the experimental ones by adjusting the force constants in an automatic least-squares fitting program. We found that a simple valence force field is inadequate for a reasonable fit to some of the important bands. The fit was markedly improved by allowance for repulsive action between non-bonded atoms (UB force field).

---

[1]) Krijgslaan 281-S1, B-9000 Gent, Belgium.

[2]) Present address: Université du Burundi, Faculté des Sciences, B.P. 2700, Bujumbura, République du Burundi.

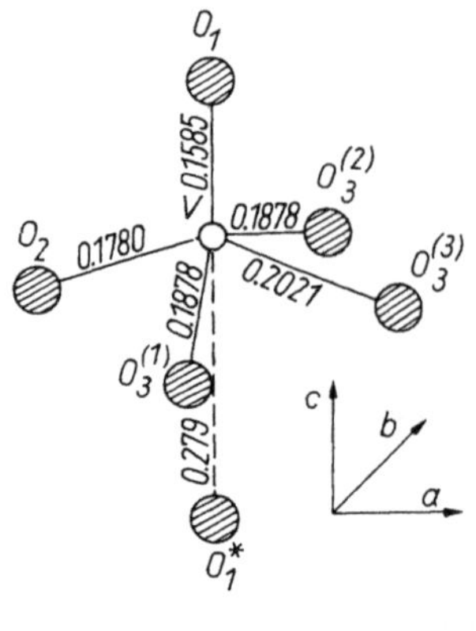

Fig. 1. Coordination of vanadium in the V_2O_5 lattice following the data of Bachmann et al. [12]; bond distances in nm

2. Factor-Group Analysis

We refer to Bachmann et al. [12] for a detailed description of the V_2O_5 structure and the list of atomic positions. The elementary structural unit, with the labels of the different atomic species used in this paper, is given in Fig. 1. The $V–O_3$ bonds form puckered chains in the *b*-direction, which are linked to each other in the *a*-direction by the bridging oxygen O_2. The layers so formed are held together by weak $V–O_1^*$ bonds.

According to Bachmann and Barnes [13], the short and intermediate bonds are mainly covalent, whereas the long $V–O_3^{(3)}$ bonds are mainly ionic. Using the empirical Pauling relation [14] between bond length d_n and multiplicity n,

$$d_n = d_1 - 2k \lg n \ ,$$

Wilhelmi et al. [15] derived estimates for the bond multiplicities as appearing in Table 1 (with $d_1 = 0.1789$ nm and $k = 0.039$ nm).

Table 1

Bond multiplicities in V_2O_5 as derived by Wilhelmi et al. [15]

bond	$V–O_1$	$V–O_2$	$V–O_3^{(1,2)}$	$V–O_3^{(3)}$	$V–O_1^*$
n	1.83	1.03	0.77	0.50	0.05

The V_2O_5 space group is $P_{mmn}(D_{2h}^{13})$. The unit cell (Fig. 2) contains two formula units (V_4O_{10}), yielding a total number of 39 optical zone-centre modes. According to factor-group analysis, the optical fundamentals have symmetries and IR or Raman activity as summarized in Table 2. Owing to the inversion symmetry, the modes split up into IR (odd) and Raman (even) modes. Note that only three modes are inactive according to symmetry.

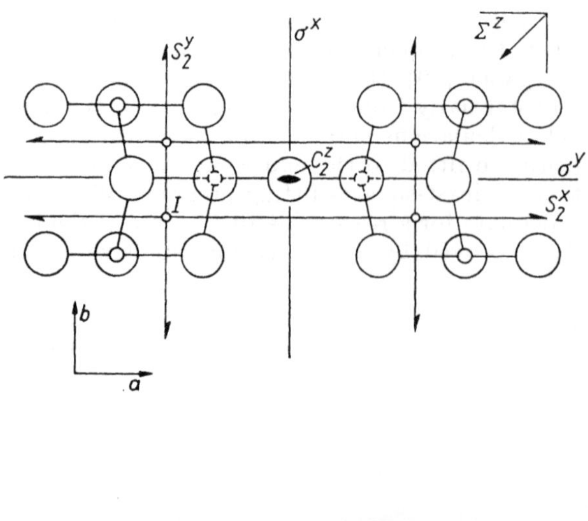

Fig. 2. Symmetry elements in the V_2O_5 unit cell

Table 2

Factor-group analysis of V_2O_5. The atomic positions and the symmetry elements in the unit cell are given in Fig. 2

representation of D_{2h}	A_g	B_{1g}	B_{2g}	B_{3g}	A_u	B_{1u}	B_{2u}	B_{3u}
number of modes	7	3	7	4	3	6	3	6
IR	—	—	—	—	—	τ_z ($\boldsymbol{E} \parallel \boldsymbol{c}$)	τ_y ($\boldsymbol{E} \parallel \boldsymbol{b}$)	τ_x ($\boldsymbol{E} \parallel \boldsymbol{a}$)
Raman	$\alpha_{xx}, \alpha_{yy}, \alpha_{zz}$	α_{xy}	α_{xz}	α_{yz}	—	—	—	—

It was indicated by Beatty and Gilson [3] that a convenient way to describe vibrations in binary oxide lattices is to treat the oxygen atoms as vibrating against an array of immobile metal atoms. Each oxygen species is then considered first as vibrating according to its site symmetry and following this the unit cell modes are generated by symmetry operations. If the bond strengths are sufficiently different, the group frequency concept applies for the stronger bonds and the vibrations of the corresponding oxygen species may be treated independently of the other vibrations. If we apply this to the V_2O_5 lattice, we obtain 30 oxygen modes as summarized in Table 3. This result is equivalent to the one appearing in the paper of Gilson et al. [8]. In Table 3 the representations have been grouped according to symmetry or antisymmetry with respect to the σ_y mirror plane. Note that multiplets which are derived from the same vibration type belong entirely to one of the two blocks. The high number of point group representations and accordingly the seven different IR and Raman polarizations should be most helpful in the assignment of the experimental frequencies to oxygen modes. It must however be emphasized that the approximate descriptions in Table 3 are not expected to hold for vibrations of the weaker bonds. For such vibrations normal-mode calculations should be carried out. The method of Beattie and Gilson is however very instructive for the understanding of the general features in the spectra. Taking the stretching of the strong $V-O_1$ bond as an example, Table 3 shows that this vibration will give rise to doublets in the IR and the Raman spectrum which are situated at the high-frequency side and well separated from other bands.

3. Infrared and Raman Data

We have already discussed the IR-reflectance data of V_2O_5 single crystals in an earlier paper [10]. The data were analyzed by the Kramers-Kronig method, classical-oscillator dispersion analysis, and four-parameter dispersion analysis, the three methods giving very similar results for the TO frequencies. For the purpose of this paper, we have summarized in Table 4 the ν_{TO} values as they follow from the four-parameter dispersion, together with values of an alternatively defined mode strength (see Section 5.1 for the definition of the latter). The data are visualized by the ε'' plots in Fig. 3 (ε'' is the imaginary part of the dielectric constant). For B_{3u} and B_{2u}, the full number of allowed resonances as predicted in Table 2 was observed. As to B_{1u} only two of the six allowed modes were found. Weak resonances in the latter polarization would be invisible due to a low S/N ratio, as the reflectance was necessarily recorded from a plane cut perpendicularly to the natural cleavage plane.

Table 3

Symmetry classification of oxygen modes in V_2O_5. C_s is the point group (E, σ_y). The Roman numerals refer to the classification hierarchy in Table 7

type of vibration			site symmetry		D_{2h}							
					symmetric to σ_y				anti-symmetric to σ_y			
					A_g	B_{2g}	B_{1u}	B_{3u}	B_{1g}	B_{3g}	A_u	B_{2u}
I	$V{-}O_1$	stretching ‖ ***c***	C_s	$A'(2)$	1	1	1	1	—	—	—	—
VI		deformation ‖ ***a***		$A'(1)$	1	1	1	1	—	—	—	—
IX		deformation ‖ ***b***		A''	—	—	—	—	1	1	1	1
II	$V{-}O_2{-}V$ (bridging)	stretching ‖ ***a***	C_{2v}	B_1	—	1	—	1	—	—	—	—
VIII		deformation ‖ ***b***		B_2	—	—	—	—	—	1	—	1
V		deformation ‖ ***c***		A_1	1	—	1	—	—	—	—	—
III	$V{-}O_3(V)(V)$	stretching ‖ ***b***	C_s	A''	—	—	—	—	1	1	1	1
IV		deformation ‖ ***a*** and stretching $V{-}O_3^{(3)}$		$A'(1)$	1	1	1	1	—	—	—	—
VII		deformation ‖ ***c***		$A'(2)$	1	1	1	1	—	—	—	—

Table 4

Fundamental IR frequencies of V_2O_5 obtained from reflection data [10]. The labels correspond with Fig. 3. The experimental and calculated intensities are defined in Section 5.1

polar-ization	No.	ν_{TO} (cm^{-1})	intensities in units of $10^4 cm^{-2}$	
			$I^k_{x,exp}$	$I^k_{x,calc}$
$\boldsymbol{E} \parallel \boldsymbol{a}$ (B_{3u})	1	72.4	0.06	0.52
	2	261.0	0.24	0.45
	3	303.0	6.1	19.7
	4	411.0	17.5	20.0
	5	767.5	33.1	16.6
	6	980.5	0.29	0.01
$\boldsymbol{E} \parallel \boldsymbol{b}$ (B_{2u})	1	212.0	0.57	0.35
	2	284.0	1.7	22.9
	3	506.5	45.3	24.4
$\boldsymbol{E} \parallel \boldsymbol{c}$ (B_{1u})	5	473.0	1.64	10.3
	6	975.5	12.6	30.7

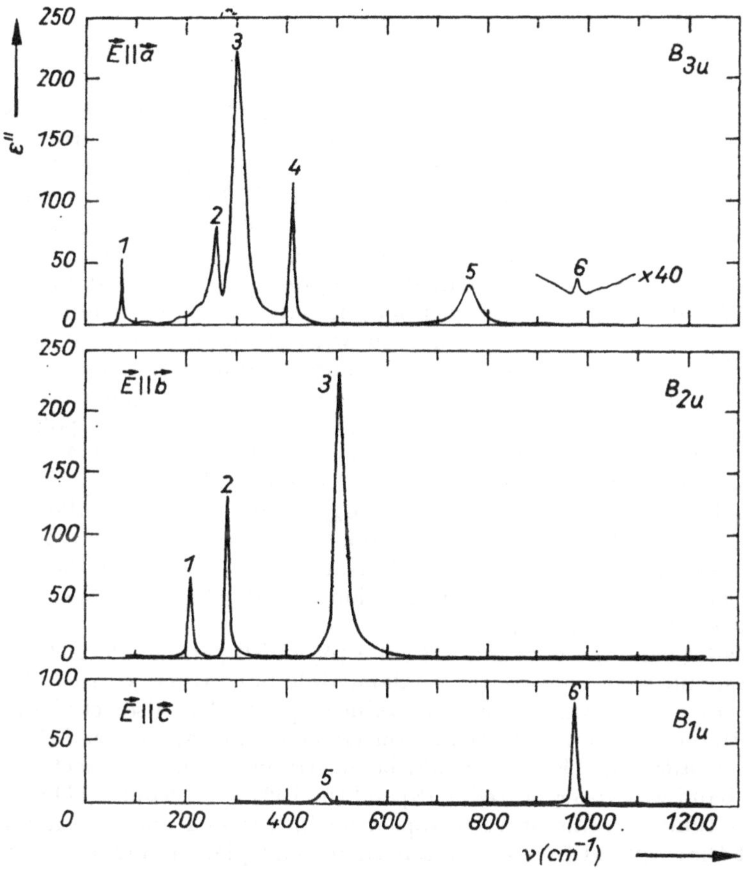

Fig. 3. ε'' spectrum of V_2O_5 obtained from reflection data [10]. The corresponding TO wave numbers are listed in Table 4

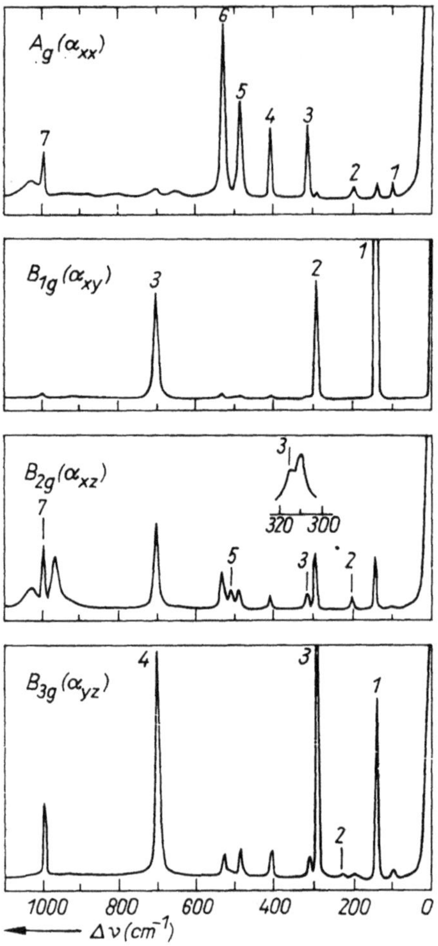

Fig. 4. Raman spectrum of V_2O_5 at 5 cm^{-1} resolution. The insert in the B_{2g} spectrum is recorded with a resolution of 2 cm^{-1}. The peaks carrying labels belong to the indicated polarization; the corresponding wave numbers are listed in Table 5

We have carried out Raman experiments on V_2O_5 single crystals in the 90° scattering configuration. We used the Ar^+ 514.5 nm laser line and a SPEX 1401 double monochromator. The crystal was cooled to 77 K as this enhanced the intensity of the Raman lines appreciably. The shift in the line positions is only a few wave numbers as compared with room temperature. The spectra are summarized in Fig. 4. It is seen that strong lines from one polarization break through into spectra of other polarizations. This was only troublesome for the weak B_{2g} spectrum. In Fig. 4 only those lines are labeled which belong to the appropriate polarization. The corresponding wave numbers are listed in Table 5. For A_g, B_{1g}, and B_{3g} the full number of allowed resonances (see Table 2) was detected. As for B_{2g}, some resonances may have been obscured by strong break-throughs from other polarizations. The labels of the observed B_{2g} resonances have been assigned by inspection of Table 3 and comparison with the list of calculated mode frequencies.

The origin of the broader bands at 1024 and 963 cm^{-1} (A_g and B_{2g}) is not clear. They probably do not correspond with zone-centre fundamentals but rather with combination bands. Our Raman data are very close to the values reported by Gilson et al. [8]. The only significant difference is a line reported by the latter authors at 357 cm^{-1} (B_{2g}) which we did not detect, although the line should be clearly visible in our spectra if present. The 963 cm^{-1} band was also reported in [8] and the 1024 cm^{-1} band in [11].

The list of observed fundamentals of V_2O_5 appearing in Tables 4 and 5 is fairly complete. It turns out that several frequencies appear in multiplets as predicted by Table 3. This is most obvious for the multiplet 992, 992, 975.5, and 980.5 cm^{-1} which belongs to the V–O_1 stretching. The next lower frequency at 767.5 cm^{-1} (B_{3u}) must

Table 5
Experimental Raman shifts of V_2O_5 fundamentals. The labels correspond with Fig. 4

symmetry	No.	Raman shift (cm^{-1})	symmetry	No.	Raman shift (cm^{-1})
A_g	1	107	B_{2g}	2	201
	2	200		3	315
	3	310		5	507
	4	404			(963)
	5	483		7	992
	6	528			(1024)
	7	992			
		(1024)			
B_{1g}	1	147	B_{3g}	1	149
	2	290		2	232
	3	702		3	291
				4	702

be assigned to $V-O_2-V$ but its B_{2g} counterpart is not observed. The $V-O_3-V$ stretching yields a doublet at 702 cm^{-1} in the Raman spectrum; the reason why the corresponding IR band appears only at 507 cm^{-1} is treated in the next section.

4. Calculation of the Normal Modes

4.1 Calculation method

We have adopted a calculation method that is based on the approach developed by Beattie et al. [16] for polyatomic molecules and crystals. We refer to the latter paper for a full description of the method and will give here a summary of the major steps involved in our calculations.

The first step is the construction of the matrix B defining internal displacement coordinates R in relation to the Cartesian coordinates X, $R = BX$. The atoms involved in this definition are shown in Fig. 5. Atoms 1 to 14 constitute the zeroth primitive cell; these atoms display the $3N - 3 = 39$ vibrational degrees of freedom corresponding with the zone-centre modes. It was found most convenient to take the origin at the inversion point at the centre of the figure. Following the method of

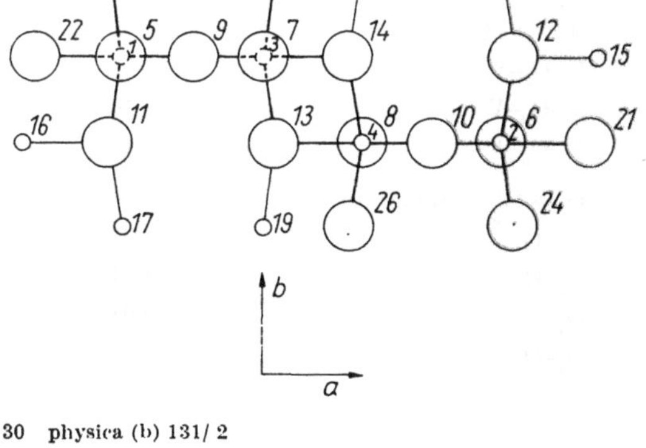

Fig. 5. Atoms involved in the definition of internal coordinates of V_2O_5. Small circles vanadium, large circles oxygen. 1 to 14 zeroth cell atoms; 15 to 26 atoms outside zeroth cell. The thick lines represent the bond-stretching coordinates that were actually used in the calculation (vanadyl bonds are directed perpendicularly to the plane of the paper). Thin lines are drawn to atoms that serve only for the definition of angle-deformation coordinates

Beattie et al., additional atoms are considered in the adjacent cells (atoms 15 to 26) so as to define the full number of internal coordinates in which primitive-cell atoms are involved. As the non-primitive atoms must vibrate identically as their primitive equivalents, appropriate summations over B_{ij} elements [16] permit restriction of the X vector to the 14 primitive atoms. Interactions with the neighbouring cells in the c-direction are neglected as this allows a serious reduction of the dimension of the problem. This approximation is thought to be justified by the weakness of the bonds between (a, b) layers in the V_2O_5 lattice. We have taken into account four types of V–O bond stretching coordinates and eight types of angle-deformation V–O–V and O–V–O coordinates. The definitions and numbering are found in Table 6. Accordingly the total number of internal coordinates is 66, giving a B matrix of dimension (66×42).

The second step is the calculation of symmetrized displacements by diagonalization of $M^{-1/2}\tilde{B}BM^{-1/2}$ (M is the diagonal matrix of the atomic masses). The eigenvectors P define a new matrix $W = BM^{-1/2}P$ that yields symmetrized displacements S_i in terms of the internal displacements: $R = WS$. The 39 columns of W are easily classified according to D_{2h} symmetry; it is the same classification as in Table 2. For V_2O_5 the $M^{-1/2}\tilde{B}BM^{-1/2}$ matrix (42×42) has the peculiarity to factorize immediately into two blocks (28×28) and (14×14) when the Cartesian coordinate vector is constructed as $x_1 \dots x_{14} z_1 \dots z_{14} y_1 \dots y_{14}$. The reason is that all the atoms of V_2O_5 are situated on σ_y mirror planes, so that vibrations are necessarily restricted to directions perpendicular or parallel to σ_y. The consequence is that the vibrations perpendicular to the mirror plane (i.e. B_{1g}, B_{3g}, A_u, B_{2u}) may be calculated separately from the

Table 6

Definition and final values of the force constant parameters appearing in the UBFF of V_2O_5

	force constant labels	type of internal or repulsive coordinate	f-values final fit
bond stretching f_i (N/cm)	f_1	$V{-}O_1$	6.64
	f_2	$V{-}O_2$	2.29
	f_3	$V{-}O_3^{(1)}$	0.90
	f_4	$V{-}O_3^{(3)}$	0.44
bond-angle deformation f_i (10^{-16} N cm/rad^2)	f_5	$O_1{-}V{-}O_2$	0.81
	f_6	$O_1{-}V{-}O_3^{(1)}$	0.40
	f_7	$O_1{-}V{-}O_3^{(3)}$	0.15
	f_8	$O_2{-}V{-}O_3^{(1)}$	0.15
	f_9	$O_3^{(1)}{-}V{-}O_3^{(3)}$	0.10
	f_{10}	$V{-}O_2{-}V$	0.48
	f_{11}	$V{-}O_3{-}V$ (2×0.1878 nm)	0.10
	f_{12}	$V{-}O_3{-}V$ (0.1878 and 0.2021 nm)	0.10
Urey-Bradley repulsion f_{ui} (N/cm)	$f_{u1}(13)$	$O_3^{(1)}{-}O_3^{(3)}$ (0.239 nm)	0.97
	$f_{u2}(14)$	$O_1{-}O_2$ (0.267 nm)	0.29
	$f_{u3}(15)$	$O_2{-}O_3^{(1)}$ (0.274 nm)	0.21
	$f_{u4}(16)$	$O_1{-}O_3^{(1)}$ (0.274 nm)	0.21
	$f_{u5}(17)$	$O_1{-}O_3^{(3)}$ (0.287 nm)	0.13

other ones. This group of vibrations is indeed independent of a whole class of internal coordinates, such as the three bond stretching coordinates that are situated in the mirror plane. Multiplets which are derived from the same V–O bond fall completely into one of the two groups (perpendicular or parallel vibrations); the same remark applies to the construction of Table 3.

The last step is the calculation of the normal modes, which follows after the introduction of the trial force-field matrix F. It involves diagonalization of $\tilde{W}FW$. The eigenvalues yield calculated frequencies, and the eigenvectors give the normal coordinates Q_j in terms of the symmetrized displacements: $S = VQ$. As only symmetrized displacements belonging to the same representation of D_{2h} combine in a given normal mode, the last diagonalization step involves small matrices; for V_2O_5 the maximum dimension is (7×7). The more elaborate calculation of W depends only on the positions and the masses of the atoms and on the choice of the internal coordinates and is carried out only once. The calculation of the frequencies is repeated for each trial force field.

The latter calculation was incorporated into an automatic force constant refinement program. Each refinement cycle consists of finding the force constant adjustments Δf_j which minimize the errors in the calculated eigenvalues of $\tilde{W}FW$ with respect to the experimental eigenvalues ($\Delta\lambda_i = \lambda_i^{calc} - \lambda_i^{obs}$). The set of m linear equations which relate the n parameters Δf_j to the m errors $\Delta\lambda_i$ (with $m > n$), is constructed following the method of Levin and Pearce [17]. The best adjustments Δf_j are then found from the least-squares solution by Householder transformation [18] of the set of equations.

After a satisfactory fit has been obtained, the program provides the following output: the list of calculated frequencies, the normal coordinates in terms of cartesian displacements of the 14 unit-cell atoms, and the potential energy distribution of each normal mode over the force field [17]. It is convenient to have both the latter listings available for a clear description of the modes, as they display which atoms and which bonds, respectively, are involved in each particular mode.

4.2 Results of the calculation

The first stage consisted of finding a set of reasonable initial force constants for the fitting program. For this purpose one calculation cycle was carried out for each trial force field. At first we adjusted the bond stretching constants f_1 to f_4 until stretching frequencies in the appropriate range were obtained. After this the values of f_5 to f_{12} were adjusted in order to obtain a better overall fit to the most prominent bands. The f-values were kept within certain limits so that the order of the bond strengths as appearing in Table 1 was respected, and on the assumption that deformations involving strong bonds have higher f-values than deformations involving weak bonds. A further help at this stage was the property that, as mentioned in Section 4.1, the vibrations belonging to (B_{1g}, B_{3g}, A_u, B_{2u}) depend only on f_3, f_6, f_8, f_9, and f_{12}.

Using this procedure a series of experimental frequencies could be quite well approximated, several of them appearing in multiplets in agreement with the oxygen-mode concept. It became however rapidly clear that the simple valence force field (SVFF) was not able to account for some phenomena such as the important splitting between the IR and Raman frequencies belonging to the V–O_3–V stretching (507 and 702 cm^{-1}, respectively). The assignment of the latter frequencies is beyond doubt, as they appear as the first member of their polarization, well separated from the deformations. The calculations however always yielded a closely spaced quartet, so that the force field was obviously too simple. The choice of the generalized force field to be used was made by noting that the even modes of the V–O_3–V stretch have higher frequency than the odd modes. In the odd modes the distance between the O_3 atoms

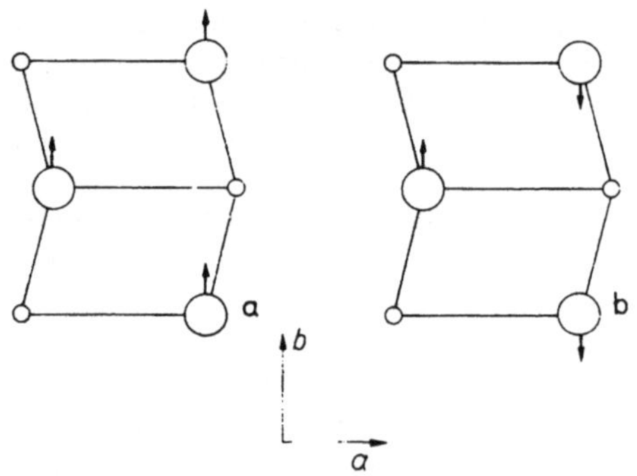

Fig. 6. Motion of the O_3 atoms in the V–O_3–V stretching (type III in Tables 3 and 7). Small circles vanadium, large circles O_3. a) Odd mode, b) even mode

remains unchanged during the vibration; in the even modes the O_3 atoms move towards each other (Fig. 6). One therefore obtains a splitting of the quartet into doublets and a shift of the even modes to higher frequencies, if one allows for repulsion between the O_3 atoms.

Repulsion between non-bonded atoms can be introduced in a straightforward manner by use of a Urey-Bradley force field (UBFF) [19]. In the UBFF the displacement coordinate between two repulsing atoms is expressed in terms of internal coordinates in which the atoms are involved. As a consequence the matrix F, which is diagonal in the case of a SVFF, obtains non-zero off-diagonal elements. A usual simplification in the UBFF is to assume that the repulsive potential energy is proportional to r^{-9}, with r the distance between the two atoms. In this way each type of repulsion introduces one additional force constant. We refer to the book of Nakamoto [20] for the calculation of the elements of the F matrix.

We have allowed for five types of repulsing non-bonded oxygen atom pairs. Their definition is found in Table 6. The $O_3^{(1)}$–$O_3^{(3)}$ pair has the shortest distance and is consequently considered as having the strongest repulsive action. As all repulsions considered are between oxygen atoms, it was thought justified to leave only f_{u1} as an adjustable parameter and to scale $f_{u2}, \dots, f_{u5}$ to the f_{u1} value by assuming the f_{ui} being proportional to r^{-11}. The introduction of the UBFF immediately yielded a better fit for the vibrations normal to the mirror plane.

Once a trial UBFF was obtained that gave a reasonable fit to the most important bands, the refinement procedure was allowed to start. Initially all 13 ($f_1, \dots, f_{12}, f_{u1}$) parameters were kept adjustable. This led to instabilities in the calculations with some parameters running to strongly negative values. It was then decided to treat only the most important force constants, i.e., f_1 to f_6, f_{10}, and f_{u1} as adjustable parameters, the five deformation force constants involving one or two weaker bonds being kept fixed at a low value: (0.15 and 0.10) $\times 10^{-16}$ N cm/rad^2. The refinement carried out in this way yields stable solutions. The final fit to the experimental data is listed in Table 7. The corresponding set of f-values is listed in Table 6.

Table 7 gives an overall summary of the calculated and experimental frequencies, the potential energy distribution in per cent, and an assignment of the modes in terms of the classification in Table 3. The latter assignment was made by taking account of the potential energy distribution over the internal coordinates and of the Cartesian displacements of the atoms as well. The latter listing is of too excessive length in order to be given in this paper. Contributions to the potential energy smaller than 2% have been omitted from the table for clarity.

5. Discussion

The UBFF seems to allow a very satifactory fit to the experimental frequencies. Especially the stretching vibrations and most of the deformations are well fitted. Large relative errors occur only for a few low-frequency chain modes for which the assumed force field is probably too simple.

5.1 IR intensities and effective charges

A separate check for the significance of the calculated results is to compare experimental and calculated mode strengths. For the IR active resonances the intensities are proportional to the square of the dipole moment change and follow directly from the eigenvectors provided the effective charges of the atoms are known. A direct comparison with the mode strength S obtained in the classical oscillator fit [10] is not straightforward for the multimode case because of the complicated dependence on the dielectric constant. A consistent analysis may in such cases be carried out as follows. It has been shown by Scott [21] that for a given polarization the total mode strength can be written as

$$\sum_j (\nu_{\mathrm{LO},j}^2 - \nu_{\mathrm{TO},j}^2) = \frac{e^2 N}{\pi c^2} \sum_i \left(\frac{Z_i^2}{m_i}\right), \tag{1}$$

where ν_{LO} and ν_{TO} are the wave numbers of the longitudinal and transverse optical modes, N is the number of unit cells per cm³, $Z_i e$ is the effective charge of the i-th ion with mass m_i. The j summation is over all the modes belonging to a given polarization, the i summation is over the atoms in the unit cell. For V_2O_5 we obtain for each polarization x, y, or z an expression

$$\sum_j (\nu_{\mathrm{LO},j}^2 - \nu_{\mathrm{TO},j}^2) = \frac{e^2 N}{\pi c^2} \left[4\,\frac{Z_{\mathrm{V}}}{m_{\mathrm{V}}} + 10\,\frac{Z_{\mathrm{O}}}{m_{\mathrm{O}}}\right] \tag{2a}$$

assuming equal effective charges for the three kinds of oxygen ions. The neutrality condition is

$$4Z_{\mathrm{V}} + 10Z_{\mathrm{O}} = 0\,. \tag{2b}$$

Using (2a) and (2b) we calculated values for Z_{V} and Z_{O}[3]) from the list of experimental ν_{LO} and ν_{TO} frequencies which resulted from the four-parameter dispersion analysis [10]:

$$\boldsymbol{E} \parallel \boldsymbol{a}: Z_{\mathrm{V}} = +3.42\,, \quad Z_{\mathrm{O}} = -1.37\,,$$

$$\boldsymbol{E} \parallel \boldsymbol{b}: Z_{\mathrm{V}} = +3.12\,, \quad Z_{\mathrm{O}} = -1.25\,.$$

The latter values may now be used for a calculation of the IR-intensities of the normal modes. The coefficient $\mu_k^x = \partial\mu_x/\partial Q_k$ is a measure of the dipole moment change in the polarization x due to the k-th mode [22]. The dipole moment is given by $\mu_x = e \sum_{i=1}^{14} Z_i X_i$ with X_i the Cartesian displacement coordinate. The latter coordinates are expressed in terms of the normal coordinates by $X = M^{-1/2} P V Q = CQ$ so that we obtain for the coefficients

$$\mu_x^k = e \sum_{i=1}^{14} Z_i C_i^k \tag{3}$$

[3]) Values for the effective charges (Z_{V}, Z_{O}) according to energy band calculations are: (+3.56, −1.39 to −1.46) [23], (+2.6, −1.04) [24], and (+2.23, −0.76 to −0.93) [25].

Table 7

Summary of the final results of the normal mode calculation. Contributions to the potential energy are arranged in columns according to the 17 force constants (see Table 6) and normalized to 100 per mode (contributions below 2% omitted). n.o.: not observed

mode label		wave number (cm^{-1})		potential energy distribution																	assignment		remark
repr.	No.	ν_{calc}	ν_{exp}	1	2	3	4	5	6	7	8	9	10	11	12	13	14	15	16	17	type	description	
A_g	7	985.6	992	95															2		I	V–O_1 stretching ‖ *c*	
B_{2g}	7	984.4	992	95															2				
B_{1u}	6	985.6	975.5	95															2				
B_{3u}	6	984.4	980.5	95															2				
B_{2g}	6	778.2	n.o.		86			2									4	8			II	V–O_2–V stretching ‖ *a*	
B_{3u}	5	776.4	767.5		87			2									4	7					
B_{3g}	4	681.4	702			45										43		4	4		III	V–O_3<(V, V) stretching ‖ *b*	
B_{1g}	3	680.7	702			45										44		4	4				
A_u	3	532.6	—			79			3						4			5	6				
B_{2u}	3	533.4	506.5			79			3						4			6	5				
A_g	6	536.4	528			4	16				3	2		2	2	59		9		3	IV	V–O_3<(V, V) stretch. V–$O_3^{(3)}$ + deform. ‖ *a*	
B_{2g}	5	533.7	507		2	4	14				3	2		2	2	59		8		3			
B_{1u}	4	432.1	n.o.		15	10	30	3			3	4		6	3	9		8	4	4			
B_{3u}	4	422.6	411			12	31	2	3		3	4		7	3	9		12	7	6			
A_g	5	488.4	483		16			33					24				24				V	V–O_2–V deform. ‖ *c*	
B_{1u}	5	487.9	473		16			34					24				24						
A_g	4	365.7	404		55		9	19		5	2					2	2			2	VI	V–O_1 deform. ‖ *a*	vanadium
B_{2g}	4	314.3	n.o.			8	4	34	6	8				4		4	16		10	5			+ VII
B_{1u}	3	355.2	n.o.		25	3		14	15	10				2			2		15	11			+ VII
B_{3u}	3	320.4	303				2	21	17	13	3						10	5	15	11			+ VII
A_g	3	292.8	310			13			32	2	2			7		3			36	2	VII	V–O_3<(V, V) deform. ‖ *c*	
B_{2g}	3	287.0	315			5	4	17	28		3			3			8	4	27				+ VI
B_{1u}	2	235.8	n.o.		25		8	5	13		7				2			24	12				
B_{3u}	2	257.8	261			2	5	36	10		7				2		17	11	9				+ VI

B_{3g}	3	297.1	291		13					11					7		16	15		VIII	$V-O_2-V$	+ IX
B_{2u}	2	278.9	284					30		16							35	19			deform. ‖ ***b***	+ IX
B_{3g}	2	243.2	232					18		20							47	13		IX	$V-O_1$	+ VIII
B_{1g}	2	287.4	290		12			50		2					9			23			deform. ‖ ***b***	
A_u	2	262.7	—					57										40				
B_{2u}	1	240.5	212					28		15							32	23				+ VIII
A_g	2	177.5	200	2		9	3		18		2	24		2	4	2			33	X	chain	(B_{3u}:
B_{2g}	2	159.3	201		4	34			6	3	4		2	4	7		9		25		transl. ‖ ***a***	acoustic)
B_{1u}	1	123.3	n.o.	7	3		5	2	6			58				4	2	3	9			
B_{3g}	1	159.0	149		26			6			5			5	36		6	16		XI	chain	(B_{2u}:
B_{1g}	1	171.3	147		26			10		5	4			4	26		4	21			transl. ‖ ***b***	acoustic)
A_u	1	75.8	—							30							66					
A_g	1	59.3	107	4		4	8		8			51			2	5			14	XII	chain	(B_{1u}:
B_{2g}	1	116.5	n.o.			4	13		27	6					4	7	14		23		transl. ‖ ***c***	acoustic)
B_{3u}	1	33.8	72.4		6	4	6	5	25				3	2		3	4	7	37			

with the C_i^k following directly from the normal mode calculation. The IR intensity of the k-th mode is then calculated consistently with (1) as

$$I^k_{x,\mathrm{calc}} = \frac{e^2 N}{\pi c^2} (\mu^k_x)^2 \tag{4}$$

which may now be compared with the experimental intensity

$$I^k_{x,\mathrm{exp}} = \nu^2_{\mathrm{LO},k} - \nu^2_{\mathrm{TO},k} \,. \tag{5}$$

The internal consistency of this intensity analysis is illustrated by the property that both summations $\sum_k I^k_{x,\mathrm{calc}}$ and $\sum_k I^k_{x,\mathrm{exp}}$ yield exactly the same original experimental result given by the left-hand side of (1). This is obviously a result of the PV matrix being orthogonal.

The individual IR intensities are summarized in the last two columns of Table 4. For $\boldsymbol{E} \parallel \boldsymbol{a}$ and $\boldsymbol{E} \parallel \boldsymbol{b}$ the values of Z_V and Z_O were used as given above. For $\boldsymbol{E} \parallel \boldsymbol{c}$ the set of observed resonances is incomplete, so we used the Z values obtained for $\boldsymbol{E} \parallel \boldsymbol{a}$. The comparison of the IR intensities in Table 4 should not be taken too strictly but merely as an indication for strong or weak resonances. First the calculated intensities are expected to be less accurate than the eigenvalues and secondly the assumption of point charges is an obvious simplification. Accordingly the results in Table 4 may be regarded as fair. The correspondence is the best for the B_{3u} modes, which is in agreement with the excellent fit of the frequencies belonging to B_{3u}. It is interesting to see that the calculation predicts a very weak IR band in the $\boldsymbol{E} \parallel \boldsymbol{a}$ polarization for the V–O_1 stretching mode at 980.5 cm^{-1} as is indeed observed. The correspondence is somewhat worse for B_{2u}. The results as a whole may nevertheless be regarded as satisfactory and as a support for the mode description as appearing in Table 7.

5.2 *Assignment of the frequencies*

The assignment of the observed resonances to modes of a certain type has been summarized in the last three columns of Table 7.

For modes of type I to V the assignment (which is based on symmetry considerations (Table 3), the potential energy distribution, and the Cartesian displacement co-ordinates) is straightforward. These modes may indeed be considered as genuine "oxygen modes" as described in Section 2.

Modes of type VI to IX are less easily described in terms of simple "oxygen modes": many of them are of mixed origin with two oxygen types or V–O_i bonds involved. For these cases the mode has been classified according to the oxygen type or bond displaying the largest amplitude and the type admixed is then given in the last column. "Vanadium" indicates an important amplitude of the vanadium atoms.

Modes of type X, XI, and XII are no longer derived from vibrations of the oxygens on their own. Drawings of the Cartesian coordinate displacements reveal that these modes may best be described as translations of whole layers consisting of single $O_3^{(1)}$–V–$O_3^{(2)}$–V chains with the O_1 atoms on top of the vanadiums. The site symmetry of the layers (or chains) is C_s and unit cell vibrations of the layers against each other are generated in the same way as was displayed in Table 3. For each type there is one mode that corresponds with a collective translation of the lattice and accordingly is acoustic.

In practice the absorption spectrum of V_2O_5 is often recorded using powdered samples. In our previous paper [10] such a spectrum was shown and the correspondence with the single crystal bands was discussed. Considering the possible practical importance, a separate assignment of the powder absorption bands is given in Table 8.

Table 8
Assignment of IR absorption bands in the spectrum of polycrystalline V_2O_5 pellets [10]

peak wave number (cm^{-1})	mode label		assignment (see Table 7)	
	repr.	No.	type	description
1023	B_{1u}	6	I	$V-O_1$ stretching ‖ ***c***
982	B_{3u}	6	I	$V-O_1$ stretching ‖ ***c***
813	B_{3u}	5	II	$V-O_2-V$ stretching ‖ ***a***
605	B_{2u}	3	III	$V-O_3\langle^{V}_{V}$ stretching ‖ ***b***
472	B_{3u}	4	IV	$V-O_3\langle^{V}_{V}$ stretching + deform. ‖ ***a***
370	B_{3u}	3	VI	$V-O_1$ deformation ‖ ***a***
294	B_{2u}	2	VIII	$V-O_2-V$ deformation ‖ ***b***
262	B_{3u}	2	VII	$V-O_3\langle^{V}_{V}$ deformation ‖ ***c***
217	B_{2u}	1	IX	$V-O_1$ deformation ‖ ***b***

References

[1] L. D. Frederickson and D. M. Hausen, Anal. Chem. **35**, 821 (1963).
[2] Y. Kera and K. Hirota, J. phys. Chem. **73**, 3973 (1969).
[3] I. R. Beattie and T. R. Gilson, J. Chem. Soc. (A), 2322 (1969).
[4] A. A. Abdullaev and L. D. Kislovskii, Kristallografiya **15**, 988 (1970).
[5] G. W. Anderson and W. D. Compton, J. chem. Phys. **52**, 6166 (1970).
[6] I. Hevesi, A. A. Abdullaev, and V. S. Lebedev, Soviet Phys. — Cryst. **16**, 275 (1971).
[7] I. I. Reshina, Soviet Phys. — Solid State **14**, 287 (1972).
[8] T. R. Gilson, O. F. Bizri, and N. Cheetham, J. Chem. Soc. (A), 291 (1973).
[9] B. Bootz, H. Finkenrath, G. Franz, and N. Uhle, Solid State Commun. **13**, 1477 (1973).
[10] P. Clauws and J. Vennik, phys. stat. sol. (b) **76**, 707 (1976).
[11] G. W. Anderson and J. L. Verble, J. appl. Phys. **50**, 2765 (1979).
[12] H. G. Bachmann, F. R. Ahmed, and W. H. Barnes, Z. Krist. **115**, 110 (1961).
[13] H. G. Bachmann and W. H. Barnes, Z. Krist. **115**, 215 (1961).
[14] L. Pauling, J. Amer. Chem. Soc. **69**, 542 (1947).
[15] K. A. Wilhelmi, K. Waltersson and L. Kihlborg, Acta chem. Scand. **25**, 2675 (1971).
[16] I. R. Beattie, N. Cheetham, M. Cardner, and D. E. Rogers, J. Chem. Soc (A), 2240 (1971).
[17] I. W. Levin and R. A. R. Pearce, in: Vibrational Spectra and Structure, Vol. 4, Ed. J. L. Durig, Elsevier Publ. Co., Amsterdam 1975 (Chap. 3).
[18] P. Businger and G. H. Holub, in: Handb. for Automatic Computation, Vol. 2, Ed. J. H. Wilkinson and C. Reinsch, Springer Verlag, Berlin 1971.
[19] T. Shimanouchi, J. chem. Phys. **17**, 245 (1949).
[20] K. Nakamoto, Infrared Spectra of Inorganic and Coordination Compounds, John Wiley & Sons, New York 1963 (p. 57).
[21] J. F. Scott, Phys. Rev. B **4**, 1360 (1971).
[22] E. B. Wilson, J. C. Decius, and P. C. Cross, Molecular Vibrations, Mc.Graw-Hill Publ. Co., New York 1955.
[23] W. Lambrecht, B. Djafari-Rouhani, M. Lannoo, P. Clauws, L. Fiermans, and J. Vennik, J. Phys. C **13**, 2503 (1980).
[24] D. W. Bullett, J. Phys. C **13**, L595 (1980).
[25] W. Lambrecht, private communication.

(Received June 18, 1985)

phys. stat. sol. (b) **131**, 475 (1985)

Subject classification: 6 and 13.4; 14.3; 20.1; 22.1.1

Max-Planck-Institut für Festkörperforschung, Stuttgart[1])

Physical Properties of Ion-Implanted Laser Annealed n-Type Germanium

By

G. CONTRERAS[2]), L. TAPFER, A. K. SOOD[3]), and M. CARDONA

Dedicated to Prof. Dr. Dr. h. c. Dr. E. h. P. GÖRLICH on the occasion of his 80th birthday

An investigation is made of the impurity distribution and carrier concentration in n-type germanium layers doped with P, As, and Sb by ion implantation and recrystallized by laser annealing. The impurity distribution is determined with SIMS. The electron concentration is obtained from Hall effect and from ir reflectivity measurements. Impurity concentrations as high as $2 \times \times 10^{21}$ cm^{-3} are reached, with carrier concentrations N_e as high as 2.4×10^{20} cm^{-3}, much higher than in bulk doped samples ($N_e \leqq 4 \times 10^{19}$ cm^{-3}). For the P-doped samples the implanted layers are found to have a lattice constant larger than the substrate for small implantation doses and smaller for larger doses. An analysis of these data leads to the conclusion that most of the ions are implanted substitutionally and, at high concentrations, they are compensated by pairing defects, probably vacancies. With this hypothesis it is also possible to interpret the dependence of the strength of the phosphorus local mode on implantation dose, as found with Raman spectroscopy. The softening of the Raman phonon with dose observed experimentally is also compatible with this hypothesis. A softening of the acoustic phonons with dose found with Brillouin spectroscopy is also reported and discussed.

Es wird die Störstellenverteilung und die Ladungsträgerkonzentration in n-leitenden, durch Ionenimplantation P-, As- und Sb-dotierten und mittels Laserausheilung rekristallisierten Germaniumschichten untersucht. Die Störstellenverteilung wird mit SIMS bestimmt und die Elektronenkonzentration aus Hall-Effekts- und aus IR-Reflexionsmessungen erhalten. Störstellenkonzentrationen bis zu 2×10^{21} cm^{-3} werden erreicht mit Ladungsträgerkonzentrationen N_e bis zu $2{,}4 \times 10^{20}$ cm^{-3}, die viel höher als in volumendotierten Proben ($N_e \leqq 4 \times 10^{19}$ cm^{-3}) sind. Für P-dotierte Proben wird gefunden, daß die implantierten Schichten eine Gitterkonstante aufweisen, die für geringe Implantationsdosen größer ist als die des Substrats und kleiner für große Dosen. Eine Analyse dieser Werte führt zu dem Schluß, daß die meisten der Ionen substitutionell implantiert sind und bei hohen Konzentrationen durch paarbildende Defekte, wahrscheinlich Leerstellen, kompensiert werden. Mit dieser Hypothese ist es auch möglich, die Abhängigkeit der Stärke der lokalen Phosphormode von der Implantationsdosis zu erklären, wie mit Ramanspektroskopie gefunden wird. Die Erweichung des Ramanphonons mit der Dosis, die experimentell beobachtet wird, ist mit dieser Hypothese ebenfalls kompatibel. Eine Erweichung des akustischen Phonons mit der Dosis, die mit Brillouin-Spektroskopie gefunden wird, wird ebenfalls mitgeteilt und diskutiert.

---

[1]) Heisenbergstr. 1, D-7000 Stuttgart 80, FRG.

[2]) DAAD Fellow, on leave from E.S.F.M.-I.P.N., Edif. 6, U.P.A.L.M., Deleg. G. A. Madero, 07300 México, D.F. México.

[3]) Present address: Materials Science Laboratory, Reactor Research Centre, Kalpakkam 603102, India.

1. Introduction

Laser annealing of ion-implanted semiconductors, especially Si [1 to 6], Ge [7], and GaAs [8] offers the possibility of reaching higher carrier concentrations than those governed by solubility limits in the bulk. In this sense, ion implantation and laser annealing have been particularly successful for Si [9]. In this case ions such as P, As, and B are incorporated substitutionally to very high concentrations ($\approx 10\%$). They are electrically active and the ion-implanted, laser-annealed (IILA) layers seem to have a relatively small number of defects. In the case of IILA germanium considerable coarse damage, in the form of craters and voids, has been observed [7, 10]. Because of this fact, and of the technological applications of Si, much less work has been published for IILA Ge than for similarly treated Si.

In this paper we present investigations of IILA samples implanted with either phosphorus, arsenic, or antimony and annealed with a pulsed excimer laser. The IILA layers are single crystalline with the orientation of the substrate, as shown by Raman scattering. The implantation profiles were investigated by secondary ion mass spectroscopy (SIMS). The free-electron concentration was obtained from the infrared reflection and the Hall effect. It was found to be as high as 2.4×10^{20} cm^{-3}, although, in this case, a much larger concentration of electrically inactive ions existed. These ions were shown, in the case of phosphorus, to occupy substitutional sites by Raman measurements of the local vibrational modes induced by the phosphorus. A model for this electrical inactivity, based on compensation or passivation of the donor impurities by neighboring vacancies, was developed.

The presence of active and inactive implanted impurities is also revealed by measurements of the change in the lattice constant of the IILA layers implanted with P with concentration: at low P concentrations the lattice constant is larger than that of the intrinsic substrate while at high concentrations it is smaller.

Shifts and broadenings of the Raman phonon are also observed in the IILA samples. They can be attributed to the presence of impurities. The effect of doping on the LA phonons has also been investigated by means of Brillouin spectroscopy. A large softening is observed. It can be partly attributed to electron–phonon interaction and partly to a decrease in refractive index produced by the presence of voids.

2. Experiment

2.1 Sample preparation and determination of the carrier concentration

Instrinsic Ge substrates were implanted with different donor ions (P, As, and Sb) and varying doses N_s (2×10^{15} cm$^{-2} \leqq N_s \leqq 6 \times 10^{16}$ cm^{-2}). The implanted samples were laser annealed with a XeCl laser ($\lambda = 308$ nm, 10 ns pulse duration, 0.8 J/cm^2 energy density). The carrier concentrations of the samples were determined by ir reflectivity [11 to 13] and Hall effect measurements (van der Pauw geometry [14]). The Hall measurements were performed at room temperature with the sample on the original substrate (≈ 1 mm thick) and with this substrate thinned down to 0.6 and 0.3 mm. The data were then linearly extrapolated to zero substrate thickness so as to remove the influence of the intrinsic Ge substrate ($\approx 10^{14}$ carriers/cm^3) on the measurements. The carrier concentration N_e is related to the Hall constant R through

$$N_e = f \frac{1}{Rec}, \qquad (1)$$

where e is the electron charge and c the speed of light. The scattering and mass anisotropy factor f was taken equal to unity for simplicity. The scattering part of it is

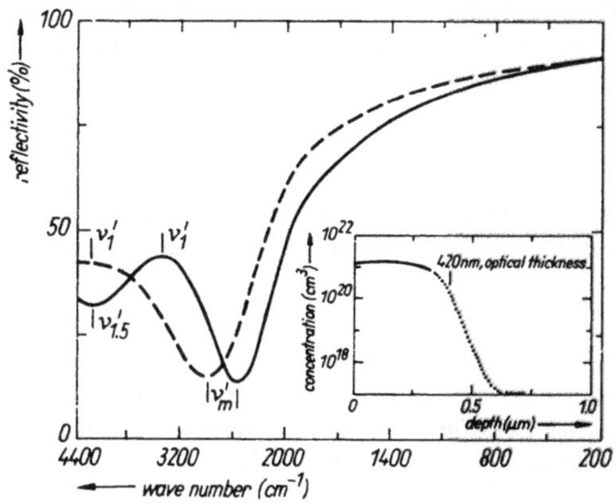

Fig. 1. Ir reflectivity spectra measured at room temperature for two IILA samples; – – – Ge:P, 1×10^{16} cm^{-2}, (110) face; ——— Ge:P, 6×10^{16} cm^{-2}, (110) face. The insert shows the SIMS profile of phosphorus atoms obtained for one of these samples (Ge(P), 6×10^{16} cm^{-2}, (110) face). The ir spectra show the reflectivity minima ν'_m and the first interference extrema ν'_1, $\nu'_{1.5}$. At the inset read cm^{-3} instead of cm^2. The wave-number scale has to be interupted left to 2000 cm^{-1}

uncertain. The mass anisotropy alone would give $f = 0.8$ [15]. Recent data for silicon [16] suggest that the value of f should be close to that calculated from the mass anisotropy alone for dopings in the 10^{20} cm^{-3} range (cf. [16] and references therein).

The determination of the Hall constant R from the van der Pauw data requires the knowledge of the sample thickness. Three avenues are open for its approximate determination: measurement of optical interferences in the implanted layer, SIMS, and Rutherford backscattering (RBS). The latter can only be used to determine thicknesses of layers implanted with atoms much heavier than Ge, in our case Sb. RBS results for our Sb-implanted Ge layers yield thicknesses of about 300 nm, in reasonable agreement with the other measurements (see Table 1) [17].

A typical SIMS doping profile obtained with 12 keV O_2^+ primary ions for a P-doped Ge sample is shown in the insert of Fig. 1. The ion concentration scale was calibrated by equating the integrated surface density to the total implantation dose. The depth at which the density decreases to one-half of its maximum value is seen to be 380 nm.

We have also shown in Fig. 1 the infrared reflectivity spectra of two P-doped samples including that for which the SIMS profile is given. These measurements were performed with a Nicolet M10 (2.0 μm $< \lambda <$ 50 μm) and a Cary 17D (0.6 μm $< \lambda <$ $<$ 2.0 μm) spectrometers. We derive the carrier concentration N_e from the wave number of the reflectivity minimum ν'_m by assuming that this minimum occurs in our layers at approximately the same position as for bulk material [11]. The correctness of this assumption is supported by detailed studies performed for IILA layers of n-type Si [18]. The qualitative justification is that only the front surface, with the sharp refractive index discontinuity, should have nearly zero reflectivity when its refractive index equals one. The interface between the implanted layer and the substrate shows a gradual decrease in the impurity concentration and is expected to have a small reflectivity, nearly independent of wave number. The back surface between intrinsic Ge and air should only contribute a weak, incoherent background to the reflection of the front surface. Within the region of our measurement (ν'_m between 1500 and 3500 cm^{-1}) we found that the experimental data which exist for the bulk can be written as a function of the plasma wave number ν'_p as

$$\nu'_m \approx 1.01 \nu'_p \quad \text{with} \quad \nu'_p = \left(\frac{N_e e^2}{\pi m^* \varepsilon_0 c^2}\right)^{1/2}, \tag{2}$$

where m^* is the optical effective mass of the free electrons ($m^* = 0.12m_e$)[4]) and $\varepsilon_0 = 16$ the dielectric constant of the intrinsic semiconductor. The values of N_e so obtained for our samples are listed in Table 1.

Table 1

List of the samples used with their corresponding carrier and impurity concentrations and the thickness determined from interferences in the near ir (optical thickness), see text

sample	dopant	surface	N_s implantation dose (10^{16} cm^{-2})	N_e carrier concentration (10^{20} cm^{-3})	N_i impurity concentration (10^{20} cm^{-3})	optical thickness d (nm)
1	phosphorus	[100]	1	1.5	2.9	353.2
2			2	1.8	5.3	380.0
3			3	1.7	7	432.0
4			4	1.5	8.46	473.0
5			6	1.1	12	489.0
6		[110]	0.2	0.55	0.6	332.0
7			0.5	0.96	1.5	340.0
8			1	1.5	2.9	340.0
9			6	1.2	14	421.8
10		[111]	1	1.6	3	335.1
11			2	2	5.1	392.1
12	arsenic	[111]	1	1.8	3.3	302.2
13			2	2.4	6.3	316.1
14			6	2.2	19	319.6
15		[100]	6	2.1	19	319.7
16	antimony	[100]	1	2.1	3.2	312.0
17			2	2.2	6.2	321.7

Our samples also exhibit interference fringes for wave numbers above ν'_m. Fig. 1 shows fringes of order $m = 1$ and $1.5 (\nu'_1, \nu'_{1.5})$. From these fringes the "optical thickness" d of the implanted layer can be determined by means of the expression

$$2d \left[\varepsilon_0 \left(1 - \frac{\nu_p'^2}{\nu_m'^2}\right)\right]^{1/2} = \frac{m}{\nu'_m}. \tag{3}$$

Using the plasma wave numbers ν'_p obtained from ν'_m with (2) we found the "optical thicknesses" listed in Table 1. The "optical thickness" is indicated by a mark in the insert of Fig. 1. We see that it is only about 10% higher than that determined by SIMS at the point where the ion density is half that at the maximum. This deviation is found systematically for all samples measured. The reason for it may lie in uncertainties in the SIMS work related to initial variations of the sputtering rate. We thus take the optical thickness to be the nominal thickness of our samples as listed in Table 1. With these thicknesses, the values of N_e listed in Table 1 were determined from the van der Pauw measurements.

[4]) The optical effective mass is given by $1/m^* = \frac{1}{3}(2/m_t + 1/m_l)$; values of m_t and m_l at room temperature are found in [19].

2.2 X-ray diffraction measurements

An X-ray double-crystal diffractometer in (+1, −1) non-dispersive Bragg geometry [20] was used to measure the difference between the lattice constant of the IILA samples and that of the substrate. The X-ray rocking curves on IILA samples were obtained with $CuK\alpha_1$ radiation ($\lambda = 0.1514$ nm) on the net planes (004), (333), and (044), (022) for the [001], [111], and [011] sample orientations, respectively. The X-ray beam was collimated and rendered nearly monochromatic by asymmetric reflection on a dislocation-free single crystal of Ge.

2.3 Raman and Brillouin scattering measurements

The Raman spectra of our IILA samples were measured at room temperature in the backscattering geometry with a Spex 1 m double monochromator in the photon counting mode. As exciting radiation we used several laser lines of Kr^+ and Ar^+ lasers. Brillouin measurements were made in backscattering geometry from samples with [111] orientation at $\lambda = 647.1$ nm (≈ 200 mW) using a piezoelectrically scanned five-pass Fabry-Perot interferometer [21] having an instrumental contrast of $\geqq 10^9$ and finesse ≈ 40. The spectra were recorded in a single, slow scan.

3. Results

3.1 X-ray diffraction

A typical X-ray diffractogram, taken for sample 4 (Ge:P), [100] orientation, $N_s = 4 \times 10^{16}$ cm^{-2} is shown in Fig. 2 for the (400) reflection as a function of scattering angle θ. This diffractogram shows two peaks which are easy to attribute to the diffraction from the substrate (sharp peak taken to be the origin of θ) and to the IILA layer (broad peak). This assignment is easily accomplished by considering the extinction length of the (400) reflection for $CuK\alpha_1$ radiation [20]. The double-peak structure was observed for all P-doped Ge IILA samples investigated, regardless of orientation, but neither for Sb-doped nor for As-doped Ge. For N_s smaller than 10^{16} cm^{-2} the diffraction peak of the IILA layer shifted to negative values of $\Delta\theta$, this fact indicates that the distance between lattice planes parallel to the layer is larger than that of the substrate for $N_s \leqq 10^{16}$ cm^{-2} and smaller for $N_s \geqq 10^{16}$ cm^{-2}. The relative change in d for the IILA layer, as compared with the substrate, is obtained with [22]

$$\frac{\Delta d}{d} = -\frac{1}{\tan \theta_B} \Delta\theta\,, \tag{4}$$

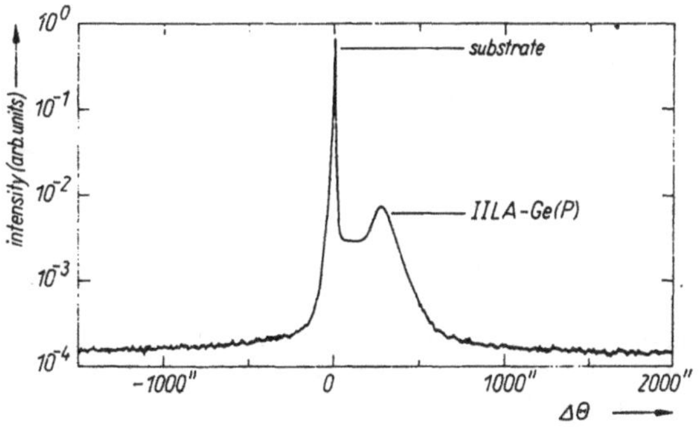

Fig. 2. X-ray diffractogram of a P-IILA Ge sample showing the (400) peak of the substrate (pure Ge) and that of the IILA layer (4×10^{16} cm^{-2}, (100) face)

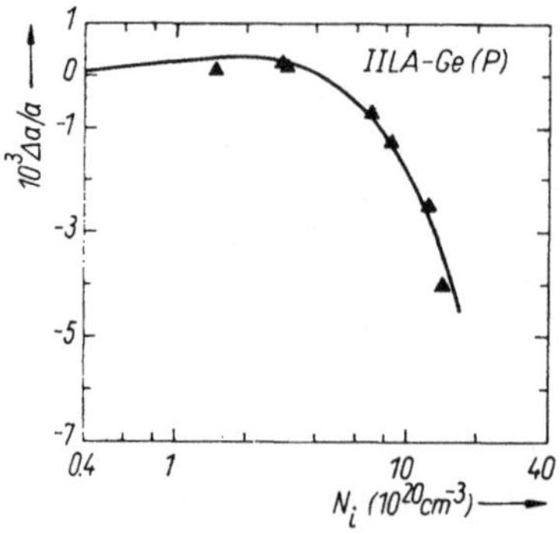

Fig. 3. Relative change in the lattice constant as a function of the impurity concentration N_i. The samples measured were those with the numbers 1, 3, 4, 5, 7, 9, and 10 with (400), (440), and (333) as reflection planes (see Table 1). The solid line represents a fit to the experimental points with (6) for $\beta_e = 7.4 \times 10^{-24}$ cm^3, $\beta_i = -3.0 \times 10^{-24}$ cm^3

where θ_B is the kinematic Bragg angle. The values of $\Delta d/d$ obtained with (4) for the (400), (440), (220), and (333) reflexes of correspondingly oriented samples have to be converted into values of $\Delta a/a$ for bulk material by taking into account the fact that the lattice constant parallel to the substrate remains the same as that of the substrate. This conversion is made with

$$\frac{\Delta a}{a} = \frac{\Delta d}{d}\left(1 + \frac{2C_{12}}{C_{11}}\right)^{-1} \tag{5}$$

for (100) surfaces. For (110) and (111) surfaces the elastic stiffness constants C_{12} and C_{11} must be replaced by

$$C_{11} \to \tfrac{1}{2}(C_{11} + C_{12} + 2C_{44}), \quad C_{12} \to \tfrac{1}{4}(C_{11} + 3C_{12} - 2C_{44}) \quad \text{for (110) surfaces},$$

$$C_{11} \to \tfrac{1}{3}(C_{11} + 2C_{12} + 4C_{44}), \quad C_{12} \to \tfrac{1}{3}(C_{11} + 2C_{12} - 2C_{44}) \quad \text{for (111) surfaces}.$$

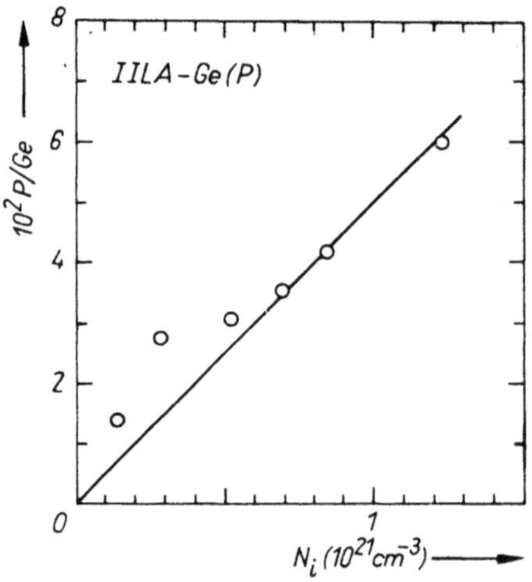

Fig. 4

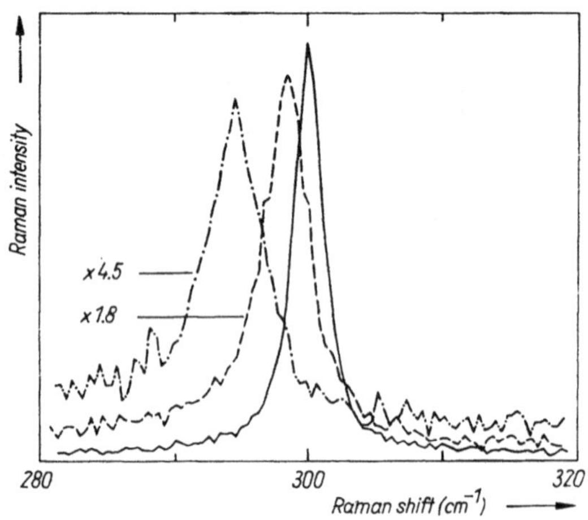

Fig. 5

Fig. 4. Integrated intensities of the local vibrational modes of P implanted in Ge vs. impurity concentration; $T = 300$ K, $\lambda = 647.1$ nm. The intensities are normalized to those of the Raman phonon of Ge

Fig. 5. Room temperature spectra of the Raman phonon in intrinsic Ge and in two IILA samples showing phonon softening and broadening of the latter. ——— pure Ge, – – – Ge:As ($N_e = 1.8 \times 10^{20}$ cm^{-3}), —·— Ge:As ($N_e = 2.1 \times 10^{20}$ cm^{-3})

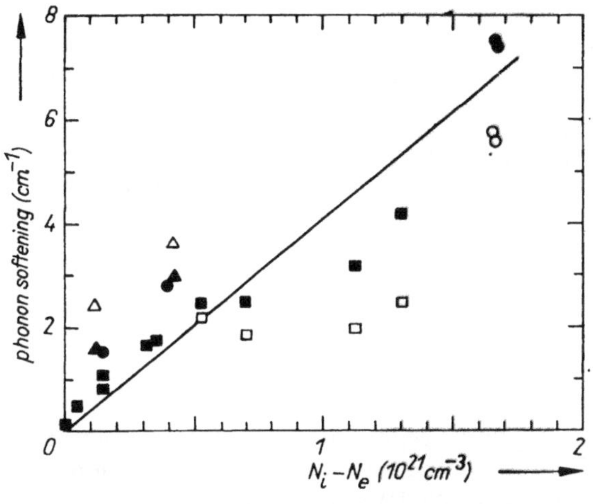

Fig. 6. Raman phonon softening observed for IILA Ge samples at room temperature. The hollow points are as measured, the solid points after correction for changes in lattice constant, $\Delta a/a$, induced by the doping. The solid line was calculated as induced by vacancies (see text). Squares P, circles As, triangles Sb doping

The values of $\Delta a/a$ obtained with (5) are shown in Fig. 3 as a function of N_i (from Table 1). This figure shows clearly the reversal in the sign of $\Delta a/a$ referred to above.

3.2 Raman measurements

Raman measurements have been shown to reveal the localized vibrational modes of the phosphorus impurities in Ge (at $\approx 336\ cm^{-1}$) [23]. Their integrated strength, normalized to that of the Raman phonon, is shown in Fig. 4. Within the estimated error in the determination of these strengths, the intensity ratios of Fig. 4 are seen to vary linearly with impurity concentration N_i.

Fig. 5 shows the Stokes spectra of the Raman phonon of intrinsic Ge and two As–IILA Ge samples. A softening and a broadening of the phonons of the IILA sample, which seems to depend on N_i rather than N_e, is seen. Similar softenings were observed for P- and Sb-implanted layers, as displayed in Fig. 6. The hollow points in this fi-

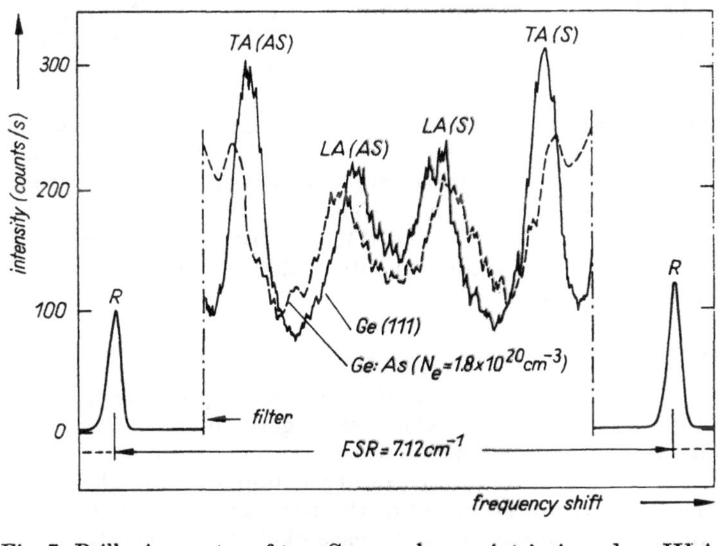

Fig. 7. Brillouin spectra of two Ge samples, an intrinsic and an IILA one, showing the softening of the LA phonons of the latter. R Rayleigh line, FSR free spectral range, S and AS in parentheses Stokes and anti-Stokes

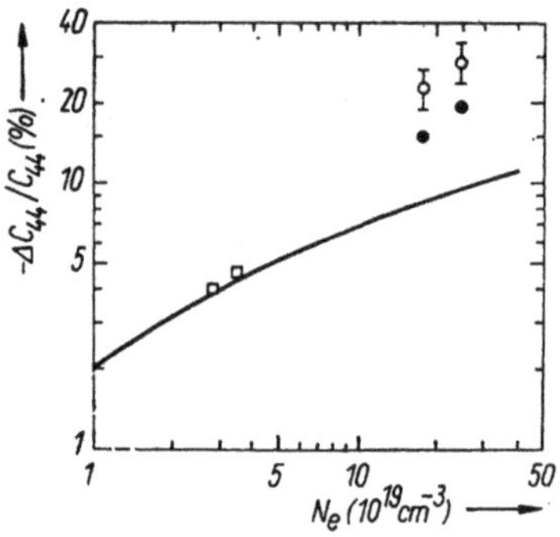

Fig. 8. Softening in C_{44} induced by electron–phonon interaction in n-Ge at room temperature. The squares represent ultrasonic data for a bulk sample [24, 25] and the line the theoretical prediction (Keyes theory). The circles (○ as measured, ● corrected for n (Drude model)) were obtained from Brillouin data for IILA samples. They can be brought to agree with theory by assuming that the samples have ≈ 3% of spherical voids (see text)

gure represent raw data while the filled points represent data after correction for changes in lattice constant induced by the doping (see Fig. 3 and discussion). The abscissa represent the number of electrically inactive impurities $N_i - N_e$. Within the estimated scatter of the experimental points ($\approx \pm 1$ cm^{-1}) the dependence of the frequency softening on $N_i - N_e$ can be assumed to be linear.

3.3 Brillouin scattering

Fig. 7 shows the Brillouin spectra of longitudinal acoustic phonons (LA) obtained for intrinsic Ge and for an As-doped IILA layer. A softening of these LA phonons is seen. A similar softening has also been found from ultrasonic propagation measurements for less doped samples [24, 25] (see Fig. 8).

4. Discussion

The above results show that the IILA n-Ge samples have much higher electron concentrations (N_e up to 2.4×10^{20} cm^{-3}, Table 1) than similar bulk doped samples (the N_e limit is quoted to be (4 to 5) $\times 10^{19}$ cm^{-3} for bulk samples [11]). Nevertheless the most heavily implanted samples must contain a large number of electrically inactive impurities according to the data of Table 1. The electrically active impurities can be assumed to be substitutional. One of the main questions to be approached below is the structure of the electrically inactive impurities, in particular whether they are substitutional (with their electrical activity compensated by other defects, such as vacancies) or simply correspond to precipitated large clusters of the dopant atoms. No evidence of large impurity segregation at the surface or at the interface between the IILA layer and the substrate was seen in either SIMS or RBS [17] measurements. We should point out that heating of the samples to only ≈ 200 °C led to the lowering of N_e to the standard values for bulk samples.

4.1 X-ray diffraction data

Data for the dependence of the lattice constant of silicon on doping are quite extensive [26 to 28] but much less extensive in the case of germanium [28, 29]. It is known [26] that two contributions to this dependence exist: one due to the free carriers themselves and the other due to the hard cores of the dopant ions. The former can be estimated from the deformation potential of the free carriers for hydrostatic deformations, which is known to be negative in the case of electrons in Ge [26, 30].[5])

[5]) The magnitude of this deformation potential is the subject of some controversy, nevertheless we believe the sign of [26] should be correct, see [30].

Hence the presence of free electrons should cause the lattice constant to *expand* so as to reduce their free energy [31]. By substitutional doping, however, we not only introduce free carriers but also change atomic cores. This is expected to lead, in the case of Sb, to an increase in the lattice constant because of the larger covalent radius of Sb as compared to Ge. For As this effect should be small while in the case of P a decrease of the lattice constant should be obtained.

In view of the discussion below (Section 4.2) we assume that all of the dopant atoms are substitutional, compensated, in the case of the $N_i - N_e$ fraction of electrically inactive ones, by defects such as vacancies or dangling bonds at the internal surfaces of voids. Donor–vacancy pairs have indeed been shown to lead to electrically inactive defects [32]. We further assume that these vacancies do not lead to significant changes in the lattice constant a. We can thus write

$$\frac{\Delta a}{a} = \beta_e N_e + \beta_i N_i \,. \tag{6}$$

Using the relationship between N_e and N_i given in Table 1 we have determined β_e and β_i by fitting (6) to the experimental data of Fig. 3 (solid line). We find

$$\beta_e = 7.4 \times 10^{-24}\ \text{cm}^3 \,, \qquad \beta_i = -3 \times 10^{-24}\ \text{cm}^3 \,. \tag{7}$$

The value of β_i can be estimated from the covalent radius of Ge and the ionic radius of P obtained by extrapolation between those of P^0 (covalent radius) and P^{-3} [33]. In this manner one finds $\beta_i = -5.7 \times 10^{-24}$ cm^3 [26], in reasonable agreement with the results of our fit. The value of β_e has been theoretically estimated in [26] to be $\beta_e \approx 2.8 \times 10^{-23}$ cm^3 although the value obtained from experimental data for As, Sb [28], and P [29] seems to be smaller ($\approx 4 \times 10^{-24}$ cm^3 on the average) and more in line with the one found here. We note that a value $\beta_e + \beta_i \approx 10^{-24}$ cm^3 was obtained in [29] for P-doped Ge.

We should also note that the bond lengths of As-doped IILA silicon also shortened when the dopant becomes electrically inactive through furnace annealing [34].

In view of the excellent data and the theoretical interpretation given above for $\Delta a/a$ we must try to answer the question of why we do not observe separate X-ray diffraction peaks in the case of As and Sb. In the case of As the answer may lie in the smallness of the $\Delta d/d$ shifts expected. In the case of Sb it should be quite large. Variation of the lattice constant in the neighborhood of the impurities may lead to rather large widths for the diffraction lines, making them inobservable.

4.2 Raman measurements

We believe that the data of Fig. 4 give strong evidence of substitutional doping by most of the N_i atoms. If the intensities of the local mode are plotted against N_e the simple linear dependence of Fig. 4 is not obtained. Interstitials and other segregated clusters of P should have local vibrational frequencies very different from those of the local modes.

Softenings of Raman phonons upon doping have been observed in heavily doped n-Si [35, 36] and in p-Si [37, 38] and p-Ge [39] and interpreted as due to electron–phonon interaction. The theory used for this interpretation, however, requires the existence of a very small excitation gap for the free carriers [40] which applies to n-Si, p-Si, and p-Ge but not to n-Ge. We have calculated the softening due to phonon-induced intraband transitions in n-Ge [41] and found it to be about an order of magnitude smaller than the results of Fig. 5 and 6. Hence another explanation for this data has to be found.

We should point out that the data of Fig. 6 are shown as "raw data" (hollow points) and also as data corrected for changes in lattice constants of the type discussed in Section 4.1. For this correction we have considered only the hydrostatic component of the strain (one third of $\Delta d/d$). The uniaxial component, which produces splittings of the phonons, should be negligible in most measurements performed with unpolarized light [42]. For this correction we have used the mode Grüneisen parameter for the Raman phonons of Ge [43]. In the case of P doping we have used experimental values of $\Delta d/d$ (Fig. 3). For Sb and As we have used the β_e and β_i coefficients as given by [26], in order to calculate $\Delta a/a$ with the values of N_e and N_i as given in Table 1. Note that the corrected data of Fig. 6 vary more linearly with $N_i - N_e$ than the uncorrected ones.

One may also ask whether one should apply a correction to take into account the different atomic masses of the dopants. It is easy to see along the lines of [44] that the mass correction is unnecessary. In fact, the eigenvectors of optical phonons tend to avoid the atoms with large differences in mass. Hence the effect of such differences is negligible.

We attribute the softening of Fig. 6 mainly to broken bonds of the vacancies or voids which compensate the donor impurities. The frequency of the Raman phonons ω_R can be written as [45]

$$\omega_R^2 = 4\,\frac{\alpha + \beta}{M}, \tag{8}$$

where α and β are Keating's bond stretching and bending force constants, respectively, and M is the atomic mass. The factor 4 in (8) corresponds to the four bonds attached to each atom. A vacancy breaks one of the four bonds and three of the six bond angles of four atoms, hence a vacancy concentration $N_v = N_i - N_e$ should give a frequency softening

$$\frac{\Delta\nu_R'}{\nu_R'} = \frac{\Delta\omega_R}{\omega_R} = \frac{N_v}{2N_{Ge}}\,\frac{\alpha + 2\beta}{\alpha + \beta} = 0.6\,\frac{N_v}{N_{Ge}}, \tag{9}$$

where we have made use of the fact that $\beta \approx 0.3\alpha$ for Ge [45]. The straight line of Fig. 6 represents (9) with $\nu_R' = 300\ \text{cm}^{-1}$. It accounts remarkably well for the observed softening.

The softening of the LA modes of Fig. 7 can be attributed, at least in part, to electron–phonon interaction [46, 47]. Since the measurements were performed in the backscattering configuration we can relate the observed Brillouin shift to the stiffness constants C_{11}, C_{12}, and C_{44} through [48]

$$\omega_L = \frac{4\pi n}{\lambda}\sqrt{\tfrac{1}{3}\,(C_{11} + 2C_{12} + 4C_{44})/\varrho}\,, \tag{10}$$

where λ is the laser wavelength, n the refractive index of the material at this wavelength, and ϱ its density. The electron–phonon interaction should soften C_{44} in n-type Ge but not $C_{11} + 2C_{12}$ [46]. Hence the observed softening must be related, at least in part, to a softening in C_{44}. From (10) we easily find

$$\frac{\Delta C_{44}}{C_{44}} = 3.66\left(\frac{\Delta\omega_L}{\omega_L} - \frac{\Delta n}{n}\right). \tag{11}$$

The change in n, Δn can be estimated from the Drude contribution to the dielectric constant (radicand of (3)). Using the values of ν_p' found from ir reflectivity measurements (cf. (2)) we obtain $\Delta n/n = -0.021$ and -0.024 for $N_e = 1.8 \times 10^{20}$ and 2.5×10^{20} cm^{-3}, respectively. Inserting these values and the experimental $\Delta\omega_L$ into (11) we find the values of $\Delta C_{44}/C_{44}$ shown by full circles in Fig. 8. The open circles represent the softening obtained for $\Delta n/n = 0$, for comparison. These values are more than twice as large as those calculated from the electron–phonon interaction [46] (solid line in Fig. 8). Hence we conclude that the electron–phonon interaction does not suffice to explain the observed softening.

We next consider the possibility that vacancies or voids also contribute to the lowering of the Brillouin frequency. If such defects are present, they should not, to a first approximation, lower the radicand in (10) since both ϱ and C_{ij} should be affected in roughly the same way. Following Landau and Lifschitz [49] the effective dielectric constant $\bar{\varepsilon}$ of a composite medium with spherical voids ($\varepsilon_2 = 1$) imbedded uniformly in the host medium ($\varepsilon_1 = 16$) is given by

$$\bar{\varepsilon} = \varepsilon_1 \left(1 - 3C \frac{\varepsilon_1 - 1}{\varepsilon_2 + 2\varepsilon_1}\right), \qquad (12)$$

where C is the volume fraction of the voids. Using the $\Delta n/n$ corresponding to (12) we can fit the difference between the full circles and the calculated curve in Fig. 8 provided we take $C = 0.03$ for sample 12 and $C = 0.04$ for sample 13. Such a volume fraction of voids is compatible with reports of the existence of voids in IILA Ge which are found in the literature [7].

5. Conclusions

It has been shown that electron concentrations as high as 2.4×10^{20} cm^{-3} can be obtained in IILA Ge implanted with column-five atoms. In this case, however, a larger number of electrically inactive donors is also present. A model, involving donor–vacancy pairs, is proposed for the electrically inactive donors. This model explains the anomalous dependence of the lattice constant on implantation dose, the intensities of the local vibrational modes of phosphorus, and the softening of the Raman phonon. The observed softening of the LA Brillouin peaks observed for (111) surfaces requires the existence of $\approx 3\%$ of voids in the IILA layer.

Acknowledgements

We would like to thank H. Hirt, M. Siemers, and P. Wurster for expert technical assistance and W. Stolz for the Hall measurements, H. Breitschwerdt and W. König for help with the ir measurements, and Dr. P. Eichinger and Mrs. Elisabeth Frenzel of the Fraunhofer-Institut für Festkörpertechnologie for performing the SIMS measurements. Thanks are also due to A. Axmann for the ion implantation which was performed at the Fraunhofer-Institut für Angewandte Festkörperphysik in Freiburg, Federal Republic of Germany.

References

[1] G. K. Celler, CRC Crit. Rev. Solid State and Mater. Sci. 12, 193 (1985).
[2] S. D. Ferris, H. J. Leamy, and J. M. Poate (Ed.), Laser–Solid Interactions and Laser Processing, AIP Conf. Proc. No. 50, Amer. Inst. Phys., New York 1979.

[3] C. W. WHITE and P. S. PEERCY (Ed.), Laser and Electron Beam Processing of Materials, Academic Press, New York 1980.
[4] J. F. GIBBONS, L. D. HESS, and T. W. SIGMON (Ed.), Laser and Electron-Beam Solid Interaction and Material Processing, Elsevier/North-Holland, New York 1981.
[5] B. R. APPLETON and G. K. CELLER (Ed.), Laser and Electron-Beam Interactions with Solids, Elsevier/North-Holland, 1982.
[6] J. NARAYAN, W. L. BROWN, and R. A. LEMONS (Ed.), Laser-Solid Interactions and Transient Thermal Processing of Materials, Elsevier/North-Holland, New York 1983.
[7] O. W. HOLLAND, J. NARAYAN, C. W. WHITE, and B. R. APPLETON, ibid. (p. 297).
[8] J. S. WILLIAMS, ibid (p. 621).
[9] J. NARAYAN, R. T. YOUNG, and C. W. WHITE, J. appl. Phys. 49, 3912 (1978).
[10] B. R. APPLETON, O. W. HOLLAND, J. NARAYAN, O. E. SCHOW III, J. S. WILLIAMS, K. T. SHORT, and E. LAWSON, Appl. Phys. Letters 41, 711 (1982).
[11] V. I. FISTUL, Heavily Doped Semiconductors, Plenum Press, New York 1969.
[12] P. A. SCHUMANN, Jr., Solid State Technol. 13, 50 (1970).
[13] A. A. KUKHARSKII and V. K. SUBASHIEV, Soviet Phys. – Solid State 8, 603 (1966).
[14] L. J. VAN DER PAUW, Philips Res. Rep. 13, R334 (1958).
[15] J. D. QUIRT and J. R. MARKO, Phys. Rev. B 7, 3842 (1973).
[16] J. A. DEL ALAMO and R. M. SWANSON, J. appl. Phys. 57, 2314 (1985).
[17] A. COMPAAN, private communication.
[18] M. MIYAO, T. MOTOOKA, N. NATSUAKI, and T. TOKUYAMA, Solid State Commun. 37, 605 (1981).
[19] LANDOLT-BÖRNSTEIN Tables, Vol. 17A, Springer-Verlag, Berlin/Heidelberg/New York 1982 (p. 93).
[20] Z. G. PINSKER, Dynamical Scattering of X-Rays in Crystals, Springer-Verlag, Berlin/Heidelberg/New York 1978.
[21] J. R. SANDERCOCK, Phys. Rev. Letters 28, 237 (1972).
[22] B. G. COHEN, Solid State Electronics 10, 33 (1967).
[23] G. CONTRERAS, A. COMPAAN, and A. AXMANN, J. Physique 44, C5-193 (1983).
[24] J. J. HALL, Phys. Rev. 137, A960 (1965).
[25] L. J. BRUNER and R. W. KEYES, Phys. Rev. Letters 7, 55 (1961).
[26] J. A. VERGÉS, D. GLÖTZEL, M. CARDONA, and O. K. ANDERSEN, phys. stat. sol. (b) 113, 519 (1982).
[27] M. MIHARA, T. HARA, M. ARAI, M. NAKAJIMA, and S. NAKAMURA, Appl. Phys. Letters 29, 1 (1976).
[28] V. T. BUBLIK, S. S. GORELIK, and A. N. DUBROVINA, Soviet Phys. – Solid State 10, 2247 (1969).
[29] F. I. LEDNEVA, V. M. USTINOV, G. G. AKIMOV, S. G. DVORNIKOVA, and B. G. ZAKHAROV, Kristallografiya 25, 815 (1979); Soviet Phys. – Cryst. 25, 466 (1980).
[30] R. M. MARTIN and C. G. VAN DE WALLE, Bull. Amer. Phys. Soc. 30, 226 (1985).
[31] J. YOKOTA, J. Phys. Soc. Japan 19, 1487 (1964).
[32] T. V. MASHOVETS and V. V. EMTSEV, Proc. Internat. Conf. Lattice Defects in Semiconductors, Freiburg (FRG), 1974 Adlard & Son, London 1975 (p. 103).
[33] L. PAULING, The Nature of the Chemical Bonding, Cornell University Press, New York 1967.
[34] A. ERBIL, G. S. CARGILL, and R. F. BOEHME, Mater. Res. Soc. Meeting, Boston 1984.
[35] M. CHANDRASEKHAR, J. B. RENUCCI, and M. CARDONA, Phys. Rev. B 4, 1623 (1978).
[36] G. CONTRERAS, A. K. SOOD, M. CARDONA, and A. COMPAAN, Solid State Commun. 49, 303 (1984).
[37] F. CERDEIRA, T. A. FJELDLY, and M. CARDONA, Phys. Rev. B 8, 4734 (1973).
[38] M. BALKANSKI, K. P. JAIN, R. BESSERMAN, and M. JOUANNE, Phys. Rev. B 12, 4328 (1975).
[39] D. OLEGO and M. CARDONA, Phys. Rev. B 23, 6592 (1981).
[40] G. ABSTREITER, M. CARDONA, and A. PINCZUK, in: Light Scattering in Solids IV, Ed. M. CARDONA and G. GÜNTHERODT, Springer-Verlag, Heidelberg 1984 (p. 127).
[41] F. CERDEIRA and M. CARDONA, Phys. Rev. B 5, 1440 (1972).
[42] F. CERDEIRA, C. J. BUCHENAUER, F. H. POLLAK, and M. CARDONA, Phys. Rev. B 5, 580 (1972).

[43] B. A. WEINSTEIN and R. ZALLEN, see [40] (p. 463).
[44] G. BENEDEK and G. F. NARDELLI, Phys. Rev. 155, 1004 (1967).
[45] R. M. MARTIN, Phys. Rev. B 1, 4005 (1970).
[46] R. W. KEYES, Solid State Phys. 20, 37 (1967).
[47] L. PINTSCHOVIUS, J. A. VERGÉS, and M. CARDONNA, Phys. Rev. B 26, 5658 (1982).
[48] C. KITTEL, Introduction to Solid State Physics, 3rd edition, John Wiley & Sons, New York 1967.
[49] L. D. LANDAU and E. M. LIFSHITZ, Electrodynamic of Continuous Media, Pergamon Press, Oxford 1963.

(Received June 11, 1985)

phys. stat. sol. (b) **131**, 489 (1985)
Subject classification: 4 and 13; 22

Zentralinstitut für Optik und Spektroskopie der Akademie der Wissenschaften der DDR, Berlin[1])

Ionicity and Static Ionic Charge

By
P. ENDERS

Dedicated to Prof. Dr. Dr. h. c. Dr. E. h. P. GÖRLICH on the occasion of his 80th birthday

Recently, Falter et al. have proposed an ionicity scale basing on the longitudinal effective charge e_L (that is wrongly identified with the static ionic charge q). It is shown that ionicities that vary linearly with q or e_L are, however, not appropriate to construct structural maps, and it is concluded that nor q, nor e_L is very relevant or even a measure for structural properties.

Kürzlich haben Falter u. a. eine Ionizitätsskala auf der Basis der longitudinalen effektiven Ladung e_L vorgeschlagen (wobei e_L fälschlicherweise mit der statischen Ionenladung q identifiziert wird). Es wird gezeigt, daß Ionizitäten, die linear mit q oder e_L variieren, jedoch nicht geeignet sind, Kristallstrukturdiagramme zu konstruieren, und es wird daraus geschlossen, daß weder q noch e_L allzu wesentlich oder gar ein Maß für strukturelle Eigenschaften ist.

1. Introduction

The concept of ionicity of chemical bonding has been proved to be very useful in the characterization of molecular and solid state properties, and there exist many ionicity scales proposed to correlate with a wide range of chemical and physical properties, as there are heats of formation and cohesive energies, crystal structure and bond forces, polarization properties (linear and non-linear polarizabilities, dynamic effective charges), static ionic charges, and band structure parameters [1 to 6]. The dielectric ionicity scale [2] seems to work most successfully, at least for average valence-4-compounds.

Among these properties, the static ionic charge seems to be most problematic. An immediate connection exists only to MO-LCAO-based ionicities [2, 3, 6], and there is a crude correlation between Pauling's ionicity and dipole charges in diatomic molecules [1]. The identification of the ionicity of the bond with the static ionic charge divided by the chemical valence has been accepted also by Urusov [4]. But this seems to be less useful owing to the possibility of a negative charge on a formal cation, as, for example, in SnTe [7].

Hübner [5] has proposed the following linear relation between Phillips's ionicity f_i [2] and static cation charge q for octet compounds:

$$q = 4(f_i - 1) + Z\,, \tag{1}$$

where Z is the chemical valence (cation core charge). This formula leads, however, to an unphysical electron transfer from the halogen to the group-Ib atom in their tetrahedral compounds, in contradiction to the experiment [8]. This can be removed using Harrison's polarity α_p [9] or $f_i^{1/2}$ [10] instead of f_i in (1).

[1]) Rudower Chaussee 6, DDR-1199 Berlin, GDR.

In order to obtain (1), Falter et al. [11] have constructed an ionicity scale from the valence charges, Z, and longitudinal effective charges, e_L. The question now arising is, which is the significance of a so-defined ionicity scale? This is the subject of the present paper, in other words, we examine the problem, which role do the static ionic charges play for the chemical bonding, especially, concerning structural properties (crystal structures).

2. Ionicity and Static Charge Transfer

Despite of the very great success of the dielectric theory in systematizing a wide range of properties of many — mainly average valence-4-compounds [2, 5] it seems to be less able to be related immediately to static ionic charges. (The well-known problem of assigning a static charge to an individual species in a compound is disregarded here, cf. [4, 12 to 14].) Especially, (1) being a linear scaling assumption basing on the correct boundary values for $f_i = 0$ and $f_i = 1$, respectively, on principle cannot account for the charge transfer in tetrahedral I–VII compounds [8, 9]. This difficulty does not exist [9] within the framework of the bond orbital model [3], where

$$q = 4(\alpha_p - 1) + Z\,. \tag{2}$$

It is also left when using the formula [10]

$$q = 4(f_i^{1/2} - 1) + Z\,. \tag{3}$$

Note that one can show that the MO ansatz sketched by Bazhenov et al. [15] leads also to (3).

In contrast to the spectroscopically and MO-theoretically defined f_i and α_p, respectively, Falter et al. [11] have proposed an ionicity — called here f_F — basing on the static charge transfer. Consequently, it is a matter of *definition*, that (1) can be retained in terms of f_F [11, 16]. It is worth noting, however, that the identification of e_L with q is incorrect, because the former is a dynamic charge and therefore principally different from the latter [12, 14], although their numerical values are generally of the same order.

In Table 1 we list experimental values of q as obtained from several methods (all being approximate, of course) [4] and relate them to various ionicity scales using as reference a "charge ionicity"

$$f_q = \frac{q - Z}{N} + 1\,, \tag{4}$$

where N is the mean valence electron number per atom.

3. Ionicity and Structure

As can be seen from Table 1, there is no satisfactory overall correlation or even agreement between the experimental q-values (despite of the indetermination of their definition and measurement) and those corresponding to (1), (2), or (3), except for special groups of compounds. On the other hand, the dielectric ionicity f_i distinguishes exactly four-fold coordination from six- and eight-fold one [2]. Consequently, we have to investigate the relevance of f_F for structural properties (crystal structure).

Table 1

Static cation charge q and charge ionicity f_q in comparison with other ionicity scales and $f_i^{1/2}$

a) NaCl structure (R)

R	q			f_q			f_F	f_i	$f_i^{1/2}$
	th.[a]	ch.[b]	at.[c]	th.	ch.	at.			
LiF	0.73	—	0.986	0.93	—	0.996	0.887	0.915	0.957
LiCl	—	—	0.968	—	—	0.992	—	0.903	0.950
LiBr	—	—	0.972	—	—	0.993	—	0.899	0.948
LiI	—	—	0.971	—	—	0.993	—	0.890	0.943
NaF	—	0.6	—	—	0.9	—	0.899	0.946	0.973
NaCl	0.79	0.9	0.966	0.95	0.98	0.992	0.869	0.935	0.967
NaBr	—	0.8	0.971	—	0.95	0.993	0.859	0.934	0.966
NaI	—	0.8	0.972	—	0.95	0.993	—	0.927	0.963
KF	—	(0.3[d])	0.972	—	—	0.993	—	0.955	0.977
KCl	0.83	—	0.969	0.96	—	0.992	0.879	0.953	0.976
KBr	—	—	0.977	—	—	0.994	0.869	0.952	0.976
KI	0.79	—	0.979	0.95	—	0.995	0.857	0.950	0.975
RbBr	0.84	—	—	0.96	—	—	0.873	0.957	0.978
AgCl	0.77	0.7	—	0.94	0.92	—	0.835	0.856	0.925
AgBr	0.75	—	—	0.94	—	—	0.830	0.850	0.922
			th.[e]			th.[e]			
MgO	1.08	1.0	1.02	0.77	0.75	0.775	0.669	0.841	0.917
MgS	—	—	0.96	—	—	0.740	—	0.786	0.887
CaO	1.15	—	1.1	0.79	—	0.775	0.707	0.913	0.956
CaS	—	—	1.1	—	—	0.775	—	0.902	0.950
SrO	1.13	—	—	0.78	—	—	0.655	0.926	0.962
BaO	1.06	—	1.12	0.76	—	0.780	—	—	—
BaS	—	—	1.14	—	—	0.785	—	—	—
CdO	0.75	0.7	0.84	0.69	0.68	0.71	0.635	0.785	0.886

a) From thermochemical data, [4], Tables V.2 and V.18.
b) From chemical shift, [4], Table V.16.
c) From energy of atomization, [4], Table VI.7.
d) From electron scattering, [17].
e) From thermochemical data, upper bounds, [4], Table VI.14.

Table 1 (continued)

b) Tetrahedral structure (Zb, W)

Zb, W	q		f_q		f_F	f_i [11]	$f_i^{1/2}$	α_p[i]
	th.[f]	ch.[g]	th.	ch.				[3]
CuCl	0.78	0.25	0.945	0.812	0.826	0.746/0.77	0.864/0.877	0.71
CuBr	0.79	0.29	0.948	0.822	0.822	0.735	0.857	0.72
CuI	0.80	0.15	0.950	0.788	0.802	0.692	0.832	0.69
AgI	0.79	0.22	0.948	0.805	0.822	0.770	0.877	0.73
ZnS	0.72	0.5	0.68	0.65	0.605	0.623	0.789	0.56
HgS	0.56	—	0.64	—	—	—	—	—
HgSe	0.54		0.64		—	—	—	—

Table 1 (continued)

Zb, W	q		f_q		f_F	f_i [11]	$f_i^{1/2}$	α_p [i])
	th. [f])	ch. [g])	th.	ch.				[3]
		X [h])		X				
HgTe	—	0.60 ± 0.05	—	0.65	0.533	—/0.65	—/0.81	—
AlP	—	0.80 ± 0.20	—	0.45	0.326	0.307	0.554	0.34
AlAs	—	0.60 ± 0.15	—	0.40	0.314	0.274	0.523	0.34
AlSb	—	0.45 ± 0.08	—	0.36	0.297	0.425/0.250	0.652/0.5	0.30
GaP	—	0.75 ± 0.20	—	0.44	0.310	0.374/0.327	0.612/0.572	0.32
GaAs	—	0.51 ± 0.10	—	0.38	0.300	0.31	0.56	0.32
GaSb	—	0.43 ± 0.07	—	0.36	0.287	0.261	0.511	0.27
InP	—	0.58 ± 0.15	—	0.40	0.317	0.421	0.649	0.40
InAs	—	0.35 ± 0.07	—	0.34	0.302	0.357	0.597	0.40
InSb	—	0.18 ± 0.05	—	0.30	0.289	0.321	0.567	0.34
BeO	1.15	—	0.79	—	0.653	0.602	0.776	0.46
ZnO	0.88	—	0.72	—	0.639	0.616	0.785	0.60
CdS	0.64	0.5	0.66	0.62	0.599	0.685	0.828	0.63
CdSe	0.47	0.4	0.62	0.6	0.589	0.699	0.836	0.63

f) Upper bounds from thermochemical data, [4], Table V.3.
g) Chemical shift, [8] and [4], Table V.18.
h) From X-ray scattering, [4], Table V.20.
i) The earlier values [18] quoted in [11] are higher.

Table 1 (countinued)
c) Ten-electron compounds (⟨5⟩)

⟨5⟩	q	f_q		α_p [k])	f_F
	th. [a])	[j])	[e])	[7]	
TlCl	0.65	0.53	0.88	—	—
TlBr	0.65	0.53	0.88	—	0.851
TlI	0.68	0.54	0.89	—	0.845
PbS	0.84	0.37	0.61	0.59	—
PbSe	0.75	0.35	0.58	0.53	—
PbTe	0.77	0.35	0.59	0.44	—
	ch. [g])				
GeS	0.9	0.38	0.63	—	—
SnO	0.9	0.38	0.63	—	—
SnS	0.8	0.36	0.60	—	—
SnSe	0.4	0.28	0.47	—	—
SnTe	0.6	0.32	0.53	0.29	—

j) $Z = 3$ and 4, respectively, $N = 5$.
k) $Z = 1$ and 2, respectively, $N = 3$.

Assuming $0 \leqq e_L \leqq 1$ (cf. [11, 19]; also, $q \leqq 1$, cf. Table 1 and Pauling's electroneutrality principle [1]), we obtain the following ranges of f_F for octet compounds (cf. [11]):

Z	f_F
4	0 to 0.25
3	0.25 to 0.5
2	0.5 to 0.75
1	0.75 to 1

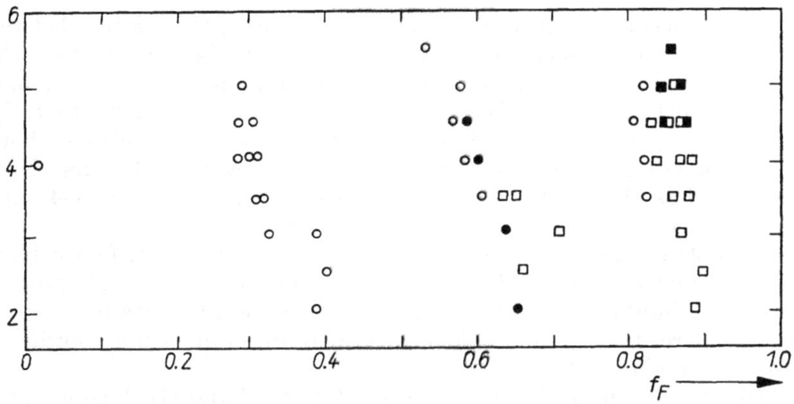

Fig. 1. Mooser-Pearson plot $\bar{n}$ vs. f_F for the 47 AB compounds listed in Table 1 of [11]. □ NaCl, ■ CsCl, ○ ZnS(Zb), ● ZnO(W)

That means, f_F classifies rather according to the valence charge number, Z, than according to structures (coordination number). This is clearly depicted in a Mooser-Pearson plot [20] $\bar{n}$ versus f_F, where $\bar{n} = \frac{1}{2}(n_A + n_B)$ is the mean row number of the constitutents in the Periodic Table (see Fig. 1).

The situation is somewhat improved when using an effective valence charge $Z_{eff} = 1.5$ [19] (cf. [21]) for the group-Ib halides accounting for the cation d-electrons. This lowers their f_F-values to about 0.7, but a separation of the rocksalt structure (AgCl, AgBr) from the zincblende structure (CuCl, CuBr, CuI, AgI) is still not achieved. Also, the f_F-values for TlBr and TlCl are clearly too high in view of their mixed ionic-covalent character [22], and $Z = 1$ is certainly not appropriate owing to the cation s^2-cores. $Z_{eff} = 2$ [19] lowers f_F by an amount of 0.2 and shifts it into the domain of the II–VI compounds, what is more satisfactory.

4. Discussion

The facts collected in the previous sections show the following picture: Ionicities that are closely related to structural properties of solids, as f_i and f_{Pe} (f_{Pe} is derived from Pettifor's [23] structural electronegativities, see [24]), are less related to static ionic charges, q, while those basing on q, f_q, or e_L, f_F, are less able to describe structural properties. Consequently, neither q, nor e_L alone are very relevant or even a measure for structural properties, at least not in the sense of a linear correlation. In other words, if one wishes to systematize structural properties by means of an ionicity (or electronegativity) scale, one should not define it through these charges.

Several factors might account for this: (i) the crystal structure of non-metallic compounds is determined not only by Coulombic and covalent forces, but also through geometrical and volume factors (as well as temperature and pressure, not considered here) [1, 4, 13]; (ii) the stability of crystal structures, especially with respect to small lattice distortions, is often given through the stiffness of certain phonon modes, i.e. determined by dynamical porperties (but not e_L, as can be seen from ferroelectric phase transitions), not static parameters.

Obviously, the close correlation between ionicity and charge depends upon the very definition of ionicity. Thus, f_i scales well with $e_L^2\varepsilon_\infty$ [2], where ε_∞ is the static electronic dielectric constant, while the LCAO-based ones by Coulson [6] and by Harrison [3] are by definition connected with q, as f_F is with e_L. Therefore, the LCAO-based ionicity parameters contain structural properties as far as the relation between ionic (Coulombic) and covalent bond forces is concerned (cf. [4]), while the other ones should be more related to structural stability against dynamic forces (cf. [2, 3,5]). In reality, of course, both aspects of bonding are connected with each other. For example, all three: f_i, α_p, and f_F, scale linearily with the relation between bond-stretching and bond-bending force constants in tetrahedral semiconductors [2, 11]. Further, Harrison [3] has sketched a structural map for average valence-4 solids using α_p and metallicity α_m (for IV–VI compounds, see Suski et al. [25]).

None of the known structural maps seems to be able to separate satisfactorily, i.e. in topologically connected domains, *all* binary compounds within a *single* plot (cf. [2, 23, 24]). Thus, one should prefer to construct such maps for certain groups or classes of solids, and here, the St. John-Bloch orbital radii as quantum coordinates seem to work most successfully [26 to 28].

Equation (3) works best for group-I halides, but not for II–VI and III–V compounds. The latter is probably due to the fact that the optical properties are determined there predominantly by the p-like valence states [3]. But the objection that (3) is a "non-necessary manipulation" [16] is unfounded.

References

[1] L. Pauling, The Nature of the Chemical Bond, Cornell University Press, Ithaca (N. Y.) 1960.
[2] J. C. Phillips, Rev. mod. Phys. **42**, 317 (1970).
[3] W. A. Harrison, Electronic Structure and the Properties of Solids, Freeman & Co., San Francisco 1980.
[4] V. S. Urusov, Energeticheskaya kristallokhimiya, Izd. Nauka, Moscow 1975.
[5] K. Hübner, in: Festkörperphysik, Ed. K. Unger and G. Schneider, Chap. 13, Akademische Verlagsgesellschaft Geest & Portig, Leipzig 1979, and references therein.
[6] C. A. Coulson, Valence, Oxford University Press, Oxford 1961.
[7] P. Enders, phys. stat. sol. (b) **120**, 735 (1983).
[8] A. Goldmann, J. Tejeda, N. Shevchik, and M. Cardona, Phys. Rev. B **10**, 4388 (1974).
[9] S. T. Pantelides and W. A. Harrison, Phys. Rev. B **11**, 4049 (1975).
[10] P. Enders, phys. stat. sol. (b) **124**, K21 (1984).
[11] C. Falter, W. Ludwig, M. Selmke, and W. Zierau, Phys. Letters A **105**, 139 (1984).
[12] J. Suchet, C. R. Acad. Sci. (France) **258**, 2305 (1964).
[13] W. B. Pearson, The Crystal Chemistry and Physics of Metals and Alloys, Wiley-Interscience, New York 1972.
[14] K. B. Tolpygo, in: Khimicheskaya svyaz v kristallakh poluprovodnikov i polumetallov, Ed. N. N. Sirota, Izd. Nauka i Tekhnika, Minsk 1973 (p. 196).
[15] V. K. Bazhenov, V. A. Presnov, T. I. Soboleva, and M. G. Foigel, see [14] (p. 18).
[16] C. Falter, M. Selmke, and W. Zierau, Solid State Commun. **52**, 949 (1984).
[17] A. P. Margolina and A. G. Buntar, see [14] (p. 85).
[18] W. A. Harrison, Phys. Rev. B **10**, 767 (1984).
[19] G. Lucovsky, R. M. Martin, and E. Burstein, Phys. Rev. B **4**, 1367 (1971).
[20] E. Mooser and W. B. Pearson, Acta cryst. **2**, 1015 (1959).
[21] P. Lawaetz, Phys. Rev. B **5**, 4039 (1972).
[22] J. Treusch, Phys. Rev. Letters **34**, 1343 (1975).
[23] D. G. Pettifor, Solid State Commun. **51**, 31 (1984).
[24] P. Enders, to be published.
[25] T. Suski, J. Karpinski, K. L. I. Kobayashi, and K. F. Komatsubara, J. Phys. Chem. Solids **42**, 479 (1981).
[26] J. C. Phillips, Festkörperprobleme **17**, 109 (1977).
[27] A. Zunger, Phys. Rev. B **22**, 5839 (1980).
[28] P. B. Littlewood, CRC Comments Solid-State Phys. **11**, 229 (1983).

(Received June 3, 1985)

phys. stat. sol. (b) 131, 495 (1985)

Subject classification: 4; 1.3; 21.1; 21.2

Polytechnical Institute, Vologda[1])

Many-Body Contribution to the Energy of Metals with B.C.C. Lattice

By

V. I. BOGDANOV and D. L. FUKS

Dedicated to Prof. Dr. Dr. h. c. Dr. E. h. P. GÖRLICH on the occasion of his 80th birthday

The Fourier transforms of covalent charge density and the contribution of the third and fourth order of perturbation series for the energy (E_{bs}^{cov}) of alkali and some transition metals are calculated on the basis of self-consistent Brillouin-Wigner pseudopotential perturbation theory. The scattering of electrons on the planes of the first and second Brillouin zones in the b.c.c. lattice is taken into account. It is shown that the value E_{bs}^{cov} is of the same order as the heat of polymorphic transition. The convergence of perturbation series in pseudopotential theory for the calculation of transition metal energy with Animalu's form-factors is investigated.

На основе самосогласованной теории возмущений Бриллюэна-Вигнера рассчитаны фурье-образы плотности ковалентного заряда и вклад третьего и четвертого порядков теории возмущений E_{bs}^{cov} в энергию связи щелочных и некоторых переходных металлов. Учтено рассеяние электронов на гранях первой и второй зон Бриллюэна в ОЦК решетке. Показано, что величина E_{bs}^{cov} даже в щелочных металлах сравнима или на порядок больше теплоты перехода при полиморфных превращениях. Аналогичные результаты получены для переходных металлов. Исследована сходимость ряда теории возмущений по псевдопотенциалу при расчетах энергии этих металлов с формфакторами Анималу.

1. Introduction

Paper [1] was seemingly the first one, where attention has been attracted by the role of nonpair indirect interaction in describing the properties of metals. This interaction plays a principal role in the formation of phonon spectra and predetermines the appearance of covalent forces in metals. The effects of "covalency" (many-ion interactions) were discussed in [2] on the basis of the Brillouin-Wigner perturbation theory while analysing the stability of crystal structures of Si, Ge, and α-Sn. Still, the direct calculations of high-order perturbation series contributions to the binding energy of b.c.c. metals are absent, while these calculations may become necessary in the description of the energy difference of crystal structures. So, in alkali metals the heats of h.c.p.–b.c.c. transitions are very small ($\approx 10^{-5}$ Ryd/ion) and the band structure energy (the second order of perturbation series) is of the order 10^{-2} to 10^{-3} Ryd/ion.

As for the transition metals with b.c.c. lattice, the application of pseudopotential theory for the investigation of their properties is still a matter of discussion. At the same time the model form-factor [3] is used for calculating the binding energy. In such calculations the characteristic s–d interaction is taken into account mainly by the diagonal matrix elements, the non-diagonal matrix elements, contributing to the band structure energy, are small. Moreover, the band structure energy is an integral over the electronic spectrum and its value is not sensitive to the delicate peculiar features of the spectrum. But the problem of calculating the value of high-order perturbation series in the band structure energy of such metals is not cleared up yet.

[1]) Ul. Voroshilova 3, 160600 Vologda, USSR.

The aim of this paper is to calculate the covalent contribution to the binding energy of b.c.c. metals in connection with the theoretical description of polymorphism in metals and the analysis of perturbation series convergence in determining the binding energy of transition metals. For this purpose we shall use the Brillouin-Wigner perturbation theory. It gives the possibility to calculate the contributions of third and fourth order (in electron–ion interaction) by means of a direct, self-consistent approach not resorting to the help of diagram techniques. Exceeding the limits of the second-order perturbation theory results in the appearance of an additional contribution to the electron density. The inhomogeneity of this contribution is a reason for nonlinear effects in pseudopotential screening.

2. Formalism

As is shown in [2] the Fourier transforms of the covalent charge density $\varrho^{\text{cov}}(g)$ and the many-body contribution to the band structure energy $E_{\text{bs}}^{\text{cov}}$ may be written as follows:

$$\varrho^{\text{cov}}(\boldsymbol{g}) = \frac{2}{\Omega} \sum_{\boldsymbol{k}} \frac{\tilde{W}(\boldsymbol{g}, \boldsymbol{k}) - W(\boldsymbol{g})}{E_{\boldsymbol{k}} - E^0_{\boldsymbol{k}+\boldsymbol{g}}}, \tag{1}$$

$$E_{\text{bs}}^{\text{cov}} = \frac{\Omega}{2} \sum_{\boldsymbol{g}}{}' \frac{W^0(\boldsymbol{g})\, \varrho^{\text{cov}}(\boldsymbol{g})}{\varepsilon^*(\boldsymbol{g})}, \tag{2}$$

where

$$\tilde{W}(\boldsymbol{g}, \boldsymbol{k}) = W(\boldsymbol{g}) + \sum_{\boldsymbol{g}' \neq \boldsymbol{g}}{}' \frac{W(\boldsymbol{g}')\, W(\boldsymbol{g}' - \boldsymbol{g})}{E_{\boldsymbol{k}} - E^0_{\boldsymbol{k}+\boldsymbol{g}'}} + \sum_{\boldsymbol{g}' \neq \boldsymbol{g}}{}' \sum_{\boldsymbol{g}'' \neq \boldsymbol{g}'}{}' \frac{W(\boldsymbol{g}'')\, W(\boldsymbol{g}'' - \boldsymbol{g}')\, W(\boldsymbol{g}' - \boldsymbol{g})}{(E_{\boldsymbol{k}} - E^0_{\boldsymbol{k}+\boldsymbol{g}'})\,(E_{\boldsymbol{k}} - E^0_{\boldsymbol{k}+\boldsymbol{g}''})}. \tag{3}$$

Here Ω is the atomic volume, $E^0_{\boldsymbol{k}}$ the free electron energy, the summation in (1) is carried out over all occupied states, in (2) and (3) over the reciprocal lattice, $W(\boldsymbol{g})$ is the total screened matrix element of the pseudopotential,

$$W(\boldsymbol{g}) = \frac{1}{\varepsilon^*(\boldsymbol{g})} \left\{ W^0(\boldsymbol{g}) + \frac{4\pi\chi(\boldsymbol{g})}{\boldsymbol{g}^2}\, \varrho^{\text{cov}}(\boldsymbol{g}) \right\}, \tag{4}$$

where $\varepsilon^*(\boldsymbol{g})$ is the modified dielectric function allowing for exchange-correlation effects, $W^0(\boldsymbol{g})$ is the matrix element of the "bare"-ion pseudopotential, the function $\chi(\boldsymbol{g})$ takes into consideration the exchange-correlation effects in screening,[1])

$$\varepsilon^*(\boldsymbol{g}) = 1 + \frac{1}{2\pi k_f \eta^2} \left(\frac{1 - \eta^2}{2\eta} \ln \left| \frac{1 + \eta}{1 - \eta} \right| + 1 \right) \chi(\boldsymbol{g}); \quad \eta = \frac{g}{2k_f}.$$

In the sums (1) to (3) we take into account only the terms, which give the "main" contribution, i.e. those for which the energy denominators in (1), (3) are tending to zero. The method of their selection is described in [2, 4]. The expressions for matrix elements are quite cumbersome. So, for example, for $\tilde{W}(112, 001)$

$$\tilde{W}(112, 001) = W(112) - \frac{a^2}{(2\pi)^2} \Big[(W(110))^2 + \frac{5}{4}\, W(110)\, W(200) + \frac{7}{5}\, W(200)\, W(112) \Big] + \frac{a^4}{(2\pi)^4} \Big[\frac{53}{20}\, (W(110))^2\, W(200) +$$

[1]) In this paper $\chi(\boldsymbol{g})$ is taken in the form, suggested by Geldart and Vosko.

$$+\frac{7}{2}W(110)\,(W(200))^2+\frac{47}{10}(W(110))^2W(112)+$$

$$+\frac{201}{100}(W(200))^2W(112)+\frac{7}{2}(W(110))^3+\frac{12}{5}(W(112))^3\Big]-$$

$$-\frac{a^2}{(2\pi)^2}[3(W(110))^2W(200)+(W(200))^2W(112)+$$

$$+3\,(W(110))^2W(112)]\,\frac{1}{E_{[001]}-E^0_{[010]}}\,,$$

where a is the lattice parameter of the b.c.c. lattice.

As the valency of the metals under consideration is $Z=1,2,3,5$, one must take into account the scattering of electrons on the planes of the first and second Brillouin zones. So, for the alkali metals the Fermi surface does not cross the planes of the first Brillouin zone, while for other metals it intersects the planes of (110) and (200) types. This circumstance must be considered while calculating the dispersion law near these planes.

The following expression is used for the determination of energy gaps in (1), (3):

$$E_{k_j}-E^0_{k_j}=\sum_{g}{}'\frac{W(g)\,\tilde W(g,k_j)}{E_{k_j}-E^0_{k_j+g}}\,, \tag{5}$$

where k_j are the wave vectors giving the main contribution to (1), (3) and the summation is performed over such g which make the energy denominators minimal or equal to zero.

In contrast to [4], here the calculation of electron reflection at the planes of Brillouin zones leads to the necessity of using the many-band model. Really for $k_j=(2\pi/a)\,(\bar{1}00)$ one must take into account the contribution to the sum from the vectors $g=(2\pi/a)\times\times(110)$, (101), $(1\bar{1}0)$, $(10\bar{1})$, (200); for $k_j=(2\pi/a)\,(\frac{\bar 1}{2}\,\frac{\bar 1}{2}\,\bar{1})$ the corresponding vectors g are $(2\pi/a)\,(110)$, (112). After substitution of the expressions for the energy gaps calculated according to this procedure into the matrix elements $\tilde W(g,k)$ as described in [2] and writing down formulas for the Fourier transforms of "covalent charge distribution" $\varrho^{\text{cov}}(g)$ one can get the equations for $\varrho^{\text{cov}}(g)$. This fact is connected with the dependence of the matrix elements $W(g)$ entering $\tilde W(g,k)$ upon $\varrho^{\text{cov}}(g)$ (see (4)). These equations may be linearized because one must keep in $\varrho^{\text{cov}}(g)$ only the terms of second and third order in the pseudopotential form-factor if the energy is formally written up to the fourth order of perturbation theory. After realizing this scheme the analytical expressions for $\varrho^{\text{cov}}(g)$ may be written as follows:

$$\varrho^{\text{cov}}(110)=\frac{1}{\Omega}\left\{\frac{3Z}{48\varepsilon_f}\,\frac{a^2}{(2\pi)^2}\left\{\left(3\cdot 2^{2/3}(3.667\overset{①}{\pm}11)-8\overset{②}{\pm}\frac{51}{S_1}+\frac{1}{2S_2}\right)A+\right.\right.$$

$$+\left(3\cdot 2^{2/3}\left(4.167\overset{①}{\pm}\frac{23}{3}-2\right)-12\overset{②}{\pm}\frac{30}{S_1}+\frac{12}{S_2}\right)B-1.864C+$$

$$+\left(\frac{6}{S_2}\overset{②}{\pm}2^{2/3}\cdot 13\overset{②}{\pm}\frac{54}{S_1}\right)D+\left(\overset{②}{\pm}\frac{18}{S_1}-\frac{9}{S_2}\right)E+\left(\overset{②}{\pm}\frac{15}{S_1}-\frac{27}{8S_2}\right)F+$$

$$+\left(\overset{②}{\pm}\frac{24}{S_1}-\frac{87}{8S_2}\right)G+\frac{a^2}{(2\pi)^2}\Big[211.767W_1^3+327.522W_1^2W_2+$$

$$+ 155.354 W_1^2 W_3 + 123.265 W_1 W_2^2 + 76.489 W_2^2 W_3 +$$

$$+ 155.620 W_1 W_3^2 \overset{③}{\pm} \frac{3 \cdot 2^{2/3}}{2} \delta_1 \left(11 W_1^2 + \frac{23}{3} W_1 W_2 + \frac{13}{3} W_1 W_3 \right) \overset{③}{\pm}$$

$$\overset{③}{\pm} 3\delta_2 \left(\frac{17}{2} W_1^2 + 5 W_1 W_2 + 9 W_1 W_3 + \frac{5}{2} W_2^2 + 3 \frac{W_2^2 W_3}{W_1} + 4 W_3^2 \right) +$$

$$+ \frac{3}{4} \delta_3 \left(\frac{1}{3} W_1^2 + 8 W_1 W_2 + 4 W_1 W_3 ! - \frac{9}{4} W_2^2 - 6 \frac{W_2^2 W_3}{W_1} - \frac{29}{4} W_3^2 \right) \Bigg] \Bigg\} \Bigg\} . \qquad (6)$$

The signs marked by the circles must be chosen as follows: ① stands for "+". if $Z \leqq 1.481$, ② stands for "+" if $Z \leqq 4.18878$, ③ stands for "−" if $Z \leqq 4.18878$. These values of valency are taken in such a way because (6) and the following formulas for $\varrho^{\text{cov}}(200)$ and $\varrho^{\text{cov}}(112)$ can be used for binary substitutional alloys if matrix elements $W(\boldsymbol{g})$ are replaced by $C W_A(\boldsymbol{g}) - (1 - C) W_B(\boldsymbol{g})$ and Z is replaced by $C Z_A + (1 - C) Z_B$ where C is the concentration of elements of sort A. These values of Z correspond to the Fermi spheres touching the planes of Brillouin zones. So, the increase of the valency leads to the crossing between Fermi sphere and planes, and changing of the formulas for $\varrho^{\text{cov}}(\boldsymbol{g})$. This fact is taken into account in calculating the energy gaps from (5).

$$\varrho^{\text{cov}}(200) = \frac{1}{\Omega} \left\{ \frac{3Z}{48\varepsilon_f} \frac{a^2}{(2\pi)^2} \left\{ \left(-3 \cdot 2^{2/3} - \frac{2}{3} \pm \frac{24}{S_3} \right) A + \right. \right.$$

$$+ \left(-12 \pm \frac{12}{S_3} - \frac{2}{3} \right) G \pm \frac{48}{S_3} H \pm \frac{36}{S_3} D \pm \frac{48}{S_3} L +$$

$$+ \frac{a^2}{(2\pi)^2} \Big[76.001 W_1^3 + 38.413 W_1^2 W_2 + 45.223 W_1^2 W_3 +$$

$$+ 24.387 W_1 W_2 W_3 + 25.976 W_2 W_3^2 \mp$$

$$\mp 3\delta_4 \left(8 \frac{W_1^3}{W_2} + 4 W_1^2 + 6 W_1 W_3 + 8 \frac{W_1^2 W_3}{W_2} + 2 W_3^2 \right) \Bigg] \Bigg\} \Bigg\} . \qquad (7)$$

Here the top sign corresponds to $Z \leqq 4.18878$.

$$\varrho^{\text{cov}}(112) = \frac{1}{\Omega} \left\{ \frac{3Z}{48\varepsilon_f} \frac{a^2}{(2\pi)^2} \left\{ \left(-3 \cdot 2^{-1/3} + 6 + \frac{9}{S_4} \right) A - 6C + \right. \right.$$

$$+ \frac{9}{4S_4} M + \frac{15}{2S_4} F - \frac{3}{S_4} N - \frac{3}{S_4} P + \frac{3}{2S_4} G +$$

$$+ \frac{a^2}{(2\pi)^2} \Big[38.039 W_1^3 + 5.973 W_1^2 W_2 + 25.508 W_1^2 W_3 +$$

$$+ 13.762 W_1 W_2^2 + 19.241 W_2^2 W_3 + 11.35 W_3^3 +$$

$$+ \frac{3}{4} \delta_5 \left(- \frac{3 W_1 W_2^2}{2 W_3} - 5 W_2^2 + 2 \frac{W_1^3}{W_3} + \right.$$

$$\left. + 2 \frac{W_1^2 W_2}{W_3} + 6 W_1^2 + W_3^2 \right) \Bigg] \Bigg\} \Bigg\} . \qquad (8)$$

The symbols used in (6) to (8) are: W_1, W_2, W_3 are screened matrix elements of the pseudopotential in the reciprocal lattice at the first, second, and third coordination spheres, ε_f is the Fermi energy,

$$A = \frac{1}{\varepsilon_1^{*2}}\left[|W_1^0|^2 + \frac{8\pi\chi_1}{g_1^2} W_1^0 R_1\right], \tag{9}$$

$$B = \frac{1}{\varepsilon_1^*\varepsilon_2^*}\left[W_1^0 W_2^0 + \frac{4\pi\chi_1 W_2^0}{g_1^2} R_1 + \frac{4\pi\chi_2 W_1^0}{g_2^2} R_2\right]. \tag{10}$$

C may be found from B by substituting index 1 by index 2, and D may be found from B by substituting index 2 by 3,

$$F = \frac{1}{\varepsilon_2^{*2}}\left[|W_2^0|^2 + \frac{8\pi\chi_2}{g_2^2} W_2^0 R_2\right], \tag{11}$$

$$E = \frac{\varepsilon_1^*}{\varepsilon_2^{*2}\varepsilon_3^* W_1^0}\left[|W_2^0|^2 W_3^0 + \frac{4\pi\chi_3}{g_3^2}|W_2^0|^2 R_3 + \right.$$
$$\left. + \frac{8\pi\chi_2}{g_2^2} W_2^0 W_3^0 R_2 - \frac{4\pi\chi_1}{g_1^2}\frac{|W_2^0|^2 W_3^0}{W_1^0} R_1\right], \tag{12}$$

$$H = \frac{\varepsilon_2^*}{\varepsilon_1^{*3} W_2^0}\left[|W_1^0|^3 + \frac{12\pi\chi_1}{g_1^2}|W_1^0|^2 R_1 - \frac{4\pi\chi_2}{g_2^2}\frac{|W_1^0|^3}{W_2^0} R_2\right]. \tag{13}$$

G may be found from F by such substitution of indexes $2 \to 3$; L from E by substitution $2 \to 1$, $1 \to 2$; P from E by substitution $2 \to 1$, $3 \to 2$, $1 \to 3$; M from E by substitution $3 \to 1$, $1 \to 3$; N from H by substitution $2 \to 3$.

In (9) to (13) W_1^0, W_2^0, W_3^0, ε_1^*, ε_2^*, ε_3^* are unscreened form-factors and modified dielectric functions allowing exchange-correlation effects at the reciprocal lattice vectors $\boldsymbol{g}_1 = (2\pi/a)\,(110)$, $\boldsymbol{g}_2 = (2\pi/a)\,(200)$, $\boldsymbol{g}_3 = (2\pi/a)\,(112)$. The function $\chi(\boldsymbol{g})$ takes into consideration the exchange-correlation interaction of conduction electrons in screening and in these sets has the values χ_1, χ_2, χ_3.

$$R_1 = -\frac{1}{\Omega}\left\{\frac{3Z}{4\varepsilon_f}\frac{a^2}{(2\pi)^2}\left[7.785 W_1^2 + 4.683 W_1 W_2 + 0.115 W_2 W_3 + \right.\right.$$
$$\left.\left. + 5.72 W_1 W_3 + 1.531 W_2^2 + 2.906 W_3^2 + 2.25\frac{W_2^2 W_3}{W_1}\right]\right\},$$

$$R_2 = \frac{1}{\Omega}\left\{\frac{3Z}{4\varepsilon_f}\frac{a^2}{(2\pi)^2}\left[2.452 W_1^2 + 2.056 W_3^2 + 4\frac{W_1^3}{W_2} + 3 W_1 W_3 + 4\frac{W_1^2 W_3}{W_2}\right]\right\},$$

$$R_3 = \frac{1}{\Omega}\left\{\frac{3Z}{4\varepsilon_f}\frac{a^2}{(2\pi)^2}\left[1.052 W_1^2 - \frac{1}{2} W_2 W_3 + 0.187\frac{W_1 W_2^2}{W_3} + \right.\right.$$
$$\left.\left. + 0.625 W_2^2 - \frac{W_1^3}{4 W_3} - \frac{W_1^2 W_2}{4 W_3} + W_3^2\right]\right\}.$$

Besides the given denotations in (6) to (8) the following ones are used:

$$\delta_1 = 3.667 W_1 + 4.167 W_2 + 2.286\frac{W_2 W_3}{W_1},$$

$$\delta_2 = \frac{-2W_1 - W_2 - 2\dfrac{W_2W_3}{W_1} - \dfrac{1}{2}\dfrac{|W_3|^2 W_2}{|W_1|^2}}{2\left(1 + \dfrac{|W_2|^2}{4|W_1|^2}\right)^{3/2}},$$

$$\delta_3 = \frac{2.667W_1 - 4W_2 - \dfrac{W_2W_3}{2W_1} + \left|\dfrac{W_3}{W_1}\right|^2 \left(4\dfrac{|W_1|^2}{W_3} - 4W_2\right)}{\left(1 + \left|\dfrac{W_3}{W_1}\right|^2\right)^{3/2}},$$

$$\delta_4 = \frac{-2\dfrac{|W_3|^2}{W_2} - 4\left|\dfrac{W_1}{W_2}\right|^2 \left(2W_1 + W_2 + 2\dfrac{W_2W_3}{W_1}\right)}{\left(1 + 4\left|\dfrac{W_1}{W_2}\right|^2\right)^{3/2}},$$

$$\delta_5 = \frac{4\dfrac{|W_1|^2}{W_3} - 4W_2 + \left|\dfrac{W_1}{W_3}\right|^2 \left(2.667W_1 - 4W_2 - \dfrac{W_2W_3}{2W_1}\right)}{\left(1 + \left|\dfrac{W_1}{W_3}\right|^2\right)^{3/2}},$$

$$S_1 = 2\left(1 + \left|\frac{W_2}{2W_1}\right|^2\right)^{1/2}, \qquad S_2 = -\left(1 + \left|\frac{W_3}{W_1}\right|^2\right)^{1/2},$$

$$S_3 = \left(1 + \left|\frac{2W_1}{W_2}\right|^2\right)^{1/2}, \qquad S_4 = \left(1 + \left|\frac{W_1}{W_3}\right|^2\right)^{1/2}.$$

Substituting $\varrho^{\text{cov}}(\boldsymbol{g})$ from (8) to (10) in (2) one can calculate the covalent contribution to the band structure energy of metals with b.c.c. lattice.

3. Numerical Results and Discussion

It is of interest to compare the results of the calculation of the covalent (many-body) contribution to the binding energy of metallic hydrogen according to (2), (6) to (8) with the data of [7] where the techniques described in [1] of perturbation theory in electron–ion interaction is used. The advisability of such a comparison is caused by the fact that the "exact" form-factor

$$W^0(g) = -\frac{4\pi e^2}{\Omega g^2}$$

is used for the calculation of this many–ion contribution to the binding energy. It is seen from Table 1 that the agreement of our calculation with the results from [7] is good.

Table 1

Covalent contributions to the binding energy of metallic hydrogen

$r_s = (3\Omega/4\pi Z)^{1/3}$	$E_{\text{bs}}^{\text{cov}} = E^{(3)} + E^{(4)}$ [7] (Ryd)	$E_{\text{bs}}^{\text{cov}}$ (Ryd)
1.05	−0.02650	−0.02101
1.45	−0.03907	−0.02859

Table 2

Fourier transforms of covalent charge density (in (at. units)$^{-3}$) and covalent contribution to the band structure energy of metals with b.c.c. lattice

	Na	K	Rb	Cs	Ca	Be
$\varrho^{cov}(110) \times 10^4$	0.26	−0.018	0.95	0.34	0.06	−0.53
$\varrho^{cov}(200) \times 10^4$	0.03	−0.024	0.16	0.11	−0.49	0.35
$\varrho^{cov}(112) \times 10^5$	0.28	0.04	0.21	0.12	0.06	0.50
$E_{bs}^{cov} \times 10^3$	0.715	−0.040	−5.933	−1.02	1.582	−2.730
Ω	254.5	481.4	587.9	745.5	303.5	54.4

The values of $\varrho^{cov}(\boldsymbol{g})$ and E_{bs}^{cov} for alkali metals, calcium, and berillium are represented in Table 2. The numerical calculations are made with model form-factors from [5]. Screening functions are taken in the form used in [6]. It is seen from Table 2 that $\varrho^{cov}(110)$ in Na is one order higher than $\varrho^{cov}(200)$ and $\varrho^{cov}(112)$, and in K the value $\varrho^{cov}(200)$ is higher than $\varrho^{cov}(110)$. Rubidium has the maximal values of the covalent charge densities and potassium has the minimal ones in the row of alkali metals. This fact causes the largest contribution of the third- and fourth-order terms of perturbation series to the binding energy of rubidium. Comparing the values of $\varrho^{cov}(\boldsymbol{g})$ and E_{bs}^{cov} in Na and K one can conclude that in potassium the role of many-body effects is smaller than in sodium. The obtained results correlate with the results of a calculation of the covalent contribution to the elastic constants of these metals from [1]. The expressions for E_{bs}^{cov} and $\varrho^{cov}(\boldsymbol{g})$ in the h.c.p. lattice may be obtained in an analogous way and will be published in another paper. The results of numerical calculations of E_{bs}^{cov} with the model potential [5] in the h.c.p. structure of alkali metals are as follows:

$$E_{bs}^{cov}(\mathrm{Na}) = 1.96 \times 10^{-4}\,\mathrm{Ryd}\,, \qquad E_{bs}^{cov}(\mathrm{K}) = 2.38 \times 10^{-5}\,\mathrm{Ryd}\,,$$
$$E_{bs}^{cov}(\mathrm{Rb}) = -7.88 \times 10^{-4}\,\mathrm{Ryd}\,, \qquad E_{bs}^{cov}(\mathrm{Cs}) = -1.23 \times 10^{-5}\,\mathrm{Ryd}\,.$$

Comparing these results with the data from Table 2 one can easily see that the difference of covalent contributions to the binding energy of alkali metals only in potassium is of the same order as the enthalpy of the phase transition ($\approx 10^{-5}$ Ryd/ion). In other metals this difference is approximately two orders greater. It means that in spite of the small contribution of E_{bs}^{cov} to the binding energy of alkali metals the theoretical calculation of the energy differences of crystal structures in polymorphic transitions cannot be performed without taking into consideration the high orders of perturbation theory. In a subsequent paper we shall show that these contributions allow to obtain good results for the temperatures of the f.c.c. → b.c.c. transition in Ca and the h.c.p. → b.c.c. transition in Be. Fig. 1 and 2 illustrate the results of the calculation of the density of covalent charge in the b.c.c. lattice of calcium. It is seen from Fig. 1

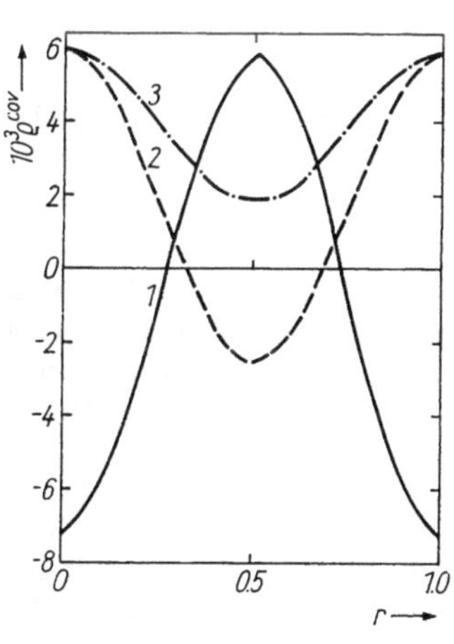

Fig. 1. The density of covalent charge in the b.c.c. lattice of calcium: ——— direction [111], – – – direction [110], — · — direction [100]. For (1) r is in $a\sqrt{3}$ units; for (2) r is in $a\sqrt{2}$ units; for (3) r is in a units

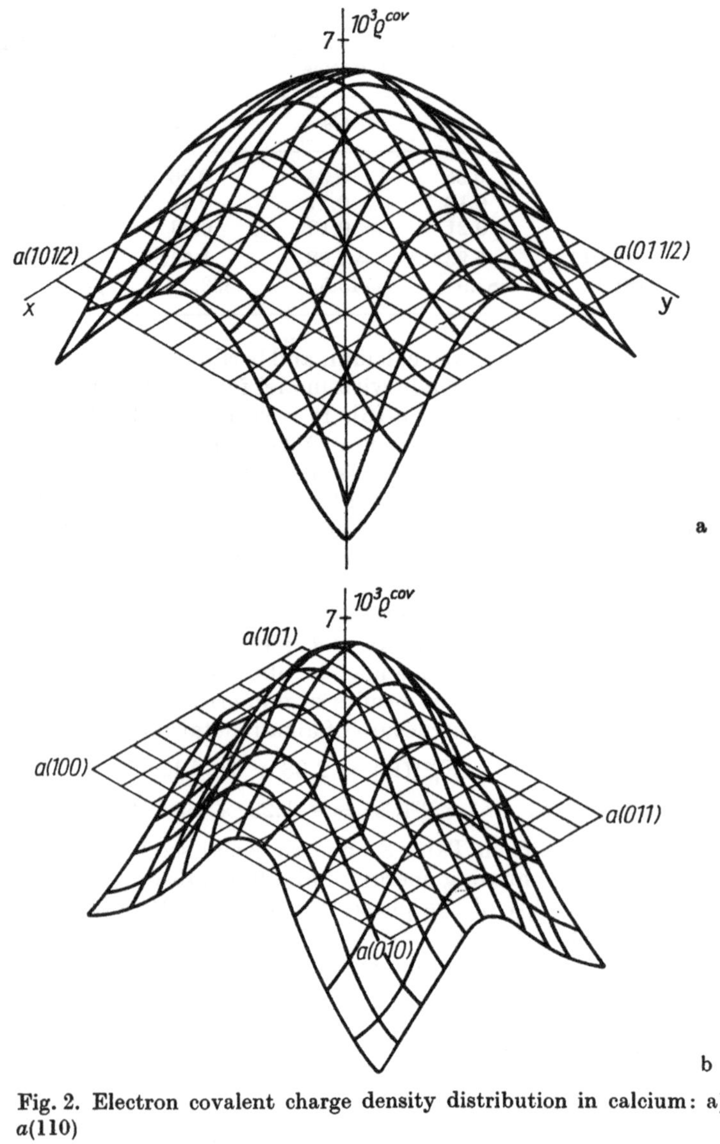

Fig. 2. Electron covalent charge density distribution in calcium: a) plane $(a/2)(001)$; b) plane $a(110)$

that in the direction [111] there is an increase of the charge density in the neighbourhood of the ion, situated in the centre of the cubic cell, and in the directions [100] and [110] there is a decrease of $\varrho^{cov}(\boldsymbol{r})$ between ions. Fig. 2 shows the cross sections of $\varrho^{cov}(\boldsymbol{r})$ by the planes $(a/2)(001)$ and $a(110)$ and also indicates the increase of the covalent charge density in the centre of the cubic cell.

Table 3 presents $\varrho^{cov}(\boldsymbol{g})$ and E_{bs}^{cov} in transition metals calculated with form-factors from [3]. Data given in this table show that for metals with the valency $Z = 3$ the value E_{bs}^{cov} is much smaller than that in metals with $Z = 5$. At the same time it is of course greater than in alkali metals. Comparison of the value E_{bs}^{cov} in Fe with the enthal-

Table 3

Fourier transforms of covalent charge density (in (at. units)$^{-3}$), covalent contribution to the band structure energy, the second (E_2) and the fourth ($E_{bs}^{(4)}$) order of perturbation theory for transition metals

	Fe	Cr	V	Ta	Nb
$\varrho^{cov}(110) \times 10^4$	−1.23	0.12	−3.30	4.50	5.20
$\varrho^{cov}(200) \times 10^4$	−0.15	−0.08	−1.30	−0.94	−0.86
$\varrho^{cov}(112) \times 10^4$	0.20	0.05	1.24	1.64	1.69
$E_{bs}^{cov} \times 10^2$ (Ryd)	0.176	−0.019	3.460	17.479	5.553
$E_{bs}^{(4)} \times 10^2$ (Ryd)	0.034	−0.003	0.352	1.514	0.420
E_2 (Ryd)	−0.0892	−0.0230	−0.9782	−1.0990	−1.1208

py of the polymorphic transition (b.c.c. → f.c.c.) equal to 6×10^{-4} Ryd/ion [8] shows that in this case E_{bs}^{cov} is also one order greater and as in alkali metals it must be taken into account while calculating this phase transition.

The contributions of the second order (E_2) and of the fourth order ($E_{bs}^{(4)}$) of perturbation theory in the pseudopotential are also given in Table 3. Comparing the results of the calculation of the values E_2, E_{bs}^{cov}, and $E_{bs}^{(4)}$ in the above-mentioned metals one can see the convergence of the perturbation series in the electron–ion interaction in transition metals.

References

[1] E. Brovman and Yu. Kagan, Uspekhi fiz. Nauk **112**, 369 (1974).
[2] G. L. Krasko and A. B. Makhnovetskii, phys. stat. sol. (b) **80**, 713 (1977).
A. B. Makhnovetskii and G. L. Krasko, phys. stat. sol. (b) **81**, 263 (1977).
[3] A. O. E. Animalu, Phys. Rev. B 8, 3542 (1973).
[4] V. I. Bogdanov and D. L. Fuks, Fiz. Metallov i Metallovedenie **48**, 903 (1979).
[5] A. O. E. Animalu and V. Heine, Phil. Mag. **12**, 1249 (1965).
[6] D. J. W. Geldart and S. H. Vosko, Canad. J. Phys. **44**, 2137 (1966).
[7] Yu. Kagan, V. V. Pushkarev, and A. Holas, Zh. eksper. teor. Fiz. **73**, 967 (1977).
[8] V. P. Glushko (Ed.), Termicheskie konstanti veshchestv, Vol. 6, Viniti, Moscow 1972.

(Received April 17, 1985)

phys. stat. sol. (b) 131, 505 (1985)
Subject classification: 4 and 6; 14.4; 22.5.3; 22.8.1

Department of Physics, Agra University[1])

Volume Dependence of Dielectric Properties of Some Fluorite-Structure Crystals

By
N. DUTT, G. G. AGRAWAL, and J. SHANKER

Dedicated to Prof. Dr. Dr. h. c. Dr. E. h. P. GÖRLICH on the occasion of his 80th birthday

The experimental data on first and second order pressure derivatives of dielectric constants are used to make an analysis of the volume dependence of polarizabilities, effective charge parameter, and transverse-optic-mode frequencies in CaF_2, SrF_2, BaF_2, PbF_2, $Sr(NO_3)_2$, $Ba(NO_3)_2$, and $Pb(NO_3)_2$. The short range force constant responsible for the ion-displacement polarizability is calculated with the help of Lundqvist's three-body interaction model. Reasonable values for the volume derivatives of various dielectric parameters are obtained. Values of optic-mode-Grüneisen parameters and their volume derivatives are also calculated and compared with available experimental data.

Die experimentellen Werte der Druckableitungen der Dielektrizitätskonstanten erster und zweiter Ordnung werden benutzt, um eine Analyse der Volumenabhängigkeit der Polarisierbarkeiten, der effektiven Ladungsparameter und der Frequenzen der transversalen optischen Mode in CaF_2, SrF_2, BaF_2, PbF_2, $Sr(NO_3)_2$, $Ba(NO_3)_2$ und $Pb(NO_3)_2$ durchzuführen. Die kurzreichweitige Kraftkonstante, die für die Ionenverschiebungspolarisation verantwortlich ist, wird mit Hilfe des Lundqvist-Drei-Körperwechselwirkungsmodells berechnet. Vernünftige Werte für die Volumenableitungen der verschiedenen dielektrischen Parameter werden erhalten. Die Werte der Grüneisenparameter der optischen Mode und ihre Volumenableitungen werden ebenfalls berechnet und mit vorhandenen experimentellen Werten verglichen.

1. Introduction

There have been many attempts [1 to 5] for studying the volume dependence of dielectric properties of alkali halides. However, desirable attention in this field has not been paid in case of fluorite-structure crystals. Adequate potential models based on Lundqvist's three-body interactions have been developed recently [6 to 8] for calculating the third-order elastic constants of fluorite-structure crystals. Using the interionic potentials [8] we calculate the short-range force constant responsible for ion-displacement polarizability in CaF_2, SrF_2, BaF_2, PbF_2, $Sr(NO_3)_2$, $Ba(NO_3)_2$, and $Pb(NO_3)_2$ crystals. Various order derivatives of these force constants with respect to volume or interionic distance make it possible to analyse the volume derivatives of polarizabilities, effective charge parameter, and optic mode frequencies using the experimental data on dielectric constants and their pressure derivatives [9 to 11]. The method of analysis is presented in Section 2. Results are discussed and compared with available experimental data in Section 3.

2. Methods of Analysis

The dielectric behaviour of ionic crystals can be described in terms of low-frequency or static dielectric constant (ε_0), high-frequency or electronic dielectric constant (ε_∞), corresponding polarizabilities α_0 and α_∞, effective ionic charges (ze), short-range force

[1]) Agra 282004, India.

constant (A), transverse-optic-mode frequency (ω_{TO}) and longitudinal-optic-mode frequency (ω_{LO}). These quantities are related to each other through the Lorentz-Lorenz, the Clausius-Mossotti, the Szigeti, and the Lyddane-Sachs-Teller relations given as follows [1]:

$$\frac{\varepsilon_0 - 1}{\varepsilon_0 - 2} = \frac{4\pi}{3V} \alpha_0 , \tag{1}$$

$$\frac{\varepsilon_\infty - 1}{\varepsilon_\infty + 2} = \frac{4\pi}{3V} \alpha_\infty , \tag{2}$$

$$\alpha_0 = \alpha_\infty + \alpha_i , \tag{3}$$

$$\alpha_i = \frac{(ze)^2}{A}, \tag{4}$$

$$\mu\omega_{TO}^2 = \frac{4\pi}{9V} \frac{(\varepsilon_\infty + 2)^2}{(\varepsilon_0 - \varepsilon_\infty)} (ze)^2 , \tag{5}$$

and

$$\left(\frac{\omega_{LO}}{\omega_{TO}}\right)^2 = \frac{\varepsilon_0}{\varepsilon_\infty}, \tag{6}$$

where α, V, and μ are the polarizability, volume, and reduced mass per ion pair, respectively. First- and second-order volume derivatives of dielectric constant (ε_0 or ε_∞) obtained from (1) and (2) are written as follows:

$$V \frac{d\varepsilon}{dV} = \frac{4\pi}{9} (\varepsilon + 2)^2 \frac{d\alpha}{dV} - \frac{(\varepsilon + 2)(\varepsilon - 1)}{3} \tag{7}$$

and

$$V^2 \frac{d^2\varepsilon}{dV^2} = \frac{4\pi}{9} (\varepsilon + 2)^2 V \frac{d^2\alpha}{dV^2} + \frac{2}{(\varepsilon + 2)} \left(V \frac{d\varepsilon}{dV}\right)^2 - 2V \frac{d\varepsilon}{dV}, \tag{8}$$

Volume derivatives of polarizabilities are deduced from (3) and (4) which yield

$$\frac{d\alpha_0}{dV} = \frac{d\alpha_\infty}{dV} + \frac{r\alpha_i}{3V} \left[2 \frac{z^I}{z} - \frac{A^I}{A}\right] \tag{9}$$

and

$$V \frac{d^2\alpha_0}{dV^2} = V \frac{d^2\alpha_\infty}{dV^2} + \frac{r^2\alpha_i}{9V} \left[2 \frac{z^{II}}{z} - \frac{A^{II}}{A} + 2\left(\frac{z^I}{z} - \frac{A^I}{A}\right)^2 - \frac{2}{r}\left(2 \frac{z^I}{z} - \frac{A^I}{A}\right)\right], \tag{10}$$

where z^I, z^{II}, A^I, and A^{II} are the first- and second-order derivatives of z and A with respect to interionic distance r. Values of $d\alpha_\infty/dV$ are derived from experimental photoelastic data [9, 12]. $d^2\alpha_\infty/dV^2$ are then obtained using the relation [13]

$$\alpha_\infty = kV^\lambda , \tag{11}$$

where k and λ are constants for a given crystal. For evaluating the derivatives of the effective charge parameter z^I and z^{II}, we need to know the derivatives of the short-range force constants A^I and A^{II}. These are calculated using the expression

$$A = \frac{1}{3} \left[\Phi^{II}(r) + \frac{2}{r} \Phi^I(r)\right], \tag{12}$$

where $\Phi^I(r)$ and $\Phi^{II}(r)$ are the first- and second-order derivatives of the short-range potential energy. Using the interionic potentials from [8] we can write

$$\Phi(r) = 8b\beta_{+-} \exp \frac{(r_+ + r_- - r)}{\varrho_{+-}} - \frac{8\alpha_M \varepsilon f(r) e^2}{a} + \\ + 6b\beta_{--} \exp \frac{(2r_- - r')}{\varrho_{--}} - \frac{6C_{dd}}{r'^6}. \tag{13}$$

The potential parameters for repulsive and van der Waals interactions have been derived from the fitting of equilibrium condition and bulk modulus. The three-body charge transfer parameter f is calculated from the Cauchy derivation $(C_{12} - C_{44})$ [8]. The expressions for A^I and A^{II} derived from (12) are written as follows:

$$A^I = \frac{1}{3}\left[\Phi^{III}(r) + \frac{2}{r}\Phi^{II}(r) - \frac{2}{r^2}\Phi^I(r)\right] \tag{14}$$

and

$$A^{II} = \frac{1}{3}\left[\Phi^{IV}(r) + \frac{2}{r}\Phi^{III}(r) - \frac{4}{r^2}\Phi^{II}(r) + \frac{4}{r^3}\Phi^I(r)\right], \tag{15}$$

where Φ^{III} and Φ^{IV} are the third- and fourth-order derivatives of Φ with respect to r.

Calculated values of A and its derivatives are reported in Table 1. Data on dielectric constants and polarizabilities [9 to 11] are also given in Table 1. Values of effective

Table 1

Values of the short-range force constant A [10^{-8} N/nm] and its derivatives A^I in [10^{-7} N/nm^2] and A^{II} [10^{-6} N/nm^3], and static and electronic dielectric constants ε_0 and ε_∞, ionic polarizability α_i [10^{-3} nm^3], and effective ionic valency z

crystals	A	A^I	A^{II}	ε_0	ε_∞	α_i	z
CaF_2	19.50	−68.80	237.29	6.81	2.06	3.84	1.80
SrF_2	16.48	−54.81	181.37	6.48	2.07	4.42	1.78
BaF_2	13.92	−43.55	99.63	7.34	2.18	5.58	1.84
PbF_2	14.35	−46.40	152.76	29.30	3.08	6.14	1.95
$Sr(NO_3)_2$	10.51	−34.20	107.43	5.17	2.46	7.24	1.82
$Ba(NO_3)_2$	7.64	−22.14	64.28	4.59	2.47	6.87	1.51
$Pb(NO_3)_2$	9.59	−30.74	88.92	11.90	3.18	10.47	2.09

charge parameter z deduced from (4) are listed in the last column of Table 1. First- and second-order derivatives of dielectric constants and polarizabilities derived from the pressure derivatives of dielectric constants [9, 10] are reported in Table 2 along with the calculated values of the derivatives of the effective charge parameter.

Table 2

First- and second-order volume derivatives of ε_∞, ε_0, and α_i, and calculated values of rz^I and r^2z^{II} derived using experimental data from [9, 10, 12]

crystal	$V\frac{d\varepsilon_\infty}{dV}$	$V^2\frac{d^2\varepsilon_\infty}{dV^2}$	$V\frac{d\varepsilon_0}{dV}$	$\frac{d\alpha_i}{dV}$	$V^2\frac{d^2\varepsilon_0}{dV^2}$	$\frac{d^2\alpha_i}{dV^2}$	rz^I	r^2z^{II}
CaF_2	−0.57	+0.97	18.7	0.29	112	0.02	0.79	− 5.14
SrF_2	−0.88	+1.72	16.0	0.29	85	0.01	1.01	−11.76
BaF_2	−1.07	+2.31	18.4	0.29	106	0.01	0.73	−24.68
PbF_2	−1.66	+3.51	142.9	0.27	—	—	−1.43	—

Expressions for γ_{TO} and γ_{LO}, and their volume derivatives are obtained from (5) and (6) as follows:

$$\gamma_{TO} = \frac{1}{2} + \frac{1}{2(\varepsilon_0 - \varepsilon_\infty)}\left[V\frac{d\varepsilon_0}{dV} - V\frac{d\varepsilon_\infty}{dV}\right] - \frac{V}{z}\frac{dz}{dV}, \tag{16}$$

$$\gamma_{LO} = \gamma_{TO} - \frac{1}{2}\left[\frac{V}{\varepsilon_0}\frac{d\varepsilon_0}{dV} - \frac{V}{\varepsilon_\infty}\frac{d\varepsilon_\infty}{dV}\right], \tag{17}$$

$$\begin{aligned} V\frac{d\gamma_{TO}}{dV} = {} & \left(\frac{V}{z}\frac{dz}{dV}\right)^2 - \frac{V}{z}\left(\frac{dz}{dV}\right) - \frac{V^2}{z}\left(\frac{d^2z}{dV^2}\right) - \frac{V}{(\varepsilon_\infty + 2)}\frac{d\varepsilon_\infty}{dV} - \\ & - \frac{V^2}{(\varepsilon_\infty + 2)}\frac{d^2\varepsilon_\infty}{dV^2} + \left(\frac{V}{\varepsilon_\infty + 2}\frac{d\varepsilon_\infty}{dV}\right)^2 + \frac{1}{2(\varepsilon_0 - \varepsilon_\infty)} \times \\ & \times \left[V^2\frac{d^2\varepsilon_0}{dV^2} + V\frac{d\varepsilon_0}{dV} - V^2\frac{d^2\varepsilon_\infty}{dV^2} - V\frac{d\varepsilon_\infty}{dV}\right] - \\ & - \frac{1}{2(\varepsilon_0 - \varepsilon_\infty)^2}\left[V\frac{d\varepsilon_0}{dV} - V\frac{d\varepsilon_\infty}{dV}\right]^2, \end{aligned} \tag{18}$$

$$\begin{aligned} V\frac{d\gamma_{LO}}{dV} = \frac{1}{2}\Bigg[& 2V\frac{d\gamma_{TO}}{dV} - \frac{V}{\varepsilon_0}\frac{d\varepsilon_0}{dV} + \frac{V}{\varepsilon_\infty}\frac{d\varepsilon_\infty}{dV} - \frac{V^2}{\varepsilon_0}\frac{d^2\varepsilon_0}{dV^2} + \\ & + \frac{V^2}{\varepsilon_\infty}\frac{d^2\varepsilon_\infty}{dV^2} + \left(\frac{V}{\varepsilon_0}\frac{d\varepsilon_0}{dV}\right)^2 - \left(\frac{V}{\varepsilon_0}\frac{d\varepsilon_\infty}{dV}\right)^2\Bigg]. \end{aligned} \tag{19}$$

3. Results and Discussion

We note from Table 1 that values of ion displacement polarizabilities α_i derived from dielectric constants are significantly larger for nitrates than the corresponding values for fluorides. The polarizability α_i is inversely proportional to the short-range force constant A. This is verified independently from the values of A derived from elastic constants. Values of A are found to be larger for fluorides than those for nitrates. Calculated values of z are in the range 1.5 to 2. In fact z represents the highest common factor of the charges on both types of ions. The nominal value of z for the crystals under study should be 2. Values of the effective charge parameter, defined as the ratio of calculated and nominal values of z, turn out to be in between 0.75 and 1. These estimates suggest that the crystals under investigation are essentially ionic as far as the nature of their chemical bond is concerned [11].

Calculated values of the derivatives of effective charge (z^I and z^{II}) are listed in Table 2. These calculations have been possible only for fluorides for which experimental data on pressure derivatives of dielectric constants are available. Values of z^I are positive for CaF_2, SrF_2, and BaF_2, but negative for PbF_2. This distinct behaviour of PbF_2 crystal is also apparent from the fact that the dielectric constant ε_0 and its volume derivative $V(d\varepsilon_0/dV)$ for PbF_2 are, respectively, about four and eight times larger than the corresponding values for alkaline earth fluorides. Positive values of z^I or dz/dr imply that z decreases with increasing pressure. This prediction can be accounted for on the basis of the deformation dipole model [14] following the similar approach as used for alkali halides [15, 16].

Values of z^I and z^{II} are utilized to calculate the transverse and longitudinal optical mode Grüneisen parameters and their volume derivatives using equations (16) to (19).

Table 3

Transverse and longitudinal optical mode Grüneisen parameters and their volume derivatives

crystal	γ_{TO}		γ_{LO}		average γ		$\frac{V}{\gamma_{TO}} \frac{d\gamma_{TO}}{dV}$	$\frac{V}{\gamma_{LO}} \frac{d\gamma_{LO}}{dV}$
	calc.	expt.	calc.	expt.	optical mode (21)	thermo-elastic (22)		
CaF_2	2.52	3.2	1.01	1.69	1.36	1.74	2.24	−0.12
SrF_2	2.43	3.1	0.99	1.66	1.34	1.62	1.89	−0.03
BaF_2	2.51	2.4	1.01	0.90	1.35	1.61	2.41	+0.88
PbF_2	3.81	—	1.01	—	1.28	—	—	—

The calculated values are found to present good agreement with experimental data available for fluorides (Table 3). For performing such calculations in nitrates we assume that z^{I} and z^{II} for nitrates are equal to those for fluorides with the same cation. This assumption gets support from the previous work [2, 17] where it has been found that the volume derivatives of effective charge parameter remain nearly unchanged for a group of ionic crystals with the same cation. This assumption enables us to make an approximate estimation of the optical mode Grüneisen parameters, γ_{TO} and γ_{LO}, volume derivatives of dielectric constants and volume derivatives of γ_{TO} and γ_{LO} in case of nitrates. Values thus estimated are reported in Table 4.

Table 4

Calculated values of first- and second-order derivatives of static dielectric constant, optical mode Grüneisen parameters, and their volume derivatives

crystal	$V \frac{d\varepsilon_0}{dV}$	$V^2 \frac{d^2\varepsilon_0}{dV^2}$	γ_{TO}	γ_{LO}	average γ	$\frac{V}{\gamma_{TO}} \frac{d\gamma_{TO}}{dV}$	$\frac{V}{\gamma_{LO}} \frac{d\gamma_{LO}}{dV}$
$Sr(NO_3)_2$	11.86	67.92	2.74	1.44	1.86	1.82	−0.02
$Ba(NO_3)_2$	6.99	17.46	2.32	1.39	1.72	0.78	+0.27
$Pb(NO_3)_2$	38.25	—	3.33	1.51	1.89	—	—

We note from Tables 3 and 4 that values of γ_{TO} are larger than γ_{LO}. Average values of γ can be obtained from γ_{TO} and γ_{LO} using a suitable averaging process as suggested by Madan [18],

$$\gamma = \frac{\gamma_{TO}\omega_{TO}^2 + \gamma_{LO}\omega_{LO}^2}{\omega_{TO}^2 + \omega_{LO}^2}. \tag{20}$$

In view of the Lyddane-Sachs-Teller relation (6) we can rewrite (20) as

$$\gamma = \frac{\gamma_{TO}\varepsilon_\infty + \gamma_{LO}\varepsilon_0}{\varepsilon_\infty + \varepsilon_0}. \tag{21}$$

Values of average γ obtained from (20) or (21) can be compared with the macroscopic Grüneisen parameter which in terms of thermoelastic data is defined as [19]

$$\gamma = \frac{V\beta B_T}{C_v}, \tag{22}$$

where β is the thermal expansion coefficient, B_T the isothermal bulk modulus, and C_v the specific heat at constant volume. These data are available in the literature for alkaline earth fluorides [20, 21]. A comparison of the average optical mode Grüneisen parameter and the thermoelastic Grüneisen parameter has been made in Table 3 for alkaline earth fluoride. They are found to compare well with each other.

Values of volume derivatives of the Grüneisen parameters γ_{TO} and γ_{LO} calculated from (18) and (19) are compared in Tables 3 and 4. It is noted that the values of $q_{TO} = (V/\gamma_{TO})\,(d\gamma_{TO}/dV)$ are positive and larger than the corresponding values of $q_{LO} = (V/\gamma_{LO})\,(d\gamma_{LO}/dV)$. Small positive or in some cases negative values of q_{LO} suggest that either γ_{LO} decreases very slowly or it increases with pressure. On the other hand, q_{TO} is positive for all the crystals and thus revealing that γ_{TO} decreases with pressure. These conclusions are similar to those for alkali halides [1]. It should also be mentioned that values of q_{TO} obtained in the present work are nearly equal to those obtained from thermoelastic properties [19]. The present study thus helps to understand the dielectric and thermoelastic behaviour of fluorite-structure crystals.

References

[1] A. V. SINGH, J. C. SHARMA, and J. SHANKER, Physica (Utrecht) **94B**, 331 (1978).
[2] J. SHANKER, V. P. GUPTA, and O. P. SHARMA, Phys. Rev. B **18**, 5869 (1978).
[3] J. C. SHARMA and J. SHANKER, Phys. Rev. B **19**, 6604 (1979).
[4] A. BATANA and E. R. GONZALEZ, phys. stat. sol. (b) **76**, K105 (1976).
[5] A. BATANA and C. HENSE, J. Phys. Chem. Solids **41**, 863 (1980).
[6] S. S. BEDI and M. P. VERMA, phys. stat. sol. (b) **94**, 297 (1979).
[7] S. S. BEDI and M. P. VERMA, phys. stat. sol. (b) **97**, 203 (1980).
[8] J. SHANKER and J. P. SINGH, phys. stat. sol. (b) **110**, 697 (1982).
[9] C. ANDEEN, D. SCHUELE, and J. FONTANELLA, Phys. Rev. B **6**, 591 (1972).
[10] G. A. SAMARA, Phys. Rev. **B13**, 4529 (1976).
[11] L. SIRDESHMUKH, Current Sci. (India) **44**, 690 (1975).
[12] E. D. D. SCHMIDT and K. VEDAM, J. Phys. Chem. Solids **27**, 1563 (1966).
[13] E. E. HAVINGA and A. J. BOSMAN, Phys. Rev. **129**, 1593 (1963).
[14] J. R. HARDY, Phil. Mag. **6**, 27 (1960).
[15] A. V. SINGH, H. P. SHARMA, and J. SHANKER, Solid State Commun. **21**, 643 (1977).
[16] C. K. KIM, A. FELDMAN, D. HOROWITZ, and R. M. WAXLER, Solid State Commun. **25**, 397 (1978).
[17] J. FONTANELLA, C. ANDEEN, and D. SCHUELE, Phys. Rev. **B6**, 582 (1972).
[18] M. P. MADAN, Physica (Utrecht) **51**, 526 (1971).
[19] J. SHANKER and K. SINGH, phys. stat. sol. (b) **112**, 615 (1982).
[20] G. C. BENSON and E. DEMPSEY, Proc. Roy. Soc. **A266**, 344 (1962).
[21] S. ALTEROVITZ and D. GERLICH, Phys. Rev. **B1**, 4136 (1970).

(Received October 16, 1984)

phys. stat. sol. (b) **131**, 511 (1985)

Subject classification: 6; 10.2; 20.1; 22.1.2

Shanghai Institute of Metallurgy, Academy of Sciences of China[1])

The Nature of Two Intense Si–H IR Stretching Bands in FZ-Si:H

By

T. S. Shi, L. M. Xie, G. R. Bai, and M. W. Qi

Dedicated to Prof. Dr. Dr. h. c. Dr. E. h. P. Görlich on the occasion of his 80th birthday

Based on the isotope effect, two H-defect complexes: tetrahedral interstitial silane $(SiH_4)_{T_d}$ and hydrogenated ⟨100⟩ split interstitials $(I_{sp} \cdot 2\,H)$ are proposed to account for the two strong Si–H IR stretching bands in FZ-Si grown in H atmosphere (2210 cm^{-1}, 1946 cm^{-1}). A series of experimental facts, including the vibration frequency, the correlation with Si–H IR bands in the bending-wagging region, the temperature dependence of band-width and the annealing behaviour are presented, which support the proposed model. The results coincide with the view that the self-interstitial is the dominate thermal defect in crystalline silicon at high temperature.

На основе результатов изотопного эффекта, два водород-дефектного комплекса: междузельного атома Si в тетроэдрической позиции и ⟨100⟩ расколького междузельного атома Si, связанных с атомами водорода были предложены для обяснения двух интенсивных ИК пиков в кристаллическом Si, выращенных зонной плавкой в водородном атмосфере (2210 cm^{-1} и 1946 cm^{-1}). Ряд экспериментальных фактов, включающих частоту ИК абсорбционных пиков, их корреляцию с Si–H ИК пиками в области колебания типа изгиба и качения, температурную зависимость ширины пиков и поведение пиков при отжиге, был представлен и поддерживал предложенные модели. Эти результаты согласны с точкой зрения что междузельный атом является главном термодинамически равновесным точечным дефектом в кристаллическом Si при высокой температуре.

1. Introduction

In floating zone melted silicon crystals grown in hydrogen atmosphere (FZ-Si:H) a number of infrared absorption bands corresponding to the local vibration modes of silicon–hydrogen bonds (Si–H IR bands) were observed. They attracted much attention because of their thermal stability and connection with Si–H IR bands in other more complicated systems such as proton-implanted crystalline silicon (c-Si) or neutron transmutation doped (NTD) FZ-Si:H.

Cui et al. [1], who first observed the Si–H IR bands in FZ-Si:H, assigned the three major bands at 2210, 2124, and 1949 cm^{-1} to the interstitial H atom located at hexagonal site (Hex), tetrahedral site (T_d), and antibonding site (Ab), respectively, based on the EHT calculation of Singh et al. [2].

But the experimental results and theoretical calculations obtained later rendered this interpretation doubtful. Experimentally the observed number of Si–H IR peaks in the bond stretching region increased gradually from 3—4 to 7—8, which exceeds the number of possible interstitial sites in c-Si. Moreover, the Si–H IR peaks in the bond bending-wagging region were also observed, suggesting strongly that the Si–H local vibration modes in FZ-Si:H resemble that of silane molecules [3, 4]. Theoretically, the calculations performed with the more sophisticated MNDO method [5] predicted that

[1]) 865 Chang Ning Road, Shanghai 200050, People's Republic of China.

the equilibrium position of the H atom in c-Si is not at T_d, Hex, or Ab sites, the H atom is quite mobile at room temperature (activation energy of diffusion $E < 0.5$ eV) and strongly inclines to combine with each other, forming the IR inactive hydrogen molecule (H_2), and the single H atom, which saturates the Si dangling bond at defect site, forming the normal Si–H bond, is stable and IR active. So the Si–H IR peaks in FZ-Si:H and in H-implanted c-Si [6] both are related to defects and correspond to various H-defect complexes, sometimes incorporating impurity atoms [7]. Therefore, it is necessary to reconsider the nature of the Si–H IR peaks in FZ-Si:H. Some attempts have already been made [8, 9], but the common understanding is still lacking.

This work aims at exploring the nature of the two intense Si–H IR bands (2210 and 1946 cm^{-1}) in FZ-Si:H, based on the experimentally observed isotope effect [10]. The corresponding models of H-defect complexes have been suggested and supported by additional experimental results.

2. The Model of H-Defect Complexes Corresponding to 2210 and 1946 cm^{-1} Bands

The two intense Si–H IR absorption peaks 2210 cm^{-1} and 1946 cm^{-1} (shifted to 2223 cm^{-1} and 1952 cm^{-1} at 10 K) have always been observed in the stretching vibration region of the IR spectrum of FZ-Si:H and are not related to other impurities in Si. Recently the isotope effects of these two peaks provided important information. When the H was replaced by D, the 2223 cm^{-1} (2210 cm^{-1}) and 1952 cm^{-1} (1946 cm^{-1}) shifted to 1617 cm^{-1} and 1421 cm^{-1}, respectively, at 10 K in good accordance with the reduced mass increasing upon isotopic substitution. It proved that they are local vibrational modes of the Si–H bond resulting from the interaction between H and a defect

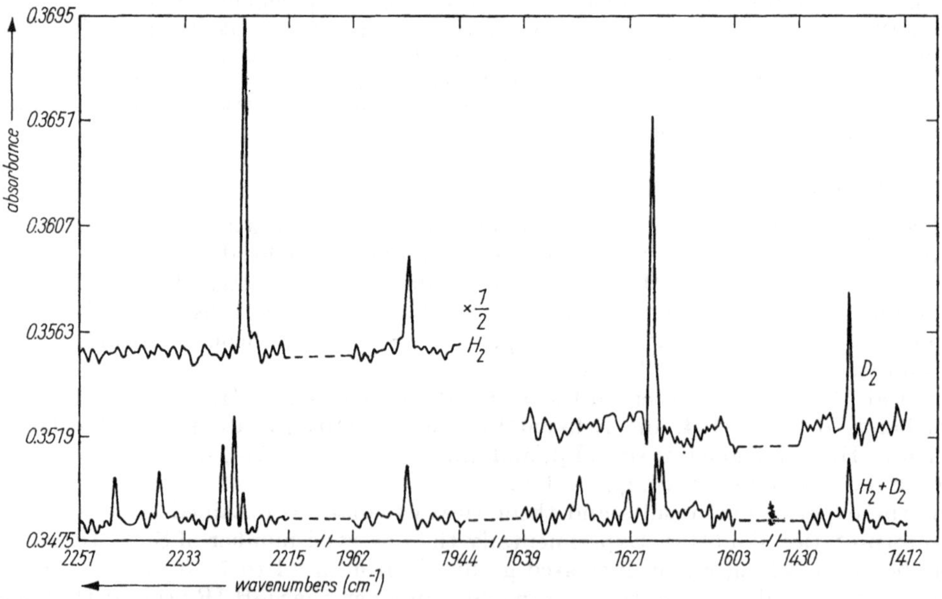

Fig. 1. The spectra of the FZ silicon crystal grown in the hydrogen (H_2), deuterium (D_2), and the mixed $H_2 + D_2$ atmosphere at 10 K, which show the isotope shift of the 2223 cm^{-1} (2210 cm^{-1}) and 1952 cm^{-1} (1946 cm^{-1}) bands upon the substitution of D for H to 1617 cm^{-1} and 1421 cm^{-1} respectively, and the splitting of 2223 cm^{-1} and 1617 cm^{-1} bands upon the partial substitution of D for H

in the Si lattice. Furthermore, for the FZ-Si grown in mixed atmosphere 36% H and 64% D, both the 2223 cm^{-1} and 1617 cm^{-1} peaks showed fine structure, splitting into 5 peaks, while the 1952 cm^{-1} and 1421 cm^{-1} peaks remained unchanged (Fig. 1). Having analysed these results via group theory, we came to the conclusions that the 2223 cm^{-1} complex contains more than one H atom, (actually four H atoms) and has T_d symmetry, while the 1952 cm^{-1} complex contains only one H atom, or more than one, which are not connected to the same Si atom and very weakly interact between each other [10]. The lowering of symmetry from T_d to C_{3v} and C_{2v} upon the partial substitution of D for H in the 2223 cm^{-1} complex results in the observed splitting of the 2223 cm^{-1} peak as well as the 1617 cm^{-1} peak. For more detail we refer to the original paper [10].

Now we consider the structure of these H-defect complexes in detail.

2.1 The 2210 cm^{-1} band

The T_d symmetry of the 2210 cm^{-1} complex excludes all suggested models of other symmetry [7, 8, 11]. One possible H-defect complex with T_d symmetry in c-Si is the Si self-interstitial located at T_d site with four H atoms, saturating its dangling bonds along four ⟨111⟩ directions, e.g. the T_d interstitial silane, $(SiH_4)_{T_d}$, as shown in Fig. 2.

In the Si lattice the diameter of the T_d interstitial site is 0.235 nm and it is just suitable to hold a Si interstitial of the same diameter in it. The Si–H bond length equals 0.148 nm, which exceeds the distance between T_d site and Hex site (0.118 nm) by 0.03 nm.

The diameter of the Hex site is 0.225 nm, close to twice the Van der Waals radius of H (0.24 nm). So $(SiH_4)_{T_d}$ is just embedded in the Si lattice. Here the possible relaxation of surrounding Si atoms is ignored.

Naturally the stretching vibration frequency of $(SiH_4)_{T_d}$ should be close to that of the gaseous silane (2189 cm^{-1}) [7, 12]. The small discrepancy between them may be brought about by the environment effect [12]. The four H atoms of $(SiH_4)_{T_d}$ are located near the centre of the Hex site, surrounded by six Si atoms. It is quite probable that the additional potential fields imposed by these Si atoms increase the force constant of the Si–H bond, and raise the vibration frequency by several ten wave numbers.

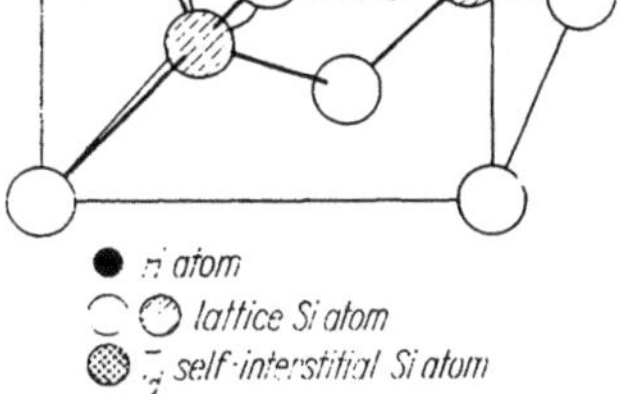

Fig. 2

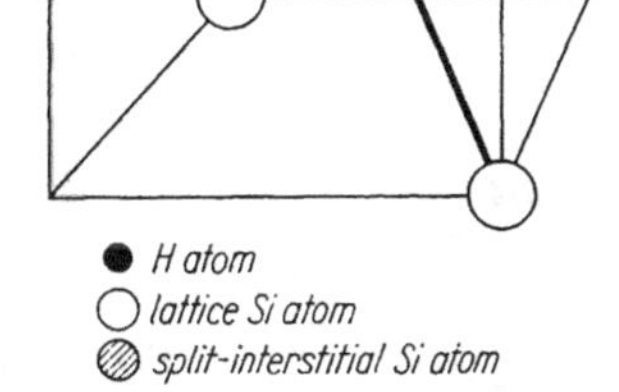

Fig. 3

Fig. 2. The T_d interstitial silane (hydrogenated T_d self-interstitial) in the silicon lattice

Fig. 3. The hydrogenated ⟨100⟩ split-interstitial Si atom

2.2 1946 cm^{-1} band

According to the correlation between the Si–H stretching vibration frequency and the structure of the vibration centre in c-Si proposed by Shi et al. [7], the possible structure of the vibration centre with frequency near 1946 cm^{-1} in undoped c-Si (without participation of other impurities such as O, C, B, P) are (Sid)SiH_2, (SiSid)SiH, and (SiSiSi)SiH, where d represents a dangling bond. Owing to the thermal stability of the 1946 cm^{-1} peak, the last structure is more reasonable. The hydrogenated ⟨100⟩ split self-interstitial (I_{sp} · 2 H) may give rise to these kind of centres [7, 13] (Fig. 3).

The two Si atoms of the split self-interstitial aligned along ⟨100⟩ and together occupying one lattice point, are subjected to a compressive stress as their nearest neighbours. Both these Si atoms have only three nearest neighbours to bond with. All these three Si–Si bonds are on the same (110) plane and are of sp^2-character. Another bond with nearly p-character, perpendicular to this plane is dangling and may be saturated by a H atom.

In the hydrogenated split self-interstitial, (I_{sp} · 2 H), after saturating these two dangling bonds by H atoms, the relaxation of the whole complex may occur, so that the p-character of the Si–H bond should decrease somehow. But the remaining effect still exists, which may give rise to a shift of the Si–H vibration frequency from that of the normal (SiSiSi)SiH structure ($\approx$ 1993 cm^{-1}) to the experimentally observed value 1946 cm^{-1} [14].

3. Some Experimental Results Supporting the Above-Mentioned Models

Some additional experimental results have been obtained which support the above-mentioned models. The experimental procedure used is the same as in [10].

3.1 Correlation of 2210 cm^{-1} and 1946 cm^{-1} with the Si–H IR bands in the bending-wagging region

There always appear two strong bands related to the local vibration modes of Si–H bond in the bending-wagging region in the IR absorption spectrum of FZ-Si:H at 791, 812 cm^{-1} [3] (shifted to 794, 814 cm^{-1} at 10 K). The correlation of 2210 cm^{-1} and 1946 cm^{-1} with these peaks is investigated in detail by measuring the IR spectrum of the differently thermal-treated samples and shown in Fig. 4.

It is obvious from Fig. 4 that the peaks 791, 812 cm^{-1} are correlated linearly with 1946 cm^{-1} but not with 2210 cm^{-1}. And furthermore there are no other strong peaks in the bending-wagging region, which may be correlated with 2210 cm^{-1}.

The fact is unexpected because of the disagreement with the first impression obtained from the room temperature experiments [8], which are less accurate due to the strong back-

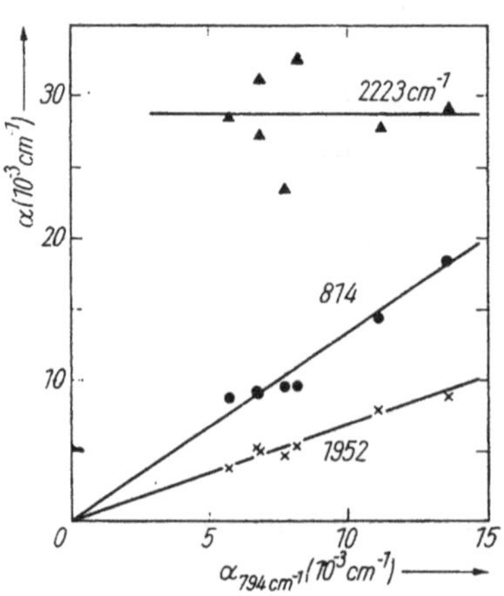

Fig. 4. The correlation between the 2223 cm^{-1} (2210 cm^{-1}), 1952 cm^{-1} (1946 cm^{-1}) and the two intense Si–H IR peaks in the bond-bending-wagging region 794 cm^{-1} (791 cm^{-1}), 814 cm^{-1} (812 cm^{-1})

ground absorption, and seems even more strange because the two bending-wagging vibration modes are correlated with the 1946 cm^{-1} band with SiH structure, rather than with the 2210 cm^{-1} band with SiH_4 structure.

But it is reasonable if the environment of the $(SiH_4)_{T_d}$ is considered. The $(SiH_4)_{T_d}$ is embedded in the Si lattice and its bending-wagging vibration modes are hindered, as compared to the SiH_4 molecule in free space. The amplitude, and hence the intensity of the corresponding IR band, may be lowered, so that no IR peak correlated with 2210 cm^{-1} can be observed in the bending-wagging region experimentally.

The frequency of the wagging mode of the (SiSiSi)SiH centre in a-Si is about 630 cm^{-1} [12]. In the $I_{sp} \cdot 2$ H structure the two Si atoms, which share the same lattice point, are in a compressed state. Their Si–Si bond and the bonds between them and their nearest-neighbour Si atoms are different from each other and from the normal Si–Si bond. Therefore, the degenerate HSiSi wagging mode splits into two components and shifts to higher frequency as observed experimentally.

Thus the experimental results on IR peak correlation are explained by the above-mentioned models quite naturally and satisfactorily.

3.2 The variation of peak width with temperature

As shown in Fig. 5 the change of peak width with temperature for 2210 cm^{-1} in comparison with other intense peaks (e.g. 1946 cm^{-1}) is characterized by

(i) the high sensitivity of the peak width to temperature. At a resolution of 0.24 cm^{-1} the peak width at 10 K, $\omega_{10} = 0.28$ cm^{-1}, and at 300 K $\omega_{300} = 6.0$ cm^{-1}. $\omega_{300}/\omega_{10} \geqq 20$;

(ii) the multiplicity of the peak-width variation with temperature.

There are two stages of increasing peak width centred at above 100 and 225 K, respectively, and a small plateau between them around 200 K.

The peak width of 1946 cm^{-1} is wider at 10 K ($\omega_{10} = 0.67$ cm^{-1}) and less sensitive to temperature $\omega_{300}/\omega_{10} \leqq 5$. The change is centred mainly above 225 K.

According to the proposed models these experimental facts can be qualitatively interpreted as following.

The interstitial SiH_4, as an independent molecule, is not bonded to the Si lattice, but just embedded in it, and, hence, is less tightly connected to the lattice vibration. So at low temperature (10 K) its peak width is smaller than that of other Si–H peaks.

With increasing temperature both the thermal movement of the SiH_4 molecule and the excitation of the Si lattice vibration may increase the interaction between them, which reduces the lifetime of the oscillator, and increases the peak width. It may be

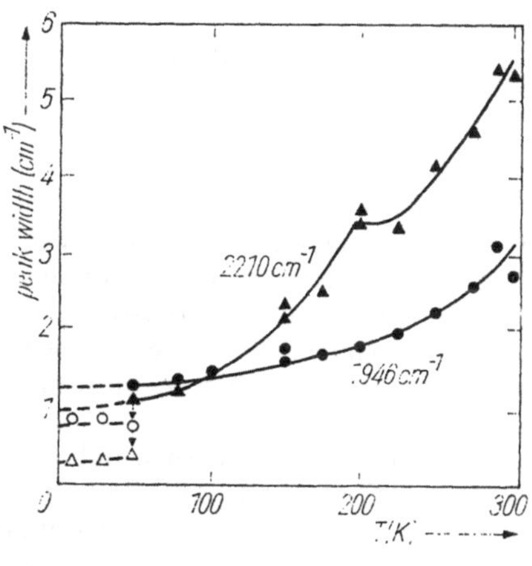

Fig. 5. The variation of peak width with temperature for 2210 cm^{-1} and 1946 cm^{-1} peaks. The arrows indicate the change of resolution in order to obtain the natural peak width at lowest temperatures. ▲, ● resolution 1 cm^{-1}, △, ○ resolution 0.24 cm^{-1}

suggested that the first stage of increasing the peak width above 100 K, which is specific for 2210 cm^{-1}, is due to the thermal movement of the $(SiH_4)_{T_d}$ molecule, while the second stage above 225 K, which is common to all Si–H peaks, is introduced by the excitation of the Si lattice vibration. The additional contribution to the increase of the peak width of 2210 cm^{-1} may arise from the disturbance of its T_d symmetry, caused by the molecule–lattice interaction. It explains the higher sensitivity of the 2210 cm^{-1} peak width to temperature as compared to 1946 cm^{-1}.

Thus it is probable that the specific character of the 2210 cm^{-1} peak width variation with temperature is dictated by the unique property of $(SiH_4)_{T_d}$ as an interstitial molecule.

More work is needed to properly understand the origin of small plateau around 200 K, which has been repeatedly observed in our experiment.

3.3 Thermal stability and annealing behaviour

The self-interstitial atom in Si is extremely unstable, its formation energy is rather high and the equilibrium concentration is quite low, while the migration energy is negligibly small, so it can migrate through the Si lattice even at the liquid helium temperature [15, 16]. Experimentally the individual self-interstitial has not yet been observed in Si. Likewise the H atom in c-Si has a low migration energy ($<$ 0.5 eV) and can move easily at room temperature. But when these two species bond to each other, the situation is entirely different. In c-Si the Si–H bond energy is about 2 to 3 eV [7]. Two to four H atoms combined with the interstitial Si atom may greatly reduce the total energy of the complex and increase its thermal stability. It is why the 2210 and 1946 cm^{-1} does not start to anneal until 500 °C and disappear completely only at 650 °C [8].

Another remarkable character in the annealing behaviour of these peaks is that the 2210 cm^{-1} starts to decrease monotonically from 500 °C, while the 1946 cm^{-1} increases first at 500 °C to a maximum and then decreases and disappears eventually at 650 °C [8]. A reasonable explanation to this fact is that the complexes responsible for these two peaks are of the same type, namely both complexes are formed by the hydrogen with two kinds of self-interstitials, I_{T_d} and $\langle 100\rangle$-I_{sp}. So they can transform from one into another, e.g. during annealing $(SiH_4)_{T_d}$ transforms first to $I_{sp}\cdot 2H$, then to I_{sp}, which eventually disappears at the appropriate sink,

$$(SiH_4)T_d \rightarrow (I_{sp}\cdot 2H) + H_2 \rightarrow I_{sp}\downarrow + 2H_2\ .$$

4. Summary

(i) Based on the isotope effect two H-defect complexes are proposed to account for the two intense SiH IR stretching bands of FZ-Si:H:

Tetrahedral interstitial silane $(SiH_4)_{T_d}$-2210cm^{-1}.

Hydrogenated $\langle 100\rangle$ split self-interstitial $(I_{sp}\cdot 2\,H)$ — 1946 cm^{-1}.

(ii) A series of experimental results including the vibration frequency, correlation with IR peaks in bending-wagging region 791, 812 cm^{-1}, variation of the peak-width with temperature, and annealing behaviour of these peaks are presented and interpreted according to the proposed models, which strongly indicate their reasonability. With other possible structure of T_d symmetry, such as hydrogenated vacancy $(V\cdot 4\,H)$, and $(V\cdot H)$ [9] it is hard to explain all the experimental results. Nevertheless more work is needed for their unambiguous exclusion.

The (iii) intrinsic defects in the as-grown c-Si are the thermal defects which have survived from the melting temperature during cooling. What is the dominant type of the thermal defects in Si — vacancy or self-interstitial — is a question of debate

[16]. Clarifying the nature of 2210 and 1946 cm^{-1} may provide direct experimental results.

We argue that the results of this work favour the view put forward by Seeger and Chik [17], having analysed the diffusion data in c-Si that the dominant thermal defect in c-Si at high temperature is the self-interstitial.

Acknowledgement

We would like to express our acknowledgement to Prof. Wu Zi-liang for his guidance and encouragement during this work.

References

[1] Cui Shu-fan, Ge Pei-wen, Zhao Ya-qin, and Wu Lan-sheng, Acta phys. Sinica 28, 791 (1979) (in Chinese).
[2] Vijay A. Singh, C. Weigel, J. W. Corbett, and L. M. Roth, phys. stat. sol. (b) 81, 637 (1977).
[3] Cui Shu-fan, Mai Zhen-hong, Ge Pei-wen, and Sheng De-yan, Kexue Tongbao 27, 382 (1982).
[4] Mai Zhen-hong, Cui Shu-fan, Ge Pei-wen, and Tsien Ling-chao, Acta cryst. A37, Suppl. c-254 (1981).
[5] Shi Tian-sheng, S. N. Sahu, J. W. Corbett, and L. C. Snyder, Scientia Sinica A27, 98 (1984).
[6] H. J. Stein, J. electronic Mater. 4, 159 (1975).
[7] T. S. Shi, S. N. Sahu, G. S. Oehrlein, A. Hiraki, and J. W. Corbett, phys. stat. sol. (a) 74, 329 (1982).
[8] Cui Shu-fan, Mai Zhen-hong, and Tsien Ling-chao, Scientia Sinica A27, 213 (1984).
[9] Gu Ben-yuan, Xu Zheng-yi, and Ge Pei-wen, Scientia Sinica A28, 740 (1985).
[10] G. R. Bai, M. W. Qi, L. M. Xie, and T. S. Shi, to be published in Solid State Commun.
[11] B. N. Mukashev, K. H. Nussupov, M. F. Tamendarov, and V. V. Frolov, Phys. Letters A 87, 376 (1982).
[12] M. Cardona, phys. stat. sol. (b) 118, 463 (1983).
[13] R. L. Kleinhenz, Y. H. Lee, Vijay A. Singh, P. M. Mooney, A. Jaworowski, L. M. Roth, J. C. Corelli, and J. W. Corbett, Inst. Phys. Conf. Ser. No. 46, 200 (1979).
[14] S. N. Sahu, Tian-sheng Shi, Pei-wen Ge, A. Hiraki, T. Imura, M. Tashira, V. A. Singh, and J. W. Corbett, J. chem. Phys. 77, 4330 (1982).
[15] S. M. Hu, Atomic Diffusion in Semiconductors, Ed. D. Shaw, Plenum Press, London/New York 1973.
[16] R. Car, P. J. Kelly, Atsushi Oshiyama, and S. T. Pantelides, Phys. Rev. Letters 52, 1814 (1984).
[17] A. Seeger and K. P. Chick, phys. stat. sol. 29, 455 (1968).

(Received June 27, 1985)

phys. stat. sol. (b) 131, 519 (1985)

Subject classification: 11; 22.1.2

Sektion Physik der Humboldt-Universität zu Berlin[1]),
Bereich Atomstoßprozesse der Festkörperphysik

Measurements of Proton Channeling Energy Losses in Silicon in the Intermediate Energy Region

By

P. GEHRMANN, K. LENKEIT, and R. STOLLE

Dedicated to Prof. Dr. Dr. h. c. Dr. E. h. P. GÖRLICH on the occasion of his 80th birthday

The energy loss of protons in the energy region from 40 to 350 keV is measured for (110), (100), and (111) planar channeling, for ⟨110⟩, ⟨111⟩, ⟨112⟩ axial channeling, and for random directions in silicon. The investigations are carried out by means of the transmission experiment using silicon targets of several thicknesses. Stopping power values are derived from the peak and the leading edge of energy spectra for channeled particles. Special procedures are applied to correct the obtained stopping power data with respect to the target thickness. In the case of leading edge the stopping power data depend strongly on the target thickness. It is shown that only the extrapolation to zero thickness leads to correct stopping power values for best channeled particles. It is found that the ratio of channeled to random stopping power shows a maximum in the energy region at about 70 keV. Here the stopping power for channeling and random directions reach their largest value, too.

In der vorliegenden Arbeit werden die Energieverluste von Protonen im Energiebereich von 40 bis 350 keV für die (110)-, (100)- und (111)-Flächenkanalisierung für die axiale ⟨110⟩-, ⟨111⟩-, ⟨112⟩-Kanalisierung und für die Randomrichtung in Silizium gemessen. Die Messungen werden in Transmissionsexperimenten an dünnen Siliziumtargets verschiedener Dicke durchgeführt. Bremsquerschnitte werden mit Hilfe des Maximums und der extrapolierten hochenergetischen Kante der Energieverteilungen der kanalisierten Teilchen bestimmt. Die erhaltenen Bremsquerschnitte werden mit Hilfe spezieller Verfahren bezüglich des Targetdickeneinflusses korrigiert. Im Fall der Extrapolation der hochenergetischen Kante der Energieverteilungen hängen die Bremsquerschnittswerte stark von der Targetdicke ab. Es wurde gezeigt, daß nur die Extrapolation auf die Schichtdicke Null zu korrekten Bremsquerschnittswerten für die am besten kanalisierten Teilchen führt. Das Verhältnis von Kanalisierungs- zu Random-Bremsquerschnitt weist ein Maximum bei einer Energie von ungefähr 70 keV auf, bei der auch die Bremsquerschnitte für Kanalisierungs- und Randomrichtung ihre Maximalwerte erreichen.

1. Introduction

Energy loss of ions penetrating matter has been widely investigated in the last few decades. Most of the investigations in this field have been done in the case of amorphous solids.

For the interpretation of experimental results based on the interaction of ions with single-crystal materials, as for instance channeling ion implantation profiles or Rutherford backscattering spectra, the knowledge of channeling electronic energy loss is essential. Whereas several measurements of channeling stopping power of ions with higher energies have been carried out [1 to 5], for the energy region corresponding to the Bohr velocity (maximum of the stopping power) there is still a deficiency of

---

[1]) Invalidenstr. 110, DDR-1040 Berlin, GDR.

experimental data. Some theoretical investigations in this field have been done [6 to 9] and a few results of measurements of the channeling energy loss of protons in the intermediate energy region have been published [4, 10, 11]. However, for best channeled protons and in the case of planar channeling experimental data for this energy region have not been published until now.

The purpose of this paper is to perform accurate measurements of the energy loss of protons with energies in the intermediate energy region for different crystallographic directions of single-crystal silicon films. The transmission experiment with ions passing through a thin target is one of the direct methods for measuring the energy loss of particles. The experiment will be described in Section 2. After numerical correction of our measured data the stopping power of the best channeled particles is extracted (Section 3) and the obtained ratio of the channeled to random stopping power is compared with experimental data and theoretical statements given in the literature.

2. Experimental

The experiment was carried out with the ion accelerator of the Humboldt-University Berlin. The detailed experimental setup was described in [12, 13], where it was applied to investigate some angle-dependent effects of channeling. Thin foils of single-crystal silicon were bombarded with protons of energies E_0 between 40 and 350 keV. The particle beam was collimated to a diameter of 0.2 mm and an angular divergence of 0.015°. The intensity of the primary proton beam was chosen to be in the picoampere region in order to prevent the damage of the foils. The E_0-energy distribution gives an energy spread smaller than 300 eV. The energies of the transmitted particles were measured by an electrostatic energy analyzer mounted behind the target in straight-through position. The particles were detected with a channeltron behind the analyzer. The energy resolution and the angular acceptance of the analyzer were $\Delta E/E_0 \leqq 0.01$ and 0.09°, respectively. Therefore in our experimental setup the acceptance angle of the analyzer was always smaller than the critical angle for channeling. The consideration of the energy dependence of the resolution of the electrostatic energy analyzer and the contribution of the neutral particle fraction to the energy spectrum leads in the region of our lowest proton energies to a correction of the measured most probable energy loss data ΔE_{mp} to higher values up to 5%.

The targets were single-crystal silicon foils with thicknesses in the range of 250 to 810 nm. They were prepared as described in [14] and had a $\langle 111 \rangle$ orientation normal to the surface. To determine the thickness Δx of the silicon foils we have measured the energy loss ΔE of protons with energies $E_0 = 300$ to 350 keV transmitted through the crystals in various random directions. The thickness was calculated using the numerical integration of the stopping power data previously obtained for amorphous silicon [13, 15]. The uncertainties of this method introduce a systematic error for the stopping power data of 5%.

The different crystallographic directions were adjusted by a three-axis goniometer with an angular resolution of 0.05°. Based on the described method we carried out a series of proton channeling energy loss measurements for several axes and planes as a function of beam energy.

3. Stopping Power Results

We consider protons of the initial energy E_0 transmitted through a thin silicon foil. While passing the foil the protons interact with the electrons and the nuclei of the target. During these interactions a small amount of energy is transferred to the target

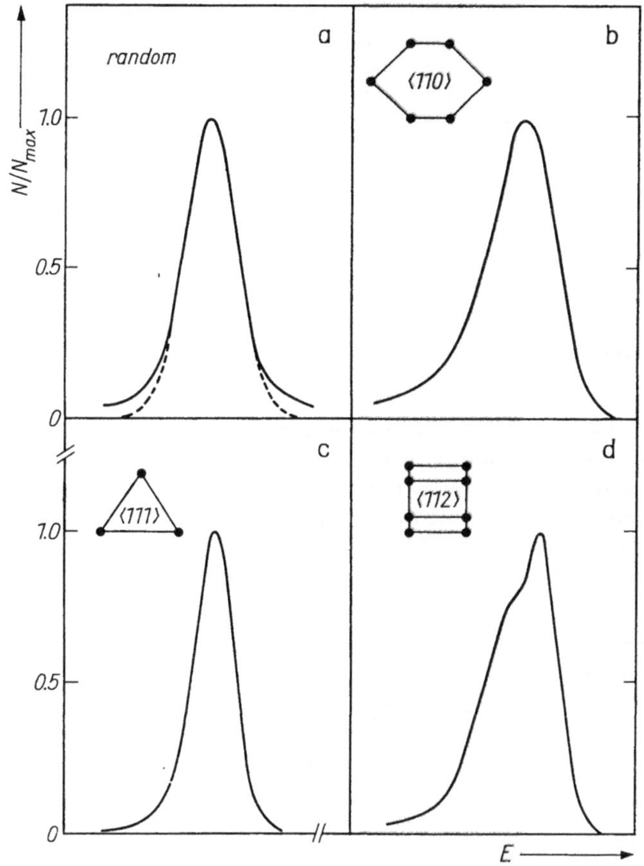

Fig. 1. Some typical energy distributions for a) the random case and b) to d) various channeling directions; dashed line Gaussian distribution

nuclei and electrons, i.e. the projectiles lose energy. In our energy region we can assume a pure electronic stopping and neglect the nuclear one. As a result of these interactions the energy of the protons after traversing the foil will be statistically distributed. Such a distribution reflects in someway the nature of the interactions inside the target. For a crystalline foil the energy distribution strongly depends on the direction of the proton velocity with respect to the crystal. So we found for a random direction in correspondence with previous papers a nearly Gaussian shape whereas in the different axial or planar cases one obtains more or less strong deviations from the Gaussian distribution [17, 19]. Fig. 1 shows some typical distributions which have been obtained for the random case and three channeling ones.

We found the halfwidths of the channeling distributions to depend obviously on the energy as expected, but independent of the target thickness. This fact is interpreted by the assumption that for not too small [16] or too large target thicknesses the channeling energy distribution is in an equilibrium state. This state is characteristic of each axial or planar channel, and it must be considered as the result of the contribution of several structural and dynamical factors. So the energy distribution reflects the features of the electron distribution of the various channels in connection with

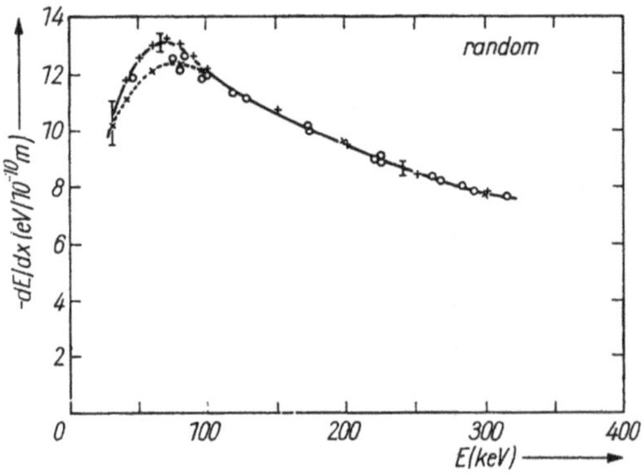

Fig. 2. Random stopping power for protons in silicon. ○ our results in approximation (1), solid curve: our corrected results, + experimental data of [13, 15], × data of Andersen and Ziegler [18]

the multiple scattering on electrons, the impact parameter distribution in the channel, and the impact parameter dependence of energy loss.

Analysing our experimental material we have carried out a correction with respect to the target thickness. Firstly we made a simple numerical approximation to extract from our measured data $\Delta E/\Delta x$ the expression dE/dx. For the first correction we used in a zeroth approximation to the dependence of stopping power on energy the follow-

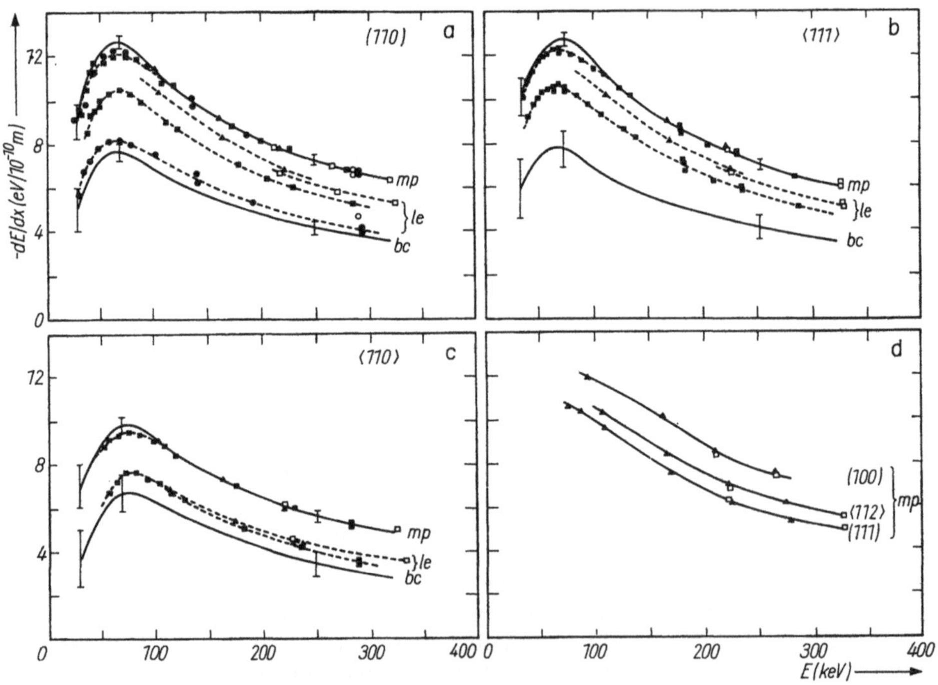

Fig. 3. Proton channeling stopping power in silicon. a) (110) planar direction, b) ⟨111⟩ axial direction, c) ⟨110⟩ axial direction, and d) ⟨112⟩ axial and (100) and (111) planar directions. The different symbols show the measured data in approximation (1) for several target thicknesses: ● 250, ○ 405, ■ 575, □ 800, ▲ 810 nm. Full drawn lines: corrected values (dE/dx) corresponding to the most probable energy loss (mp) and the energy loss of the best channeled particles (bc)

ing representation:

$$\frac{dE}{dx} \approx \frac{\Delta E}{\Delta x} = f(E^*) \tag{1}$$

with $E^* = E_0 - \Delta E/2$, ΔE and Δx are the measured energy loss and the target thickness, respectively, and E^* is the mean energy of the projectile in the target foil. Fig. 2 and 3 show our experimental data obtained in this way for random stopping power (Fig. 2) and for some channeling cases (Fig. 3). The dependence of the random stopping power $(dE/dx)_r$ on the proton energy is compared to the semiempirical results of Andersen and Ziegler [18] and with previously obtained data in amorphous silicon [13, 15]. One can see that within the limits of experimental error the $(dE/dx)_r$ values in random direction of a crystalline target are identical with those in amorphous material. For the channeling cases we derived stopping power values from the most probable energy loss ΔE_{mp} and from the leading edge of the proton energy distribution ΔE_{le}. For the further thickness correction we used a simple iterative procedure. Based on the first approximation (1) dE/dx was determined by means of numerical iteration.

The iteration procedure was repeated until the agreement between the approximation data and experimental results was satisfactory. For higher energies ($E \geqq 150$ keV) the approximation (1) is in good agreement with the real dE/dx dependence. But it leads to significant deviations from the correct curvature in the maximum of the stopping power curve (see Fig. 2 and 3).

From Fig. 3 one can see that the data $(\Delta E/\Delta x)_{mp}$ depend only weakly on target thickness. The thickness-induced deviations lie inside the error limits and are only noticeable in the stopping power maximum. This negligible thickness dependence agrees with the notion that the particles move at stable trajectories through the axial or planar channels. Under these conditions it is possible to consider the most probable energy loss data as to be thickness independent. With this consideration and the above-mentioned correction method to the experimental results we obtained the stopping power values $(dE/dx)_{mp}$ as a function of energy (see solid curves in Fig. 3). In the following we determined the stopping power of the best channeled particles $(dE/dx)_{bc}$ from our leading edge energy loss measurements $(\Delta E/\Delta x)_{le}$. In this case the obtained data $(dE/dx)_{le}$ depend strongly on the target thickness (see Fig. 4 and 3) and thereby the simple extraction of $(dE/dx)_{bc}$ values for the best channeled particles from the experimental data $(\Delta E/\Delta x)_{le}$ (as in [17]) is not possible. From Fig. 4 one can see that for our not too large target thicknesses a linear extrapolation to an

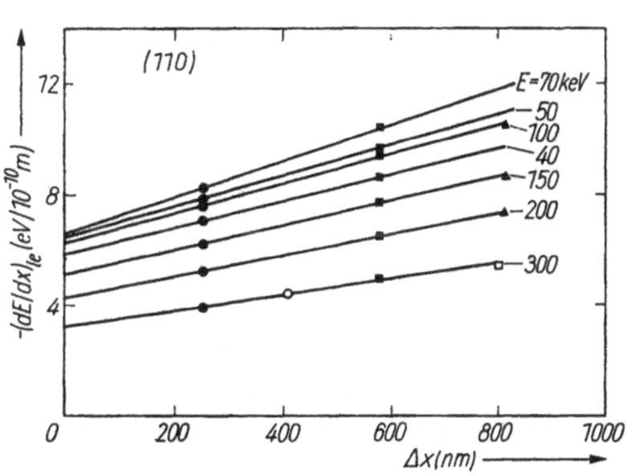

Fig. 4. Thickness dependence of leading edge stopping power $(dE/dx)_{le}$ for (110) planar direction

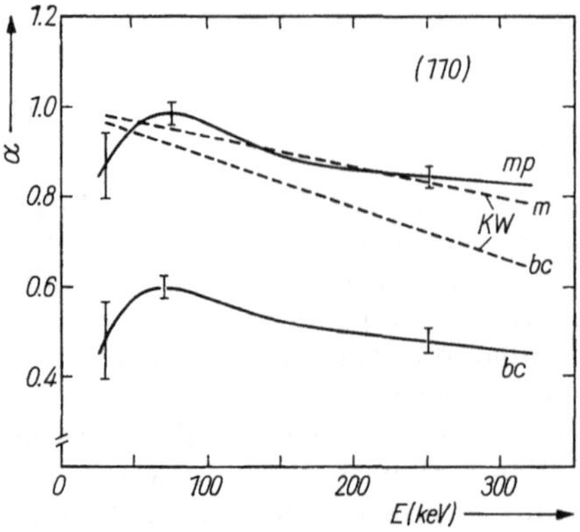

Fig. 5. Ratio between channeling and random stopping power for (110) direction. Dashed lines: theoretical results by Kührt and Wedell [9]. α_{mp} and α_{bc} are values for the particles with the most probable energy loss and the best channeled particles, respectively

infinitesimal target thickness is possible. Therefore, the $(dE/dx)_{le}$ data extrapolated to zero thickness correspond to the stopping power of the best channeled particles,

$$\left(\frac{dE}{dx}\right)_{bc} = \lim_{\Delta x \to 0} \left(\frac{dE}{dx}\right)_{le}. \tag{2}$$

The results of this procedure are shown also in Fig. 3 (solid curves with the index bc). The distinct differences in the error range for the various favoured crystallographic directions result from the different number of targets used for the various measuring series. The results for the three cases ⟨112⟩, (100), and (111) will be completed in further measurements.

The ratio between the channeled stopping power $(dE/dx)_c$ and the random stopping power $(dE/dx)_r$ is determined from the corrected data (solid curves in Fig. 2 and 3),

$$\alpha = \frac{(dE/dx)_c}{(dE/dx)_r}. \tag{3}$$

The energy dependences of α for channeling along the (110) planar and the ⟨111⟩ and ⟨110⟩ axial directions are shown in Fig. 5 and 6. In Table 1 all obtained results are

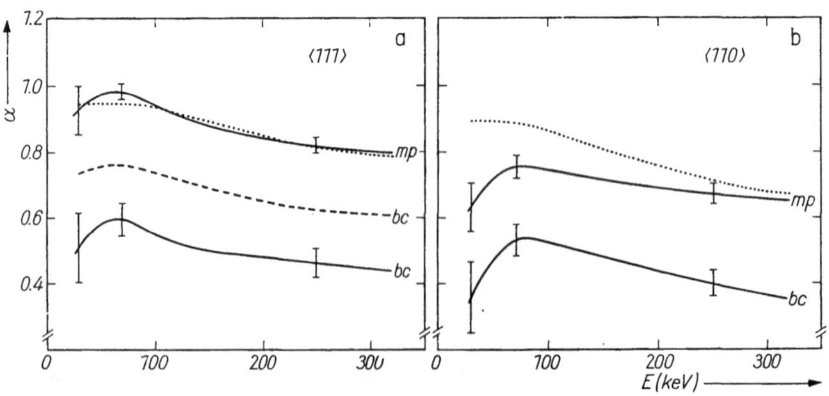

Fig. 6. Ratio between channeling and random stopping power. a) ⟨111⟩ axial direction, b) ⟨110⟩ axial direction. Dashed line: theoretical results by Komarov and Kumakhov [6], dotted line: experimental data of Carnera et al. [11]

Table 1

Experimental results of the ratio between channeling and random stopping power for protons in silicon

E (keV)	$(-dE/dx)_r$ ($eV/10^{-10}m$) [13, 15]	$\alpha_{(110)}$		$\alpha_{\langle 111\rangle}$		$\alpha_{\langle 110\rangle}$		$\alpha_{(100)}$	$\alpha_{\langle 112\rangle}$	$\alpha_{(111)}$
		mp	bc	mp	bc	mp	bc	mp	mp	mp
30	10.2	0.89	0.49	0.96	0.55	0.66	0.36	—	—	—
40	11.8	0.91	0.53	0.95	0.57	0.67	0.41	—	—	—
50	12.6	0.94	0.57	0.96	0.58	0.71	0.45	—	—	—
60	13.0	0.96	0.58	0.97	0.58	0.73	0.50	—	—	—
70	13.3	0.96	0.59	0.96	0.58	0.74	0.52	—	—	—
80	13.1	0.96	0.58	0.95	0.56	0.74	0.52	—	—	—
100	12.3	0.93	0.56	0.92	0.54	0.73	0.50	0.94	0.84	0.80
120	11.7	0.90	0.53	0.89	0.52	0.71	0.49	0.94	0.83	0.78
150	10.8	0.89	0.52	0.87	0.50	0.70	0.46	0.94	0.81	0.75
200	9.6	0.85	0.49	0.83	0.48	0.68	0.43	0.90	0.77	0.69
250	8.5	0.85	0.48	0.82	0.47	0.66	0.40	0.88	0.75	0.67
300	7.9	0.82	0.46	0.79	0.45	0.64	0.36	—	0.72	0.64

presented. For protons it is firstly shown that the ratio between the channeling (for the most probable energy loss and for the best channeled particles, α_{mp} and α_{bc}) and random stopping power reaches a maximum in the energy region about 70 keV where the stopping powers for channeling and random directions reach their largest values, too (see Fig. 2 and 3). This is in good agreement with the results for helium ions reported in [17].

We compared our results to former measurements given in [11] and to calculations by Komarov and Kumakhov [6] and Kührt and Wedell [9]. There is good agreement with the experimental data of [11], in the higher energy region (Fig. 5, 6), whereas for energies below the maximum strong deviations appear. It should be noted that the measuring error of the results presented here in this energy region is lower than that given in [11].

The experimental α_{mp} data for the $\langle 110\rangle$ channel obtained by Cembali and Zignani [10] are up to 15% larger than the data of the present paper. Theoretical results extracted from the literature differ considerably from our measurements. The values calculated by [6] describe qualitatively well the energy dependence with a maximum in the same energy region as in experiment, whereas the theory [9] did not provide such a dependence below the maximum.

4. Discussion

The method described above is useful for the analysis of energy loss spectra of channeled particles for all energies of interest, especially for energies in the vicinity or below the electronic stopping power maximum. The obtained results show that in transmission experiments only a correction with respect to the target thickness (iterative procedure to derive correct values of dE/dx from the measured data $\Delta E/\Delta x$ and extrapolation to zero thickness) leads to correct stopping power values for channeled particles. In particular, our results show that the stopping power of best channeled particles is not directly available from the leading edge data of a given transmission spectrum as assumed in earlier work [17].

The measured thickness dependence of $(dE/dx)_{le}$ (see (Fig. 4) can be explained by the increasing minimum oscillation amplitude of channeled particles due to their multiple scattering on the target electrons. Therefore, the energy loss of these ions increases with penetration depth, since the electron density close to the atomic strings is higher than in the channel centre. In coincidence with these considerations we found the lowest value of the best channeled particle energy loss in the ⟨110⟩ channel, where the electron density in the channel centre is smallest. These effects do not influence the energy distribution of a great part of channeled particles with energy losses near the most probable energy loss ΔE_{mp}. From these considerations the measured thickness independence of $(dE/dx)_{mp}$ can be explained.

Summarizing we can state that the results of our investigations once more show the features of electronic stopping of ions in favoured directions of a crystalline target, in particular at lower energies. The research and interpretation of all these processes must be continued by further experimental and theoretical work.

Acknowledgements

We wish to thank Dr. H. Kerkow for collaboration in carrying out the experiments, Dr. K. Gärtner and Dr. R. Wedell for helpful discussions in the interpretation of experimental data, and Mrs. G. I. Gorbunova for preparing of single-crystal targets.

References

[1] G. DELLA MEA, A. V. DRIGO, S. LO RUSSO, P. MAZZOLDI, and G. G. BENTINI, Radiat. Eff. **13**, 115 (1972).
[2] G. GÖTZ, K. D. KLINGE, and F. SCHWABE, Exper. Tech. Phys. **23**, 167 (1975).
[3] A. S. RUDNEV, V. I. ROSLYAKOV, E. I. SIROTININ, and A. F. TULINOV, phys. stat. sol. (a) **35**, K23 (1976).
[4] K. LENKEIT, R. STOLLE, and P. GEHRMANN, Proc. XIII. All-Union Conf. Physics of Interaction of Charged Particles with Single Crystals, Moscow 1984 (p. 75) (in Russian).
[5] J. D. MELVIN and T. A. TOMBRELLO, Radiat. Eff. **26**, 113 (1975).
[6] F. F. KOMAROV and M. A. KUMAKHOV, Radiat. Eff. **22**, 1 (1974).
[7] K. DETTMANN and M. T. ROBINSON, Phys. Rev. B **10**, 1 (1974).
[8] V. V. BELOSHITSKY and M. A. KUMAKHOV, Radiat. Eff. **35**, 209 (1978).
[9] E. KÜHRT and R. WEDELL, phys. stat. sol. (b) **116**, 585 (1983).
[10] F. CEMBALI and F. ZIGNANI, Radiat. Eff. **31**, 169 (1977).
[11] A. CARNERA, G. DELLA MEA, A. V. DRIGO, S. LO RUSSO, P. MAZZOLDI, and G. G. BENTINI, Phys. Rev. B **17**, 3492 (1978).
[12] YU. V. BULGAKOV, K. LENKEIT, and R. STOLLE, phys. stat. sol. (b) **116**, K23 (1983).
[13] K. LENKEIT, Thesis, Moscow 1982.
[14] H. J. A. VAN DIJK, J. Electrochem. Soc. **117**, 553 (1970).
[15] E. KÜHRT, K. LENKEIT, and F. TÄUBNER, phys. stat. sol. (a) **66**, K131 (1981).
[16] P. SIGMUND and K. B. WINTERBORN, Nuclear Instrum. and Methods B, to be published (1985).
[17] F. H. EISEN, G. J. CLARK, J. BØTTIGER, and J. M. POATE, Radiat. Eff. **13**, 93 (1972).
[18] H. H. ANDERSEN and J. F. ZIEGLER, Hydrogen Stopping Powers and Ranges in All Elements, Pergamon Press, New York 1977.
[19] G. HÖGBERG, H. NORDEN, and R. SKOOG, phys. stat. sol. **42**, 441 (1970).

(Received July 17, 1985)

phys. stat. sol. (b) 131, 527 (1985)

Subject classification: 1.4 and 10

Sektion Physik der Humboldt-Universität zu Berlin[1])

Dislocations and Energy of Grain Boundaries in Crystal Structures with Covalent Bonds

By

NGUYEN AN, R. HERRMANN, TH. SCHURIG, and G. WORM

Dedicated to Prof. Dr. Dr. h. c. Dr. E. h. P. GÖRLICH on the occasion of his 80th birthday

In covalent bounded crystal structures the construction of grain boundaries with low lattice deformation is possible by ring dislocations. The present work deals new aspects of the formation of binding rings. There will be introduced structure units of the grain boundaries as binding tetrahedrons enabling the formation of 5–7 and 5–7–n · 8-ring dislocations. Due to these dislocations grain boundaries with rotation angles θ between 0° and 180° can be constructed. The dislocation density in the grain boundaries does not increase monotonously with the tilt angle θ, but reaches maxima for discrete rotation angles (θ^{i}_{max} = 26.53°, 109.47°, and 148.49°). It can be shown, that dangling bonds in these grain boundaries exist only in special angular ranges.

In kovalent gebundenen Kristallstrukturen ist die Konstruktion von Korngrenzen mit geringer Gitterspannung durch Ringversetzungen möglich. Die vorliegende Arbeit beschäftigt sich mit neuen Aspekten der Bildung von Bindungsringen, wobei als Struktureinheiten der Korngrenze Bindungstetraeder eingeführt werden, mit denen 5–7- und 5–7–n · 8-Ringversetzungen begründet werden. Durch diese Versetzungen sind Korngrenzen mit Drehwinkeln θ zwischen 0° und 180° möglich. Dabei wächst die Versetzungsdichte der Korngrenze nicht monoton mit dem Winkel θ, sondern es werden für spezielle Winkel (θ^{i}_{max} = 26,53°, 109,49°, 148,49°) Maxima in der Versetzungsdichte erreicht. Es kann gezeigt werden, daß dangling bonds in diesen Korngrenzen nur in bestimmten Winkelbereichen existieren.

1. Introduction

For covalently bound crystal structures Hornstra [1] and later Holt [2] have developed dislocation models by constructing 5–7 binding rings which make it possible to integrate half planes into the lattice with only low lattice deformation. By means of such dislocations tilt boundaries may be formed. Möller [3] has applied the CSL conception to this model and has developed special dislocation structures for the {110} plane of the diamond structure. Hornstra [4] has shown that an arbitrary tilt axis in the crystal may be considered as the sum of steps in ⟨110⟩ directions. Therefore, the only simple dislocations which have to be studied are those along the ⟨110⟩ directions.

In the present work the conception of ring dislocations will be further developed and structure units of the grain boundary will be introduced, namely binding tetrahedra and their projections onto symmetry planes, respectively, enabling the formation of 5–7 and 5–7–n · 8 ring dislocations (n = 1, 2, 3 ...).

These dislocations may be understood as elements of lattice adaptation which cause an essentially lower lattice deformation than the "classical" dislocations. The grain boundary energy will be computed and our results will be compared with other known calculations from literature.

---

[1]) Invalidenstr. 110, DDR-1040 Berlin, GDR.

2. Formation of Ring Dislocations in Grain Boundaries with Rotation Axis $c = \langle 110 \rangle$

2.1 *Binding triangles*

The diamond structure is characterized by covalent bonds between the atoms. Each atom is surrounded by four nearest neighbours positioned at the corners of a regular tetrahedron. The Bravais cell contains four of such binding tetrahedra arranged in such a manner that the structure is composed of two f.c.c. lattices translated by 1/4 the length of space diagonal of the elementary cell. In the case of zincblende structure these two f.c.c. lattices are occupied by different atoms. Moreover, the covalent binding is superimposed by an ionic part because of the different charges of these atoms (e.g. $A^{III}B^{V}$ compounds).

The edges of tetrahedra are formed by $\langle 110 \rangle$ directions and the bonds between the atoms are parallel to $\langle 111 \rangle$ directions. Accordingly, the surfaces of tetrahedra are $\{111\}$ planes (Fig. 1).

The projection of a tetrahedron on the $\{110\}$ plane leaves two bonds DI and CI unchanged, whereas the bonds AI and BI result in the same projection; we designate this projection as binding triangle.

If we consider the lattice along the $\langle 110 \rangle$ directions the tetrahedra are connected with each other in such a manner that they form infinite chains and we see a sequence of binding triangles in the plane. In order to investigate the relative position of the tetrahedron chains or tetrahedron layers in the (110) plane we can use the binding triangles. The edges of the binding triangles are directed along $[\bar{1}10]$, $[\bar{1}1\bar{2}]$, and $[\bar{1}12]$; in these directions the chains of binding triangles are built up (Fig. 2).

2.2 *Deformation of the binding tetrahedra in the grain boundary*

Theoretical and experimental investigations of the structure of f.c.c. and b.c.c. metals [5, 6] have demonstrated that low energy grain boundaries may be characterized by the formation of closed atom polyhedra. The diamond and zincblende structures

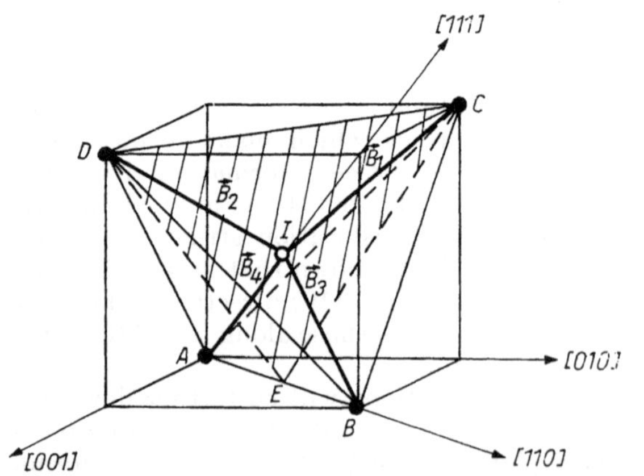

Fig. 1

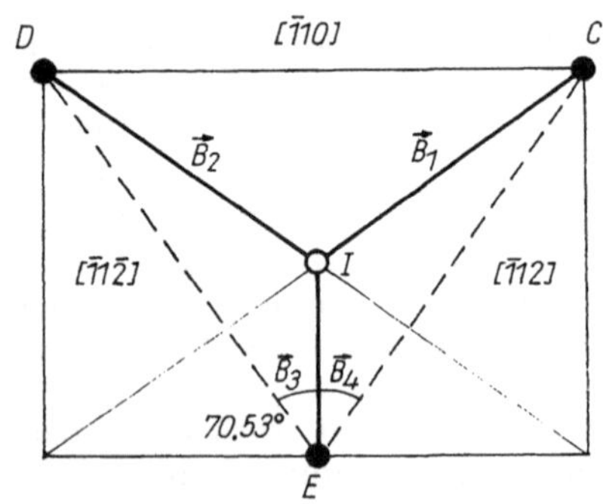

Fig. 2

Fig. 1. Position of a binding tetrahedron of sphalerite structure in the cubic lattice. The bonds between the central atom I and the four nearest neighbours A, B, C, D are represented by the vectors $\boldsymbol{B}_l$ ($l = 1, 2, 3, 4$). (Read [100] instead of [001])

Fig. 2. Projection of the binding tetrahedron from Fig. 1 on the (110) plane, resulting in the so-called binding triangle

differing in their structure from that of the metals mentioned above are characterized by binding tetrahedra. In forming grain boundaries in diamond or zincblende structures atoms can only enter a position corresponding to a low grain boundary energy if the shape of binding tetrahedra is preserved.

Our further investigations will be limited to symmetrical grain boundaries. With this supposition the parts of the binding tetrahedron on both sides of the grain boundary are equal.

In order to investigate the relative position of the binding tetrahedra with respect to the symmetrical grain boundary, the length B_l' of the projections of the bonds on the grain boundary normal is given as follows:

$$B_l' = \boldsymbol{u}\boldsymbol{B}_l = \frac{(u_{11}\boldsymbol{a}_1 + u_{12}\boldsymbol{a}_2 + u_{13}\boldsymbol{a}_3) \sum_{i=1}^{3} (1/4)\,(1 - 2\delta_{il})\,\boldsymbol{a}_i}{(\sum_{i=1}^{3} u_{1i}^2)^{1/2}}$$

$$= \frac{(1/4) \sum_{i=1}^{3} (1 - 2\delta_{il})\, u_{1i}}{(\sum_{i=1}^{3} u_{1i}^2)^{1/2}}; \tag{1a}$$

$(l = 1, 2, 3)$,

$$B_4' = \boldsymbol{u}\boldsymbol{B}_4 = \frac{-(1/4) \sum_{i=1}^{3} u_{1i}}{(\sum_{i=1}^{3} u_{1i}^2)^{1/2}}; \tag{1b}$$

u_{11}, u_{12}, u_{13} are indices of grain boundary planes (*hkl*). The vectors corresponding to the bonds $\boldsymbol{B}_i$ are shown in Fig. 1.

a) If the distance Δu between the atom positioned in the centre of the tetrahedron and the grain boundary plane is approximative B_l', for example, an atom C or D can get into the boundary and forms a common corner with two tetrahedra on both sides of the boundary (see Fig. 3a).

b) If $\Delta u \approx 1/2 B_l$, a corner of the first tetrahedron coincides with the centre of the second tetrahedron at the other side of the boundary (and vice versa) and we achieve two interpenetrating tetrahedra as shown in Fig. 3b.

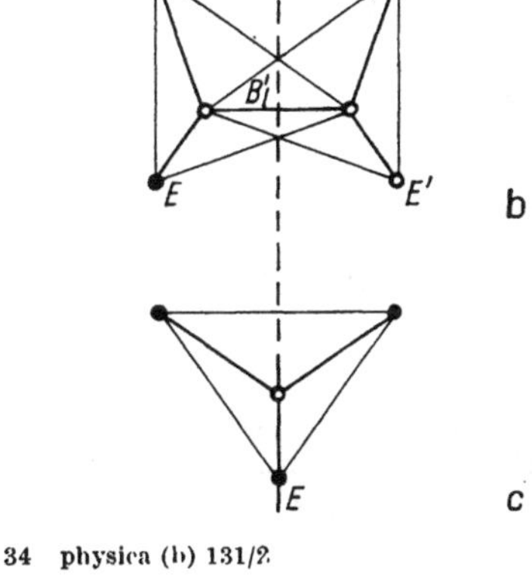

Fig. 3. Schematic representation of the construction of the binding tetrahedra at a symmetrical grain boundary (in the (110) plane). a) $\Delta u \approx B_l'$: two tetrahedra on both sides of the boundary with a common corner D are formed. b) $\Delta u \approx$ $\approx (1/2)\, B_l'$: a corner of the first tetrahedron coincides with the centre of the second tetrahedron at the other side of the boundary. c) $\Delta u \ll B_l'$: both tetrahedra contribute only half or less to the formation of a "reduced" tetrahedron (For explanation of Δu and B_l' see text.)

c) If $\Delta u \ll B'_l$, the binding tetrahedra at each side of the boundary contribute only half or less to the formation of the boundary. Two halves, neighbouring in such a manner in the boundary, form a new deformed tetrahedron, as shown in Fig. 3c.

For these three cases we have still to distinguish if the atoms lying in the boundary are positioned in the plane of projection or if they are above and below the mentioned plane. The projections of the latter coincide in point E of the boundary.

2.3 Formation of dislocations in the grain boundary

The edges of the binding tetrahedron are always parallel to three distinct directions. Therefore, in dependence upon the position of the binding tetrahedra relative to the grain boundary we must consider different dislocations.

If a boundary segment consists of three pairs or of four and five pairs, respectively, of tetrahedra with two common corners each, we find there a so-called 5–7 ring dislocation and a 5–7–8 or a 5–7–2 · 8 ring dislocation (Fig. 4 and 5).

The position of a 5–7 ring in the crystal association is shown in Fig. 6. After translation of one of the crystals neighbouring the boundary against the other by the vector $\boldsymbol{t} = (1/4)$ [111] we see from Fig. 4 and 5: The 5–7 ring is transformed into a pair of 8 rings; on the other hand, the 8 rings immediately neighbouring the 5–7 ring dislocation (Fig. 5) will be transformed into pairs of 6 rings representing the undisturbed structure.

Pairs of 8 ring dislocations in the range $70.53° < \theta < 110.40°$ and 5–7 ring dislocations in the range between $0° < \theta < 70.53°$ have been described by Hornstra [7]. Such dislocations also have been observed experimentally [8, 9].

2.4 Dislocation structure of grain boundaries with rotation axis [110]

The structure of symmetrical grain boundaries with rotation axis [110] can be analyzed by means of the binding tetrahedra. There exist six ranges in the plane of rotation (110). Each of them can be described by one distinct dislocation element (Fig. 7).

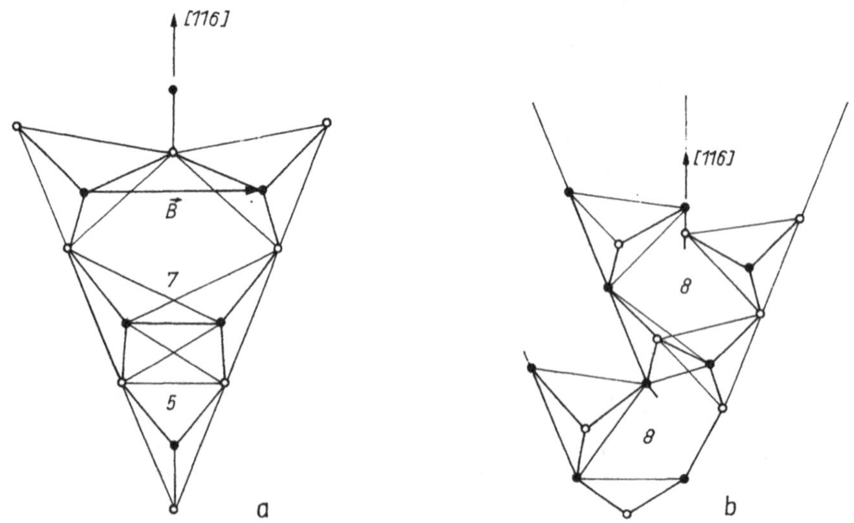

Fig. 4. a) 5–7 ring dislocation with Burgers vector $\boldsymbol{B} = (1/2)$ [110]; $|\boldsymbol{B}| = a_0/\sqrt{2}$ (a_0 lattice constant). b) Transformation of a 5–7 ring dislocation into a dislocation with a pair of 8 rings after a translation of the crystals neighbouring the grain boundary against each other by $\boldsymbol{t} = (1/4)$ [111]

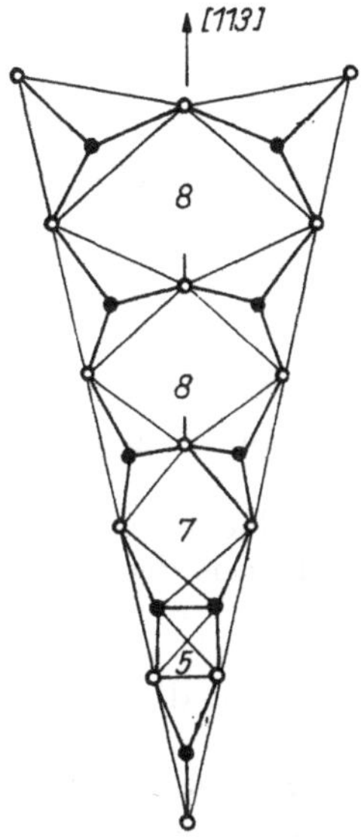

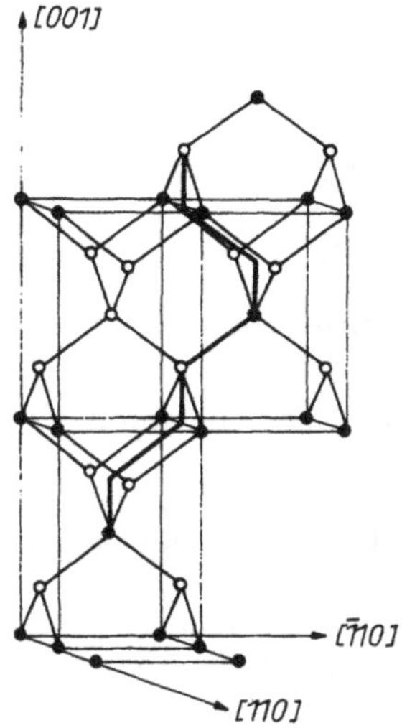

Fig. 5 Fig. 6

Fig. 5. Representation of a so-called 5–7–2 · 8 ring dislocation

Fig. 6. Position of a 5–7 ring in the crystal association

(1) The range $0° \leqq \theta \leqq 26.53°$

Along the direction $\boldsymbol{b}$[2]) = [116], corresponding to $\Sigma = 19$ and 26.53°, the boundary consists only of 5–7 ring dislocations (Fig. 7). At the distance $|\boldsymbol{b}| = \sqrt{38}\, a_0$ (a_0 is the lattice constant) there are altogether four dislocations. Consequently, the distance

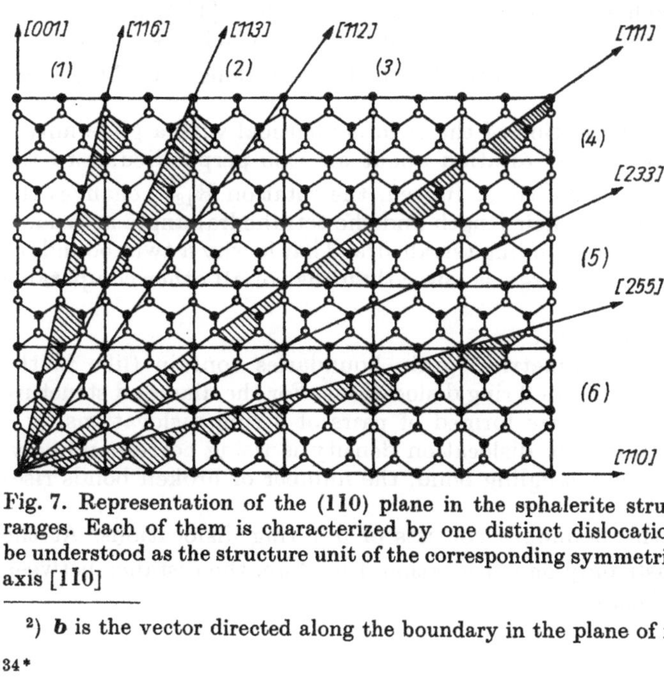

Fig. 7. Representation of the (1$\bar{1}$0) plane in the sphalerite structure subdivided in six angular ranges. Each of them is characterized by one distinct dislocation element (hatched), which can be understood as the structure unit of the corresponding symmetrical grain boundary with rotation axis [1$\bar{1}$0]

---

[2]) $\boldsymbol{b}$ is the vector directed along the boundary in the plane of rotation.

between two dislocations is

$$D = \frac{\sqrt{38}}{4}\, a_0 = 1.54 a_0 \,. \tag{2}$$

On the other hand, the Burgers vector of the 5–7 ring dislocation amounts to

$$|\boldsymbol{B}| = \frac{\sqrt{2}}{2}\, a_0 = 0.71 a_0 \,. \tag{3}$$

By comparing both quantities there follows

$$D = 2.17\, |\boldsymbol{B}| \,. \tag{4}$$

Therefore, one can suggest that the single dislocation cores do not touch each other in the boundary.

If θ is lowered from 26.53° to 0°, the distance between two dislocations D_i, at which at the same time the dislocation density in the grain boundary is decreasing, is increasing.

(2) The range $26.53° < \theta \leqq 70.53°$.

In this range a symmetrical grain boundary is built up from 5–7–$n \cdot 8$ ring dislocations ($n = 1, 2, 3, \ldots$) (Fig. 7).

If θ arises from 26.53° to 70.53° the number of 8 rings is rising, too. As will be shown later such a boundary structure is not stable in the diamond structure. Consequently, by a translation directed parallel to a ⟨111⟩-bond the 5–7 ring dislocation will be transformed into a pair of 8 ring dislocations and, on the other hand, each 8 ring into a pair of undisturbed 6 rings. Because every 8 ring dislocation has one dangling bond, the boundaries in this range subject to dangling bonds correspond to the density of 8 rings.

Consequently, a stable grain boundary structure may be achieved by a glide mirror operation [10], because the glide mirror symmetrical boundary consists of pairs of 8 ring dislocations.

If θ arises from 26.53° to 70.53° the density of pairs of 8 ring dislocations is lowered. Finally, at $\theta = 70.53°$ we achieve a twin boundary with $\boldsymbol{b} = [112]$ and $\Sigma = 3$. Moreover, it has to be taken into consideration that we cannot achieve a twin only by the rotation $R^{[0\bar{1}1]}_{70.53°}$, but an additional translation directed along the ⟨111⟩-bond is necessary.

Consequently, the symmetry operation of the twin is identical with a glide mirror operation, being equivalent to a 180° rotation about an axis perpendicular to the twinning plane, i.e. (111), at the same time. Indeed, the rotation $R^{[111]}_{180°}$ changes the direction of the bonds, but it leaves their length and the tetrahedral angle invariant, i.e. this rotation does not deform the binding tetrahedra. That is why a twin boundary consists of modified 6 rings and its energy is lower than that of all other comparable grain boundaries.

(3) The range $70.53° < \theta \leqq 109.47°$.

It is valid for this range that symmetrical grain boundaries contain (like in the previously discussed range), 5–7 and $n \cdot 8$ ring dislocations. For the diamond structure boundaries should be stable if they are formed of pairs of 8 ring dislocations. If θ arises from 70.53° to 109.47° also the dislocation density is rising correspondingly. Because every 8 ring contains one dangling bond, the number of broken bonds rises in the same proportion.

For $\theta = 109.47°$ ($\Sigma = 3$) the boundary consists of 5–7 ring dislocations. At the distance $|\boldsymbol{b}| = \sqrt{3}\, a_0$ there exists only one dislocation, therefore, the distance between two dislocations is $D = 1.732\, a_0$, i.e.

$$D = 2.44\, |\boldsymbol{B}| \,. \tag{5}$$

As a result the single dislocation cores do not touch each other at the boundary.

(4) The range $109.47° < \theta \leqq 129.52°$.

In this range a symmetrical grain boundary is built up of 5–7 ring dislocations with increasing distance between them, if θ is growing.

If θ reaches the value 129.52° ($\Sigma = 11$ and $\boldsymbol{b} = [233]$) the boundary consists of modified 6 rings. In this case, however, the binding tetrahedron will be deformed and the grain boundary energy is higher than that for the twin. No dangling bonds are present in this range.

(5) The range $129.52° < \theta \leqq 148.49°$.

For the symmetrical case we find 5–7 ring dislocations. With increasing θ the distance between the dislocations becomes lower. At $\theta = 148.40°$ ($\boldsymbol{b} = [255]$ and $\Sigma = 7$) a symmetrical grain boundary is built up of 5–7 ring dislocations of which four dislocations are at the distance $|\boldsymbol{b}| = \sqrt{54}a_0$. Therefore, the distance between the single dislocations is

$$D = 1.84\, a_0 = 2.59\, |\boldsymbol{B}| \,. \tag{6}$$

From that follows again that the dislocation cores do not touch each other at the boundary. There exist no dangling bonds all over the range mentioned.

(6) The range $148.49° < \theta \leqq 180°$

The corresponding grain boundaries consist of 5–7 and $n \cdot 8$ ring dislocations. With increasing θ also the number of 8 rings is increasing. Like in the case (2) and (3) the grain boundary structure is stable for the diamond structure, if a translation takes place in the boundary by (1/4) [111] and if the 5–7–$n \cdot 8$ ring dislocations change to pairs of 8 ring dislocations.

With growing θ also the distance between the pairs of 8 ring dislocations is increasing. All statements mentioned above suggest the following conclusions:

a) The dislocation density in the grain boundary does not increase monotonously, if the rotation angle θ is growing. It reaches maxima for discrete rotation angles θ^i_{max}, at which the angles of absolute maximum dislocation density are as follows:

$$\theta^1_{max} = 26.53°\,,$$
$$\theta^2_{max} = 109.47°\,,$$
$$\theta^3_{max} = 148.49°\,.$$

b) For boundaries corresponding to the maximum angles θ^i_{max} the distance between the dislocations is always larger than twice the Burgers vector, i.e. $D_{max} > 2|\boldsymbol{B}|$. That is why the individual disloaction cores do not touch each other. Otherwise, for boundaries with $\theta \neq \theta^i_{max}$ for the distance between the dislocations $D > D_{max}$ is valid. Consequently, the dislocation model for the description of grain boundaries may be used not only for low angle grain boundaries, but also for bondaries with arbitrary angles. Moreover, the grain boundary energy is not expressed as a function of the rotation angle, but in dependence on $1/D$. No doubt we can evaluate D for any grain boundary, but we cannot state at present an analytical expression describing the dependence $(1/D) = f(\theta)$.

c) As has been shown earlier [11] the facets corresponding to the rotation angles θ and $\theta' = 180° - \theta$ are perpendicular to each other. Furthermore it has been demonstrated in [10] that for the diamond structure one of the facets being perpendicular to each other is a symmetrical grain boundary and the other, on the contrary, a glide mirror facet.

These orthogonal facets represent the so-called joined facets of a grain boundary. In this connection the angular ranges (1) for θ and (6) for θ' and the ranges (2) for θ and (4), (5) for θ' are so-called joined ranges. In range (3) we can find for a facet corresponding to the angle θ at the same time the facet corresponding to θ'. Therefore, range (3) is a self-joined range. Furthermore we can establish that for the diamond structure the structure of the facets connected with ranges (1) and (4), (5) is stable if the facets are symmetrical. In ranges (2), (3), and (6) the stable facets have glide mirror symmetry.

d) The grain boundaries containing 8 ring dislocations have also one dangling bond per 8 ring. Consequently ranges (2), (3), and (6), in accordance with the dislocation density, also have a density of dangling bonds. By the way, for the zincblende structure we have to pay attention to the fact that the 5–7 ring dislocation has a wrong bond (bond between two atoms of the same kind) [2].

3. Energy of Grain Boundaries with Rotation Axis [011]

3.1 Energy of 5–7 and 8–8 ring dislocations

In order to estimate the energy of grain boundaries we will proceed from the energy of typical 5–7 and 5–7–$n \cdot 8$ ring dislocations and of 8–8 ring dislocations, respectively.

In general the energy of a grain boundary in a semiconductor may be calculated from two components, namely the deformation energy E_d and the energy of dangling bonds $m_0 E_b$, i.e.

$$E_{dis} = E_d + m_0 E_b \tag{7}$$

at which E_b is the average energy of a dangling bond and m_0 the number of broken bonds per unit length in the dislocation line.

The 5–7 ring dislocation is an edge dislocation with Burgers vector $\boldsymbol{B}_{5-7-} = (1/2) \times \times [110]$ and has no dangling bonds in the diamond structure (also valid for Si and Ge). According to [12] for the energy results

$$E_{dis} = E_d = \frac{\mu b^2}{4\pi(1-\nu)} \ln \frac{r_1}{r_0}. \tag{8}$$

A rough estimation for Ge yields an energy of 1.4×10^{-10} J/m (μ torsion modulus, ν Poisson ratio, r_1 crystal radius, r_0 radius of the dislocation core).

The 8 ring dislocation is an edge dislocation with Burgers vector $\boldsymbol{B}_{8-} = \frac{1}{2}[110]$. Besides the deformation energy (equation (8)) also the energy of dangling bonds has to be considered. According to Hornstra [1] there are on the average 2.83 dangling bonds per lattice constant. Consequently their number per unit length is

$$m = \frac{2.83}{a_0}. \tag{9}$$

From that we can estimate the value $mE_b = 1.3 \times 10^{-11}$ J/m, i.e. the energy of dangling bonds is one tenth of the deformation energy. After a mutual translation of both lattices along a binding direction in the boundary transforming the 5–7 ring dislocation into $a/2 \cdot 8$ ring dislocation there follows for the most stable Burgers vector

$$\tfrac{1}{2}[110] + \tfrac{1}{4}[\bar{1}\bar{1}1] = \tfrac{1}{4}[11\bar{1}], \tag{10}$$

because the energy is proportional to $|\boldsymbol{B}|^2$ and $(1/4)[11\bar{1}]$ is the shortest possible Burgers vector after a translation as mentioned above. With (8)

$$\frac{E_{d\,8-8-}}{E_{d\,5-7-}} = \frac{|\boldsymbol{B}_{8-8-}|^2}{|\boldsymbol{B}_{5-7-}|^2} = \frac{3}{8} \tag{11}$$

is valid. It has to be considered, however, that the pair of 8 ring dislocations has still $2m_0$ dangling bonds per unit length the energy of which amounts to 2.6×10^{-11} J/m $\approx$ $\approx (1/5)\, E_d$. Consequently, the 8 ring dislocations are more stable than the 5–7 and the 5–7–$n \cdot 8$ ring dislocations in the diamond structure. The 5–7 and the 8–8 ring dislocations, hence, represent the structure units of the grain boundary.

This situation approximately remains preserved, if the ⟨111⟩ or ⟨211⟩ directions are the rotation axes.

3.2 The energy of periodic dislocation rows

According to Möller [3] we describe the rows of periodically arranged dislocations as a sequence of ring dislocations now. Doing this we start from the results shown in Fig. 7.

For a row of dislocations the number of dangling bonds is increasing by the number of dislocation lines in the row: $N = n_0 m_0$, where n_0 is the number of dislocations per unit length in the direction of $\boldsymbol{b}$; for m_0 see above. The energy of the grain boundary may now be calculated in the following form:

$$E = E_d^r + NE_b\,. \qquad (12)$$

Table 1

Parameters of the dislocation model and estimated energy values for grain boundaries in Ge. $\boldsymbol{b}$ direction of the grain boundary in the rotation plane, $\boldsymbol{B}$ Burgers vector of the dislocation, n number of dislocations per $|\boldsymbol{b}|$, n_0 number of dislocations per unit length, E_d^r deformation energy of the grain boundary, NE_b energy of the dangling bonds in the boundary, E grain boundary energy

θ	$\boldsymbol{b}$	$\boldsymbol{B}$	n	$10^{-8} n_0$	E_d^r (10^{-3} J/m)	NE_b (10^{-3} J/m)	E (10^{-3} J/m)
11.10°	[1611]	(1/2) [110]	4	4.40	1.30	—	1.30
11.54°	[1411]		4	5.00	1.69	—	1.69
13.44°	[1211]		4	5.90	2.67	—	2.67
16.10°	[1011]		4	7.00	3.33	—	3.33
20.05°	[811]		4	8.70	5.06	—	5.06
26.52°	[611]		4	11.5	6.35	—	6.35
31.59°	[511]	(1/4) [111]	3	10.2	0.98	26.5	27.5
38.94°	[411]		2	8.30	1.75	21.6	23.3
50.47°	[311]		1	5.30	2.43	13.4	16.2
70.53°	[211]		—	—	—	—	—
86.63°	[322]		2	8.60	1.87	22.4	24.2
93.37°	[433]		2	6.10	0.60	15.9	16.5
109.5°	[111]	(1/2) [110]	1	10.2	5.33	—	5.33
129.52°	[233]	—	—	—	—	—	—
141.06°	[122]	(1/2) [110]	2	11.8	6.15	—	6.15
148.41°	[255]		4	9.60	5.00	—	5.00
153.47°	[133]	(1/4) [111]	2	8.10	1.89	21.1	22.9
159.95°	[144]		2	6.10	1.23	15.9	17.1
163.90°	[155]		2	4.90	0.14	12.7	12.8
166.56°	[166]		2	4.10	0.19	10.7	10.8
168.46°	[177]		2	3.50	0.52	9.10	9.62
169.90°	[188]		2	3.10	0.66	8.06	8.72

For ranges (1) and (4), (5) the grain boundary only consists of 5–7 ring dislocations, therefore $m_0 = 0$ and

$$E = E_d^r . \tag{13}$$

Otherwise, for ranges (2), (3), and (6) the boundary consists of pairs of 8 ring dislocations and (12) is valid. In accordance with [13] E_d^r may be calculated for the dislocation row.

In Table 1 the parameters of our dislocation model and the energies estimated by means of our model for the considered grain boundaries for germanium in the individual angular ranges (1) to (6) are summarized.

From this table one can see that there exists only a maximum of energy at $\Sigma = 27$ in the angular range $0° \leqq \theta \leqq 70.53°$, but no minima of energy at $\Sigma = 19$ and $\Sigma = 9$, as in [3].

The essential difference between the method applied in this paper and the investigations carried out in literature up to now is the following: In our paper the new results described in Section 2.4 and Fig. 7 signify that the dislocation density does not increase monotonously as in the case of low angle grain boundaries, but reaches maxima for the angles $\theta_{max}^1 = 26.53°$, $\theta_{max}^2 = 109.47°$, and $\theta_{max}^3 = 148.49°$ and between them decreases to zero (all considerations are valid for plane $\{110\}$).

References

[1] J. HORNSTRA, Physica **25**, 409 (1959).
[2] D. B. HOLT, J. Phys. Chem. Solids **25**, 1385 (1964).
[3] H. J. MÖLLER, J. Phys. C **43**, 1 (1982).
[4] J. HORNSTRA, J. Phys. Chem. Solids **5**, 129 (1958).
[5] V. VITEK, A. P. SUTTON, D. A. SMITH, and R. C. POND, Phil. Mag. **39**, 213 (1979).
[6] V. VITEK, D. A. SMITH, and R. C. POND, Phil. Mag. **41**, 649 (1980).
[7] J. HORNSTRA, Physica **26**, 198 (1960)
[8] O. L. KRIVANEK, S. ISOLDA, and K. KOBAYSHI, Phil. Mag. **36**, 931 (1977).
[9] A. BOURRET and J. DESSEAUX, Phil. Mag. **39**, 405 (1979).
[10] R. HERRMANN, NGUYEN AN, and G. WORM, Cryst. Res. Techn. **19**, 1607 (1984).
[11] NGUYEN AN, G. WORM, and R. HERRMANN, phys. stat. sol. (b) **114**, 349 (1982).
[12] J. FRIEDEL, Dislocations, Pergamon Press 1964.
[13] J. P. HIRTH and S. LOTHE, Theory of Dislocations, Mc Graw-Hill Publ. Co., New York 1968.

(Received July 17, 1985)

phys. stat. sol. (b) **131**, 537 (1985)

Subject classification: 4 and 10.1; 21.1

M. I. Kalinin Leningrad Polytechnical Institute[1])

Small Vacancy Clusters in Nickel

By

A. A. VASILYEV, V. V. SIROTINKIN, and A. I. MELKER

Dedicated to Prof. Dr. Dr. h. c. Dr. E. h. P. GÖRLICH on the occasion of his 80th birthday

The method of molecular dynamics is used to compute binding energies of different configurations of vacancy clusters in nickel. The interatomic pair potential is obtained according to the pseudopotential theory. It is found that in the range from divacancy to hexavacancy various atomic configurations are formed which can be considered as nuclei of dislocation loops, voids, and new crystal lattices.

Die Methode der Molekulardynamik wird benutzt, um die Bindungsenergien verschiedener Konfigurationen von Leerstellenclustern in Nickel numerisch zu berechnen. Das interatomare Paarpotential wird entsprechend der Pseudopotentialtheorie erhalten. Es wird gefunden, daß im Bereich von der Doppelleerstelle bis zur Hexavakanz verschiedene atomare Konfigurationen gebildet werden, die als Keime für Versetzungsschleifen, Hohlräume und neue Kristallgitter angesehen werden können.

1. Introduction

The point defect clusters are known to belong to the basic elements of the radiation damage structure of solids [1], which significantly affect the physical and mechanical properties of irradiated materials. For example, swelling of metals and alloys irradiated to a large dose by energetic particles results from the coagulation of vacancies leading to void formation.

Lately, large attention was attracted to the mechanisms of small vacancy cluster formation, these clusters can be the nuclei of voids, dislocation loops, or new phases [2]. In the given paper, the binding energies and equilibrium configurations of the small vacancy clusters in nickel are calculated by the molecular dynamics method and the evolution of these clusters in the process of single vacancy attachment is also studied.

2. Interatomic Potential

The calculation procedure in many respects is similar to that described previously [3]. The interatomic pair potential was calculated according to the pseudopotential theory [4]. The model Heine-Abarenkov-type pseudopotential was used, for which the parameters for transition metals were tabulated by Animalu [5]. The Animalu form-factors describe well the phonon dispersion curves of the transition metals, including nickel [6]. Besides, these form-factors were applied with success for the calculation of the short-range order parameters of some transition metal alloys, e.g. Ni–Pt [7].

The correction to the Hartree dielectric function, taking into account the effects of electron-gas exchange and correlation, was chosen in the form suggested by Singwi et al. [8]. The calculated interatomic pair potential $\varphi(r)$ is shown in Fig. 1. To evalu-

[1]) Ul. Politekhnicheskaya 29, 195251 Leningrad, USSR.

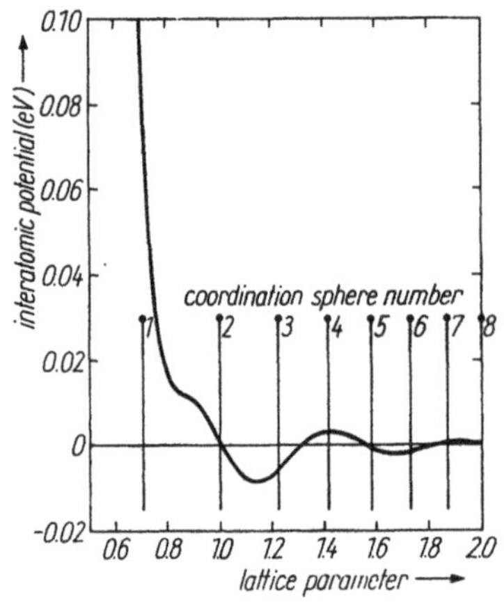

Fig. 1. Interatomic pair potential $\Phi(r)$ in nickel. The lattice parameter is taken as a unit

ate the applicability of this potential for studying defect properties, the vacancy formation and migration energies were calculated in accordance with the methods described in [9, 10] and [11], respectively. The results obtained are given in Table 1. As can be seen from the table, these values are in good agreement with the experimental ones and results reported by other authors.

Table 1
Energy parameters (in eV) of vacancy in nickel

	[12]		present results
	experiment	theory	
formation	1.54 to 1.80	1.4	1.59
migration	1.2 to 1.27	1.04 to 1.25	0.9

3. Calculation Procedure

The f.c.c. crystallite had the form of a sphere, the interatomic interactions being taken into account up to the sixth coordination sphere. The rigid boundary conditions for the atoms fixed in spherical shells having a width of 1.75 lattice parameters were used. Such boundary conditions can be acceptable [3], if the displacements of the atoms nearest to the boundary in a defect crystallite are small. The control calculations have shown that this condition was fulfilled for the clusters having up to seven vacancies, if the crystallite radius was larger than $\approx 4.0a$, where a is the lattice parameter. Such crystallite contains 1205 atoms, 369 ones being fixed. Further increasing the crystallite radius does not change practically the results.

The binding energy of the cluster, which contains n vacancies, can be written in the following form [3]:

$$E^{\mathrm{B}}_{n\mathrm{V}} = -\tfrac{1}{2} \sum_{i=1}^{n} \sum_{j \neq i}^{n} \varphi(r_{ij}) + E^{\mathrm{R}}_{n\mathrm{V}} - nE^{\mathrm{R}}_{\mathrm{V}}\,. \tag{1}$$

Here r_{ij} is the distance between the i- and j-vacant sites, $E^{\mathrm{R}}_{n\mathrm{V}}$, $E^{\mathrm{R}}_{\mathrm{V}}$ the cluster and single-vacancy relaxation energies, respectively. These energies were obtained as the difference between the potential energy of a defect-containing crystallite before and after lattice relaxation. The binding energy calculated characterizes the cluster configuration which is formed athermally (e.g. in displacement cascades) by removing a required number of atoms from a perfect lattice and allowing it to relax. (Such configurations will be called athermal ones.)

During cluster growth, each next vacancy joins to the equilibrium configuration of an already existing cluster. This process can be both diffusional and athermal (e.g.

at large irradiation doses). The cluster configurations formed under these conditions (growth configurations) could not always be compared with the athermal ones, which is necessary for the binding energy calculation by (1), so we have calculated at first the binding energy of the athermal vacancy cluster configuration relaxed using (1). The value obtained was taken as a mark for defining the binding energies of the growth clusters containing the same vacancy number. All the computations were made on a ES 1040 computer. The computing time was less than 45 min for an athermal tetravacancy configuration.

4. Results and Discussion

4.1 Divacancy

It is convenient to classify different divacancy configurations in a f.c.c. lattice as follows. One vacancy is placed in the coordinate origin, half a lattice parameter being taken as a unit. For example, a [110] configuration consists of two vacancies being at the nearest neighbour distance. The calculation results are given in Table 2. The binding energy of the most stable configuration is about 0.07 eV. Increasing the distance between the vacancies results in the monotonic reduction of the binding energy which does not exceed several thousandths of an eV beyond the fourth nearest neighbour distance.

Table 2
Binding energies of divacancy

coordination sphere	configuration	binding energy (eV)
1	110	0.066 ± 0.014
2	200	0.026 ± 0.013
3	211	0.002 ± 0.022
4	220	−0.015 ± 0.029
5	310	−0.004 ± 0.014
6	222	−0.002 ± 0.025
7	321	−0.008 ± 0.013
8	400	−0.003 ± 0.012

We do not know the experimental data on the divacancy binding energy in nickel. According to the calculations with the empirical short-range Johnson potential [13] the [110] configuration binding energy is equal to 0.25 eV which differs significantly from our results. The reasons for this discrepancy are probably similar to those considered in our previous paper [3] and can be attributed to truncating the Johnson potential at too small distances.

Two circumstances should be noted here. Numerous investigations lead to the conclusion (e.g. see Konobeev's review [1]) that under definite irradiation conditions vacancy loops in metals and alloys are formed and grow simultaneously with the interstitial ones. As a rule, the Burgers vectors and planes containing vacancy and interstitial loops coincide. Sometimes (e.g. at electron irradiation) only vacancy loops are formed but no voids. The loop formation is usually attributed to a collapse of displacement peak zones having excess vacancies. However, this mechanism is not universal. In particular, it cannot explain the experiments in which the irradiation by ≈ 1 MeV electrons was used, since such electrons generate only one or two Frenkel pairs.

On the contrary, a more general mechanism can be proposed which is not connected with the collapse. According to [14] the distortion region around the dumb-bell interstitial configuration is such that this defect should be considered as a planar inclusion. Probably for this reason, the only types of interstitial clusters are plane ones because such clusters cause lesser distortion in a crystal. According to [15] the linear configurations of di-, tri-, and tetravacancies in copper directed along $\langle 111 \rangle$ produce such atomic displacement that after relaxation the atomic configurations obtained are similar to the Guinier-Preston zones, where the vacancies play the part of atoms of smaller dimension. It is known [16] that the migration energy of a divacancy in a f.c.c. lattice is some times smaller than that of a single vacancy, so at low temperatures divacancies begin to migrate earlier than monovacancies. If we suppose that divacancies can create mainly plane clusters and take into account that at low temperatures these divacancies but not single vacancies are only mobile, we can explain consistently the following experimental data.

It is well known [1] that vacancy loops in nickel are observed at irradiation temperatures 250 to 450 °C, which is less than the lower temperature boundary of swelling. Moreover, from [17] follows that decreasing the primary knock-out atom (PKA) energy results in the increase of vacancies entering athermal divacancies. In particular, if PKA can create only two Frenkel pairs, then half of the appearing athermal vacancies will be of the form of divacancies. It means that the vacancy-loop formation probability is largest under electron irradiation.

The second remark is connected with the divacancy influence on the annealing kinetics of defects. As a rule, divacancies are not included in annealing schemes [18, 19]. In rare cases [20 to 24] to make better the agreement with experimental data one takes into account either the thermal divacancy formation, as a result of single vacancy collisions, or varies the monovacancy energy parameters to compensate for divacancy presence. In both cases, they suppose divacancies to have only one configuration.

However, in metals there are several stable divacancy configurations, e.g. two in nickel. According to Table 1 the first configuration is stable up to $kT \leqq 0.07$ eV, i.e. below ≈ 530 °C, the second one up to $kT \leqq 0.03$ eV, i.e. below ≈ 80 °C (this temperature corresponds to the third stage of annealing [23]). It is known [25] that in copper the athermal divacancies $V_2(1)$ and $V_2(2)$ are generated in the ratio $\approx 4:1$, here the index in brackets indicates the number of the coordination sphere. Approximately the same ratio takes place, as our calculations showed, in case of nickel. Heating realizes the following formation–dissociation reactions:

$$V_2(1) \leftrightarrow 2V_2(2) + 4V_2(3) + V_2(4)\,,$$

$$V_2(2) \leftrightarrow 4V_2(1) + 4V_2(3) + 4V_2(4)\,.$$

Apparently, the theories of radiation swelling and adjacent phenomena must include the athermal divacancy formation and changing of their configuration spectrum with temperature. Paying no attention to these phenomena is one of the main reasons which leads to the discrepancy between theories and experiments [26].

4.2 Trivacancy

Four trivacancy configurations and their binding energies in eV are shown in Fig. 2. All the configurations, except the linear or 180° trivacancy, were obtained from the initial [110] divacancy (so-called growth configurations). The linear trivacancy was formed athermally by removing three atoms at once from the crystallite. The configuration shown in Fig. 2a was obtained by getting away the atom which had two nearest vacant sites. As a result of this process, a 60° trivacancy was formed. During

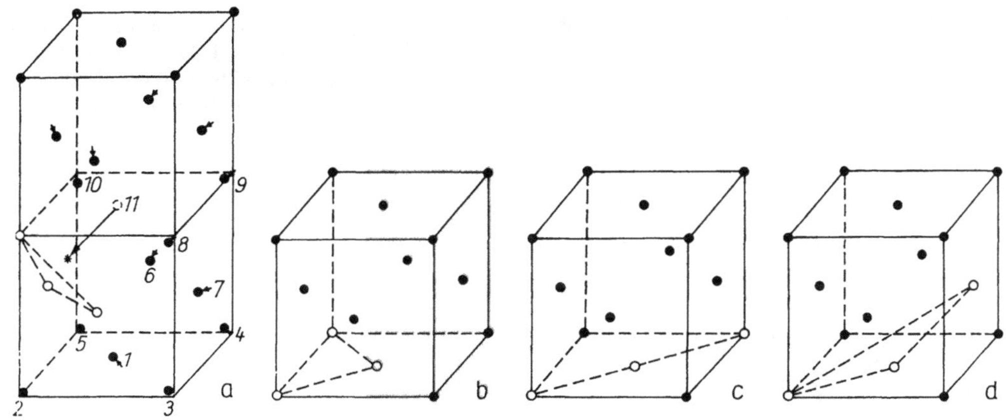

Fig. 2. Four trivacancy configurations. Strongly displaced atoms are shown by an asterisk, the displacement direction by an arrow, small atomic displacements are not represented. ● atom, ○ vacancy. a) $E^B_{3V} = 0.71 \pm 0.03$, b) 0.26 ± 0.02, c) 0.15 ± 0.01, d) 0.11 ± 0.01 eV

the following crystallite relaxation the atom which became equidistant to three vacant sites displaces strongly to the centre of the created 60° trivacancy. The configuration relaxed has the largest binding energy 0.71 eV and represents the tetrahedron consisting of four vacancies whose centre is occupied by the displaced atom. Such lattice relaxation can be called tetrahedral one. As will be shown, the tetrahedral relaxation is observed whenever any lattice atom loses three nearest neighbours in the process of vacancy cluster growth.

The defect structure shown in Fig. 2a (tetrahedral trivacancy) has been first discovered in copper [16], the empirical Morse and Born-Mayer potentials were being used. However, in nickel a 60° trivacancy with a binding energy 0.76 eV was found to be the most stable one, whereas the tetrahedral trivacancy having the binding energy 0.59 eV was only the second in stability [13]. These calculations were made with the empirical Johnson potential so that we suppose the reasons for the discrepancy with our results to be the same as in the case of a divacancy.

4.3 Tetravacancy

Five configurations of this complex are shown in Fig. 3. Among them configurations a, b, c are the growth ones which were obtained from the tetrahedral trivacancy by removing atoms No. 1, 2, and 11, respectively. Configuration c was formed after removing the atom which lay on the 90°-trivacancy plane. The linear tetravacancy was obtained athermally.

Configurations a and b have the largest binding energy being equal in the limits of calculation error to 1.23 eV. Both configurations are combinations of two adjacent tetrahedra, the displaced atom being placed at the centre of each of them. Each tetrahedron was created as a result of the tetrahedral relaxation of atoms which had three nearest vacant sites (Fig. 3a). The analysis of atom displacements near the tetrahedral trivacancy and symmetry considerations lead to the conclusion that the configuration shown in Fig. 3a can be obtained by removing any of eight atoms which have positions equivalent to atom No. 2. Exactly the same configuration was earlier obtained athermally [3] from a vacancy rhomb lying in the (111) plane.

The binding energy of the vacancy tetrahedron configuration shown in Fig. 3e is negative, therefore this configuration is unstable. However, we have not observed

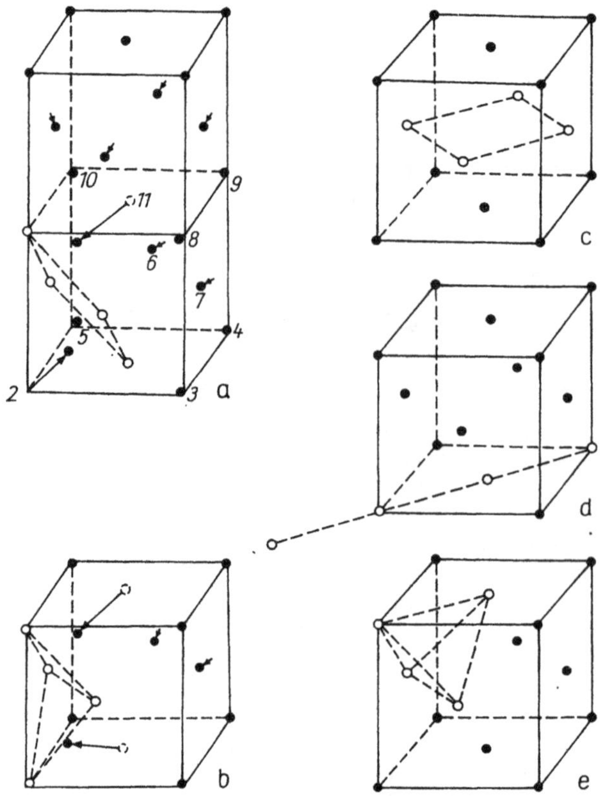

Fig. 3. Five tetravacancy configurations. a) $E^B_{V4} = 1.23 \pm 0.05$, b) $1.23 \pm \pm 0.05$, c) 0.76 ± 0.04, d) $0.22 \pm \pm 0.04$, e) -0.24 ± 0.04 eV

spontaneous transformation of this cluster. Since all the calculations made correspond to absolute zero, it means the existence of a potential barrier for this configuration transformation. However, if this configuration arises athermally, for example at temperature different from zero in collision cascades, it can transform to one of the stable configurations.

4.4 Pentavacancy

Let us consider the evolution of the most stable tetravacancy configuration which joins a single vacancy. The pentavacancy configuration given in Fig. 4a was formed

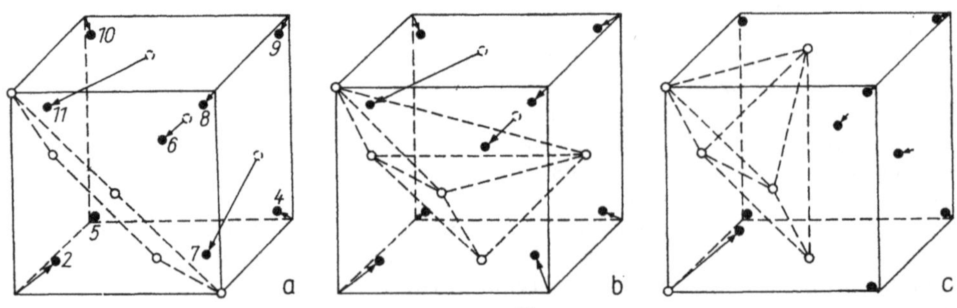

Fig. 4. Three pentavacancy configurations. a) $E^B_{5V} = 2.32 \pm 0.08$, b) 2.28 ± 0.08, c) $1.01 \pm \pm 0.11$ eV

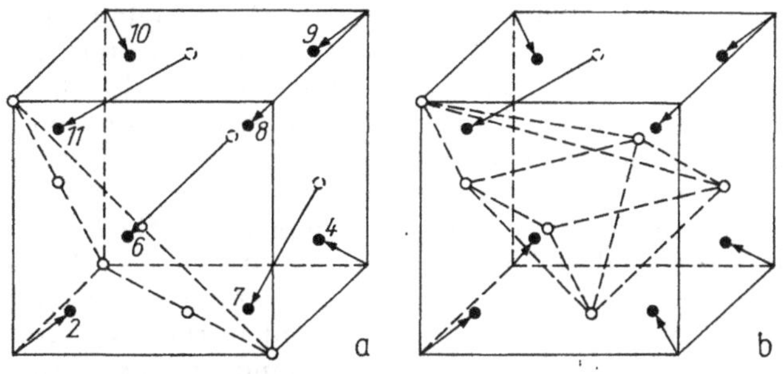

Fig. 5. Two hexavacancy configurations. a) $E^{B}_{6V} = 2.90 \pm 0.09$, b) 2.63 ± 0.08 eV

after removing atom No. 1 which had not two nearest neighbours and lay at the vacancy rhomb plane shown by a dotted line. As a result of removing, three nearest sites near atom No. 7 become vacant, so this atom displaces to the vacancy cluster plane. Atom No. 6 also loses its stability and moves to the cube centre. The arising atomic configuration represents a distorted elementary cell of a b.c.c. lattice having the parameter equal to ≈ 0.76 parameter of the nickel f.c.c. lattice.

If to get away atom No. 7 instead of atom No. 3, then the lattice relaxation leads to a configuration analogous to that considered above (Fig. 4 a, b). Both configurations have binding energies 2.3 eV. It should be noted that the b.c.c. lattice elementary cell formed by displaced atoms has been found earlier in copper [16] from the tetravacancy configuration consisting of two parallel [200] divacancies which lie in a (100) plane.

Fig. 4c shows the pentavacancy obtained by removing atom No. 11. This configuration consists of two adjacent vacancy tetrahedra, one of which contains a displaced atom. Its binding energy is about three times smaller than that of the two previous configurations. One should note that the formation of such a configuration during diffusional growth is unlikely, because in this case a vacancy must diffuse into the stretched region around atom No. 11. However, this configuration can be formed athermally during irradiation.

4.5 Hexavacancy

Two configurations of this cluster are shown in Fig. 5. The first, configuration a, was obtained by removing atom No. 5 from the vacancy disk plane. In this case, atom No. 6 displaces from the cubic cell centre in direction to the new vacancy and locates up the centre of the vacancy triangle newly created in the vacancy cluster plane. Atom No. 9, which now has no three neighbours in the (111) plane nearest to the vacant cluster plane, undergoes in turn a tetrahedral relaxation. Thus, the vacancy cluster collapse is going on, the defect formed under these conditions may be considered as the nucleus of a Frank dislocation loop.

The volume configuration of a hexavacancy (Fig. 5b) can be obtained from the pentavacancy shown in Fig. 4b, if to get away atom No. 6 located in the b.c.c. cell centre. This hexavacancy is a cubic void. Exactly the same configuration arises after removing from a perfect f.c.c. crystallite simultaneously six atoms which form a regular octahedron.

5. Conclusion

It is accepted to assume [1] that during vacancy cluster growth by means of adding single vacancies there is a bifurcation point, which corresponds to a trivacancy. Beyond this point, the vacancy cluster can grow either as a plane one (i.e. vacancy loop nucleus) or as a volume one (i.e. void nucleus). Besides, the alloying which stabilizes one of these forms can change the growth evolution of vacancy clusters [27].

The results of our calculations discussed in this paper and elsewhere [2, 3, 15] compel to refuse this simplified picture. The most stable trivacancy configuration in nickel is a vacancy tetrahedron with an atom in the centre. The most stable relaxed tetravacancy configuration may be considered, in principle, as a nucleus of a Frank dislocation loop. However, the evolution of this configuration depends on the location of a removed neighbouring atom and leads to new transitional structures. On the hexavacancy stage, there form either a cubic void nucleus or a Frank loop continues to grow. Therefore, in nickel there is no definite bifurcation point in the range from divacancy to hexavacancy. Besides, the lattice relaxation around clusters creates a variety of atomic configurations, and it is difficult to foresee what configuration will appear if we take into account only geometrical reasons.

References

[1] YU. V. KONOBEEV, in: Modelirovanie na EVM defektov v kristallakh, FTI Akad. Nauk SSSR, Leningrad 1979 (p. 40).
[2] A. I. MELKER, V. V. SIROTINKIN, and N. L. TARASENKO, Trudy LPI, to be published.
[3] A. I. MELKER and A. A. VASILYEV, Metallofiz. **6**, 3 (1984).
[4] M. L. COHEN and V. HEINE, Solid State Phys. **24**, 37 (1970).
[5] A. O. E. ANIMALU, Phys. Rev. B **8**, 3542 (1973).
[6] A. O. E. ANIMALU, Phys. Rev. B **8**, 3555 (1973).
[7] V. M. SILONOV, M. M. KHRUSHCHEV, and A. A. KATSNELSON, Fiz. Metallov i Metallovedenie **41**, 698 (1976).
[8] K. S. SINGWI, A. SJÖLANDER, M. P. TOSI, and R. H. LAND, Phys. Rev. B **1**, 1044 (1970).
[9] G. NEUMANN, V. TÖLLE, and W. HIRSCHWALD, phys. stat. sol. (b) **54**, 519 (1972).
[10] R. H. RAUTIOAHO, phys. stat. sol. (b) **115**, 95 (1983).
[11] M. W. FINNIS and M. SACHDEW, J. Phys. F **6**, 965 (1976).
[12] A. N. ORLOV and YU. V. TRUSHIN, Energies of Point Defects in Metals, Energoatomizdat, Moscow 1983 (in Russian).
[13] R. A. JOHNSON, Phys. Rev. **152**, 629 (1966).
[14] J. R. BEELER, JR., Adv. Mater. Res. **4**, 295 (1970).
[15] A. I. MELKER, S. N. ROMANOV, and N. L. TARASENKO, Trudy LPI, to be published.
[16] A. C. DAMASK and G. G. DIENES, Point Defects in Metals, Gordon & Breach, New York/London 1963.
[17] A. I. MELKER and S. N. ROMANOV, phys. stat. sol. (b) **126**, 133 (1984).
[18] Z. K. SARALIDZE, Vopr. atom. Nauki i Tekh., Ser. Fiz. rad. Povrezh. rad. Mater. No. 1 (29)/2(30), 158 (1984).
[19] YU. V. KONOBEEV, Vopr. atom. Nauki i Tekh., Ser. Fiz. rad. Povrezh. rad. Mater. No. 1(29)/2(30), 172 (1984).
[20] A. N. ORLOV and YU. V. TRUSHIN, in: Radiatsionnye defekty v metallicheskikh kristallakh Ed. SH. SH. IBRAGIMOV, Izd. Nauka, Alma-Ata 1978 (p. 30).
[21] M. H. YOO, Scripta metall. **13**, 635 (1979).
[22] V. A. PECHENKIN and A. M. STROGOVA, Zh. tekh. Fiz. **50**, 582 (1980).
[23] M. H. YOO, J. Phys. F **11**, L65 (1981).
[24] W. G. WOLFER and A. SIAHMED, Phil. Mag. **A46**, 723 (1982).
[25] A. I. MELKER and S. N. ROMANOV, phys. stat. sol. (b) **126**, 517 (1984).
[26] S. I. GOLUBOV and YU. V. KONOBEEV, Zh. tekh. Fiz., Pisma **7**, 1420 (1981).
[27] A. I. MELKER, A. A. VASILYEV, and N. L. SHISHKIN, in: EVM i modelirovanie defektov v kristallakh, FTI Akad. Nauk SSSR, Leningrad 1982 (p. 116).

(Received May 13, 1985)

phys. stat. sol. (b) 131, 545 (1985)

Subject classification: 13.5.4 and 20.1; 18.2; 18.4

Institute of Physics, Academy of Sciences of the Azerbaidzhan SSR, Baku[1])

Theory of Absorption and Scattering of Light by Singlet–Triplet Magnetic Systems

By

YU. M. SEIDOV and A. M. SULEIMANOV

Dedicated to Prof. Dr. Dr. h. c. Dr. E. h. P. GÖRLICH on the occasion of his 80th birthday

By the multi-time Green's function method the light absorption and scattering in the singlet–triplet systems are investigated by means of the pseudospin formalism. The role of the crystal field for the two-magnon light absorption in the singlet–triplet ferro- and antiferromagnets and for the one-magnon light scattering in the singlet–triplet ferromagnets is elucidated.

Методом многовременных функций Грина исследовано поглощение и рассеяние света в синглет–триплетных системах с помощью псевдоспинового формализма. Выяснена роль кристаллического поля в двухмагнонном поглощении света в синглет–триплетных ферро- и антиферромагнетиках и в одномагнонном рассеянии света в синглет–триплетных ферромагнетиках.

1. Introduction

Interesting observations have been reported on optical processes related with magnons. They are two-magnon absorption [1, 2], magnon sidebands [3, 4], magnon–phonon absorption [5, 6], and scattering of light due to one and two magnons [7, 8], in certain antiferromagnets. An interesting feature in all these measurements is the fact that the electric component of the radiation field couples to the spin system more strongly than its magnetic component. In other words, we have electric dipole moments associated with the spins, with the spins and phonons, and with the spins and excitons, which give rise to the absorption of the radiation mentioned above. Also we have a spin-dependent electric polarizability which gives rise to the scattering of light by spins. Since the dynamical properties of the spin system reflect the spin-dependent electric dipole moments and polarizabilities, one can study the properties of spin systems by means of the above-mentioned optical methods.

The purpose of the present paper is to develop the theories of light absorption and scattering in special magnetically ordered systems, singlet–triplet ferro- and antiferromagnets. Such magnetic orders occur in rare-earth compounds with group-V anions, such as TbSb, TmN, etc.. In these systems the crystal field is comparable to or even dominant over the exchange interaction between the rare-earth ions [9 to 13]. The multi-time Green's function method [14] and the pseudopsin formalism [13] are used to solve the mentioned problem. The role of the crystal field in the two-magnon light absorption in singlet–triplet ferro- and antiferromagnets and in one-magnon light scattering in singlet–triplet ferromagnets is elucidated.

2. Two-Magnon Absorption of Light in Singlet-Triplet Ferromagnets

The Hamiltonian of the singlet–triplet ferromagnet under the external uniform electric field $\boldsymbol{E}$ may be written as

$$H = H_0 - \boldsymbol{E} \cdot \boldsymbol{p} \,,$$

[1]) Prospekt Narimanova 33, 370143 Baku, USSR.

where

$$H_0 = \sum_i V_{\mathrm{cr},i} - \sum_{i,j} J(i-j)\, \boldsymbol{J}_i \cdot \boldsymbol{J}_j\,, \tag{1}$$

$$\boldsymbol{p} = \sum_{i,j} \boldsymbol{\pi}_{ij} \cdot \boldsymbol{J}_i \cdot \boldsymbol{J}_j\,. \tag{2}$$

Here $V_{\mathrm{cr},i}$ is the single-ion crystal field potential which gives a singlet ground state (Γ_1) and a triplet excited state (Γ_4) separated by an energy gap Δ. $\boldsymbol{J}_i$ is the total angular-momentum operator of the rare-earth ion at the i-th site and $J(i-j) > 0$ is the effective exchange interaction between ions at i and j. $\boldsymbol{p}$ is the electric dipole momentum associated with the i, j pair of ionic spins. $\boldsymbol{\pi}_{ij}$ is the electric polarizability vector for the singlet–triplet ferromagnets.

We calculate and analyse the dielectric susceptibility due to the electric dipole moment $\boldsymbol{p}$ in singlet–triplet ferromagnets. To do this, we use the double-time Green function technique. In the first approximation in $\boldsymbol{E}$ the dielectric susceptibility is expressed in terms of the Green function in the following form [14]:

$$\epsilon(\omega) = -i \int\limits_{-\infty}^{+\infty} \mathrm{d}t'\, \mathrm{e}^{i\omega(t-t')} \langle\langle p(t) \mid p(t') \rangle\rangle\,. \tag{3}$$

In the representation of the pseudospin formalism [13], the Hamiltonian (1) and the electric dipole moment (2) take the following form:

$$\begin{aligned} H_0 &= \Delta \sum_i \boldsymbol{S}_i \cdot \boldsymbol{T}_i - \sum_{i,j} J(i-j)\, (\alpha \boldsymbol{S}_i + \beta \boldsymbol{T}_i)\, (\alpha \boldsymbol{S}_j + \beta \boldsymbol{T}_j)\,, \\ \boldsymbol{p} &= \sum_{i,j} \boldsymbol{\pi}_{ij} (\alpha \boldsymbol{S}_i + \beta \boldsymbol{T}_i)\, (\alpha \boldsymbol{S}_i + \beta \boldsymbol{T}_j)\,, \end{aligned} \tag{4}$$

where $\boldsymbol{S}$ and $\boldsymbol{T}$ are the spin $-\frac{1}{2}$ operators introduced by the transformation of $\boldsymbol{J} = \alpha \boldsymbol{S} + \beta \boldsymbol{T}$ with the real coefficients α and β. In particular, for a cubic crystal field α and β are equal to $\frac{1}{2}\left(1 + \frac{4}{3}\sqrt{5}\right)$, $\frac{1}{2}\left(1 - \frac{4}{3}\sqrt{5}\right)$ for praseodymium Pr^{3+}, $\frac{1}{2}\left(1 + 2\sqrt{14}\right)$, $\frac{1}{2}\left(1 - 2\sqrt{14}\right)$) for thulium Tm^{3+} and terbium Tb^{3+} ions, respectively.

Expressing the Hamiltonian (4) in terms of the second-quantized operators by means of the Holstein-Primakoff transformation for the spin operators $\boldsymbol{S}$ and $\boldsymbol{T}$, passing to the momentum representation, and selecting the appropriate diagonalizing transformation for the Hamiltonian (4), for H_0 and $\boldsymbol{p}$ we obtain the following expressions:

$$H_0 = \sum_k (\omega_{k1} c_k^+ c_k + \omega_{k2} d_k^+ d_k)\,,$$

$$\boldsymbol{p} = \sum_k \boldsymbol{\Pi}(\boldsymbol{k})\, (c_k d_{-k} + c_k^+ d_{-k}^+)\,.$$

Here ω_{k1} and ω_{k2} are the energies of the elementary excitations of the singlet–triplet ferromagnets, and $\boldsymbol{\Pi}(\boldsymbol{k})$ assumes the form

$$\boldsymbol{\Pi}(\boldsymbol{k}) = 2(T\beta^2 - S\alpha^2)\, u_k v_k \boldsymbol{\pi}(\boldsymbol{k})\,,$$

where

$$\boldsymbol{\pi}(\boldsymbol{k}) = \sum_{i,j} \boldsymbol{\pi}_{ij}\, \mathrm{e}^{-i\boldsymbol{k}(\boldsymbol{R}_i - \boldsymbol{R}_j)}$$

is the Fourier transform of the vector $\boldsymbol{\pi}_{ij}$, u_k, v_k are the coefficients of the unitary diagonalizing transformation for the Hamiltonian H_0.

For simplicity we introduce the notations

$$\boldsymbol{p} = \sum_k \boldsymbol{\Pi}(\boldsymbol{k})\, p_k\,; \qquad p_k = c_k d_{-k} + c_k^+ d_{-k}^+\,. \tag{5}$$

Substituting (5) into (3), the dielectric susceptibility can be written as

$$\epsilon(\omega) = -2\pi i \sum_k \boldsymbol{\Pi}(\boldsymbol{k})\, G(p_k;\omega)\,, \tag{6}$$

where $G(p_k;\omega)$ is the Fourier transform of the double-time Green function,

$$G\big(p_k(\tau_1);\ p_k(\tau_2)\big) = -i\theta(\tau_1-\tau_2)\,\langle[p_k(\tau_1), p_k(\tau_2)]\rangle\,.$$

From the equation of motion for the Green function $G(p_k;\omega)$, we obtain

$$G(p_k;\omega) = \frac{(2\pi)^{-1}\boldsymbol{\Pi}(\boldsymbol{k})\,\langle c_k^+c_k + d_k d_k^+\rangle}{\omega - (\omega_{k1}+\omega_{k2})}. \tag{7}$$

Substituting (7) into (6), we have

$$\epsilon(\omega) = -i\sum_k \frac{\boldsymbol{\Pi}(\boldsymbol{k})\cdot\boldsymbol{\Pi}(\boldsymbol{k})\,\langle c_k^+c_k + d_k d_k^+\rangle}{\omega-(\omega_{k1}+\omega_{k2})}. \tag{8}$$

From (8) we see that absorption of light by singlet–triplet ferromagnet is due to the simultaneous emission of two magnons with opposite $\boldsymbol{k}$-vectors in different branches of the spectrum of the singlet–triplet ferromagnet.

3. Two-Magnon Absorption of Light in Singlet–Triplet Antiferromagnets

The Hamiltonian of the singlet–triplet antiferromagnet under the external uniform electric field $\boldsymbol{E}$ may be written as

$$H = H_0 - \boldsymbol{E}\cdot\boldsymbol{p}\,,$$

where

$$H_0 = \sum_i V_{\text{cr},i} - \sum_{i,j} J(i-j)\,\boldsymbol{J}_i\cdot\boldsymbol{J}_j - D\Big[\sum_i (J_i^z)^2 + \sum_j (J_j^z)^2\Big] \tag{9}$$

is the energy operator for the system under study,

$$\boldsymbol{p} = \sum_{i,j} \boldsymbol{\pi}_{ij}\boldsymbol{J}_i\cdot\boldsymbol{J}_j \tag{10}$$

is the electric dipole moment associated with the i, j pair of ionic spins. Here D is the anisotropy constant, $\boldsymbol{\pi}_{ij}$ is the electric polarizability vector for the singlet–triplet antiferromagnet. In this case $J(i-j) < 0$, i.e. the system orders antiferromagnetically.

We calculate and analyse the dielectric susceptibility due to the electric dipole moment $\boldsymbol{p}$ in the singlet–triplet antiferromagnet. In the representation of the pseudo-spin formalism the Hamiltonian (9) and the electric dipole moment (10) take the following form:

$$\begin{aligned} H_0 &= \Delta\Big(\sum_i \boldsymbol{S}_i\boldsymbol{T}_i + \sum_j \boldsymbol{S}_j\boldsymbol{T}_j\Big) - \sum_{i,j} J(i-j)\,(\alpha\boldsymbol{S}_i+\beta\boldsymbol{T}_i)(\alpha\boldsymbol{S}_j+\beta\boldsymbol{T}_j) - \\ &\quad - D\Big[\sum_i (\alpha S_i^z+\beta T_i^z)^2 + \sum_j (\alpha S_j^z + \beta T_j^z)^2\Big]\,, \end{aligned} \tag{11}$$

$$\boldsymbol{p} = \sum_{i,j}\boldsymbol{\pi}_{ij}(\alpha\boldsymbol{S}_i+\beta\boldsymbol{T}_i)(\alpha\boldsymbol{S}_j+\beta\boldsymbol{T}_j)\,.$$

Expressing (11) in the second-quantized operators through the use of the Holstein-Primakoff transformation for spin operators $\boldsymbol{S}$ and $\boldsymbol{T}$ (taking into account that the $\boldsymbol{S}$- and $\boldsymbol{T}$-spins on different sites are ordered antiferromagnetically), turning to the momentum representation and selecting the appropriate diagonalizing transformation for the Hamiltonian (11), for H_0 and $\boldsymbol{p}$ we obtain the following expressions:

$$\begin{aligned} H_0 &= \sum_k (\omega_{k1}\alpha_k^+\alpha_k + \omega_{k2}\beta_k^+\beta_k + \omega_{k3}j_k^+j_k + \omega_{k4}\xi_k^+\xi_k)\,, \\ \boldsymbol{p} &= \sum_k (\boldsymbol{\Pi}_{k1}\boldsymbol{p}_{k1} + \boldsymbol{\Pi}_{k2}p_{k2} + \boldsymbol{\Pi}_{k3}p_{k3} + \boldsymbol{\Pi}_{k4}p_{k4})\,. \end{aligned} \tag{12}$$

Here ω_{k1}, ω_{k2}, ω_{k3}, and ω_{k4} are the energies of elementary excitations of the singlet–triplet antiferromagnets. Coefficients $\boldsymbol{\Pi}_{k1}, \boldsymbol{\Pi}_{k2}, \boldsymbol{\Pi}_{k3}, \boldsymbol{\Pi}_{k4}$ and operators $p_{k1}, p_{k2}, p_{k3}, p_{k4}$ take the form

$$\boldsymbol{\Pi}_{k1} = \boldsymbol{\Pi}_{k2} = [(\alpha^2 S + \beta^2 T)\,(u_{1k}u_{2k} + v_{1k}v_{2k}) + \alpha\beta\sqrt{ST}\,(u_{1k}^2 + u_{2k}^2 + v_{1k}^2 + v_{2k}^2)]\,\boldsymbol{\pi}(\boldsymbol{k})\,,$$

$$\boldsymbol{\Pi}_{k3} = [2\alpha\beta\sqrt{ST}\,(u_{1k}u_{2k} + v_{1k}v_{2k}) + \alpha^2 S(u_{2k}^2 + v_{1k}^2) + \beta^2 T(u_{1k}^2 + v_{2k}^2)]\,\boldsymbol{\pi}(\boldsymbol{k})\,,$$

$$\boldsymbol{\Pi}_{k4} = [2\alpha\beta\sqrt{ST}\,(u_{1k}u_{2k} + v_{1k}v_{2k}) + \alpha^2 S(u_{1k}^2 + v_{2k}^2) + \beta^2 T(u_{2k}^2 + v_{1k}^2)]\,\boldsymbol{\pi}(\boldsymbol{k})\,,$$

$$p_{k1} = \alpha_k\beta_{-k} + \alpha_k^+\beta_{-k}^+\,, \qquad p_{k2} = j_k\xi_{-k} + j_k^+\xi_{-k}^+\,,$$

$$p_{k3} = j_k\beta_{-k} + j_k^+\beta_{-k}^+\,, \qquad p_{k4} = \alpha_k\xi_{-k} + \alpha_k^+\xi_{-k}^+\,,$$

where u_{1k}, u_{2k}, v_{1k}, v_{2k} are the coefficients of the unitary diagonalizing transformation for the Hamiltonian H_0.

Thus, the collective excitation spectrum ω_{kr} of the system under consideration consists of four branches, two of them are the magnon branches ($r = 1, 2$) and the other two are the "magnetic excitaton" branches ($r = 3, 4$). On the other hand, the electric dipole moment (12) describes the following four possible absorption processes of light by collective excitations in singlet–triplet antiferromagnets:

(i) absorption of light with simultaneous excitation of two magnons, one from the branch $r = 1$ and the other from the branch $r = 2$;

(ii) absorption of light with simultaneous excitation of two "excitons", one from the branch $r = 3$ and the other from the branch $r = 4$;

(iii) absorption of light with simultaneous excitation of one magnon from the branch $r = 2$ and one "exciton" from the branch $r = 3$;

(iv) absorption of light with simultaneous excitation of one magnon from the branch $r = 1$ and one "exciton" from the branch $r = 4$, with opposite wave vectors.

Thus, due to the presence of four branches in the spectrum the possibility of two-magnon light absorption in the singlet–triplet antiferromagnets is larger than that in the singlet–triplet ferromagnets.

To calculate the dielectric susceptibility corresponding to each of the above-mentioned four processes we use (3). For the sake of simplicity we introduce the following notations:

$$\boldsymbol{p}_{(1)} = \sum_k \boldsymbol{\Pi}_{k1}p_{k1}\,, \tag{13}$$

$$\boldsymbol{p}_{(2)} = \sum_k \boldsymbol{\Pi}_{k2}p_{k2}\,, \tag{14}$$

$$\boldsymbol{p}_{(3)} = \sum_k \boldsymbol{\Pi}_{k3}p_{k3}\,, \tag{15}$$

$$\boldsymbol{p}_{(4)} = \sum_k \boldsymbol{\Pi}_{k4}p_{k4}\,. \tag{16}$$

Substituting (13) into (3) the dielectric susceptibility corresponding to the light absorption process (1) in the singlet–triplet antiferromagnets, can be written as

$$\epsilon_{(1)}(\omega) = -2\pi i \sum_k \boldsymbol{\Pi}_{k1} G(p_{k1};\omega)\,, \tag{17}$$

where $G(p_{k1}; \omega)$ is the Fourier transform of the double-time Green function

$$G(p_{k1}(\tau_1); p_{k1}(\tau_2)) = -i\theta(\tau_1 - \tau_2) \langle [p_{k1}(\tau_1), p_{k1}(\tau_2)] \rangle .$$

From the equation of motion for the Green function $G(p_{k1}; \omega)$ we obtain

$$G(p_{k1}; \omega) = \frac{(2\pi)^{-1} \Pi_{k1} \langle \alpha_k^+ \alpha_k + \beta_k \beta_k^+ \rangle}{\omega - (\omega_{k1} + \omega_{k2})}. \tag{18}$$

Substituting (18) into (17), we have

$$\epsilon_{(1)}(\omega) = -i \sum_k \frac{\Pi_{k1} \Pi_{k1} \langle \alpha_k^+ \alpha_k + \beta_k \beta_k^+ \rangle}{\omega - (\omega_{k1} + \omega_{k2})}. \tag{19}$$

It is seen from (19) that two magnons take part in the given light absorption process (I) in the singlet–triplet antiferromagnet, i.e. absorption of the electromagnetic radiation is due to excitation of two magnons, one from the branch 1 (with energy ω_{k1}) and the other from the branch 2 (with energy ω_{k2}).

Now, substituting (14) into (3) yields the expression for the dielectric susceptibility $\epsilon_{(2)}(\omega)$, corresponding to light absorption process (II) in the singlet–triplet antiferromagnet,

$$\epsilon_{(2)}(\omega) = -2\pi i \sum_k \Pi_{k2} G(p_{k2}; \omega) . \tag{20}$$

From the equation of motion the Green function $G(p_{k2}; \omega)$ can be calculated,

$$G(p_{k2}; \omega) = \frac{(2\pi)^{-1} \Pi_{k2} \langle j_k^+ j_k + \xi_k \xi_k^+ \rangle}{\omega - (\omega_{k3} + \omega_{k4})}. \tag{21}$$

From (20) and (21), for the dielectric susceptibility $\epsilon_{(2)}(\omega)$ we obtain

$$\epsilon_{(2)}(\omega) = -i \sum_k \frac{\Pi_{k2} \Pi_{k2} \langle j_k^+ j_k + \xi_k \xi_k^+ \rangle}{\omega - (\omega_{k3} + \omega_{k4})}. \tag{22}$$

As seen from (22), two "excitons" take part in the given light absorption process (II) in the singlet–triplet antiferromagnet, i.e. absorption of the electromagnetic radiation is due to excitation of two "excitons", one from the branch 3 (with energy ω_{k3}) and the other from the branch 4 (with energy ω_{k4}).

By analogous calculations, expressions for the dielectric susceptibilites, corresponding to the light absorption processes (III) and (IV) in the singlet–triplet antiferromagnet, can be found as

$$\epsilon_{(3)}(\omega) = -i \sum_k \frac{\Pi_{k3} \Pi_{k3} \langle \beta_k^+ \beta_k + j_k j_k^+ \rangle}{\omega - (\omega_{k2} + \omega_{k3})}, \tag{23}$$

$$\epsilon_{(4)}(\omega) = -i \sum_k \frac{\Pi_{k4} \Pi_{k4} \langle \alpha_k^+ \alpha_k + \xi_k \xi_k^+ \rangle}{\omega - (\omega_{k1} + \omega_{k4})}. \tag{24}$$

4. One-Magnon Light Scattering by Singlet–Triplet Ferromagnets

In this section we describe a theory of one-magnon light scattering in singlet–triplet ferromagnets. Spin–orbit interaction in these systems is stronger than that in the usual magnetically ordered systems, that is why the spin–orbit mechanism is considered as the fundamental mechanism for one-magnon light scattering in this case. Within this mechanism the electric vector of the light wave interacts indirectly with the spin system via the spin–orbit interaction.

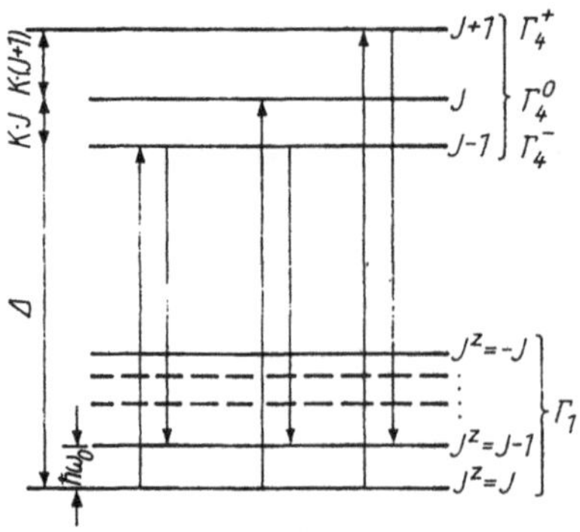

Fig. 1. Energy levels to illustrate the electric-dipole Raman mechanism. Δ is normally very much larger than KJ and $\hbar\omega_0$

Fig. 1 shows the level scheme for the rare-earth ion Tm^{3+} or Tb^{3+} in an octahedral crystal field. The ground state is split into $(2J + 1)$ components by the exchange field, and the excited state is split into three components (Γ_4^-, Γ_4^0, Γ_4^+) by the spin–orbit interaction. The collective excitations in such systems are linear combinations of the excitations of the individual ions from the $J^z = J$ to the $J^z = J - 1$ states. The Raman transition for a single ion can proceed by a pair of allowed electric–dipole transitions via the excited state. Such pairs are indicated by the vertical arrows in Fig. 1; each of the three spin–orbit-split excited states can act as the virtual intermediate state in the double transition. In one of these transitions a quantum of the excited radiation is absorbed and in the other a scattered quantum is emitted. Fig. 1 shows the Stokes scattering only. The energy difference between the $J^z = J$ and $J^z = J - 1$ states is denoted by $\hbar\omega_0$ in Fig. 1, which corresponds to the magnon excitation energy.

The electric-dipole interaction between the radiation and the electrons of the magnetic ions is given by

$$H_{\text{ed}} = -e \sum_i \boldsymbol{E}_i \cdot \boldsymbol{r}_i \,.$$

The radiative electromagnetic field in a crystal is specified by the vector potential $\boldsymbol{A}_i$,

$$\boldsymbol{E}_i = -\frac{1}{c} \frac{\partial \boldsymbol{A}_i}{\partial t} .$$

In the second-quantized notation,

$$\boldsymbol{A}_i = c \sum_k \left(\frac{2\pi h}{V \omega_k \eta_k^2} \right)^{1/2} \varepsilon_k \exp\,(i\boldsymbol{k} \cdot \boldsymbol{r}_i)\,(a_k + a_k^+) \,,$$

where η_k is the refractive index at the frequency ω_k of the radiation, $\boldsymbol{k}$ and ε_k are the wave vector and polarization vector of the radiation, respectively, V the crystal volume, and a_k^+ and a_k are the photon creation and destruction operators. Considering the three possible intermediate states according to Fleury and Loudon [15], the total matrix element for the scattering process can be written as

$$M_i = \frac{\pi h e^2 (\omega_1 \omega_2)^{1/2}\, K J^{1/2}}{\eta_1 \eta_2 V} \left[\frac{1}{(\Delta - \hbar\omega_1)^2} - \frac{1}{(\Delta + \hbar\omega_2)^2} \right] \times$$
$$\times\, (\varepsilon_1^z \varepsilon_2^+ - \varepsilon_1^+ \varepsilon_2^z)\, \langle 0, 0 |\, r_i^z \,| 1, 0 \rangle\, \langle 1, -1 |\, r_i^- \,| 0, 0 \rangle \,, \tag{25}$$

where ω_1, ω_2 and ε_1, ε_2 are the frequencies and polarization vectors of the incident and scattered light waves, respectively, η_1 is the refractive index at the incident frequency

ω_1, η_2 the refractive index at the scattered frequency ω_2, K the excited state spin-orbit coupling, Δ the energy gap between the ground state and the first excited state due to the action of the crystal field, $\varepsilon^+ = \varepsilon^x + i\varepsilon^y$, $r^- = x - iy$, in the bra and ket vectors the numbers indicate the value of J and J^z, respectively.

The matrix element M_i connects the $J^z = J$ ground state to the $J^z = J - 1$ state. It can therefore be written as an operator linear in J_i^-, using the result

$$J^- |J, J\rangle = (2J)^{1/2} |J, J - 1\rangle . \tag{26}$$

The matrix element must now be summed over all the magnetic ions in the crystal. The result is conveniently written in the form of a spin-operator Hamiltonian (Stokes scattering Hamiltonian)

$$H_{\text{Stokes}} = \Gamma \sum_i (E_1^z E_2^+ - E_1^+ E_2^z) J_i^- , \tag{27}$$

where

$$\Gamma = \frac{e^2 K}{2^{3/2}} \left[\frac{1}{(\Delta - \hbar\omega_1)^2} - \frac{1}{(\Delta + \hbar\omega_2)^2}\right] \langle 0, 0| rz |1, 0\rangle \langle 1, -1| r^- |0, 0\rangle .$$

The result analogous to (27) for the anti-Stokes scattering can be derived straightforwardly and lead to a spin-Hamiltonian

$$H_{\text{anti-Stokes}} = -\Gamma \sum_i (E_1^z E_2^- - E_1^- E_2^z) J_i^+ .$$

Now we can find the differential cross-section for Stokes scattering of light by singlet–triplet ferromagnets. Using the Hamiltonian (27) according to Loudon [16], the differential cross-section can be written as follows:

$$\left(\frac{d^2h}{d\Omega\, d\omega_2}\right)_{\text{Stokes}} = -\frac{2\hbar\omega_1\omega_2^3\eta_2}{c^4 V \eta_1} \Gamma^2 |\varepsilon_1^z\varepsilon_2^+ - \varepsilon_1^+\varepsilon_2^z|^2 \sum_{i,i'} \operatorname{Im} G(a_2 J_i^+ ; a_2^+ J_{i'}^+) , \tag{28}$$

where a_2^+ and a_2 are the creation and destruction operators of the scattered photons. The Green function $G(a_2 J_i^+ ; a_2^+ J_{i'}^-)$ can be determined using the equation of motion and the random-phase decoupling procedure of Tyablikov [14]. The Hamiltonian of the singlet–triplet ferromagnet is written in the pseudospin formalism,

$$H_0 = \Delta \sum_i \boldsymbol{S}_i \cdot \boldsymbol{T}_i - \sum_{i,j} J(i - j) (\alpha \boldsymbol{S}_i + \beta \boldsymbol{T}_i) (\alpha \boldsymbol{S}_j + \beta \boldsymbol{T}_j) -$$
$$- g\mu_B h_0 \sum_i (\alpha S_i^z + \beta T_i^z) + \hbar\omega_2 a_2^+ a_2 .$$

Here the last term represents the energy of the scattered photons. As a result, the calculations for the Green function $G(a_2 J_i^+ ; a_2^+ J_{i'}^-)$ gives

$$G(a_2 J_i^+ ; a_2^+ J_{i'}^-) =$$
$$= \frac{1}{2\pi N} \sum_k \frac{e^{ik(r_i - r_{i'})}}{(\hbar\omega_1 - \hbar\omega_2 - E_{1k}) (\hbar\omega_1 - \hbar\omega_2 - E_{2k})} \times$$
$$\times \{[\alpha^2(\hbar\omega_1 - \hbar\omega_2 - B_k) + \alpha\beta D_k] \langle S_i^+ S_{i'}^- \rangle_k +$$
$$+ [\alpha\beta(\hbar\omega_1 - \hbar\omega_2 - A_k) + \alpha^2 D_k] \langle T_i^+ S_{i'}^- \rangle_k +$$
$$+ [\alpha\beta(\hbar\omega_1 - \hbar\omega_2 - B_k) + \beta^2 D_k] \langle S_i^+ T_{i'}^- \rangle_k +$$
$$+ [\beta^2(\hbar\omega_1 - \hbar\omega_2 - A_k) + \alpha\beta D_k] \langle T_i^+ T_{i'}^- \rangle_k\} , \tag{29}$$

where

$$E_{1k} = \frac{1}{2} \{(A_k + B_k) + [(A_k + B_k)^2 + 4D_k^2]^{1/2}\} ,$$
$$E_{2k} = \frac{1}{2} \{(A_k + B_k) - [(A_k + B_k)^2 + 4D_k^2]^{1/2}\}$$

are the energies of the collective excitations of the singlet–triplet ferromagnets [17] and

$$A_k = 2\alpha^2[J(0) - J(k)] \langle S^z \rangle - \Delta \langle T^z \rangle + 2\alpha\beta J(0) \langle T^z \rangle + \alpha g \mu_B h_0 ,$$
$$B_k = 2\beta^2(J(0) - J(k)) \langle T^z \rangle - \Delta \langle S^z \rangle + 2\alpha\beta J(0) \langle S^z \rangle + \beta g \mu_B h_0 ,$$
$$D_k = \Delta \langle S^z \rangle - 2\alpha\beta \langle S^z \rangle J(k) = \Delta \langle T^z \rangle - 2\alpha\beta \langle T^z \rangle J(k) ,$$
$$J(k) = \sum_{\Delta} J(\Delta) \, e^{-ik\Delta} , \qquad J(0) = \sum_{\Delta} J(\Delta) .$$

Calculating the correlation functions $\langle S_i^+ S_{i'}^- \rangle_k$, $\langle T_i^+ S_{i'}^- \rangle_k$, $\langle S_i^+ T_{i'}^- \rangle_k$, $\langle T_i^+ T_{i'}^- \rangle_k$ and substituting these expressions into (29), for the sum $\sum_{i,i'} G(a_2 J_i^+ ; a_2^+ J_{i'}^-)$ we obtain

$$\sum_{i,i'} G(a_2 J_i^+ ; a_2^+ J_{i'}^-) =$$
$$= \frac{N}{\pi} \left[\frac{\langle \alpha^2 S^z + \beta^2 T^z \rangle D_0^2 + \alpha\beta D_0 \langle S^z + T^z \rangle [(E_{10} - (A_0 + B_0)]}{(\hbar\omega_1 - \hbar\omega_2 - E_{10}) (-E_{10} + E_{20}) [\exp(-E_{10}/k_B T) - 1]} + \right.$$
$$\left. + \frac{\langle \alpha^2 S^z + \beta^2 T^z \rangle D_0^2 + \alpha\beta D_0 \langle S^z + T^z \rangle [(E_{20} - (A_0 + B_0)]}{(\hbar\omega_1 - \hbar\omega_2 - E_{20}) (-E_{20} + E_{10}) [\exp(-E_{20}/k_B T) - 1]} \right], \tag{30}$$

where N is the number of the unit cells containing one spin each, $\langle S^z \rangle$ and $\langle T^z \rangle$ are the expectation values of the z-component of single spins $\boldsymbol{S}$ and $\boldsymbol{T}$, respectively.

Substituting (30) into (28), the differential cross-section for the Stokes scsattering of light by singlet–triplet ferromagnets is obtained in the form

$$\left(\frac{d^2 h}{d\Omega \, d\omega_2} \right)_{\text{Stokes}} =$$
$$= \hbar F_+ \left\{ \frac{\langle \alpha^2 S^z + \beta^2 T^z \rangle D_0^2 + \alpha\beta D_0 \langle S^z + T^z \rangle \left(E_{10} - (A_0 + B_0) \right)}{(E_{20} - E_{10}) [\exp(-E_{10}/k_B T) - 1]} \times \right.$$
$$\times \delta(\hbar\omega_1 - E_{10} - \hbar\omega_2) +$$
$$+ \frac{\langle \alpha^2 S^z + \beta^2 T^z \rangle D_0^2 + \alpha\beta D_0 \langle \alpha S^z + T^z \rangle \left(E_{20} - (A_0 + B_0) \right)}{(E_{10} - E_{20}) [\exp(-E_{20}/k_B T) - 1]} \times$$
$$\left. \times \delta(\hbar\omega_1 - E_{20} - \hbar\omega_2) \right\}, \tag{31}$$

where

$$F_+ = \frac{2N\omega_1\omega_2^3\eta_2}{c^4 V \eta_1} \Gamma^2 |\varepsilon_1^z \varepsilon_2^+ - \varepsilon_1^+ \varepsilon_2^z|^2 .$$

It is seen from (31) that the Stokes scattering spectrum of the singlet–triplet ferromagnets contains two peaks with the frequency shifts $E_{10}/\hbar$ and $E_{20}/\hbar$ which are independent of temperature. For the Stokes scattered light with the frequency ω_2 given by $\omega_2 = \omega_1 - (E_{10}/\hbar)$, the integrated cross-section obtained by integrating (31) over ω_2 is

$$\left(\frac{dh_1}{d\Omega} \right)_{\text{Stokes}} = \frac{F_+ S_1}{1 - \exp(-E_{10}/k_B T)} . \tag{32}$$

For the Stokes scattered light with the frequency ω_2 given by $\omega_2 = \omega_1 - (E_{20}/\hbar)$, the integrated cross-section is obtained in the form

$$\left(\frac{\mathrm{d}h_2}{\mathrm{d}\Omega}\right)_{\text{Stokes}} = \frac{F_+ S_2}{1 - \exp\left(-E_{20}/k_\mathrm{B}T\right)}, \tag{33}$$

where

$$S_1 = \frac{\langle \alpha^2 S^z + \beta^2 T^z \rangle\, D_0^2 + \alpha\beta D_0 \langle S^z + T^z \rangle\, [E_{10} - (A_0 + B_0)]}{E_{10} - E_{20}},$$

$$S_2 = \frac{\langle \alpha^2 S^z + \beta^2 T^z \rangle\, D_0^2 + \alpha\beta D_0 \langle S^z + T^z \rangle\, [E_{20} - (A_0 + B_0)]}{E_{20} - E_{10}}.$$

It is interesting to note that the integrated intensities (32) and (33) vary considerably with temperature.

In a similar way, for the anti-Stokes scattered light with the frequency ω_2 given by $\omega_2 = \omega_1 + E_{10}/\hbar$, the integrated cross-section can be found in the following form:

$$\left(\frac{\mathrm{d}h_1}{\mathrm{d}\Omega}\right)_{\text{anti-Stokes}} = \frac{F_- S_1}{\exp\left(E_{10}/k_\mathrm{B}T\right) - 1}. \tag{34}$$

For the anti-Stokes scattered light with the frequency ω_2 given by $\omega_2 = \omega_1 + E_{20}/\hbar$, the integrated cross-section is obtained as follows:

$$\left(\frac{\mathrm{d}h_2}{\mathrm{d}\Omega}\right)_{\text{anti-Stokes}} = \frac{F_- S_2}{\exp\left(E_{20}/k_\mathrm{B}T\right) - 1}, \tag{35}$$

where

$$F_- = \frac{2N\omega_1\omega_2^3\eta_2}{c^4 V \eta_1}\, \Gamma^2 \left|\varepsilon_1^z \varepsilon_2^- - \varepsilon_1^- \varepsilon_2^z\right|^2 .$$

5. Conclusion

Thus, unlike the usual ferromagnets, the possibility of two-magnon light absorption by singlet–triplet ferromagnets is related to the role of the crystal field available in these systems. On the other hand, the presence of the crystal field leads to a higher probability of two-magnon light absorption in the singlet–triplet antiferromagnets than in the singlet–triplet ferromagnets. It is probable that the analysis of (8), (19), (22) to (24) for the light absorption spectrum in the singlet–triplet systems will provide more comprehensive information on the crystal field in these systems. What is more, the crystal field in the singlet–triplet ferromagnets produces an effect on the light scattering by collective excitations of these systems. Using the analysis of (32) to (35) obtained for the Stokes and anti-Stokes light scattering spectra in the singlet–triplet ferromagnets as the basis, we can gain more detailed information on the peculiarities of the collective excitations in these systems.

References

[1] J. W. Halley and J. Silvera, Phys. Rev. Letters 15, 654 (1965).
[2] S. J. Allen, R. Loudon, and P. L. Richards, Phys. Rev. Letters 16, 463 (1966).
[3] R. L. Greene, D. D. Sell, W. M. Yen, A. L. Schawlow, and R. M. White, Phys. Rev. Letters 15, 565 (1965).

[4] P. G. Russell, D. S. McClure, and J. W. Stout, Phys. Rev. Letters **16**, 776 (1966).
[5] R. Newman and R. M. Cherenko, Phys. Rev. **114**, 1507 (1959).
[6] A. Tsuchida, J. Phys. Soc. Japan **21**, 2497 (1966).
[7] P. A. Fleury, S. P. S. Porto, L. E. Cheesman, and H. J. Guggenheim, Phys. Rev. Letters **17**, 84 (1966).
[8] P. A. Fleury, S. P. S. Porto, and R. Loudon, Phys. Rev. Letters **18**, 658 (1967).
[9] G. T. Trammel, J. appl. Phys. **31**, 362 (1960).
[10] G. T. Trammel, Phys. Rev. **131**, 932 (1963).
[11] G. J. Busch, J. appl. Phys. **38**, 1386 (1967).
[12] O. Vogt and B. K. Cooper, J. appl. Phys. **39**, 1202 (1968).
[13] M. Blume and Y. Y. Hsieh, Phys. Rev. B **6**, 2684 (1972).
[14] S. V. Tyablikov, Methods of Quantum Theory of Magnetism, Izd. Nauka, Moscow 1965 (in Russian).
[15] P. A. Fleury and R. Loudon, Phys. Rev. **166**, 514 (1968).
[16] R. Loudon, J. Phys. C **3**, 872 (1970).
[17] Yu. M. Seidov and A. M. Suleimanov, Fiz. nizk. Temp. **2**, 253 (1976).

(Received April 23, 1985)

phys. stat. sol. (b) **131**, 555 (1985)

Subject classification: 5 and 18.3; 21.1; 21.4

Institute of Nuclear Physics, Moscow State University[1])

The Magnetic Hyperfine Field for ^{119}Sn Atoms in GdMnSi Intermetallic Compound

Zero Contribution from 3d Moments

By

N. N. DELYAGIN, V. I. KRYLOV, G. T. MUJIRI,
V. I. NESTEROV, and S. I. REIMAN

Dedicated to Prof. Dr. Dr. h. c. Dr. E. h. P. GÖRLICH on the occasion of his 80th birthday

The magnetic hyperfine interaction for ^{119}Sn impurity atoms in the GdMnSi host has been studied by Mössbauer spectroscopy. The hyperfine field for Sn atoms at 5 K is $B_{hf} = (-27.6 \pm 0.3)$ T. The temperature dependences of B_{hf} and magnetization are practically identical. As follows from the comparison of the results to data for GdCoSi host and an analysis of the satellite structure of the spectrum in $Gd_{0.95}Y_{0.05}MnSi$ host, the contribution to the transferred hyperfine field from 3d moments in GdMnSi vanishes. Both experimental results and obtained empirical function of the radial dependence of the gadolinium contribution to B_{hf} emphasize the short-range nature of the hyperfine interaction for Sn impurity atoms in Gd-based magnetic alloys.

Методом мессбауэровской спектроскопии исследовано магнитное сверхтонкое взаимодействие для примесных атомов ^{119}Sn в матрице GdMnSi. При 5 K сверхтонкое поле для атомов Sn равно $B_{CT} = (-27{,}6 \pm 0{,}3)$ T. Температурные зависимости сверхтонкого поля и намагниченности практически одинаковы. Из сравнения результатов с данными для матрицы GdCoSi и из анализа сателлитной структуры спектра в сплаве $Gd_{0.95}Y_{0.05}MnSi$ следует, что в GdMnSi вклад в сверхтонкое поле от 3d-моментов равен нулю. Экспериментальные результаты совместно с найденной эмпирической функцией радиальной зависимости вкладов в B_{CT} указывают на локальный характер сверхтонкого взаимодействия для примесных атомов Sn в магнитных сплавах гадолиния.

1. Introduction

In previous papers [1, 2] we have reported on first results of the Mössbauer investigations of the magnetic hyperfine interaction (HFI) for ^{119}Sn impurity atoms in gadolinium intermetallic compounds. It has been shown that the main features of the HFI for Sn impurity atoms in Gd- and 3d-based ferromagnets are substantially different. In Mn-, Fe-, or Co-based alloys the hyperfine fields for Sn atoms can be represented as the sum of large contributions of opposite sign. The nearest magnetic neighbours of the Sn atom are responsible for the negative contribution to the hyperfine field (B_{hf}), while the magnetic atoms in the more remote spheres give the overall positive one [3]. As a result, for Sn atoms both positive and negative hyperfine fields occur equally frequently.

On the contrary, for Sn impurity atoms in the ferromagnetic Gd-based alloys only rather large negative hyperfine fields were observed ($B_{hf} = -30$ to -40 T). It has been found that there are no two contributions in this case, and B_{hf} is mostly determined by the nearest magnetic neighbours. Furthermore, with increasing Sn–Gd distance, the negative contribution to B_{hf} decreases sharply [2]. For Sn atoms in Gd-based alloys there is no positive contribution to B_{hf} which competes with the negative contribution from the nearest neighbours as in transition metal alloys. The

[1]) Moscow 117234, USSR.

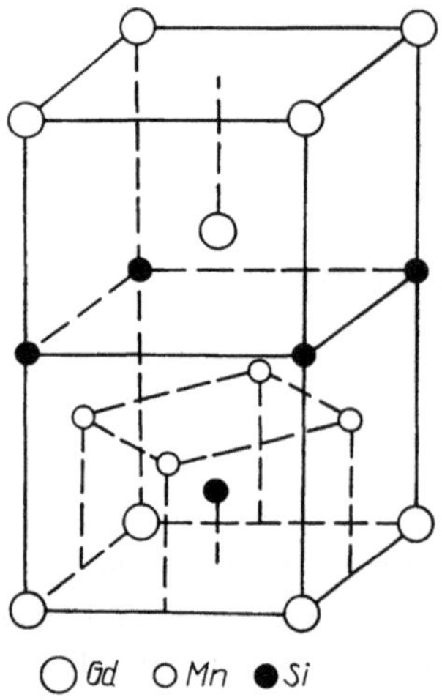

Fig. 1. Crsytal structure of GdMnSi

sharp difference between the features of the HFI for Sn in two metallic magnetic systems was unexpected and difficult to explain within the framework of the traditional HFI models.

In this connection it seems important to study the HFI for Sn impurity atoms in magnetic alloys containing both Gd ions and magnetic 3d atoms. It is reasonable to suppose that the investigation of systems like this would be very useful for an elucidation of the properties of metallic magnets which are responsible for the peculiarities of HFI described above.

To begin the research within the framework of this programme, the ordered ferrimagnet GdMnSi was chosen. The properties of this intermetallic compound have been recently studied by Kido et al. [4]. GdMnSi has a PbFCl-type tetragonal structure with lattice constants $a = 0.4009$ nm, $c = 0.7183$ nm. The Curie temperature deduced from the magnetic measurements is 320 K. The saturation moment at 77 K is equal to about 118 emu/g ($\approx 5.1\mu_B$ per formula unit). As usual the gadolinium magnetic moment may be supposed to be close to $7\mu_B$; then the manganese moment would be close to $2\mu_B$.

The atomic arrangement of GdMnSi is shown in Fig. 1. One can see that the silicon layers were sandwiched between the gadolinium and transition metal layers. In the Gd intermetallics the Sn impurity atoms only occupy the non-transition metal sites [1, 2]. Moreover, the outer electron configuration of the Si and Sn atoms is the same. One can therefore expect the impurity Sn atoms to be localized at the Si sites and the hyperfine field for Sn atoms to be determined by both manganese and gadolinium magnetic moments. However, the results of our measurements turned out to be rather surprising: B_{hf} for Sn atoms in GdMnSi is found only to be determined by the 4f moments of Gd^{3+} ions. This conclusion follows directly from an analysis of the satellite structure of the Mössbauer spectrum for ^{119}Sn in $Gd_{0.95}Y_{0.05}MnSi$ alloy and a comparison of the hyperfine fields for Sn in GdMnSi and GdCoSi isostructural hosts. We will discuss these results in Section 4 of this paper. In Section 3 the radial dependence of the partial contributions to B_{hf} for Sn in gadolinium ferromagnets will be discussed. To obtain more detailed information on this dependence, B_{hf} for ^{119}Sn impurity atoms in GdZn host also was measured.

2. Experimental Details and Results

The GdMnSi, GdCoSi, and $Gd_{0.95}Y_{0.05}MnSi$ alloys were prepared by induction melting of the components (of purity not worse than 99.9%) in an argon atmosphere. Metallic tin, enriched to 92% with the isotope ^{119}Sn was previously introduced into the silicon. The ingots were remelted several times and then annealed at 1050 K for 100 h. The sample of the GdZn intermetallic compound was prepared by directed reaction of Gd metal powder, containing a small admixture of ^{119}Sn, and Zn metall in a sealed evacuated quartz tube. The sample was heated up to 1100 K and kept at this temper-

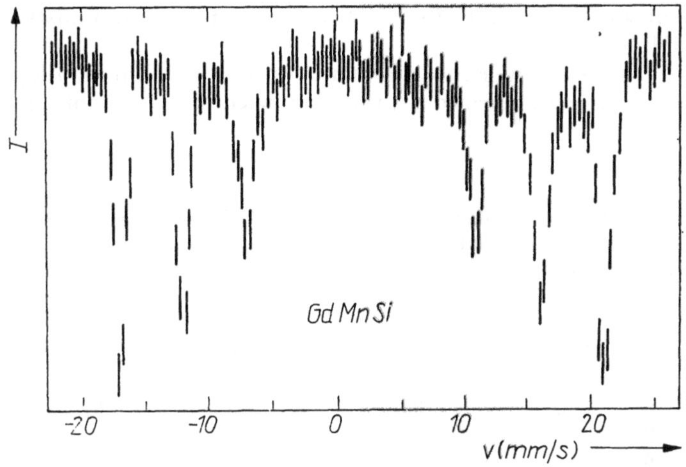

Fig. 2. Mössbauer spectrum for ^{119}Sn impurity atoms in GdMnSi host, measured at 5 K

ature for five days. The reaction product was melted and then fast cooled. The tin impurity concentration in the alloys was varied in the range from 0.3 to 0.6 at%.

Mössbauer absorption spectra of 23.9 keV ^{119}Sn γ-rays were measured with $CaSnO_3$ source, using a constant-acceleration spectrometer. The source was kept at room temperature. A resonant $CaSnO_3$ detector was used to register the Mössbauer radiation. The spectra for ^{119}Sn in GdMnSi and GdZn hosts were measured in the temperature range from 5 K to T_c, for GdCoSi and $Gd_{0.95}Y_{0.05}MnSi$ hosts the spectra were only measured at 77 K. The signs of B_{hf} were determined by measurements of the absorption spectra in an applied magnetic field of 0.9 T. To determine the hyperfine structure parameters the usual procedure of approximating the spectra by a sum of Lorentzian lines was used. The other details of the measurement technique and the spectrum reduction procedure are given in our preceding papers [1, 2].

As an example, a Mössbauer spectrum for ^{119}Sn in GdMnSi measured at 5 K is shown in Fig. 2. The magnetic hyperfine structure corresponding to B_{hf} (5 K) = $= -27.6 \pm 0.3$ T is well resolved. The temperature dependence of B_{hf} is shown in Fig. 3. This dependence appears to be practically identical with the temperature dependence of the magnetization. The Curie temperature found from the temperature dependence of B_{hf} is 321 ± 1 K, and this value agrees well with magnetization measure-

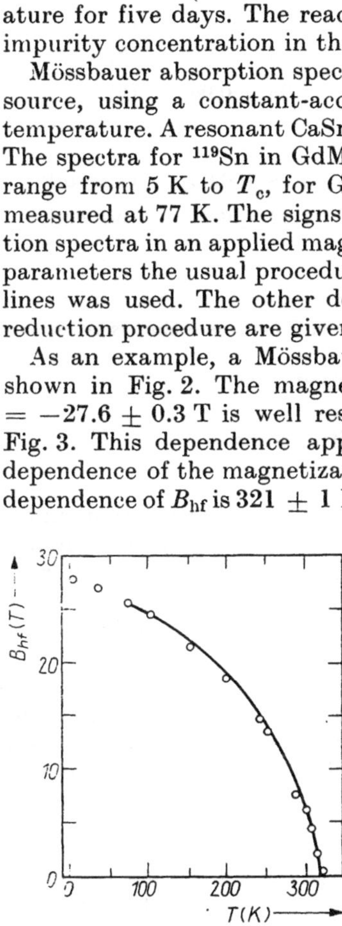

Fig. 3. Temperature dependence of magnetic hyperfine field for ^{119}Sn impurity atoms in GdMnSi. The full line shows the temperature dependence of magnetization [4] normalized to the value of the hyperfine field at 77 K

ments [4]. The shape of the absorption spectra for ^{119}Sn in GdZn and GdCoSi hosts is similar to that for Sn in GdMnSi.

The data extracted from ^{119}Sn Mössbauer measurements on the gadolinium alloys examined in the present study are summarized — together with the data obtained previously — in Table 1.

Table 1

Magnetic hyperfine fields (B_{hf}) and isomer shifts (IS) for ^{119}Sn impurity atoms in the Gd-based hosts. B_{calc} values were calculated from (5) using Gd magnetic moments (μ) and Sn–Gd distances (r) listed in this table. N is the number of the Gd atoms located at a distance r

host	$\mu(\mu_B)$**)	r (nm)	N	B_{hf} (T)*)	B_{calc} (T)	IS (mm/s)*)	ref.
Gd	7.5	0.364	6	-32.6 ± 0.3	-33.2	1.91 ± 0.02	[5, 6]
		0.357	6				
		0.510	6				
$GdAl_2$	7.1	0.327	6	-29.3 ± 0.1	-26.4	1.90 ± 0.03	[1]
		0.514	8				
$GdAg_{0.5}In_{0.5}$	7.3	0.322	8	-32.4 ± 0.6	-33.6	2.06 ± 0.03	[2]
Gd_3In	(7.3)	0.344	12	-37.0 ± 0.7	-36.7	1.93 ± 0.06	[2]
Gd_2In	(7.3)	0.312	3	-38.3 ± 0.7	-37.2	1.95 ± 0.06	[2]
		0.338	2				
		0.355	6				
GdZn	7.5	0.312	8	-36.6 ± 0.4	-40.2	1.98 ± 0.05	this work
GdMnSi	(7.0)	0.304	4	-27.6 ± 0.3	-28.8	1.97 ± 0.04	this work
		0.359	2				
		0.538	8				

*) All measurements of the B_{hf} and IS values were made at 5 K. The isomer shift is given with respect to room temperature $CaSnO_3$ source.

**) In parentheses in the second column the assumed μ values are given.

3. Radial Dependence of the HFI for Sn Impurity Atoms in Gadolinium Ferromagnetic Hosts

It follows from the results previously reported that B_{hf} for Sn impurity atoms in the Gd-based ferromagnets is mainly determined by the spins of the nearest neighbour Gd^{3+} ions [2]. It is quite probable that this conclusion has a general character for non-transition atoms in gadolinium intermetallic compounds. For example, it was found that the nearest neighbour Gd ions give 82.5% of B_{hf} for Zn in GdZn [7] and 85% of B_{hf} for Al in $GdAl_2$ [8].

Using available experimental data, one can obtain a simple mathematical expression for the radial dependence of the partial contributions to B_{hf} for Sn impurity atoms in Gd-based ferromagnets. In general form, the magnetic hyperfine field for a non-magnetic atom is the sum of the contributions from all the neighbouring magnetic atoms,

$$B_{hf} = \sum \mu(r_i)\, p(r_i)\,, \tag{1}$$

where $\mu(r_i)$ is the magnetic moment of the atom located at a distance r_i, $p(r_i)$ is a distance-dependent contribution to B_{hf} (in units of T/μ_B). Of course, knowledge of the function $p(r)$ is of prime importance for the interpretation of the experimental data [3]. For simplicity, we assume that HFI for Sn atom is isotropic. If all moments $\mu(r_i)$ are identical ($\mu(r_i) \equiv \mu$), then (1) can be written in the form

$$B_{hf} = \mu \sum N(r_i)\, p(r_i)\,, \tag{2}$$

where $N(r_i)$ is the number of the magnetic atoms located at distance r_i.

According to [2], $p(r)$ decreases sharply with increasing interatomic Gd–Sn distance. It was found that the data of Table 1 can be successfully described by a simple two-parameter function

$$p(r) = \frac{A}{r^n}\,. \tag{3}$$

where A and n are empirical constants, so that (2) takes the form

$$B_{hf} = A\mu \sum \frac{N(r_i)}{r_i^n}\,. \tag{4}$$

The parameters A and n were determined by a least-squares method using the experimental values of B_{hf} for Sn atoms in the following hosts: Gd [5, 6], $GdAl_2$ [1], $GdAg_{0.5}In_{0.5}$, Gd_3In, and Gd_2In [2], and GdZn (see Table 1). It was also supposed that the function $p(r)$ is equal to zero, then $r > r_0$, where r_0 was the third varying parameter. It was found that the best results were achieved with $r_0 = 0.55$ nm. In this case the following values for the A and n were obtained: $n = 4.8$ and $A = -157.9$ (r_i in Å, B_{hf} in T) or $A = -2.5 \times 10^{-3}$ (r_i in nm, B_{hf} in T). Thus, the hyperfine fields are given by

$$B_{hf} = -2.5 \times 10^{-3}\,\mu \sum_{(r_i \leqq 0.55\text{nm})} \frac{N(r_i)}{r_i^{4.8}} \tag{5}$$

(B_{hf} in T, μ in Bohr magnetons, r_i in nm).

The hyperfine fields calculated by (5) are presented in the sixth column of Table 1. The calculated values, B_{calc}, are in a good agreement with experimental data. It can be seen from Table 1 that the deviations of B_{calc} from the experimental values do not exceed 10%.

4. HFI for Sn Impurity Atoms in GdMnSi

According to the general approach we expected from both Gd and Mn atoms comparable contributions to B_{hf} for Sn in GdMnSi host. Assuming an additivity of the contributions to B_{hf}, we suggested to find the Mn-atom contribution by subtraction of the calculated gadolinium contribution from the observed value of B_{hf}. It turned out, however, that the gadolinium contribution calculated from (5), $B_{calc} = -28.8$ T, and the measured value, $B_{hf} = -27.6$ T, appeared to be practically identical, i.e. there was no contribution from Mn atoms within the limits of accuracy of the empirical expression (5). This result seems to be rather surprising, taking into consideration that in the GdMnSi host four nearest neighbour Mn atoms are located at shorter distance from the Sn impurity atom than the four Gd atoms [4].

It is not impossible that this is an accidental coincidence. To obtain more direct data on the Mn- and Gd-atom contributions to B_{hf}, we have investigated two other systems: $Gd_{0.95}Y_{0.05}MnSi$ alloy and GdCoSi compound. If B_{hf} in the GdMnSi host is exclusively determined by the Gd atoms, the appearance of the non-magnetic Y atoms in close vicinity of the probe Sn atom would produce a drastic reduction in B_{hf}. Using (5) and the data from Table 1, one can estimate the corresponding hyperfine field reduction. For one Y atom situated in the first neighbour shell the reduction of B_{hf} calculated using (5) is $\Delta B_{hf} = 5.3$ T. For random distribution of Y atoms in Gd sites the probability for a configuration with x yttrium atoms in the first neighbour shell is given by the binominal equation

$$W(x) = \frac{4!}{x!\,(4-x)\,!}\, c^x(I-c)^{4-x}\,,$$

where c is the concentration of Y atoms ($c = 0.05$). It follows from this equation that the ratio of the intensities of the subspectra with $x = 0$ and $x = 1$ will be 1:0.21. (These two configurations determined about 90% of the total area of the spectrum, thus all other configurations were neglected in the analysis of the spectrum.)

Fig. 4 presents the satellite structure of the Mössbauer spectrum due to substitution of Y for Gd in GdMnSi host. The satellite lines are quite well resolved. The reduction of B_{hf} and the relative intensity of the subspectrum were found to be equal to $\Delta B_{hf} = (5.0 \pm 0.6)$ and (0.24 ± 0.05) T, respectively. Both these values agree well with the calculated values quoted above.

The hyperfine field for Sn atoms in GdCoSi host at 77 K was found $B_{hf} = (-27.6 \pm 0.5)$ T, slightly larger than for Sn in GdMnSi host ($B_{hf} = (-25.5 \pm 0.5)$ T at 77 K). The interatomic distances in GdCoSi are smaller than in GdMnSi. According to Kido et al. [4] the ratio of these distances is 1.015. Then, as follows from (5), provided μ(Gd) is equal in both alloys, the ratio of the hyperfine fields for the two hosts will be $1.015^{4.8} = 1.075$. Just this value of the ratio has been observed experimentally. Thus, the difference between the B_{hf} values in GdMnSi and GdCoSi hosts is completely accounted for by the radial dependence of the contributions to B_{hf}. The substitution of Mn atoms by Co atoms (with the corresponding change of the 3d moments) in itself does not affect the B_{hf} value.

Thus, it is convincingly demonstrated by these experiments that for Sn impurity atoms in GdMnSi and GdCoSi hosts the HFI is determined by the 4f moments of Gd ions. The contribution to B_{hf} from the 3d moments of the Mn and Co atoms is smaller than the experimental error.

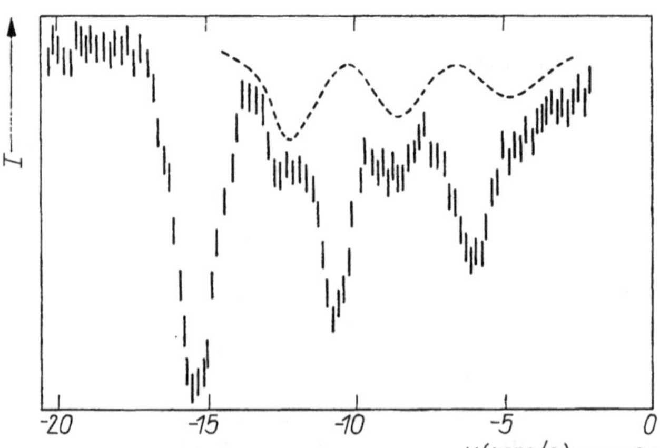

Fig. 4. Left part of the Mössbauer spectrum for ^{119}Sn in $Gd_{0.95}Y_{0.05}MnSi$ host at 77 K. The dashed line shows the subspectrum attributed to replacement of one Gd atom by an Y atom in the nearest neighbour shell

5. Conclusion

According to our data, while the Mn atoms in GdMnSi possess a magnetic moment, they produce no contribution to the transferred HFI. This result was especially unexpected in view of the fact that Mn atoms undoubtedly take part in the exchange interaction. In particular, this follows from the rather high Curie temperature of GdMnSi.

We believe that the obtained data demonstrate the strong interrelation between HFI and chemical bonding for the impurity atoms in the rare earth alloys. The results look as a puzzle if one treats the impurity atom merely as a probe immersed in the ferromagnetic medium. In a more realistic model it seems necessary to take into account the peculiarities of the chemical bonding between the non-magnetic impurity atom and the neighbouring magnetic atoms. In the present case one can suppose that both HFI and chemical bonding for Sn atoms are governed by the Gd ions. Further confirmation of this suggestion results from an examination of the isomer shift (IS) values. As is seen from Table 1, for all investigated hosts, the IS values are nearly equal and close to the IS for Sn in the pure gadolinium host. No matter which other atoms occur as the nearest neighbours of the Sn atom, the electron density at Sn nuclei corresponds to Sn–Gd interatomic bonding. For example, in the $GdAl_2$ host six Al atoms are the nearest neighbours of the Sn atom, however, the IS value is practically the same as that in the pure gadolinium host. Both the IS data and the radial dependence of the partial contributions to B_{hf} emphasize the short-range nature of the HFI for Sn atoms in Gd-based ferromagnets.

Of course, this discussion does not mean that a similar situation will take place in all other rare-earth–transition metal alloys. On the contrary, we expect a gradual transition to the situation typical for the transition metal hosts [3] as the transition metal concentration increases. More systematic studies on the other rare-earth–transition metals alloys are now in progress in our laboratory.

References

[1] N. N. Delyagin, V. I. Nesterov, and S. I. Reiman, Zh. eksper. teor. Fiz. **84**, 1580 (1983); Soviet Phys. — J. exper. theor. Phys. **57**, 922 (1983).

[2] N. N. Delyagin, G. T. Mujiri, V. I. Nesterov, and S. I. Reiman, Zh. eksper. teor. Fiz. **86**, 1016 (1984).

[3] N. N. Delyagin, V. I. Krylov, and V. I. Nesterov, Zh. eksper. teor. Fiz. **79**, 1049 (1980); Soviet Phys. — J. exper. theor. Phys. **52**, 532 (1980).

[4] H. Kido, M. Shimada, and M. Koizumi, phys. stat. sol. (a) **70**, K23 (1982).

[5] V. Gotthardt, H. S. Möller, and R. L. Mössbauer, Phys. Letters A **28**, 480 (1969).

[6] S. I. Reiman, N. I. Rokhlov, V. S. Shpinel, and E. P. Kaminskaya, Zh. eksper. teor. Fiz. **86**, 330 (1984).

[7] K. Eckrich, E. Dormann, A. Oppelt, and K. H. J. Buschow, Z. Phys. B **23**, 157 (1976).

[8] E. Dormann, K. H. J. Buschow, K. N. R. Taylor, G. Brown, and M. Issa, J. Phys. F **3**, 220 (1973).

(Received March 4, 1985)

phys. stat. sol. (b) 131, 563 (1985)

Subject classification: 18; 1.6

Sektion Physik der Humboldt-Universität zu Berlin, Bereich 04[1])

The Random Field Ising Model in 2 + ε Dimensions

Critical Behaviour from Interface Fluctuations

By

TH. NATTERMANN

Dedicated to Prof. Dr. Dr. h. c. Dr. E. h. P. GÖRLICH on the occasion of his 80th birthday

The critical behaviour of the random field Ising model is calculated in $d = 2 + \varepsilon$ dimensions, using previous findings on interface roughening of Grinstein and Ma. The critical exponents are expressed in terms of ε and the dimension β/ν of the order parameter field. ε and β/ν are assumed to be small. For nonvanishing random field strength H_R the critical behaviour is governed by a zero temperature fixed point $H_{Rc} \sim (\varepsilon + 2\beta/\nu)^{3/4}$. In particular $\mu = 3/4$ and $\nu^{-1} = \frac{2}{3}(\varepsilon + 2\beta/\nu)$ are obtained for the interface tension and the correlation length exponent, respectively. The relation between interface and bulk correlation lengths in the various regions is discussed.

Das kritische Verhalten des Ising-Modells in einem Zufallsfeld wird in ($d = 2 + \varepsilon$)-Dimensionen mit der schon früher von Grinstein und Ma gefundenen Grenzflächenrauhigkeit berechnet. Die kritischen Exponenten werden über ε und die Dimension β/ν des Ordnungsparameterfeldes ausgedrückt. Es wird angenommen, daß ε und β/ν klein sind. Für eine nichtverschwindende Stärke des Zufallsfeldes H_R wird das kritische Verhalten durch einen $T = 0$ Fixpunkt $H_{Rc} \sim (\varepsilon + 2\beta/\nu)^{3/4}$ bestimmt. Speziell für den Grenzflächenspannungs- und den Korrelationslängenexponenten werden die Werte $\mu = 3/4$ und $\nu^{-1} = \frac{2}{3}(\varepsilon + 2\beta/\nu)$ erhalten. Die Beziehung zwischen Grenzflächen- und Volumen-Korrelationslänge in verschiedenen Bereichen wird diskutiert.

1. Introduction

The drastic effects of random fields (RF) on the behaviour of Ising-like models have been debated extensively in recent years. Stability analysis of the ordered state initiated by Imry and Ma [1] yields $d_l = 2$ as the lower critical dimension (l.c.d.), below which no long-range order is present. This result has been called into question by results of a perturbation theory around the upper critical dimension $d_c = 6$, which gave a dimensionality shift by two between the RF and the pure Ising model [2]. This suggests $d_l = d_{0,l} + 2 = 3$. In an interlude people tried to understand a l.c.d. equal to three in terms of the Imry-Ma argument by domain wall roughening which might lower the domain wall energy and hence lead to a destabilization of the ferromagnetic state for $d < 3$ even at $T = 0$ [3]. Deeper consideration, however, showed that walls become actually rough, but the corresponding lowering of wall energy is not strong enough to shift the l.c.d., thus $d_l = 2$ [4, 5]. More detailed information about this development as well as references can be found in review papers [6, 7].

The present status of our knowledge of the *critical behaviour* of the RF Ising model (RFIM) is unsatisfactory. The above-mentioned perturbation theory, which works in $6 - \varepsilon$ dimensions breaks probably down below four or five dimensions because of the existence of metastable states [7].

[1]) Invalidenstr. 42, DDR-1040 Berlin, GDR.

In their paper Aharony et al. [2] also raised the possibility of a reduction in dimension such that the properties of the d-dimensional RFIM are those of the d_0-dimensional pure Ising model with

$$d_0 = d - 2 + \eta_0(d_0)\,, \tag{1.1}$$

where $\eta_0(d_0) = \eta(d)$ is the critical exponent of the pure system in d_0 or of the RFIM in d dimensions. Since $\eta_0(d_{0l} = 1) = 1$, i.e. $d_l = d_{0l} + 1$ but $\eta_0(d_{0c}) = 0$, i.e., $d_c = d_{0c} + 2$, (1.1) would reconcile domain arguments with perturbation theory. Schwartz [8] rederived (1.1) by relating the RFIM with quenched disorder to a corresponding annealed problem.

Using a non-perturbative approach, which treats basically the *local* response of the system, Shapir [9] finds a reduction

$$\begin{aligned} d_0 &= d - \frac{1}{\nu_0(d_0)} && \text{if} \quad \alpha_0(d_0) > 0\,, \\ d_0 &= \frac{1}{2}\left(d + \frac{1}{\nu_0(d_0)}\right) && \text{if} \quad \alpha_0(d_0) < 0\,. \end{aligned} \tag{1.2}$$

This yields again $d_l = 2$. More interesting, according to (1.2) the critical properties of the RFIM in $d = 3$ dimensions should be those of the two-dimensional Ising model. In particular, the predictions $\alpha = 0$ and $\nu = 1$ are in agreement with specific heat, birefringence, and heat capacitance experiments [10]. We note however, that although important for fitting the experimental data, (1.2) probably does not describe the true *equilibrium* properties of RFIM.

In the present paper, we try to extract the critical properties of the RFIM from the behaviour of interfaces close to the critical point. The approach is oriented on — although at present less profounded than — the interface calculations for pure models as invented by Wallace and Zia [11]. Our calculation is based on the results of Grinstein and Ma (GM) [4] for the fluctuation renormalized interface tension. However, in order to enter the critical region we have to alter their expressions in a self-consistent way. The results should be valid in $d = 2 + \varepsilon$ dimensions. As for pure models, we are not able to obtain the exponent β (or η) from interface calculations alone. This requires the consideration of a multi-droplet ensemble [11]. Unfortunately, contrary to the pure problem, even the determination of ν involves β in the present case. To lowest order in ε and $2\beta/\nu \ll 1$ we get

$$\nu^{-1} = \frac{2}{3}\left(d - 2 + \frac{2\beta}{\nu}\right).$$

Together with $\mu = \frac{3}{4}$ for the surface tension exponent and the modified hyperscaling relation (4.6) other exponents can be found once β is given. Despite this modest achievement we hope that the present study motivates further investigations in this direction.

2. Interface Fluctuations. The $T = 0$ Case

Grinstein and Ma [4] start from the following Hamiltonian for the distortions $f(r)$ of an interface in the RFIM:

$$\mathcal{H} = \int \mathrm{d}^{d-1}r \left[\sigma_0 \sqrt{1 + (\nabla f)^2} - \int\limits_0^{f(r)} \mathrm{d}z\, \Phi(r, z)\, H_\mathrm{R}(r, z)\right]\,. \tag{2.1}$$

σ_0 denotes the bare interface tension, including, say, interface fluctuations on a length scale $\varrho < \varrho_0$ with ϱ_0 arbitrary but fixed. A natural choice for ϱ_0 would be $\varrho_0 \approx \xi_0$ where ξ_0 denotes the interface width at vanishing disorder: $T = H_\mathrm{R} = 0$. Below we use $\varrho_0 = 1$.

$H_R(\boldsymbol{r})$, $\boldsymbol{r} = \{r, z\}$ is a Gaussian correlated random field with

$$\overline{H_R(\boldsymbol{r})} = 0\,, \qquad \overline{H_R(\boldsymbol{r})\,H_R(\boldsymbol{r}')} = H_R^2\,\delta(\boldsymbol{r} - \boldsymbol{r}')\,. \tag{2.2}$$

The bar denotes the configurational average. The bulk order parameter $\Phi(\boldsymbol{r})$ was set equal to unity in GM's original treatment. Below we use the notation $H(\boldsymbol{r}) = H_R(\boldsymbol{r})\,\Phi(\boldsymbol{r})$ etc.

Neglecting the change of the short-distance cut-off ϱ_0, the scale transformation $(b > 1)$

$$r \to br'\,, \qquad z \to bz'\,, \qquad f \to bf'\,, \qquad \varrho \to b\varrho' \tag{2.3}$$

leaves $\mathcal{H}/T$ invariant, provided we set

$$T' = b^{1-d}T\,, \qquad H' = b^{(2-d)/2}H\,. \tag{2.3'}$$

GM calculate now the change $\Delta\sigma_1$ of the interface tension by integrating over fluctuations on the scale $\varrho_0 \dots b\varrho_0$. This scale transformation and integration is repeated many times until we reach a macroscopic scale. The interface tension after the n-th renormalization group (RNG) step is

$$\sigma_n = \sigma_0 + \sum_{l=1}^{n} \Delta\sigma_l\,, \tag{2.4}$$

where for $T = 0$ $(\Delta l = 1)$ [4, 5]

$$\Delta\sigma_l = -\,C\,\frac{H_l^{4/3}}{\sigma_0^{1/3}}\,\Delta l\,; \qquad H_l = Hb^{((2-d)/2)l}\,. \tag{2.5}$$

C is a constant of order unity which could not be determined within the crude approximation scheme of GM. Below we absorb C in the definition of H_R. As an important restriction GM ignore any new terms which may be generated by the elimination of the short-wavelength degrees of freedom and so just keep the two variables T and H at each stage. Further H_l/σ_0 had to be assumed to be small. This restricts the calculation in general to H much smaller than the critical field strength H_c where the randomness disorders the system $(d > d_l = 2)$.

In order to extract critical properties from the GM results, we introduce two important alterations:

(i) It is in the spirit of the RNG calculus to make the computation self-consistent by replacing σ_0 by the true interface tension σ_l on the scale $\varrho = b_l$. For an infinitesimal scale change $\Delta l \to \mathrm{d}l = \mathrm{d}\varrho/\varrho \ln b$ we get from (2.5)

$$\frac{\partial\sigma_l}{\partial l} = -\frac{H_l^{4/3}}{\sigma_l^{1/3}} \tag{2.6}$$

provided $h_l = H_l/\sigma_l \ll 1$ now. For simplicity we have chosen $b = e$.

(ii) Close to the critical point it is important to take into account the scaling properties of the order parameter. Thus, instead of (2.5), we use

$$H_l = H_R\,\mathrm{e}^{\frac{2-d}{2}l}\,\Phi_l = H_R\,\mathrm{e}^{-l\left(\frac{d-2}{2}+\Delta\right)}\,\varphi(\mathrm{e}^l/\xi)\,, \tag{2.7}$$

where $\Delta = \beta/\nu$ with β and ν are the exponents of the order parameter and bulk correlation length ξ, $\varphi(x) \approx 1$, x^Δ for $x \to 0$, $x \to \infty$, respectively.

(2.7) can be understood as follows. H_l is essentially given by

$$\begin{aligned} H_l^2 &\approx \int_{|\boldsymbol{r}-\boldsymbol{r}'|\le 1} \mathrm{d}^d(r - r')\,\overline{H_{R,l}(\boldsymbol{r})\,\Phi_l(\boldsymbol{r})\,H_{R,l}(\boldsymbol{r}')\,\Phi_l(\boldsymbol{r}')} = \\ &= \mathrm{e}^{2l} \int_{|\boldsymbol{r}-\boldsymbol{r}'|\le 1} \mathrm{d}^d(r - r')\,\overline{H_R(\boldsymbol{r}\,\mathrm{e}^l)\,\Phi(\boldsymbol{r}\,\mathrm{e}^l)\,H_R(\boldsymbol{r}'\,\mathrm{e}^l)\,\Phi(\boldsymbol{r}'\,\mathrm{e}^l)}\,. \end{aligned} \tag{2.7'}$$

For $e^l \ll \xi$ the main contribution to H_l arises from terms on the right-hand side of (2.7′), where Φ is replaced by $\bar{\Phi} \sim \xi^{-\beta/\nu}$. Together with (2.2) this yields (2.7) for $x = e^l/\xi \gg 1$. On the contrary, for $\xi \gg e^l$ we are in the scaling region where $\Phi(\boldsymbol{r}\, e^l) = e^{l\Delta}\, \Phi(\boldsymbol{r})$, Δ denotes the dimensionality of the field. This yields (2.7) for $x \ll 1$. No further renomalization of H_{R} besides the trivial one has been assumed!

The introduction of a non-zero Δ and of ξ look rather artificial at this point. It has no counterpart in the treatment of the corresponding pure model [11]. However, for a consistent description of the RF model, it is necessary to take into account the change of the coupling strength between the RF and the interface distortion due to fluctuations right and left from the interface (e.g. domains within domains).

From (2.6), (2.7) we get for $h_l = H_l/\sigma_l$

$$\frac{\partial h_l}{\partial l} = -\left[\frac{d-2}{2} + \Delta - \frac{\partial \ln \varphi\,(e^l/\xi)}{\partial l}\right] h_l + h_l^{7/3}\,. \tag{2.8}$$

(2.8) has an ultraviolet stable fixpoint,

$$h_{\mathrm{c}} = \left(\frac{d-2}{2} + \Delta\right)^{3/4}. \tag{2.8′}$$

For $h_l < h_{\mathrm{c}}$, h_l is iterated to zero under the RNG flow. On the contrary, for $h_l > h_{\mathrm{c}}$, h_l increases, leaving finally the domain of validity of our approach. Since h_{c} is infrared unstable, it is tempting to identify it with a critical point. Because of the condition $h_l \ll 1$ our treatment is restricted to dimensionalities $d - 2 + 2\,\Delta \ll 1$. Since we expect that the critical point vanishes for $d \to 2$, Δ should vanish as well. Thus, we have to assume $d - 2 = \varepsilon \ll 1$.

Integration of (2.6) between l^* and l yields

$$\sigma_l^{4/3} = \sigma_{l^*}^{4/3} - h_{\mathrm{c}}^{-4/3}(H_l^{4/3} - H_{l^*}^{4/3})\,. \tag{2.9}$$

Actually, H_l on the right-hand side of (2.9) is now given by (2.7) with φ replaced by a new cross-over function $\varphi_1(e^l/\xi)$ with

$$\varphi^{4/3}(x) = \varphi_1^{4/3}(x)\left[1 - h_{\mathrm{c}}^{-4/3}\,\frac{\partial \ln \varphi_1(x)}{\partial \ln x}\right].$$

$\varphi_1(x)$ has essentially the same properties as $\varphi(x)$. The macroscopic surface tension σ_∞ can then be written as

$$\sigma_\infty = \sigma_{l^*}\left[1 - \left(\frac{H_{\mathrm{R}}}{H_{\mathrm{c}}}\right)^{4/3}\right]^{\mu}; \qquad \mu = \frac{3}{4}, \tag{2.10}$$

where the critical field

$$H_{\mathrm{c}} = \tilde{H}(l^*, \sigma_{l^*}) = h_{\mathrm{c}}\sigma_{l^*}\, e^{l^*(d-2+2\Delta)/2}\, \varphi_1^{-1}(e^{l^*}/\xi) \tag{2.11}$$

has been introduced. For $H_{\mathrm{R}} \to H_{\mathrm{c}}$ a transition to the disordered phase occurs.

We may extract a characteristic length ξ_{I} from (2.9). To this aim it is convenient to go back to the original length scale $\varrho = e^l$, using the notation $\sigma_l = \sigma(\varrho)$, etc.. (2.9) can then be written as

$$\sigma(\varrho) = \sigma_\infty\, g(\varrho/\xi, \varrho/\xi_{\mathrm{I}})\,, \tag{2.12}$$

$$\xi_{\mathrm{I}} = \varrho^*\left[\left(\frac{H_{\mathrm{c}}}{H_{\mathrm{R}}}\right)^{4/3} - 1\right]^{-\nu_{\mathrm{I}}} \varphi_1^{-4/3\nu_{\mathrm{I}}}(\varrho^*/\xi)\,,\qquad \nu_{\mathrm{I}} = \frac{3}{2(d-2+2\Delta)}\,, \tag{2.13}$$

$$g(x, y) = [1 + \varphi_1^{4/3}(x)\, y^{-1/\nu_{\mathrm{I}}}]^{3/4}\,. \tag{2.14}$$

Since we are restricted so far to $T = 0$, it is natural to choose $\varrho^* = 1$, i.e. $\sigma_{l^*} = \sigma_0$ in (2.9) to (2.13). Note that for $d = 2$ (2.8) yields a characteristic length $\xi_\mathrm{I} \sim \exp C(\sigma_0/H_\mathrm{R})^{4/3}$, in agreement with earlier findings [6, 7].

ξ_I denotes the scale on which the interface is *locally fuzzy*. To show this, we consider the expression for the mean-square interface width on the scale ϱ, $w^2(\varrho) = \overline{f^2}$ (equation (3.29) of [4]). To be consistent with our above approach, we replace again σ_0 by $\sigma(\varrho)$ and H by $H\varrho^{-\Delta}\,\varphi(\varrho/\xi)$. $w(\varrho)$ can then be written as

$$\frac{w(\varrho)}{\lambda(\varrho)} \sim \left[\frac{\varrho}{\lambda(\varrho)}\right]^{(5-d)/3}, \tag{2.15}$$

where we have introduced the scale $\lambda(\varrho)$,

$$\lambda(\varrho) = \left[\frac{H_\mathrm{R}\varrho^{-\Delta}\,\varphi(\varrho/\xi)}{\sigma(\varrho)}\right]^{2/(d-2)}. \tag{2.16}$$

For $\varrho \leqq \lambda$, the typical gradient of the interface is of order unity, or larger, i.e. the interface is fuzzy. In particular, for $\varrho \approx \xi_\mathrm{I}$ we get from (2.10) to (2.12)) and (2.16)

$$\lambda(\xi_\mathrm{I}) \approx h_\mathrm{c}^{4\nu_\mathrm{I}/3}\xi_\mathrm{I}\,, \tag{2.17}$$

i.e. ξ_I sets indeed the fuzzyness scale. Note that $\xi_\mathrm{I} \to 0$ for $H_\mathrm{R} \to 0$ (or, more generally, for vanishing disorder), whereas ξ remains finite.

In the critical region only a single length scale should exist and hence $\xi_\mathrm{I} \approx \xi$ has to be expected. We come back to this point in Section 5.

On scales $\varrho \gg \xi_\mathrm{I}, \xi$, $\lambda(\varrho) \to \lambda_H$ with

$$\lambda_H = \left[\frac{H_\mathrm{R}\xi^{-\beta/\nu}}{\sigma_\infty}\right]^{2/(2-d)} \sim \xi_\mathrm{I}\left(\frac{\xi_\mathrm{I}}{\xi}\right)^{2\Delta/(d-2)}. \tag{2.18}$$

λ_H controls the large-scale interface roughening. In the opposite case $\varrho \ll \xi_\mathrm{I}, \xi$ we get

$$\lambda(\varrho) = \left(\frac{H_\mathrm{R}\xi_\mathrm{I}^{-3/4\nu_\mathrm{I}}}{\sigma_\infty}\right)^{2/(d-2)}\varrho\,, \tag{2.19}$$

i.e. $w(\varrho)/\varrho$ approaches a constant value for $\varrho \to 0$. In analogy to the pure case we expect that the interface has a fractale structure on this scale.

3. Interface Fluctuations at Finite Temperatures

Grinstein and Ma [4] extended their treatment to finite temperatures. Introducing again self-consistency in their equations (3.23), (3.24), we get for infinitesimal scale changes

$$\frac{\partial\sigma_l}{\partial l} = -\frac{H_l^{4/3}}{\sigma_l^{1/3}}\,(1 + Dx_l) \quad \text{if} \quad x_l \ll 1\,, \tag{3.1}$$

$$\frac{\partial\sigma_l}{\partial l} = -D'T_l\left(1 + \frac{C'}{D'}\,x_l^{-3/2}\right) \quad \text{if} \quad x_l \gg 1\,, \tag{3.2}$$

where

$$x_l = \frac{T_l\sigma_l^{1/3}}{H_l^{4/3}} = \frac{t_l}{h_l^{4/3}}; \quad T_l = T\,\mathrm{e}^{l(1-d)}\,. \tag{3.3}$$

The factors D, D' include a weak (logarithmic) dependence on T_l and σ_l which has been neglected here, C' is a constant of order unity. In order that (3.1), (3.2) are valid, $h_l \ll 1$ and $t_l = T_l/\sigma_l \ll 1$, respectively, have to be fulfilled. From (3.2), (3.3),

$$\frac{\partial t_l}{\partial l} = -(d-1)\,t_l + D't_l^2\left(1 + \frac{C'}{D'}\,x_l^{-3/2}\right); \qquad x_l \gg 1\,. \tag{3.4}$$

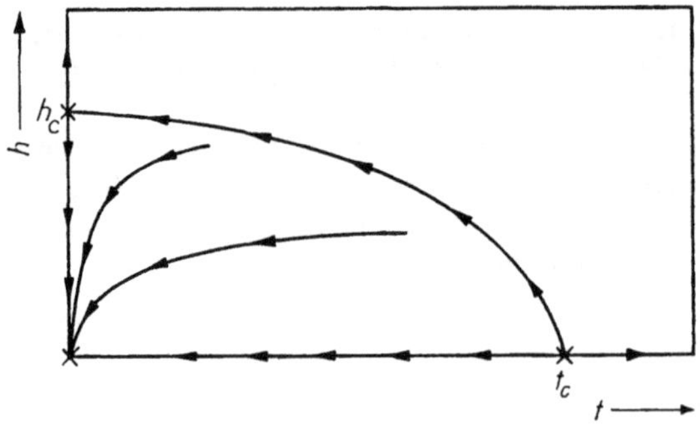

Fig. 1. Schematic renormalization-group flow for the RFIM above the lower critical dimension. The critical behaviour is governed by the fixed point h_c; $h_l = H_l/\sigma_l$; $t_l = T_l/\sigma_l$

Since x_l is large, (3.4) can be used only for $t_l < t_c \ll 1$ with

$$t_c = \frac{d-1}{D'} \tag{3.5}$$

which restricts us close to $d = 1$ dimension. For $t_l > t_c$, t_l iterates to large values if l becomes large. However, from (2.7), (3.3),

$$\frac{\partial x_l}{\partial l} = -\frac{d+1-4\Delta}{3} x_l + \frac{\partial \ln \sigma_l}{\partial l}, \tag{3.6}$$

i.e. x_l decreases strongly ($\Delta \ll 1$). The schematic RNG flow is depicted in Fig. 1. Thence there is a critical value l_c such that $x_l < \tilde{\varepsilon}$ for $l > l_c$, and $\tilde{\varepsilon} \ll 1$ fixed. On scales $l > l_c$ we may use the results of Section 2, identifying l_c with l^*.

In principle we can integrate $\partial\sigma/\partial l$ between $l = 0$ and $l = l_c$, using (3.2) and (3.1) for $x_l \gg 1$ and $x_l \ll 1$, respectively. As a result, we get σ_{l_c} with l_c, σ_{l_c} depending on T, H, σ_0 (and $\tilde{\varepsilon}$). Since the conditions $d - 2 \ll 1$ and $d - 1 \ll 1$ are not compatible, we do not try here to find the explicit T- and H-dependence of σ_{l_c} from (3.1), (3.2), but assume that the resulting

$$H_c = \tilde{H}(e^{l_c}, \sigma_{l_c}) = \tilde{h}(H_R, T) \tag{3.7}$$

(cf. (2.11)) leads to a unique solution $H_R = H_c(T)$ (or $T_c(H_R)$) of

$$H_R = \tilde{h}(H_R, T)\,. \tag{3.8}$$

With $t = [T - T_c(H_R)]/T_c(H_R)$ we get from (2.10), (2.13), and (3.7)

$$\sigma_\infty = \sigma_{l^*}[\varkappa(H_R)\,|t|]^\mu\,, \tag{3.9}$$

$$\xi_I \sim [\varkappa(H_R)\,|t|]^{-\nu_I}\,, \tag{3.10}$$

where

$$\varkappa(H_R) = -\frac{4}{3}\,\frac{T_c(H_R)}{H_R}\,\frac{\partial \tilde{h}}{\partial T}\bigg|_{T = T_c(H_R)}$$

and μ and ν_I given in (2.10), (2.13).

4. Scaling and Critical Exponents

It is the aim of this section to calculate the critical exponents of the RFIM from the results obtained above. As mentioned already in Section 2, one should expect only a single length scale $\xi \sim \xi_I$ in the critical region. To substantiate this belief, we consider the scaling behaviour of λ_H by approaching $T_c(H_R)$ for H_R fixed.

$$\sigma_\infty \sim \sigma_0\,|t|^\mu \sim \sigma_0 \Phi^2 \xi^{\eta-1} \sim \sigma_0\,|t|^{2\beta+\nu(1-\eta)}$$

yields $\mu = 2\beta + \nu(1 - \eta)$. With $\xi \sim |t|^{-\nu}$ we get

$$\lambda_H \sim \left(\frac{H_R}{\sigma_0}\right)^{2/(d-2)} |t|^{-\nu_H}\,; \qquad \nu_H = \frac{2}{d-2}\,(\beta + \nu(1 - \eta))\,. \tag{4.1}$$

Since $H_R \ll \sigma_0$, $d > 2$, $\lambda_H \ll \xi$ at low temperatures.

Considering the approach to $T_c(H)$, we find for the *Landau region* $\nu_H = 2/(d - 2)$, $\nu = 1/2$, i.e. $\nu_H > \nu$ for $d < d_c = 6$. Thus λ_H increases stronger than ξ and there is a temperature t_H, where $\lambda_H \approx \xi$,

$$|t_H| \approx \left(\frac{H_R}{\sigma_0}\right)^{4/(6-d)}. \tag{4.2}$$

t_H denotes the Levanyuk-Ginzburg critical region. Indeed, for the Ginzburg-Landau version of the RFIM,

$$\mathcal{H}_{GL} = \int d^dr\{\sigma_0\,[\tfrac{1}{2}\,t\Phi^2 + \tfrac{1}{4}\,\Phi^4 + \tfrac{1}{2}\,(\nabla\Phi)^2] - H_R(\boldsymbol{r})\,\Phi(\boldsymbol{r})\}\,, \tag{4.3}$$

the RF dominated critical region is characterized by a contribution I from the basic graph (Fig. 2a) larger than σ_0 (the bare value of the coupling constant),

$$\frac{I}{\sigma_0} \approx \frac{\sigma_0}{T}\,\frac{H_R^2}{T^2} \int d^dk \left[\frac{T}{\sigma_0(2|t| + k^2)}\right]^3 \approx \left|\frac{t_H}{t}\right|^{(6-d)/2}. \tag{4.4}$$

Thus $I/\sigma_0 > 1$ for $|t| < |t_H|$ if $d < 6$.

Incidentally we remark that for purely thermal disorder interface roughening is scaled by a length $\lambda_T = (T/\sigma_\infty)^{1/(d-1)} \sim (T/\sigma_0)^{1/(d-1)}\,|t|^{-\nu_T}$, $\nu_T = \mu/(d - 1)$ [6]. Landau theory gives $\nu_T = 3/2(d - 1)$, i.e. λ_T grows faster than ξ for $d < 4$ such that we get $\lambda_T \approx \xi$ for $|t| \sim |t_T| = (T/\sigma_0)^{2/(4-d)}$. t_T is the Ginzburg critical region of the pure model as can be easily shown from (4.3). The corresponding graph is depicted in Fig. 2b.

We conclude from these two examples that within interface models the Levanyuk-Ginzburg critical region is characterized by $\lambda \to \xi$, where λ scales the large-scale interface roughening and ξ denotes the bulk correlation length.

For weak fields such that $|t_H| \ll |t| \ll |t_T|$, exponents get renormalized by thermal fluctuations. Using the classical hyperscaling relation (we use the subscript zero for the exponents of the pure model),

$$\nu_0 d = 2 - \alpha_0 \tag{4.5}$$

and $\alpha + 2\beta + \gamma = 2$, we get $\mu = \mu_0 = \nu_0(d - 1)$, $\nu_H = [\nu_0/(d - 2)]\,(d - \eta_0) > \nu_0$. In this case $\lambda_H \approx \xi$ for $|t| \to |\tilde{t}_H| \approx (H_R/\sigma_0)^{2/\gamma_0}$. For $d < 4$, $|\tilde{t}_H| < |t_H|$ and the RF Ginzburg region is given by $|\tilde{t}_H|$.

At $|\tilde{t}_H|$, $\lambda_H \sim \xi = (\sigma_0/H_R)^{2/(2-\eta_0)}$. Villain [12] has shown that $\lambda_H(\tilde{t}_H)$ characterizes the minimal size of *metastable* domains in the vicinity of $T_c(H)$.

In the last part of this section we consider the region $|t| < |\tilde{t}_H|$. The approach has certain speculative aspects. The essential point is the assumption of the hyperscaling relation

$$\nu(d - 2 + \eta) = 2 - \alpha \tag{4.6}$$

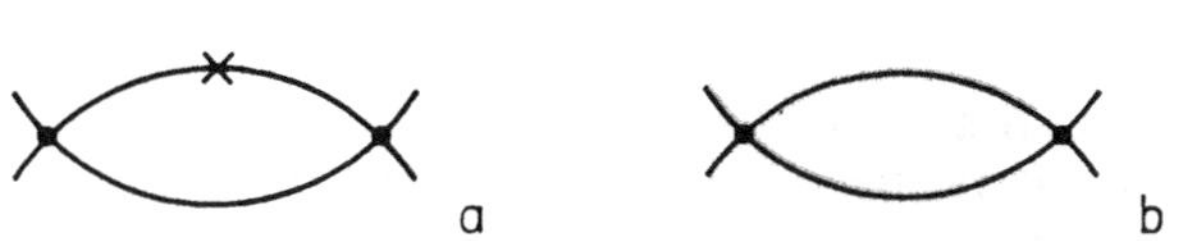

Fig. 2. Lowest-order diagrams contributing to the Φ^4 coupling constant. The cross denotes a factor H_R^2 arising from averaging over the random field

which replaces (4.5). (4.6) is usually derived from a variation of a Pippard argument [2, 6] (see also [8]). An alternative derivation goes as follows. In a homogeneous external field h, $\xi \sim h^{-\nu/(\beta+\gamma)}$ for $T = T_c$. For a correlated region ξ^d there is an average excess field $h \sim H\xi^{-d/2}$ or $\xi \sim (h/H)^{-2/d}$. Thus, $\beta + \gamma = \nu d/2$ which yields with the d-independent scaling law $2(\beta + \gamma) = 2 - \alpha + \nu(2 - \eta)$ (4.6). Inserting (4.6) into (4.1) yields

$$\mu = \nu(d - 3 + \eta)\,; \qquad \nu_H = \nu\,. \tag{4.7}$$

From (2.18) $\nu_H = \nu_I + (\nu_I - \nu)\, 2\Delta/(d - 2)$ which gives with (4.7)

$$\nu_I = \nu\,. \tag{4.8}$$

We come now back to the results of Section 2. From (2.10), (2.13), (4.7), (4.8), and $\Delta = \beta/\nu$,

$$\nu = \frac{3 - 4\beta}{2\varepsilon}; \qquad \varepsilon = d - 2 \ll 1\,. \tag{4.9}$$

The other exponents follow from scaling relations obeying (4.6), e.g.

$$\alpha = -\frac{3 - 4\beta}{2\varepsilon} + \frac{5}{4}, \qquad \eta = 1 - \frac{\varepsilon}{2}\,\frac{3 - 8\beta}{3 - 4\beta}.$$

Since the present approach does not determine β (typical for interface calculations [11]), we are even not able to determine ν (typical for the RFIM because of the special coupling between the disorder and the interface coordinate). Assuming $\beta < \infty$, for $d \to 2$ $\nu \to \infty$, $\alpha \to -\infty$, $\eta \to 1$, in agreement with previous findings [9] and the common expectations at the l.c.d.

In order to get results for $d = 3$ we have to make an assumption about the value of β. From their Monte Carlo data for the RFIM Landau and coworkers [13] find $\beta = 0.2$ in $d = 3$ dimensions. Using this value we get $\nu = 1.1$, $\eta = 0.68$, $\gamma = 1.45$, $\alpha = 0.15$. The value for γ is somewhat larger than $\gamma = 1.40$ found by Khurana et al. [14] from a high-temperature expansion. Also ν and α are larger than the experimental values $\nu \approx 1$ and $\alpha \approx 0$ [10]. However, the ε-expansion around the l.c.d. is usually believed to give less satisfactory agreement with the true exponents than other approaches potentially do. Moreover, it is rather likely that present experiments do not reveal the real equilibrium properties of the system but merely probe the behaviour in a certain metastable state.

Within the present approach the next work to be done should be the determination of β/ν from a multi-droplet theory as worked out by Bruce and Wallace [11] for the pure case. It would be also important to see, how a systematic expansion in ε could be constructed. This task includes the determination of higher-order terms in the Gell-Mann-Low function (2.8), which is probably very hard. Moreover, it would be more satisfactory if one could avoid the use of the hyperscaling relation (4.6) or, possibly, to prove it to a certain order in ε.

Acknowledgements

The author thanks Prof. W. Ebeling for providing the possibility to perform this work and A. Engel for critical remarks on the manuscript. In particular, he is much indepted to Prof. J. Villain for a discussion in the early stage of the calculation which has stimulated the further work.

References

[1] Y. IMRY and S. K. MA, Phys. Rev. Letters **35**, 1399 (1975).
[2] A. AHARONY, Y. IMRY, and S. K. MA, Phys. Rev. Letters **37**, 1367 (1976).
A. P. YOUNG, J. Phys. C **10**, L 257 (1977).

[3] E. PYTTE, Y. IMRY, and D. MUKAMEL, Phys. Rev. Letters **46**, 1173 (1981).
K. BINDER, Y. IMRY, and E. PYTTE, Phys. Rev. B **24**, 6736 (1981).
D. MUKAMEL and E. PYTTE, Phys. Rev. B **25**, 4779 (1982).
[4] G. GRINSTEIN and S. K. MA, Phys. Rev. Letters **49**, 685 (1982); Phys. Rev. B **28**, 2588 (1983).
[5] J. VILLAIN, J. Physique Letters **43**, L 551 (1982).
J. VILLAIN, B. SÉMÉRIA, F. LANCON, and L. BILLARD, J. Phys. C **16**, 6153 (1983).
[6] Y. IMRY, J. statistical Phys. **34**, 849 (1984).
J. C. TOLEDANO, Ann. Telecomm. **39**, 277 (1984).
G. GRINSTEIN, J. appl. Phys. **55**, 2371 (1984).
T. NATTERMANN, Lecture Notes in Physics, Vol. 206, Springer-Verlag, Heidelberg 1984; J. Phys. C **16**, 6407 (1983).
[7] J. VILLAIN, in: Elementary Excitations and Fluctuations in Magnetic Systems, Ed. S. LOVESEY, V. TOGNETTI, and U. BALUCANI, Springer-Verlag, Heidelberg 1985.
J. VILLAIN, in: Proc. Geilo Conf., Springer-Verlag, 1985.
[8] M. SCHWARTZ, Tel-Aviv University Report, TAUP 1139, 1983.
[9] Y. SHAPIR, Phys. Rev Letters **54**, 154 (1985); J. appl. Phys., to be published.
[10] D. P. BELANGER, A. R. KING, and V. JACCARINO, Phys. Rev. Letters **49**, 1050 (1982); J. appl. Phys. **55**, 2383 (1984).
D. P. BELANGER, A. R. KING, V. JACCARINO, and J. L. CARDY, Phys. Rev. B **29**, 2522 (1983).
I. B. FERREIRA, A. R. KING, V. JACCARINO, and J. L. CARDY, Phys. Rev. B **28**, 5192 (1983).
S. M. REZENDE, A. R. KING, and V. JACCARINO, J. appl. Phys. **55**, 2413 (1983).
[11] D. J. WALLACE, Perturbative Approach to Surface Fluctuations, Edinburgh Preprint 82/234, to be published in Proc. Ecole d' Eté de Physique Theorique, Les Houches 1982, North-Holland Publ. Co.
A. D. BRUCE and D. J. WALLACE, J. Phys. A **16**, 1721 (1983).
A. D. BRUCE, J. Phys. A **17**, 3363 (1984).
[12] J. VILLAIN, Phys. Rev. Letters **52**, 1543 (1984).
[13] D. P. LANDAU, H. H. LEE, and W. KAO, J. appl. Phys. **49**, 1358 (1978).
E. P. RASMUSSEN, M. A. NOVOTNY, and D. P. LANDAU, J. appl. Phys. **53**, 1925 (1982).
[14] J. A. KHURANA, F. J. SECO, and A. HOUGHTON, Phys. Rev. Letters **54**, 357 (1985).

(Received June 14, 1985)

Note added in proof:

After submitting this paper for publication we received two very interesting preprints by Villain [a] and Bray and Moore [b] who derived scaling relations between *three* independent exponents. In our notation, the third exponent results from a non-trivial renormalization of H_R: $\Delta = \beta/\nu - \zeta/2$ with $\zeta \geqq 0$ and likely $\zeta \to 0$ for $d \to 2$. ζ is related to the anomalous dimension $\bar{\eta}$ entering the disconnected part of the correlation function, $\bar{\eta} = 2\eta - \zeta$.

Since T is a dangerous irrelevant variable, the hyperscaling relation is modified to $\nu(d + \lambda_t) = 2 - \alpha$ where λ_t denotes the thermal eigenvalue of the RNG recursion relation linearized at the zero temperature fixed point h_c [c]. From $\partial\sigma/\partial l = H_l \cdot F(h_l, t_l)$, (2.7) and (3.3) $\lambda_t = (-d/2) + \Delta = -2 + \eta - \zeta$, irrespective of the details of the function F'. The d-independent scaling laws remain unchanged. Hence, if $\zeta > 0$, $\mu = \nu(d - 1 + \lambda_t)$, $\nu^{-1} = 2(\varepsilon + 2\Delta)/3$, the other exponents are related to ν via the modified scaling relations. Contrasting theoretical results are $\nu^{-1} = 2\varepsilon/3$ [a] and $\nu^{-1} = \varepsilon$, $\zeta = \beta = 0$ [b], respectively, whereas Birgenau et al. [d] found recently $\nu = 1.52 \pm 0.13$ for a 3-dimensional experimental realization of the RFIM. The latter would be in accord with our results if $\Delta \simeq 0$.

[a] J. VILLAIN, Journale de Physique, Nov. 1985.
[b] A. J. BRAY and M. A. MOORE, subm. to J. Phys. C.
[c] G. GRINSTEIN, Phys. Rev. Lett. **37**, 944 (1976).
[d] R. J. BIRGENEAU, R. A. COWLEY, G. SHIRANE, and Y. YOSHIZAWA, Phys. Rev. Lett. **54**, 2147 (1985).

phys. stat. sol. (b) 131, 573 (1985)

Subject classification: 18.2; 22.8

Laboratoire de Magnétisme, associé au C.N.R.S., Meudon[1])

Frequency Dependence of the AC Susceptibility of the Reentrant System $CdCr_{2x}In_{2-2x}S_4$ $(0.85 < x \leq 1)$

By

J. L. DORMANN, A. SAIFI, V. CAGAN, and M. NOGUES

Dedicated to Prof. Dr. Dr. h. c. Dr. E. h. P. GÖRLICH on the occasion of his 80th birthday

AC permeability measurements of the reentrant spin glass system $CdCr_{2x}In_{2-x}S_4$ $(0.85 < x \leq 1)$ are performed on toroidal shaped samples by using the transformer technique since the demagnetizing factor is minimized. The complex permeability observed is very close to the true initial permeability. The abrupt decrease of μ' at low temperatures is an intrinsic effect, probably related to the transition to the reentrant spin glass state. The out-of-phase permeability presents a maximum at the transition temperature T_{RSG}. The frequency dependence of T_{RSG} is studied (130 Hz $\leq \nu \leq$ 100 kHz). Its variation corresponds with the explanation of reentrant properties based on the existence of clusters of which the freezing destroys the ferromagnetic order.

En utilisant la méthode du transformateur, des mesures de perméabilité alternative ont éte effectuées sur des échantillons en forme de tore du système réentrant verre de spin $CdCr_{2x}In_{2-x}S_4$ $(0{,}85 < x \leq 1)$. Le facteur démagnétisant étant minimisé, la perméabilité complexe observée est très proche de la perméabilité initiale intrinsèque. A basse température, la décroissance rapide de μ' est un effet intrinsèque, probablement relié à la transition vers un état spin glass réentrant. La partie complexe de la perméabilité présente un maximum à la température de transition T_{RSG}. La variation de T_{RSG} avec la fréquence a été étudiée (130 Hz $\leq \nu \leq$ 100 kHz); elle est en accord avec l'interprétation des propriétés réentrantes basée sur l'existence de clusters dont le gel détruit l'ordre ferromagnétique.

1. Introduction

One of the most interesting phenomena which has been recently studied in the spin glass field is the so-called reentrant transition from a ferromagnetic (FM) to a spin glass (SG) state. Predicted by Sherrington and Kirkpatrick (SK) [1], an unbalance between both the positive and negative interactions can lead to another transition: at decreasing temperatures, the paramagnetic phase (PM) first evolves into an ordered FM phase and then into a disordered SG state. From the SK mean field model, the influence of an applied field H has been deduced. For an Ising spin glass, it appears a transition line from PM to replica symmetry broken order, i.e. the de Almeida-Thouless (AT) line [2]. In a Heisenberg model that transition corresponds to a crossover from weak to strong irreversibility and another transition line appears, i.e. the Gabay-Toulouse (GT) line which corresponds to the beginning of transverse irreversibility [3, 4]. By continuity the same lines exist for $H = 0$.

Reentrant spin glass (RSG) behaviour has been observed in a wide variety of materials such as polycrystalline metals Fe_xCr_{1-x} [5], Fe_xAu_{1-x} [6], Mn_xNi_{1-x} [7], and $Fe_{80-x}Ni_xCr_{20}$ [8], amorphous metals $(MT_xMT'_{1-x})_{75}P_{16}B_6Al_3$ where MT = Fe or Co, MT' = Mn or Ni [9], $(Fe_xNi_{1-x})_{80}P_{14}B_6$ [10], $(Fe_xNi_{1-x})_{78}Si_9B_{13}$ [11], and $Ni_{70}Fe_{10}$·

[1]) 92195 Meudon, France.

$\cdot Si_8B_{12}$ [12], crystalline semiconductors $Eu_xSr_{1-x}S$ [13], EuS_xSe_{1-x} [14], and $Eu_{1-x}Gd_xS$ [15] and insulators $KMn_xZn_{1-x}F_3$ [16] and $CdCr_{2x}In_{2-2x}S_4$ [17]. However, the nature of the low temperature state remains unclear and two possibilities seem to exist. For the concentrated magnetic materials, the neutron scattering studies on $Eu_xSr_{1-x}S$ and $(Fe_xMn_{1-x})_{75}P_{16}B_6Al_3$ [18] have not shown the coexistence of FM order at low temperature. In disagreement with the GT model these properties could be explained by the existence of random fields due to the freezing of finite clusters decoupled from the infinite network. These random fields can destroy the long-range magnetic order [19]. For the less concentrated magnetic materials, Mössbauer experiments on Fe_xAu_{1-x} [20] have shown the existence of a canted state, i.e. a superposition of FM and SG orders in agreement with the GT model. Neutron scattering experiments on $Fe_{19}Au_{81}$ [21] and $Ni_{79}Mn_{22}$ [22] have revealed well-defined spin waves at all temperatures below T_C which, together with other results, suggests that FM and SG orders coexist at low temperatures [21]. However, that latter interpretation is still controversial [18].

RSG properties have been often deduced from susceptibility measurements: when the temperature decreases below T_C, the real part χ' shows a plateau due to the demagnetizing field and then strongly falls down at a given temperature. However, such a decrease of χ' could be also due to the onset of the anisotropy and hysteresis behaviour not related to the existence of a transition [23]. A better indicator is the variation of the maximum slope of the hysteresis loop; however, it is difficult in that case to define the transition temperature T_{RSG}. This problem can be partially solved by looking at the imaginary part of the susceptibility χ''_{ac}. The presence of a χ''_{ac} cusp clearly reveals a magnetic transition and T_{RSG} can be defined as the χ''_{ac} maximum. Only a very few χ''_{ac} measurements have been performed on materials showing RSG properties.

To our knowledge, no measurement of the frequency dependence of T_{RSG} exists; such results could clear up the debated nature of the RSG state, according as the T_{RSG} time dependence is similar or not to the variation of T_F for adjoining composition showing only paramagnetic–SG transition.

In this context, we carried out χ_{ac} measurements on the $CdCr_{2x}In_{2-2x}S_4$ RSG system [17], by using a new technique which permits the macroscopic demagnetizing field to be eliminated and a wide frequency range to be covered. Preliminary results of this study have been presented at the MMM Conference [24].

2. Experimental

The synthesis and characterization of powder samples are described elsewhere [25]. For $2x = 2$, 1.93, 1.90, and 1.80 the toroid shaped samples, required for susceptibility measurements, have been prepared by pressing the powder at four tons and then by annealing the samples under a H_2S stream at 900 °C for 12 h. Powder from the same batch as for our earlier measurements has been used.

The annealing temperature and the cooling rate are the same for the synthesis, in order to avoid any change of the cation distribution; they did not permit a very low porosity to be obtained.

The transformer method used for susceptibility measurements allows the macroscopic demagnetizing factor to be eliminated, a direct calibration of the complex permeability μ^* to be eliminated and recording at various measuring times to be performed. In the present study, measurements have been limited to the frequency range 130 Hz to 100 kHz, although the method permits the range to be extended at least up to 1 GHz. The transformer technique used for complex permeability measure-

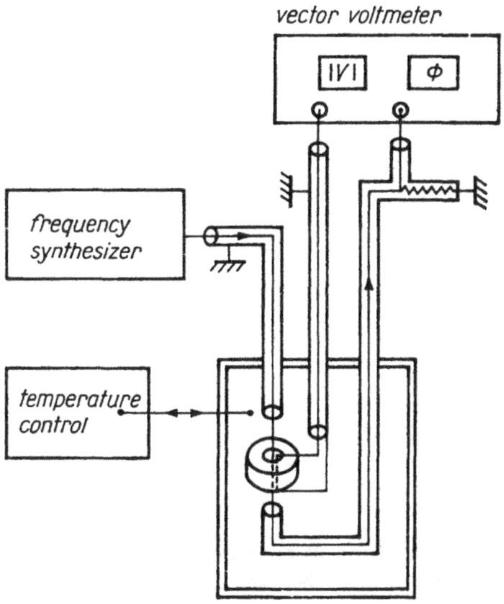

Fig. 1. Schematic diagram of the permeability measurement set-up at moderate frequency

ments has been already described [26]. Designed at first for high frequencies, it also allows easy measurements to be performed in the lower part of the frequency range, compared with the conventional technique using impedance determination. Fig. 1 shows a schematic diagram.

The toroid shaped sample is magnetized by a primary winding feeded from the ac generator. The bad resistor is used to measure the value of the magnetizing current as well as to obtain an in-phase reference signal. The flux variations inside the sample produce a complex voltage induced in the secondary winding, which is measured by a vector voltmeter. The real and imaginary parts of the permeability are then determined. In the present experimental conditions, both the primary and secondary windings have the same number of turns. To be placed in the low temperature cell, the pressed toroid shaped sample is sealed with epoxy in a plastic container; the windings are inserted inside the thickness of the two parallel flanges, so that the sample can be tighted between two brass discs thermally connected to the variable temperature arrangement.

On the other hand, hysteresis loops have been recorded. Since the purpose of such an experiment was to check the possible influence of the demagnetizing factor on the initial permeability value and on the slope of the loop, we use the same toroid shaped sample and then the same measurement cell.

A "burst" technique allows "high" fields to be applied to the sample with no excessive heating due to the winding or to the losses in the material.

The method [27] consists in feeding the primary winding by a short sequence of ac current for only a few periods. The induced voltage is integrated and then recorded by using a digital oscilloscope controlled by a microcomputer. For the present measurements, the frequency 1 kHz has been chosen; however, the geometrical and technical limitations have not permitted applied field higher than 26 Oe to be reached.

3. Results and Discussion

3.1 μ' values

Fig. 2 presents the temperature dependence of μ^* for $2x = 1.9$, and Fig. 3 the variation of μ'/μ'_{max} for $2x = 2$, 1.93, 1.90, and 1.8. Several observations can be done:

for $2x = 2$, 1.93, and 1.9, a plateau (Fig. 3) is observed, revealing the effect of a demagnetizing field. The Curie temperature T_C can be easily deduced;

for $2x = 1.8$, μ' behaves differently and does not present an abrupt decrease near T_C, showing that the FM order is already very perturbed for this dilution;

for $2x = 1.93$, 1.9, and 1.8, a rapid decrease of μ' is observed at low temperature and the curve is slightly shifted in temperature when the frequency varies (Fig. 2).

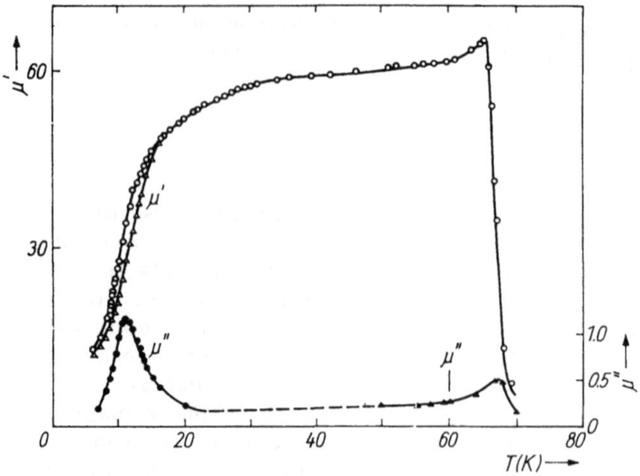

Fig. 2. Temperature dependence of the complex permeability, for $CdCr_{1.9}In_{0.1}S_4$, as function of the two extrema frequencies used: 130 Hz and 100 kHz. ○ ν = 130 Hz, △ 100 kHz, ● and ▲ 1 kHz,

The important question is to know whether the abrupt decrease of μ'_{ob} at low temperature is in our case a specific effect due to the disappearance of long-range FM order or an artefact due to the broadening of the hysteresis loop [23].

Theoretically, for FM materials, μ' diverges at T_C. If demagnetizing fields exist, the intrinsic permeability μ^*_{in} can be deduced from the observed permeability μ^*_{ob} by

$$\mu^*_{in} = \mu^*_{ob} \frac{4\pi - N}{4\pi - N\mu^*_{ob}}, \tag{1}$$

where N is the demagnetizing factor along the applied field direction ($\sum N = 4\pi$). In this case, at T_C, μ'_{ob} shall be limited by

$$\mu'_{ob}(T = T_C) = \frac{4\pi}{N}. \tag{2}$$

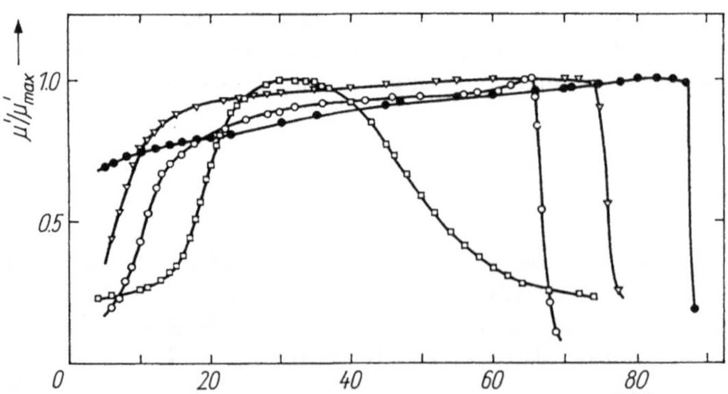

Fig. 3. Temperature dependence of the in-phase permeability in the system $CdCr_{2x}In_{2-2x}S_4$ ($x > 0.85$) for ν = 1 kHz. □ x = 0.9, ○ 0.95, ▽ 0.965, and ● 1.0

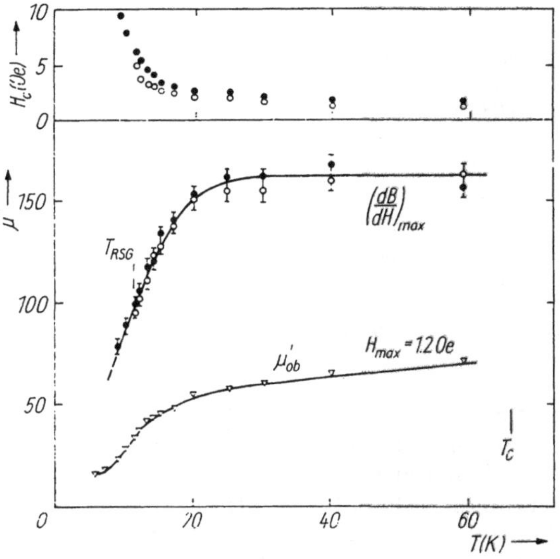

Fig. 4. Temperature dependence of the in-phase permeability μ'_{ob} (H_{max} = = 1.2 Oe), of the hysteresis loop maximum slope (H_{max} = 11.5 (○) and 26.5 Oe (●), and of the coercive force, for $CdCr_{1.9}In_{0.1}S_4$

Really, μ'_{in} does not diverge at T_C. The existence of a magnetic short-range order will round μ'_{in}. On the other hand, if the anisotropy does not vanish near T_C, a hysteresis loop will be observed. In this case the initial permeability μ'_{ob} is smaller than $4\pi/N$, which is the factor limiting the maximum slope $(dB/dH)_{max}$ of the hysteresis loop.

Fig. 4 presents the results of hysteresis loop measurements for $2x = 1.9$. Three maximum field values have been chosen. (i) $H_{max} = 1.2$ Oe which permits to obtain the initial permeability, (ii) $H_{max} = 11.5$ Oe, and (iii) $H_{max} = 26.2$ Oe which permit $(dB/dH)_{max}$ to be determined. At low temperatures ($T \leqq 10$ K), the H_{max} value we used is not high enough to determine $(dB/dH)_{max}$ due to the anisotropy; for other temperatures we obtain a good approximation. One can also determine a value of the coercive force H_c for the two fields H_{max}; since the H_{max} field values are too small to saturate the sample, the true H_c value should be much higher. The μ'_{ob} values ($H_{max} = 1.2$ Oe) are in agreement with the values obtained with the other method ($H_{max} \approx 0.1$ Oe) (Fig. 2) with an accuracy better than 3%, except at low temperature ($T \leqq 10$ K) and near T_C where the values for $H = 1.2$ Oe are higher than for $H \approx$ ≈ 0.1 Oe. At low temperature, the hysteresis effects become larger and could result in such a discrepancy.

From our results it is clear that the hysteresis effects exist in the whole temperature range below T_C. μ'_{ob} is smaller than $(dB/dH)_{max}$ limited by $4\pi/N$. For $T \geqq 20$ K, the real plateau is observed for $(dB/dH)_{max}$. For $T < 20$ K, an abrupt decrease of $(dB/dH)_{max}$ is observed, like for μ'_{ob}, ruling out for the variation of this latter parameter the possibility of an artefact. For $T > 10$ K, the ratio $(dB/dH)_{max}/\mu'_{ob}$ is approximatively constant ($\approx 2.75 \pm 0.15$), coherent with the small variation of H_c.

In our case, the macroscopic demagnetizing factor along the applied field direction is equal to zero due to the sample shape, but microscopic demagnetizing fields due to the porosity can exist. Such fields may act, roughly, as a macroscopic demagnetizing field. Similar results are obtained for $2x = 2$; no abrupt decrease of $(dB/dH)_{max}$ is observed like for μ'_{ob}. For $2x = 1.8$, our present possible maximum field is not high enough to determine $(dB/dH)_{max}$: only a minor loop can be observed. We have indicated, in Table 1, for the different samples:

(i) the μ'_{max} values corresponding to the maximum value of μ'_{ob}, (ii) the value of $(dB/dH)_{plateau}$ corresponding to $(dB/dH)_{max}$ in the plateau region, and (iii) the porosity p defined by $p = 1 - (d_{ob}/d_{th})$ where d_{ob} and d_{th} are the observed and theoretical densities, respectively. We have also indicated the ratios $4\pi/\mu'_{max}$, $4\pi/(dB/dH)_{plateau}$ which corresponds to the equivalent macroscopic demagnetizing factor N. We can see that the values of $4\pi/(dB/dH)_{plateau}$ are almost constant for porosity ranging from

Table 1

μ'_{max} values corresponding to the μ'_{ob} maximum, $(dB/dH)_{plateau}$ corresponding to the values of $(dB/dH)_{max}$ in the plateau region, porosity p (cf. text), ratios $4\pi/\mu'_{max}$ and $4\pi/(dB/dH)_{plateau}$ as function of composition x

$2x$	2	1.93	1.90	1.80
μ'_{max}	33	39.5	61	19
$(dB/dH)_{plateau}$	200	—	163	—
p	0.44	0.43	0.33	0.40
$4\pi/\mu'_{max}$	0.38	0.32	0.19	0.66
$4\pi/(dB/dH)_{plateau}$	0.063	—	0.077	—

0.33 to 0.44; we can suppose that for $2x = 1.93$ and $2x = 1.80$, $N \approx 0.08 \pm 0.02$ since the porosity values are 0.43 and 0.40, respectively.

For $2x = 1.8$, the ratio μ'_{in}/μ'_{ob} can be calculated by using the formula (1) and the values of μ'_{max} and N (Table 1). The maximum value of this ratio is 1.13. μ'_{in} is then very close to μ'_{ob} for all temperatures. The low values of μ'_{in} confirm that the FM order is perturbed even at high temperature. One must note that observing the non-divergence of χ'_{in} was not possible from classical susceptibility measurements, where the demagnetizing factor N is often near 2.5.

For the other compositions, in which FM order does not seem perturbed at high temperature, the existence of microscopic demagnetizing fields does not permit the variation of μ'_{in} to be observed, although the equivalent demagnetizing factor is approximatively equal to 0.1. Nevertheless, these low values permitted the shift with frequency to be observed (Fig. 2).

3.2 μ'' values

The variation of μ'' is represented in Fig. 5 and 6 for $2x = 1.9$ and 1.8, respectively, in the low temperature range where an anomalous decrease and frequency shift of μ' exist. μ'' presents a relatively sharp cusp which is frequency dependent, almost symmetrical for $2x = 1.9$ and clearly dissymetrical for $2x = 1.8$. A light frequency shift of μ' is also observed.

An important question is to appreciate the influence of the demagnetizing field on μ''_{ob}, in particular the difference between the temperature maximum of μ''_{ob} and μ''_{in}, $(T_{RSG})_{ob}$ and $(T_{RSG})_{in}$, respectively, and the shape modification of μ''_{ob}. The answer is difficult since μ^*_{in} is unknown. If we consider the susceptibility, usually used as measurement, we can deduce from (1)

$$\mu^*_{ob} - 1 = (4\pi - N)\,\chi^*_{ob} \tag{3}$$

and with $\chi''_{ob} \ll \chi'_{ob}$

$$\chi'_{ob} = \frac{\chi'_{in}}{1 + N\chi'_{in}} \quad \text{and} \quad \chi''_{ob} = \frac{\chi''_{in}}{(1 + N\chi'_{in})^2}. \tag{4}$$

If we suppose, as a first approximation, that χ''_{in} is Lorentzian and that χ''_{ob} varies linearly near $(T_{RSG})_{ob}$, it is possible to evaluate $(T_{RSG})_{ob} - (T_{RSG})_{in}$. For $2x = 1.93$ and 1.90, this value is near -0.2 K and is negligible for $2x = 1.80$, i.e. $(T_{RSG})_{ob}$ is lowered by N. For $N = 2.5$, the shift should be near -1.4 K for $2x = 1.93$ and 1.90 and near -0.8 K for $2x = 1.8$. These calculations are, of course, rather approximative, but the tendency is experimentally verified. For example, for $2x = 1.90$, $N \approx 2.5$

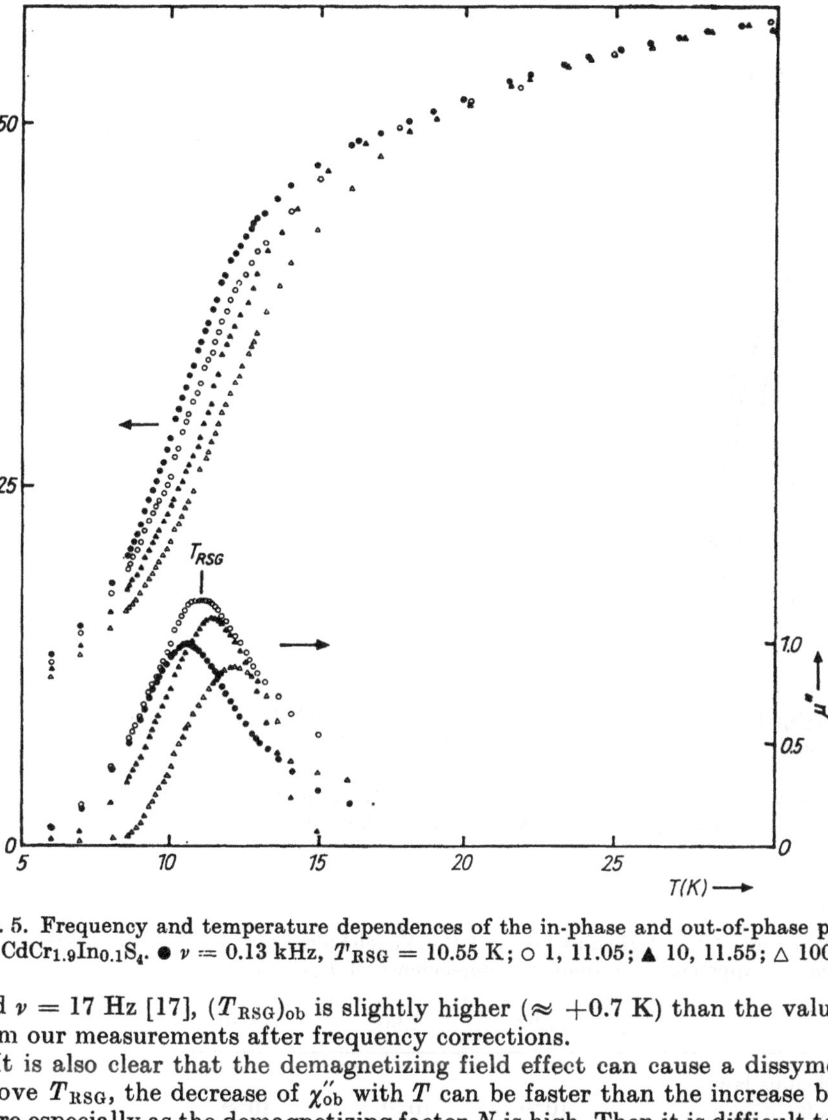

Fig. 5. Frequency and temperature dependences of the in-phase and out-of-phase permeabilities for $CdCr_{1.9}In_{0.1}S_4$. ● $\nu = 0.13$ kHz, $T_{RSG} = 10.55$ K; ○ 1, 11.05; ▲ 10, 11.55; △ 100, 12.20

and $\nu = 17$ Hz [17], $(T_{RSG})_{ob}$ is slightly higher ($\approx +0.7$ K) than the values deduced from our measurements after frequency corrections.

It is also clear that the demagnetizing field effect can cause a dissymetry of χ''_{ob}. Above T_{RSG}, the decrease of χ''_{ob} with T can be faster than the increase below T_{RSG}, more especially as the demagnetizing factor N is high. Then it is difficult to interprete the shape of χ''_{ob} obtained by conventional susceptibility measurements ($N \approx 2.5$). In our case, N is small and the shape of μ''_{ob} is near those of μ''_{in}. The clearly dissymetrical shape of μ''_{ob} for $2x = 1.8$ is probably related to the perturbed FM order existing at higher temperature. A last observation about the frequency effect is the following: if μ^*_{in} is only shifted in temperature, the same shift occurs for μ^*_{ob}. Our experimental results confirm this hypothesis since the observed shift of μ'_{ob} to low temperature is approximatively equal to the variation of $(T_{RSG})_{ob}$ with frequency.

3.3 Frequency variation

The temperature T_{RSG} of the maximum of μ''_{ob} is frequency dependent. The same tendency is observed for μ'_{ob} and the shift is approximately equal. The frequency dependence of $(T_{RSG})_{ob}$ is represented on Fig. 7 in a plot T versus $1/(\ln \nu_0 - \ln \nu)$. This

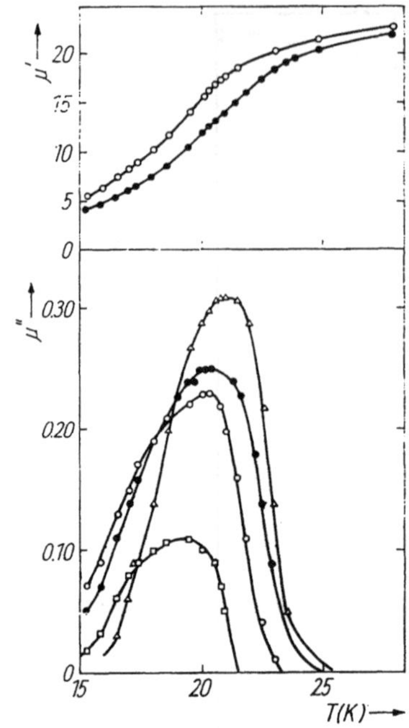

Fig. 6

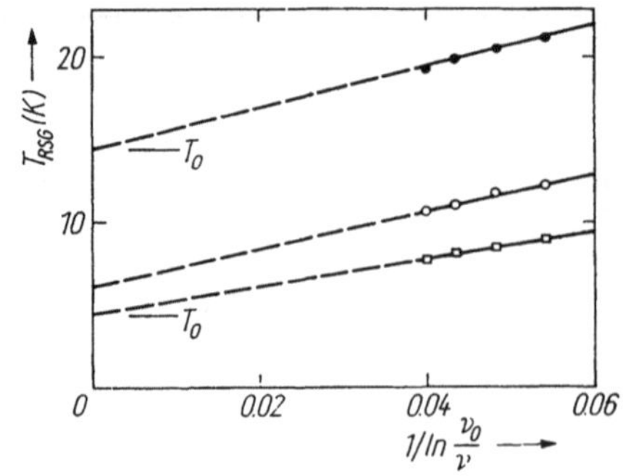

Fig. 7

Fig. 6. Frequency and temperature dependences of the in-phase and out-of-phase permeabilities for $CdCr_{1.8}In_{0.2}S_4$. For the upper part: ○ $\nu = 0.13$ and ● 100 kHz; for the lower part: △ $\nu = 100$ kHz, $T_{RSG} = 21.1$ K; ● 10, 20.5; ○ 1, 19.9; □ 0.13, 19.2

Fig. 7. Plot of T vs. $1/(\ln \nu_0 - \ln \nu)$, with $\nu_0 = 10^{13}\ s^{-1}$, giving the frequency dependence of T_{RSG} in the system $CdCr_{2x}In_{2-2x}S_4$ ($x > 0.85$). ● $x = 0.9$, ○ 0.95, □ 0.965

Table 2

Parameters of the Arrhenius law and the Vogel-Fulcher law; values of criteria which permit to appreciate the frequency dependence of T_{RSG} (see text)

$2x$	1.93	1.90	1.80
Arrhenius law			
τ_0 (s)	10^{-24}	10^{-24}	10^{-34}
Ea/k (K)	395	525	1430
Vogel-Fulcher law			
τ_0 (s)	10^{-13}	10^{-13}	10^{-13}
T_0 (K)	4.5	6.05	14.05
Ea/k (K)	81	114	132
$\dfrac{\Delta T_{RSG}}{T_{RSG} \log \nu}$	0.05	0.045	0.035
$\dfrac{Ea}{kT_{RSG}}$	10.5	10.8	6.9
$\dfrac{T_{RSG} - T_0}{T_{RSG}}$	0.42	0.43	0.27

variation can fit either an Arrhenius law or a Vogel-Fulcher law [28]. The parameters of these laws are given in Table 2, but for the Arrhenius law, the parameters have no physical meaning.

The nature of the transition from the paramagnetic phase to a SG state is still controversial. Is it either a true thermodynamic phase transition at a well defined temperature T_f, or a dynamic non-equilibrium process which is produced in an extended temperature range due in particular to a progressive freezing of cluster moments, with superparamagnetic type properties? The frequency dependence of T_f was proposed as a possible criterion of distribution. The ratio $\Delta T_f/(T_f \Delta \log \nu)$, corresponding to the relative variation of T_f per decade in a given frequency range, has been suggested [28] but this criterion does not allow a clear distinction between the two behaviours for systems with intermediate frequency dependence like $Eu_xSr_{1-x}S$. Recently another criterion has been proposed when the Vogel-Fulcher law is followed [29][2]), i.e. the parameters $(T_f - T_0)/T_f$ or $Ea/(kT_f)$, for a given frequency, which seems to permit a better distinction.

In the case of RSG state, the frequency dependence of T_{RSG} can be predicted only when this state is due to the freezing of finite clusters which destroys the FM order [19]. It must be similar to those of spin glasses of which the interpretation is based on the freezing of clusters.

In Table 2 are indicated the values of $\Delta T_f/(T_f \Delta \log \nu)$, $(T_f - T_0)/T_f$, and $Ea/(kT_f)$ with $T_f = (T_{RSG})_{ob}$ for $\nu = 130$ Hz and in Table 3 the same parameters ($\nu = 10$ to 20 Hz) for typical spin glass materials [28 to 32]. In this latter table, we have added the results obtained on small magnetically interacting particles [31] of which the variation of the blocking temperature can be represented in first approximation by a Vogel-Fulcher law.

For the calculation of Table 2, we have used the $(T_{RSG})_{ob}$ values since $(T_{RSG})_{in}$ are unknown. Nevertheless, since the difference between these two temperatures does not approximatively depend on the frequency, we can predict that $(Ea/k)_{in} \approx \approx (Ea/k)_{ob}$, $(T_0)_{in} = (T_0)_{ob} - (T_{RSG})_{ob} + (T_{RSG})_{in}$ and that the criteria $\Delta T_f/(T_f \Delta \log \nu)$, $(T_f - T_0)/T_f$, and $Ea/(kT_f)$ are multiplied by a factor ranging from 0.95 to 1, which does not change the order of magnitude.

Table 3

Values of criteria which permit to appreciate the frequency dependence of T_f for typical spin glass materials

	$\dfrac{\Delta T_f}{T_f \Delta \log \nu}$	$\dfrac{T_f - T_0}{T_f}$	$\dfrac{Ea}{kT_f}$	ref.
$La_{0.994}Gd_{0.006}Al_2$	0.11	1	30	29
$Eu_{0.2}Sr_{0.8}S$	0.04	0.5	14	28
$ZnCr_{1.6}Ga_{0.4}O_4$	0.05	—	—	30
Small interacting particles Fe/Al_2O_3	0.05	0.5	11	31
$Eu_{0.4}Sr_{0.6}S$	0.02	0.25	7	29
Au–Fe 10%	0.008	0.10	2.8	29
$CsNiFeF_6$	0.01	0.08	1.9	32
Cu–Mn	0.005	0.07	2.1	29
$Eu_{0.5}Gd_{0.5}S$	—	0.04	0.85	29

[2]) See also references therein.

Table 3

Parameters of the power law and the generalized Arrhenius law, with $\tau_0 = 10^{-13}$ s

$2x$		1.93	1.90	1.80
power law				
($T_C \neq 0$)	$Z\nu$	17	18	15
	T_C	6	8	16
generalized Arrhenius law ($T_C = 0$)				
$Z\nu$		2.2	2.1	3.1

The properties of the materials listed at the top of Table 3 are rather relevant to progressive freezing of cluster moments (this is true of course for small particles); whereas for the materials listed at the bottom, a true phase transition occurs probably at T_f. It is difficult to classify $Eu_{0.4}Sr_{0.6}S$ (intermediate properties ?). The criterion values obtained for our system correspond to the first category, very near those obtained for small interacting particles for $2x = 1.93$ and 1.90 and near those of $Eu_{0.4}$·$Sr_{0.6}S$ for $2x = 1.80$. Note that for $2x = 1.70$, a SG state is observed and the criterion values decrease strongly [33].

Our results were also analyzed with two other laws: the power law $\tau = \tau_0 |T/(T - T_C)|^{Z\nu}$ [34], and the generalized Arrhenius law $\ln (\tau/\tau_0) \sim T^{-Z\nu}$ [35]. Considering the experimental accuracy and the studied restricted frequency range, the solutions are not unique and good fits can be find in a wide range of τ_0. The results are given in Table 4 for $\tau_0 = 10^{-13}$ with the aim to compare it with the Vogel-Fulcher law (Table 2). The same trend is observed than for the criteria given above; for $2x = 1.8$, the exponents are nearer those materials showing a true phase transition. Measurements in a wider frequency range could permit to refine the parameter and to choose among these laws.

From a point of view of T_f variation with frequency, our results favour the explanation of reentrant properties based on the existence of clusters, decoupled from the infinite network, the freezing of which builds up random fields destroying the FM order [19]. When x decreases, the cluster number increases; for $2x = 1.8$, this number if sufficiently high to hardly perturb the FM order at high temperature and the state obtained is probably intermediate between a FM order with RSG properties and a SG order, coherent with the criterion values.

4. Conclusion

By using a transformer method to measure the permeability of toroidal shaped samples, we have almost eliminated the demagnetizing fields and obtained μ^* values very near the intrinsic permeability. Using this method has permitted the abrupt decrease of μ'_{ob} to be shown as an intrinsic effect, probably related to the transition to a reentrant spin glass state. In addition it has been shown that for $2x = 1.8$, i.e. near the composition where a paramagnetism–spin glass transition is observed, the ferromagnetism order is hardly perturbed even at high temperature. From μ'' measurements, the frequency dependence of the temperature T_{RSG} has the same order of magnitude than those of spin glass systems of which the properties are rather relevant to progressive freezing of cluster moments; that favours the interpretation of reentrant properties based on the existence of clusters of which the freezing destroys the ferromagnetic order.

Acknowledgements

The authors are indebted to M. Guyot for helpful discussions and hysteresis measurements and would like to thank P. Renaudin for his very skilled technical assistance.

References

[1] D. Sherrington and S. Kirkpatrick, Phys. Rev. Letters **35**, 1792 (1975).
[2] J. R. L. de Almeida and D. J. Thouless, J. Phys. A **11**, 983 (1978).
[3] M. Gabay and G. Toulouse, Phys. Rev. Letters **47**, 201 (1981).
[4] D. M. Cragg, D. Sherrington, and M. Gabay, Phys. Rev. Letters **49**, 158 (1982).
[5] S. H. Burke, R. Cywinski, J. R. Davis, and B. D. Rainford, J. Phys. **13**, 441, 451, 471 (1983).
[6] B. R. Coles, B. V. B. Sarkissian, and R. H. Taylor, Phil. Mag. **B37**, 489 (1978).
[7] R. G. Aitken, T. D. Cheung, J. S. Kouvel, and H. Hurdequint, J. Magnetism magn. Mater. **30**, L1 (1982).
[8] A. K. Majumdar and P. V. Blanckenhagen, Phys. Rev. B **29**, 4079 (1984).
[9] Y. Yeshurun, M. B. Salamon, K. V. Rao, and H. S. Chen, Phys. Rev. B **24**, 1536 (1981).
J. A. Geohegan and S. M. Bhagat, J. Magnetism magn. Mater. **25**, 17 (1981).
[10] L. M. Kistler and S. M. Bhagat, J. Phys. C **15**, L929 (1982).
[11] V. Manns, R. A. Brand, W. Keune, and R. Marx, Solid. State Commun. **48**, 811 (1983).
[12] Girish Chandra, J. L. Dormann, R. Krishnan, and P. Renaudin, to be published.
[13] H. Maletta, G. Aeppli, and S. M. Shapiro, J. Magnetism magnetic Mater. **31**–**34**, 1367 (1983).
[14] K. Westerholt and H. Bach, J. Phys. F **12**, 1227 (1982).
[15] A. Burton, C. Chaussy, J. Odin, R. Rammal, J. Souletie, J. L. Tholence, R. Tournier, F. Holtzberg, and S. von Molnar, J. appl. Phys. **52**, 1763 (1981).
[16] R. A. Cowley, G. Shirane, R. J. Birgeneau, E. C. Svensson, and H. J. Guggenheim, Phys. Rev. B **22**, 4412 (1980).
[17] M. Alba, J. Hammann, and M. Nogues, J. Phys. C **15**, 5441 (1982).
[18] G. Aeppli, S. M. Shapiro, H. Maletta, R. J. Birgeneau, and H. S. Chen, J. appl. Phys. **55**, 1628 (1984).
[19] G. Aeppli, S. M. Shapiro, R. J. Birgenau, and H. S. Chen, Phys. Rev. B **28**, 5160 (1983); **29**, 2589 (1984).
[20] F. Varret, A. Hamzic, and I. A. Campbell, Phys. Rev. B **26**, 5285 (1982).
J. Lauer and W. Keune, Phys. Rev. Letters **48**, 1850 (1982).
[21] B. Hennion, M. Hennion, F. Hippert, and A. P. Murani, Phys. Rev. B **28**, 5365 (1983).
[22] A. P. Murani, Phys. Rev. B **28**, 432 (1983).
[23] D. W. Carnegie, Jr., C. J. Tronchita, and H. Claus, J. appl. Phys. **50**, 7318 (1979).
[24] A. Saifi, V. Cagan, J. L. Dormann, and M. Nogues, MMM Conf., San Diego (USA) 1984.
[25] M. Nogues, A. Saifi, M. Hamedoun, J. L. Dormann, A. Malmanche, D. Fiorani, and S. Viticoli, J. appl. Phys. **53**, 7699 (1982).
C. Battistoni, E. Paparazzo, Y. Dumond, and M. Nogues, Solid State Commun. **46**, 333 (1983).
[26] C. Cagan and M. Guyot, IEEE Trans. Magnetics **20**, 1732 (1984).
[27] M. Guyot and V. Cagan, Mém. sci. Rev. Métall. **9**, 433 (1984).
[28] J. L. Tholence, Solid State Commun. **35**, 113 (1980); Physica (Utrecht) **108 B**, 1287 (1981).
[29] J. L. Tholence, Physica (Utrecht) **126B**, 157 (1984).
[30] D. Fiorani, S. Viticoli, J. L. Dormann, J. L. Tholence, and A. P. Murani, Phys. Rev. B **30**, 2776 (1984).
[31] J. L. Dormann, D. Fiorani, J. L. Tholence, and C. Sella, J. Magnetism magnetic Mater. **35**, 117 (1983).
J. L. Dormann, L. Bessais, D. Fiorani, and G. Villers, to be published.
[32] C. Pappa, J. Hammann, G. Jehanno, and C. Jacoboni, J. Phys. C **17**, 1303 (1984).
[33] J. Hammann, private communication.
[34] P. C. Hohenberg and B. I. Halperin, Rev. mod. Phys. **49**, 435 (1977).
[35] K. Binder and A. P. Young, Phys. Rev. B **29**, 2864 (1984).

(Received August 5, 1985)

phys. stat. sol. (b) 131, 585 (1985)

Subject classification: 13 and 18

Institute of Metal Physics, Academy of Sciences of the USSR, Sverdlovsk[1] *(a) and Ural State University, Sverdlovsk (b)*

On the Atomic Limit of the s–f Model

By

B. V. KARPENKO (a), A. V. KUZNETSOV (b), L. D. FALKOVSKAYA (a), and G. L. BUDRINA (b)

Dedicated to Prof. Dr. Dr. h. c. Dr. E. h. P. GÖRLICH on the occasion of his80th birthday

The atomic limit of the s–f model for a system in an external magnetic field is studied. By direct calculation of traces expressions are obtained for the first time for the partition function, average number of electrons, chemical potential, magnetization, susceptibility, mean energy, and specific heat at arbitrary values of electron concentration, spin, temperature, magnetic field, and exchange and Coulomb interaction parameters. It is shown that the system behaving itself in many respects as an ensemble of "quasi-particles" of four types, i.e. holes, down-units, up-units, and pairs is always paramagnetic and the susceptibility may obey both, Curie law and Curie-Weiss law. At subnormally small intra-atomic interactions Schottky anomalies may be observed in the system.

Исследована задача атомного предела s–f модели для системы, находяцейся во внешнем магнитном поле. Методом прямого вычисления шпуров впервые получены точные выражения для статистической суммы, среднего числа электронов, химического потенциала, намагниченности, восприимчивости, средней энергии и теплоемкости при произвольных значениях концентрации электронов, спина, температуры, магнитного поля и параметров обменного и кулоновского взаимодействия. Показано, что система, во многом ведущая себя как ансамбль „квазичастиц" четырех типов—дырок, левых единиц, правых единиц и двоек—всегда парамагнитна, а ее восприимчивость может следовать как закону Кюри, так и Кюри-Вейсса. При аномально малом внутриатомном взаимодействии в системе могут наблюдаться аномалии Шоттки.

1. Introduction

For a very long time the s–f (or s–d) model has remained one of the most widely used models for the study of metals. A tremendous number of papers, presenting various results for different physical effects in metals and semiconductors, have appeared since the publication of the first fundamental work [1]. However, very few exact results have been obtained as yet and the problem is far from being solved.

Since the perturbation theory is often used for investigations, it is interesting to study in detail this or that limiting case which could be considered as a zeroth approximation. For the s–f model two such limiting situations are weak and strong intra-atomic interactions. Below we consider the strong coupling, for which the zeroth approximation is the problem of intra-atomic exchange and Coulomb interactions. This problem is known as the atomic limit or zero-bandwidth case.

The atomic limit for the Hubbard model has been studied in [2], presenting, however the result only for the chemical potential at zero temperature. The excitation spectrun in the atomic limit of the s–f model for $S = \frac{1}{2}$ has been obtained in [3]. The numerical calculations of the thermodynamic averages for the atomic limit of the s–f model ac-

[1]) 620219 Sverdlovsk, GSP 170, USSR.

counting for the molecular field are presented in [4]. The thermodynamic properties of the s–f model in the atomic limit have been studied in [5].

In the present paper, for the first time, exact expressions in closed form in the atomic limit of the s–f model are presented for the chemical potential, mangetization, susceptibility, mean energy, and specific heat.

2. The Hamiltonian, Wave Functions, and Spectrum

In the atomic limit the conduction bandwidth is considered to be zero, i.e. the translational electron energy $\varepsilon_k = T_0 = \text{const}$ for any quasi-momentum k. T_0 denotes the mean energy of the band $T_0 = (1/N) \sum_k \varepsilon_k$, where N is the number of sites in the crystal. Thus, the transfer integral between sites R_i and R_j,

$$T_{ij} = \frac{1}{N} \sum_k \varepsilon_k \, e^{ik(R_i - R_j)} , \tag{2.1}$$

is $T_{ij} = T_0 \delta_{ij}$. Since T_0 is an insignificant constant, we assume $T_0 = 0$.

In accordance with the model the concentration of conduction electrons $x = N_e/N$ where N_e is the total number of electrons in the crystal, is assumed to be varying from 0 to 2.

The Hamiltonian of the system in the assumed approximation is represented by the sum of energy operators for separate atoms, therefore it is sufficient to carry out the analysis only for one arbitrary site. In this case we shall study the behaviour of the system described by the single-atom Hamiltonian

$$\hat{H} = \hat{H}_{sf} - \mu \hat{n} , \tag{2.2}$$

$$\begin{aligned} \hat{H}_{sf} = & -I[\hat{S}^+ \hat{a}_\downarrow^+ \hat{a}_\uparrow + \hat{S}^- \hat{a}_\uparrow^+ \hat{a}_\downarrow + \hat{S}^z (\hat{n}_\uparrow - \hat{n}_\downarrow)] + \\ & + U \hat{n}_\uparrow \hat{n}_\downarrow - g\mu_B h [\hat{S}^z + \tfrac{1}{2} (\hat{n}_\uparrow - \hat{n}_\downarrow)] , \end{aligned} \tag{2.3}$$

where I is the intra-atomic s–f exchange parameter; $\hat{S}^+$, $\hat{S}^-$, and $\hat{S}^z$ are the spin operators of rising, lowering, and z-projection related to the localized f-spins of the spin value S; $\hat{a}_\sigma^+$ and $\hat{a}_\sigma$ are creation and annihilation operators for the conduction electron (s-electron) at a site with spin $\sigma = \uparrow$ and $\downarrow$; $\hat{n}_\sigma$ is the operator of the number of electrons with spin σ; U the parameter for the Coulomb interaction between two s-electrons at one site; g the Landé splitting factor; μ_B the Bohr magneton; h the external magnetic field; μ the chemical potential; $\hat{n} = \hat{n}_\uparrow + \hat{n}_\downarrow$ the total electron number operator for one site. With a view to using the grand-canonical distribution we have included a term with the chemical potential into the Hamiltonian. The chemical potential μ is derived from the equation

$$\langle \hat{n} \rangle = x ; \qquad 0 \leqq x \leqq 2 , \tag{2.4}$$

where $\langle ... \rangle$ means the average over the grand-canonical distribution with the Hamiltonian (2.2).

It is not difficult to calculate exact eigenvectors and eigenvalues for the Hamiltonian (2.3). Hereafter we shall refer to an atom without an electron as a hole. It is evident that the hole spin is S. The atom with one electron and total spin $S + 1/2$ will be referred to as the up-unit, the atom with one electron and total spin $S - 1/2$ as down-unit and the atom with two electrons as a pair with spin S as for the hole.

These four states have four corresponding atomic functions which are the eigenfunctions of the Hamiltonian (2.3),

$$\psi_0 = |S, S^z\rangle\, |0\rangle\,, \tag{2.5}$$

$$\psi_{1\uparrow} = |S + \tfrac{1}{2}, m\rangle = \sum_{\sigma=-(1/2)}^{1/2} C^{S+(1/2),m}_{S,m-\sigma;(1/2)\sigma}\, |S, m - \sigma\rangle\, \hat{a}^+_\sigma\, |0\rangle\,, \tag{2.6}$$

$$\psi_{1\downarrow} = |S - \tfrac{1}{2}, m\rangle = \sum_{\sigma=-(1/2)}^{1/2} C^{S-(1/2),m}_{S,m-\sigma;(1/2)\sigma}\, |S, m - \sigma\rangle\, \hat{a}^+_\sigma\, |0\rangle\,, \tag{2.7}$$

$$\psi_2 = |S, S^z\rangle\, \hat{a}^+_\downarrow \hat{a}^+_\uparrow\, |0\rangle\,. \tag{2.8}$$

In (2.5) to (2.8): $|S_t, M\rangle$ is the wave function of the atom with the total spin of f- and s-electrons S_t and its z-projection M; $|0\rangle$ is the vacuum function of the s-electrons; $C^{jm}_{j_1m_1,j_2m_2}$ are Clebsch-Gordon coefficients,

$$C^{S+(1/2),m}_{S,m\pm(1/2);\,1/2,\,\mp1/2} = \sqrt{\frac{S + \frac{1}{2} \mp m}{2S + 1}}\,, \tag{2.9}$$

$$C^{S-(1/2),m}_{S,m\pm(1/2);\,1/2,\,\mp1/2} = \pm \sqrt{\frac{S + \frac{1}{2} \pm m}{2S + 1}}\,. \tag{2.10}$$

By operating with the Hamiltonian (2.3) on the functions (2.5) to (2.8) we obtain eigenvalues for holes, up- and down-units, and pairs, respectively,

$$E_0 = -g\mu_B h S^z; \qquad -S \leqq S^z \leqq S\,, \tag{2.11}$$

$$E_{1\uparrow} = E_p - g\mu_B h m; \qquad -S - \tfrac{1}{2} \leqq m \leqq S + \tfrac{1}{2}\,, \tag{2.12}$$

$$E_{1\downarrow} = E_a - g\mu_B h m; \qquad -S + \tfrac{1}{2} \leqq m \leqq S - \tfrac{1}{2}\,, \tag{2.13}$$

$$E_2 = U - g\mu_B h S^z; \qquad -S \leqq S^z \leqq S\,, \tag{2.14}$$

where

$$E_p = -IS\,, \qquad E_a = I(S + 1)\,. \tag{2.15}$$

Expressions (2.11) to (2.15) are used below for calculating thermodynamic averages.

For further analysis it is convenient to introduce the following operators:

$$\hat{N}_0 = 1 - \hat{n}_\uparrow - \hat{n}_\downarrow + \hat{n}_\uparrow \hat{n}_\downarrow\,, \tag{2.16}$$

$$\hat{N}_1 = \hat{n}_\uparrow + \hat{n}_\downarrow - 2\hat{n}_\uparrow \hat{n}_\downarrow\,, \tag{2.17}$$

$$\hat{N}_2 = \hat{n}_\uparrow \hat{n}_\downarrow\,. \tag{2.18}$$

The operators (2.16) to (2.18) are the operators of the number of holes, units, and pairs, respectively, which can be ascertained directly by operating on the eigenvectors (2.5) to (2.8).

3. Thermodynamics

The average statistical value of the operator $\hat{A}$ is calculated by the usual expression

$$\langle \hat{A} \rangle = \frac{\mathrm{Sp}\,(\hat{A}\, \mathrm{e}^{-\beta \hat{H}})}{\mathrm{Sp}\, \mathrm{e}^{-\beta \hat{H}}}\,, \tag{3.1}$$

where $\beta = 1/kT$, k is the Boltzmann constant, T the temperature.

The *partition function* is given by

$$Z \equiv \mathrm{Sp}\, e^{-\beta H} = \sum_{S^z=-S}^{S} e^{\beta g \mu_B h S^z} + e^{-\beta(E_p-\mu)} \sum_{m=-S-(1/2)}^{S+(1/2)} e^{\beta g \mu_B h m} + $$
$$+ e^{-\beta(E_a-\mu)} \sum_{m=-S+(1/2)}^{S-(1/2)} e^{\beta g \mu_B h m} + e^{-\beta(U-2\mu)} \sum_{S^z=-S}^{S} e^{\beta g \mu_B h S^z} \tag{3.2}$$

or

$$Z = \frac{1}{\mathrm{sh}\, t/2} \left\{ e^{-\beta(E_p-\mu)}\, \mathrm{sh}\, [t(S+1)] + e^{-\beta(E_a-\mu)}\, \mathrm{sh}\, (tS) + \right.$$
$$\left. + (e^{\beta(2\mu-U)} + 1)\, \mathrm{sh} \left[t \left(S + \frac{1}{2} \right) \right] \right\}, \tag{3.3}$$

where

$$t = \beta g \mu_B h\,. \tag{3.4}$$

The operator averages may be calculated analogously.

The *chemical potential* is determined by (2.4) where

$$\langle \hat{n} \rangle = \sum_\sigma \langle \hat{n}_\sigma \rangle = \frac{1}{Z\, \mathrm{sh}\, t/2} \left\{ e^{-\beta(E_p-\mu)}\, \mathrm{sh}\, [t(S+1)] + \right.$$
$$\left. + e^{-\beta(E_a-\mu)}\, \mathrm{sh}\, (tS) + 2 e^{\beta(2\mu-U)}\, \mathrm{sh} \left[t \left(S + \frac{1}{2} \right) \right] \right\}. \tag{3.5}$$

From (2.4) and (3.5) it follows that

$$\mu = \frac{1}{\beta} \ln \frac{F}{\mathrm{sh}\, [t(S+\frac{1}{2})]}, \tag{3.6}$$

where

$$F = [2(2-x)\, e^{-\beta U}]^{-1} \{(x-1)\, [l_1\, \mathrm{sh}\, (t(S+1)) + l_2\, \mathrm{sh}\, (tS)] + $$
$$+ \sqrt{(x-1)^2 [l_1\, \mathrm{sh}\, (t(S+1)) + l_2\, \mathrm{sh}\, (ts)]^2 + 4x(2-x)\, e^{-\beta U}\, \mathrm{sh}^2 [t(S+\tfrac{1}{2})]}\}, \tag{3.7}$$

$$l_1 = e^{-\beta E_p}, \qquad l_2 = e^{-\beta E_a}. \tag{3.8}$$

For $h = 0$ the expression for the chemical potential takes on a simplified form,

$$\mu_0 = U + \frac{1}{\beta} \times$$
$$\times \ln \frac{(x-1)\, [l_1(S+1) + l_2 S] + \sqrt{(x-1)^2\, [l_1(S+1) + l_2 S]^2 + x(2-x)(2S+1)^2\, e^{-\beta U}}}{(2-x)\, (2S+1)}. \tag{3.9}$$

Calculation of the concentration of holes x_0, units x_1, and pairs x_2 gives

$$x_0 \equiv \langle \hat{N}_0 \rangle = \frac{\mathrm{sh}^2\, [t(S+\frac{1}{2})]}{RF}, \tag{3.10}$$

$$x_1 \equiv \langle \hat{N}_1 \rangle = x_{1\uparrow} + x_{1\downarrow}, \tag{3.11}$$

$$x_{1\uparrow} = \frac{l_1\, \mathrm{sh}\, [t(S+1)]}{R}, \qquad x_{1\downarrow} = \frac{l_2\, \mathrm{sh}\, (tS)}{R}, \tag{3.12}$$

$$x_2 \equiv \langle \hat{N}_2 \rangle = \frac{F\, e^{-\beta U}}{R}, \tag{3.13}$$

where

$$R = l_1 \operatorname{sh}[t(S+1)] + l_2 \operatorname{sh}(tS) + l_3 \operatorname{sh}[t(S+\tfrac{1}{2})], \tag{3.14}$$

$$l_3 = \frac{F}{\operatorname{sh}[t(S+\frac{1}{2})]} e^{-\beta U} + \frac{\operatorname{sh}[t(S+\frac{1}{2})]}{F}. \tag{3.15}$$

$x_{1\uparrow}$ and $x_{1\downarrow}$ are concentrations of up- and down-units, respectively. The concentrations satisfy the relationships

$$x_{1\uparrow} + x_{1\downarrow} + 2x_2 = x, \tag{3.16}$$

$$x_{1\uparrow} + x_{1\downarrow} + x_0 + x_2 = 1. \tag{3.17}$$

The *magnetization* given by

$$M = g\mu_B \langle S^z + \tfrac{1}{2}(\hat{n}_\uparrow - \hat{n}_\downarrow)\rangle \tag{3.18}$$

may be represented as a sum of four summands,

$$M = M_0 + M_{1\uparrow} + M_{1\downarrow} + M_2, \tag{3.19}$$

where

$$M_0 = g\mu_B B(S)\, x_0, \tag{3.20}$$

$$M_{1\uparrow} = g\mu_B B(S+\tfrac{1}{2})\, x_{1\uparrow}, \tag{3.21}$$

$$M_{1\downarrow} = g\mu_B B(S-\tfrac{1}{2})\, x_{1\downarrow}, \tag{3.22}$$

$$M_2 = g\mu_B B(S)\, x_2. \tag{3.23}$$

In (3.20) to (3.23) $B(y)$ is the Brillouin function,

$$B(y) = (y+\tfrac{1}{2}) \operatorname{cth}[t(y+\tfrac{1}{2})] - \tfrac{1}{2}\operatorname{cth}(t/2). \tag{3.24}$$

The averages of the operators $\hat{n}_\uparrow$ and $\hat{n}_\downarrow$ taken separately are given by

$$\langle \hat{n}_\uparrow \rangle = \frac{x}{2} + \Delta, \tag{3.25}$$

$$\langle \hat{n}_\downarrow \rangle = \frac{x}{2} - \Delta, \tag{3.26}$$

where

$$\Delta = \frac{l_1 \operatorname{sh}[t(S+1)]\, B(S+\frac{1}{2}) - l_2 \operatorname{sh}(tS)\, B(S-\frac{1}{2})}{(2S+1)\{l_1 \operatorname{sh}[t(S+1)] + l_2 \operatorname{sh}(tS) + l_3 \operatorname{sh}[t(S+\frac{1}{2})]\}}. \tag{3.27}$$

In the absence of the magnetic field ($h = 0$) it follows from (3.25) to (3.27) that

$$\langle \hat{n}_\uparrow \rangle_0 = \langle \hat{n}_\downarrow \rangle_0 = \frac{x}{2}, \tag{3.28}$$

where the subscript "0" means $h = 0$.

Since at $h = 0$ the magnetization becomes zero, from (3.18) and (3.28) it follows that

$$\langle \hat{S}^z \rangle_0 = 0. \tag{3.29}$$

The *initial susceptibility* in accordance with (3.19) is also represented by a sum of four summands,

$$\chi = \chi_0 + \chi_{1\uparrow} + \chi_{1\downarrow} + \chi_2, \tag{3.30}$$

where

$$\chi_0 = \frac{g^2\mu_B^2}{3kT} S(S+1)\,(x_0)_0\,, \tag{3.31}$$

$$\chi_{1\uparrow} = \frac{g^2\mu_B^2}{3kT}\left(S+\frac{1}{2}\right)\left(S+\frac{3}{2}\right)(x_{1\uparrow})_0\,, \tag{3.32}$$

$$\chi_{1\downarrow} = \frac{g^2\mu_B^2}{3kT}\left(S-\frac{1}{2}\right)\left(S+\frac{1}{2}\right)(x_{1\downarrow})_0\,, \tag{3.33}$$

$$\chi_2 = \frac{g^2\mu_B^2}{3kT} S(S+1)\,(x_2)_0\,. \tag{3.34}$$

Here $(x_0)_0$, $(x_1)_0$, and $(x_2)_0$ are the concentrations of holes, units, and pairs in zero magnetic field,

$$(x_0)_0 = \frac{(S+\frac{1}{2})^2}{F_0 R_0}, \tag{3.35}$$

$$(x_{1\uparrow})_0 = \frac{l_1(S+1)}{R_0}, \tag{3.36}$$

$$(x_{1\downarrow})_0 = \frac{l_2 S}{R_0}, \tag{3.37}$$

$$(x_2)_0 = \frac{F_0\,e^{-\beta U}}{R_0}. \tag{3.38}$$

Here

$$F_0 = e^{\beta U} \times$$
$$\times \frac{(x-1)\,[l_1(S+1)+l_2S] + \sqrt{(x-1)^2[l_1(S+1)+l_2S]^2 + x(2-x)(2S+1)^2\,e^{-\beta U}}}{2(2-x)}, \tag{3.39}$$

$$R_0 = l_1(S+1) + l_2S + e^{-\beta U}F_0 + \frac{(S+\frac{1}{2})^2}{F_0}. \tag{3.40}$$

For the limiting cases of $x = 0$ or 2

$$\chi = \frac{g^2\mu_B^2}{3kT} S(S+1)\,, \tag{3.41}$$

i.e. the Curie law is fulfilled.

Let us assume, for definiteness, that $I > 0$. Then in the region of low temperatures (IS/kT, $U/kT \gg 1$), since the excited states give an exponentially small contribution, the Curie law is also observed,

$$\chi \approx \frac{g^2\mu_B^2}{3kT} \times \begin{cases} [S(S+1)\,(1-x) + (S+\frac{1}{2})\,(S+\frac{3}{2})\,x]\,; & x \leqq 1\,, \\ [(S+\frac{1}{2})\,(S+\frac{3}{2})\,(2-x) + S(S+1)\,(x-1)]; & x \geqq 1\,. \end{cases} \tag{3.42}$$

In the region of high temperatures (IS/kT, $U/kT \ll 1$) the susceptibility follows approximately the Curie-Weiss law

$$\frac{1}{\chi} \approx \frac{1}{C}(T - \theta), \tag{3.43}$$

where

$$C = \frac{g^2 \mu_B^2}{3k}\left[S(S+1) + \frac{3}{8}x(2-x)\right], \tag{3.44}$$

$$\theta = \frac{x(2-x)\left[IS(S+1) + \frac{3}{16}x(2-x)U\right]}{2k\left[S(S+1) + \frac{3}{8}x(2-x)\right]}. \tag{3.45}$$

The *mean energy* is given by

$$\begin{aligned} E \equiv \langle \hat{H}_{sf} \rangle = &\left\{ l_1 \operatorname{sh}\left[t(S+1)\right]\left[-IS - tkTB\left(S + \frac{1}{2}\right)\right] + \right. \\ &+ l_2 \operatorname{sh}(ts)\left[I(S+1) - tkTB\left(S - \frac{1}{2}\right)\right] + \\ &\left. + \operatorname{sh}\left[t\left(S + \frac{1}{2}\right)\right]\left[U\,e^{-U/kT}\frac{F}{\operatorname{sh}\left[t(S+\frac{1}{2})\right]} - l_3 tkTB(S)\right]\right\} \times \\ &\times \left\{ l_1 \operatorname{sh}\left[t(S+1)\right] + l_2 \operatorname{sh}(tS) + l_3 \operatorname{sh}\left[t\left(S + \frac{1}{2}\right)\right]\right\}^{-1}. \end{aligned} \tag{3.46}$$

The *specific heat* is obtained by differentiating E from (3.46) and in the general case it is given by a very cumbersome expression. In the particular case of $h = 0$ and $x = 1$ it is given by

$$\begin{aligned} C_0 = &\frac{S(S+1)(2S+1)}{kT^2}\left\{ e^{-\beta U/2}\left[Sl_1\left(I + \frac{U}{2S}\right)^2 + \right.\right. \\ &\left.\left. + (S+1)\,l_2\left(I - \frac{U}{2(S+1)}\right)^2\right] + (2S+1)\,e^{-\beta I}\right\} \times \\ &\times \{(S+1)\,l_1 + Sl_2 + (2S+1)\,e^{-\beta U/2}\}^{-2}. \end{aligned} \tag{3.47}$$

The specific heat tends to zero in the low-temperature limit as $T^{-2}\exp(-IS/kT)$, if it is assumed that $I > 0$, reaches a maximum, and goes to zero at high temperatures as T^{-2}.

4. Discussion

Above exact expressions in the atomic limit have been obtained for the spectrum, wave functions, chemical potential, concentration of holes, units, and pairs, magnetization, susceptibility, mean energy, and specific heat at arbitrary values of electron concentration, spin of f-electrons, temperature, magnetic field, and exchange and Coulomb interaction parameters.

The system has a four-level spectrum and presents in many respects an additive system of "quasi-particles" of four types: holes, down-units, up-units, and pairs, which is clearly seen from the expressions for magnetization and susceptibility (3.19) and (3.30).

The system under consideration is paramagnetic and is characterized by a very complicated field and temperature dependence of the magnetization. The susceptibility, according to (3.30) to (3.34), is exactly the susceptibility of the Langevin paramagnet for a system of particles of the four types with the corresponding effective moments but with temperature-dependent concentrations of particles.

Depending on the parameters, the susceptibility of the system obeys both the Curie law and the Curie-Weiss law. The fact that the Curie-Weiss law is fulfilled does not mean that magnetic ordering can take place in the crystal (the system of atoms in the atomic limit under consideration is always paramagnetic!). It is rather another example of the fact that this law can result only from the intra-atomic interaction. Since with an increase in I and U the average number of down-units and pairs (consequently, of holes) decreases, then, in accordance with (3.16) and (3.17), the average number of the up-units increases. As the up-units possess the largest spin $S + 1/2$ in comparison with other particles (holes, pairs, and down-units), the increase in the number of the former with an increase in the intra-atomic interaction leads to an increase in the susceptibility.

The specific heat expressed in a very complex form passes a maximum as a function of temperature (Schottky anomaly), which, however, is detected experimentally only when I or U are anomalously small.

Recently [5] the s–f atomic limit has already been the subject of analysis but in the absence of the magnetic field. The authors of [5] have obtained the excitation spectrum given by

$$\varepsilon_1 = -IS\,, \tag{4.1}$$

$$\varepsilon_2 = I(S+1)\,, \tag{4.2}$$

$$\varepsilon_3 = U - I(S+1)\,, \tag{4.3}$$

$$\varepsilon_4 = U + IS \tag{4.4}$$

(we have assumed $T_0 = 0$ in the corresponding formula from [5]). It is evident that the spectrum (4.1) to (4.4) does not coincide with our spectrum (2.11) to (2.15) at $h = 0$. The energies from (2.11) to (2.15) are related to (4.1) to (4.4) as

$$\varepsilon_1 = E_{1\uparrow} - E_0\,, \tag{4.5}$$

$$\varepsilon_2 = E_{1\downarrow} - E_0\,, \tag{4.6}$$

$$\varepsilon_3 = E_2 - E_{1\downarrow}\,, \tag{4.7}$$

$$\varepsilon_4 = E_2 - E_{1\uparrow}\,. \tag{4.8}$$

For the analysis of the thermodynamic behaviour the authors of [5] have used a method, which is, in fact, equivalent to the temperature Green's function. They obtained approximate expressions for the chemical potential (see (3.3) and (4.3) in [5])

$$\mu_0 = \begin{cases} -IS - kT \ln \dfrac{2(S+1)\,(1-x)}{x(2S+1)}\,; & x < 1 \\[2ex] U + IS - kT \ln \dfrac{(2S+1)\,(2-x)}{2(S+1)\,(x-1)}\,; & x > 1 \end{cases} \tag{4.9}$$

for $I > 0$ and

$$\mu_0 = \begin{cases} I(S+1) - kT \ln \dfrac{2S(1-x)}{x(2S+1)}\,; & x < 1 \\[2ex] U - I(S+1) - kT \ln \dfrac{(2S+1)\,(2-x)}{2S(x-1)}\,; & x > 1 \end{cases} \tag{4.10}$$

for $I < 0$, which are well satisfied at $|I|, U \gg kT$. Expressions (4.9) and (4.10) follow from our general expression (3.9), provided it is assumed that $|I|, U \gg kT$.

The authors failed to determine unambiguously the weights $\alpha_{i\sigma}$ included into the expression for $\langle \hat{n}_\sigma \rangle$, since these contain the quantity $\langle \hat{S}^z \rangle_0$, for which their method does not give any expression. Therefore, the authors of [5] consider the quantity $\langle \hat{S}^z \rangle_0$ as a variable, upon which the other functions depend. They claim that the quantity $\langle \hat{S}^z \rangle_0$ cannot be derived self-consistently using the atomic limit Hamiltonian (2.2). As regards this, it should be noted that the quantity $\langle \hat{S}^z \rangle_0$, on the contrary, is quite unambiguously determined by direct calculation of traces applied in the present paper and, according to (3.29), is always equal to zero. Since $\langle \hat{S}^z \rangle_0 = \text{const} = 0$, the quantity $\langle \hat{S}^z \rangle_0$ cannot be used as a variable. The same refers to spin polarization also considered in [5], which is proportional to the difference $\langle \hat{n}_\uparrow \rangle_0 - \langle \hat{n}_\downarrow \rangle_0$. This polarization, contrary to the opinion of the authors [5], is always equal to zero, as, according to (3.28), $\langle \hat{n}_\uparrow \rangle_0 = \langle \hat{n}_\downarrow \rangle_0 = x/2$.

References

[1] S. V. Vonsovskii and E. A. Turov, Soviet Phys. – J. exper. theor. Phys. **24**, 419 (1953).
[2] J. Hubbard, Proc. Roy. Soc. **A276**, 238 (1963).
[3] W. Nolting, phys. stat. sol. (b) **96**, 11 (1979).
[4] M. Matlak, A. Ramakanth, and K. Skrobis, phys. stat. sol. (b) **114**, 145 (1982).
[5] W. Nolting and M. Matlak, phys. stat. sol. (b) **123**, 155 (1984).

(Received April 30, 1985)

phys. stat. sol. (b) 131, 595 (1985)

Subject classification: 13; 21.1

Institute of Physics and Chemistry of Metals, Silesian University, Katowice[1])

Electronic Contribution to the Ordering Energy of a Binary Alloy

By

K. Bortel

Dedicated to Prof. Dr. Dr. h. c. Dr. E. h. P. Görlich on the occasion of his 80th birthday

The band theory of the phase transitions in the binary alloy of transition metals is investigated by means of the Landau theory and coherent potential approximation, CPA. The investigation of the mechanisms responsible for the transition from disordered to ordered phase of an alloy is the main purpose of the considerations. It is shown that the asymmetry of interactions and cnergy transfer are the primary causes of the ordering process in a binary alloy.

Théorie de bande des transitions de phase dans un binaire alliage des métaux transitoires est étudiée à l'aide de la théorie de Landau et coherent potentiel approximation CPA. L'étude des mécanismes responsables pour la transition de la phase désordonnée à ordonnée d'un alliage est le but des considérations. On a démontré que l'asymétrie des intéractions et le transfert d'énérgie sont les primaire causes du procès d'ordre dans un binaire alloy.

1. Introduction

Theories of the order–disorder phase transitions were developed on many different ways. The first group of methods considers the "geometric" causes of the ordering effects in binary alloys, posing of accent on the configurational average of the physical quantities. The renormalized group theory [1] or approaching considerations [2, 3] represent this family of phase transitions. The homogeneity of the thermodynamical potentials in the environment of the transition point $T = T_t$, $\eta = 0$, with respect to the variables temperature T, external field h, order parameter η is the fundamental assumption of the mentioned theories.

The second group of theories of phase transitions permits also to investigate the role of localized electrons in the atomic arrangement of a binary alloy of transition metals. These are the thermodynamical [3, 4] and variational [3] methods. The latter methods use an essential assumption — the differentiability of the thermodynamical potentials in the environment of the transition point with respect to the variables T, h, η. The above hypothesis is weak in comparison with the assumption of homogeneity — it means that these methods are more general.

The majority of properties of the transition metals and their alloys is determined by the electronic structure of the narrow bands formed by localized 3d electrons. The configuration of the atomic cores in a binary alloy is determined by their effective interaction. The electronic structure of metals by reason of strong screening of the conduction and narrow-band electrons determines the effective interactions between the atomic cores in a metallic alloy.

The main purpose of the present paper consists in the investigation of the influence of the electronic structure and interactions on the atomic arrangement of some binary alloy. The above considerations permit to evaluate the electronic contribution to the

[1]) Bankowa 12, 40-007 Katowice, Poland.

ordering energy of the binary alloy and indirectly to investigate the configurational contribution of the atomic cores.

In the present paper, the configurational contribution of the atomic cores is treated within the Bragg-Williams approximation [3], the phonon–electron interactions are neglected as secondary effects. In Sections 2 and 3 of this paper we discuss the model Hamiltonian of the system and the assumptions limiting our considerations. The Landau theory of the phase transition [4] is connected with the coherent potential approximation [7] in Section 4. In the last section we represent the results and finally, we summarize the observations and conclusions suggested during the performance of this work.

2. Model Hamiltonian and Band Interpretation of the Coherent Potential Approximation

The localized electrons in transition metals (or rare metals) and their alloys are well described by the Hubbard Hamiltonian [6]

$$\mathcal{H} = \sum_{\substack{\alpha,\beta,\sigma \\ \alpha\neq\beta}} t_{\alpha\beta} c^+_{\alpha\sigma} c_{\beta\sigma} + \sum_{\alpha,\sigma} E_\alpha n_{\alpha\sigma} + \frac{1}{2} \sum_{\alpha,\sigma} U_\alpha n_{\alpha-\sigma} n_{\alpha\sigma} + \frac{1}{2N} \sum_{\substack{\alpha,\beta,\sigma \\ \alpha\neq\beta}} U_{\alpha\beta} n_{\alpha\sigma} n_{\beta\sigma} , \tag{2.1}$$

where $t_{\alpha\beta}$ is the transfer integral, E_α the atomic level, $c_{\alpha\sigma}$ the creation operator of an electron in site α with spin σ, and N denotes the number of sites of the lattice.

In a binary alloy A_xB_{1-x} of transition metals U_α, $U_{\alpha\beta}$ are the Coulomb interactions of electrons appertaining to the same site (atom U_A or U_B) and to two different sites (two same atoms U_{AA}, U_{BB} or different atoms U_{AB}), respectively. The sums are developed over all the sites of the lattice.

The widths of the narrow bands formed by localized electrons of A and B atoms are comparable, so that in Bloch representation the first term of Hamiltonian (2.1) gets the form

$$\sum_{\substack{\alpha,\beta,\sigma \\ \alpha\neq\beta}} t_{\alpha\beta} c^+_{\alpha\sigma} c_{\beta\sigma} = \sum_{k\sigma} \varepsilon_k n_{k\sigma} , \tag{2.2}$$

where $(k\sigma)$ denotes the Bloch states related to the Wannier states $(\alpha\sigma)$. Then we have supposed that the two narrow bands formed by energy levels of A and B atoms constitute one narrow band of halfwidth w.

In the calculations we limit ourselves to the coherent potential approximation [7]. This method is consistent with the Hartree-Fock approximation in which the Hamiltonian (2.1) is given by

$$\mathcal{H} = \sum_{\substack{\alpha,\beta,\sigma \\ \alpha\neq\beta}} t_{\alpha\beta} c^+_{\alpha\sigma} c_{\beta\sigma} + \sum_{\alpha,\sigma} \tilde{\varepsilon}_\alpha n_{\alpha\sigma} , \tag{2.3}$$

where

$$\tilde{\varepsilon}_{\alpha\sigma} = E_\alpha + U_\alpha \langle n_{\alpha-\sigma}\rangle + \frac{1}{N} \sum_\beta U_{\alpha\beta} \langle n_{\beta\sigma}\rangle . \tag{2.3a}$$

The final considerations in the approximations made concern the system of free electrons in some effective medium field $\Sigma(z, \eta)$. This system is described by an effective medium Hamiltonian (Hamiltonian of the free electrons)

$$\mathcal{H}_{\text{eff}} = \sum_{\substack{\alpha,\beta,\sigma \\ \alpha\neq\beta}} t_{\alpha\beta} c^+_{\alpha\sigma} c_{\beta\sigma} + \sum_{\alpha,\sigma} \Sigma(z, \eta)\, n_{\alpha\sigma} \tag{2.4}$$

which in Bloch representation gets the form

$$\mathscr{H}_{\text{eff}} = \sum_{k\sigma} [\varepsilon_k + \Sigma(z, \eta)]\, n_{k\sigma}\,, \tag{2.5}$$

where η denotes the order parameter, $z = \omega + i\delta$ ($\delta > 0$) is the complex energy variable. The one-particle Green function in Bloch representation,

$$G_{kk'} = \langle k|\, \mathscr{H}_{\text{eff}}\, |k'\rangle = \frac{\delta_{kk'}}{z - \varepsilon_k - \Sigma(z, \eta)}\,, \tag{2.6}$$

is determined on the base of the last Hamiltonian. The poles $z - \varepsilon_k - \Sigma(z, \eta) = 0$ of the Green's function determine the effective energy levels of the electrons,

$$\omega_{0k} = \varepsilon_k + \Sigma_1(\omega, \eta)\,,$$

and the time of relaxation $\tau^{-1} = (2/\hbar)\, \Sigma_2(\omega, \eta)$ on these levels. The obtained Green function (2.6) permits to determine the partial density of states

$$\varrho(\omega, \varepsilon_k) = -\frac{1}{\pi}\, \text{Im}\, G_{kk} = \frac{1}{\pi}\, \frac{\Sigma_2(\omega, \eta)}{(\omega - \varepsilon_k - \Sigma_1)^2 + \Sigma_2^2}\,, \tag{2.7}$$

where

$$\Sigma(\omega, \eta) = \Sigma_1(\omega, \eta) + i\Sigma_2(\omega, \eta)\,.$$

The coherent potential $\Sigma(z, \eta)$ is defined by the autocoherence condition

$$\sum_{\alpha} \langle T_\alpha \rangle = 0\,, \tag{2.8}$$

where the single-site scattering matrix T_α is given by

$$_\alpha T = \frac{v_\alpha}{1 - v_\alpha F(z - \Sigma)} \tag{2.9}$$

with

$$F(z - \Sigma) = \sum_k [z - \varepsilon_k - \Sigma(z, \eta)]^{-1}\,, \qquad v_\alpha = \tilde{\varepsilon}_{\alpha\sigma} - \Sigma(z, \eta)\,. \tag{2.10}$$

In conformity with the above interpretations the total energy of the system is determined by the real part of the coherent potential $\Sigma(z, \eta)$, simultaneously the effective medium Hamiltonian in the following form

$$H_{\text{eff}} = \sum_{k\sigma} [\varepsilon_k + \Sigma_1(\omega, \eta)]\, n_{k\sigma} \tag{2.11}$$

defines the free energy of the alloy.

Summarizing the above considerations we conclude that the investigated system of localized electrons is represented by the system of free electrons in the narrow band determining some effective interaction of the atomic cores — the ordering energy of the alloy.

3. Configurational and Statistical Averaging the Total Free Energy of System

We suppose some binary, equiatomic alloy $A_{0.5}B_{0.5}$ ($x = 0.5$) of transition metals. The two sublattices a, b of A and B atoms with the numbers of sites N_a, N_b form the lattice of the alloy possessing the CsCl structure. The "unnatural" order parameter η of the sublattice of A atoms is defined indirectly by probabilities p_a^A, p_a^B of occupation of the sites N_a by A and B atoms,

$$p_a^A = \frac{N_a^A}{N_a} = \frac{1}{2}(1+\eta),$$
$$p_a^B = \frac{N_a^B}{N_a} = \frac{1}{2}(1-\eta); \qquad N_a + N_b = N. \tag{3.1}$$

N_a^A, N_a^B denote the number of A and B atoms in sites of a-sublattice. Symmetrically, this same order parameter determines the probabilities of occupation of the sites of the b-sublattice. In ordered state $\eta = 1$, atoms occupy the sites of the parental sublattices

$$p_a^A = p_b^B = 1, \qquad p_a^B = p_b^A = 0.$$

The above definitions are the basis of the configurational averaging.

The total free energy comprises the configurational F_{con} and electronic F_{el} contributions. The first component F_{con} is evaluated independently from the electronic contribution F_{el} and given by

$$F_{con} = F_0 - TS_{con}, \tag{3.2}$$

where F_0 is the free energy of the symmetric phase. S_{con} denotes the configurational entropy,

$$S_{con} = -N[\tfrac{1}{2}(1+\eta)\ln\tfrac{1}{2}(1+\eta) + \tfrac{1}{2}(1-\eta)\ln\tfrac{1}{2}(1-\eta)] \tag{3.2a}$$

evaluated within the Bragg-Williams approximation, it means that the variation of the configuration of the cores caused by interactions of the localized electrons determines the configurational component of the free energy. This approximated treatment of the configurational component is sufficient in a consideration of the electronic contribution. The atomic cores of the transition metals forming the binary alloy AB are strongly screened by conduction and narrow-band electrons, therefore, their direct interactions are negligible.

The electronic contribution of the free energy is determined by the effective medium Hamiltonian (2.11), which defines the statistical operator

$$\hat{\varrho} = \exp(-\beta H_{eff}); \qquad \beta = \frac{1}{T}. \tag{3.3}$$

Evaluating the statistical sum

$$Q = \mathrm{Tr}\,[\exp(-\beta H_{eff})], \tag{3.4}$$

we average over the mixed energy states characterized by the partial density of states $\varrho(\omega, \varepsilon_k)$ (2.7). Then the electronic component of the free energy is given by

$$F_{el} = -T\ln\{\mathrm{Tr}\exp[-\beta\sum_{k\sigma}(\varepsilon_k + \Sigma_1(\omega,\eta))\,n_{k\sigma}]\} =$$
$$= -2T\int_{-w}^{w}\int_{-w}^{w}\varrho_0(\omega)\,\varrho(\omega,\varepsilon)\ln\{1+\exp[-\beta(\varepsilon_k + \Sigma_1(\omega,\eta))]\}\,d\omega\,d\varepsilon, \tag{3.5}$$

where $\varrho_0(\omega)$ is the non-perturbed density of states,

$$\varrho_0(\omega) = \begin{cases} \dfrac{2}{\pi}\sqrt{1-\left(\dfrac{\omega}{w}\right)^2} & \text{for} \quad \omega \leqq w, \\ 0 & \text{for} \quad \omega > w. \end{cases} \tag{3.5a}$$

In some sense the above averaging possesses double character, statistical expressed by the density of states and configurational determined by the order parameter.

The dependence upon the configuration of atoms is expressed indirectly by the coherent potential $\Sigma(\omega, \eta)$ evaluated from the autocoherence condition (2.8). The sums over the sites of the lattice in (2.8) for the coherent potential are evaluated in conformity with the Bragg-Williams approximation.

The total free energy F_{tot} of the binary alloy AB is given by the sum of electronic F_{el} and configurational F_{con} components,

$$F_{\text{tot}} = F_{\text{con}} + F_{\text{el}} = F_0 - TS_{\text{con}} + F_{\text{el}} \,. \tag{3.6}$$

The remaining considerations concerning the existence and classification of the phase transitions are derived in thermodynamical limit [5, 8] on the basis of the free energy,

$$\mathcal{F}_{\text{tot}} = \lim_{N \to \infty} \frac{F_{\text{tot}}}{N} \,. \tag{3.7}$$

4. The Merit of Calculations

The correspondence between the thermodynamical Landau theory of phase transitions and the CPA approximation in essence consists in the manner of autocoherence of the calculations. In reality, the conditions of autocoherence for the coherent potential $\Sigma(\omega, \eta)$ are determined by the conditions of existence of the phase transitions — by energy conditions imposed on the system. Developing the total free energy (3.7) with respect to η in the environment of $\eta = 0$ we obtain

$$\mathcal{F}_{\text{tot}} = F_{0\text{s}}(T) + a(T)\,\eta + b(T)\,\eta^2 + c(T)\,\eta^3 + d(T)\,\eta^4 + O(\eta^5) \,, \tag{4.1}$$

where $F_{0\text{s}}$ denotes the free energy of the disordered state (symmetric phase) and

$$a(T) = \left.\frac{\partial \mathcal{F}_{\text{tot}}}{\partial \eta}\right|_{\eta=0} = 2 \int_{-w}^{w} \int_{-w}^{w} \varrho_0(\omega)\, \varrho(\omega, \varepsilon)\, f(\varepsilon + \Sigma_1) \left.\frac{\partial \Sigma_1}{\partial \eta}\right|_{\eta=0} \mathrm{d}\omega \, \mathrm{d}\varepsilon \,, \tag{4.2a}$$

$$b(T) = T + \int_{-w}^{w} \int_{-w}^{w} \varrho_0\, \varrho(\omega, \varepsilon)\, f(\varepsilon + \Sigma_1) \left.\left[\frac{\partial^2 \Sigma_1}{\partial \eta^2} - \beta(1 - f)\left(\frac{\partial \Sigma_1}{\partial \eta}\right)^2\right]\right|_{\eta=0} \mathrm{d}\omega \, \mathrm{d}\varepsilon \,, \tag{4.2b}$$

$$c(T) = \left(\text{terms expressed by } \frac{\partial^3 \Sigma_1}{\partial \eta^3};\ \frac{\partial^2 \Sigma_1}{\partial \eta^2}\,\frac{\partial \Sigma_1}{\partial \eta};\ \left(\frac{\partial \Sigma_1}{\partial \eta}\right)^3\right), \tag{4.2c}$$

$$d(T) = \left(2T + \text{terms expressed by } \frac{\partial^4 \Sigma_1}{\partial \eta^4};\ \frac{\partial^3 \Sigma_1}{\partial \eta^3}\,\frac{\partial \Sigma_1}{\partial \eta};\ \left(\frac{\partial^2 \Sigma_1}{\partial \eta^2}\right)^2;\ \left(\frac{\partial \Sigma_1}{\partial \eta}\right)^4\right). \tag{4.2d}$$

$f = f(\varepsilon + \Sigma_1)$ is the Fermi function. All the terms depend upon the derivatives $\partial^n \Sigma_1 / \partial \eta^n|_{\eta=0}$ of the real part of the coherent potential $\Sigma(\omega, \eta)$.

In conformity with the Landau theory of the phase transitions we obtain some conditions for the coefficients (4.2a) to (4.2d), for example:

(i) for the first-kind phase transition

$$a(T) = 0 \,, \qquad c(T) \neq 0 \,, \tag{4.3}$$

(ii) for the second-kind phase transition

$$a(T) = 0 \,, \qquad c(T) = 0 \,. \tag{4.4}$$

Simultaneously, these are the conditions of the real autocoherence for the coherent potential $\Sigma(\omega, \eta)$. The Landau considerations are phenomenological — in some sense experimental, and this theory can be treated as well motivated assumption, i.e. as hypothesis.

Conditions (4.3) or (4.4) should be satisfied for arbitrary energy intervals of integration and for arbitrary temperature, then from (4.2a), (4.2c) we get that in case of first kind of phase transitions

$$\left.\frac{\partial \Sigma_1}{\partial \eta}\right|_{\eta=0} = 0 \quad \text{and} \quad \left.\frac{\partial^3 \Sigma_1}{\partial \eta^3}\right|_{\eta=0} \neq 0$$

and in case of the second kind of phase transitions

$$\frac{\partial \Sigma_1}{\partial \eta} = 0 \tag{4.5a}$$

and

$$\left.\frac{\partial^3 \Sigma_1}{\partial \eta^3}\right|_{\eta=0} = 0\,. \tag{4.5b}$$

Finally, we limit our investigations to second-order phase transitions. In this case we possess three equations determining the coherent potential $\Sigma(\omega, \eta)$.

Equation (2.8) for the scattering matrix is treated as the definition of $\Sigma(\omega, \eta)$ and conditions (4.5) of the second kind-phase transitions treated as the autocoherence conditions for the coherent potential $\Sigma(\omega, \eta)$. In agreement to the CPA approximation, in the calculations of the derivatives $\partial^n \Sigma_1/\partial \eta^n|_{\eta=0}$ we have neglected all the terms of second order with respect to v_α^2 (2.10) in the perturbation series

$$\frac{\partial}{\partial \eta}\left(\langle n_{A,B}\rangle\right)|_{\eta=0} = 0\,, \qquad \left.\frac{\partial F(z-\Sigma)}{\partial \eta}\right|_{\eta=0} = 0\,. \tag{4.6}$$

For example, the first derivative is equal to

$$\left.\frac{\partial \Sigma}{\partial \eta}\right|_{\eta=0} = \left.\left(\frac{\partial \Sigma_1}{\partial \eta} + i\,\frac{\partial \Sigma_2}{\partial \eta}\right)\right|_{\eta=0}, \tag{4.7}$$

where

$$\left.\frac{\partial \Sigma_1}{\partial \eta}\right|_{\eta=0} = \Gamma_B + \frac{(\Gamma_A - \Gamma_B)\,(1 + V_R^2 - V_I^2)}{(1 + V_R^2 - V_I^2)^2 + 4V_R^2 V_I^2}, \tag{4.8a}$$

$$\left.\frac{\partial \Sigma_2}{\partial \eta}\right|_{\eta=0} = \frac{2V_R V_I(\Gamma_B - \Gamma_A)}{(1 + V_R^2 - V_I^2)^2 + 4V_R^2 V_I^2}\,. \tag{4.8b}$$

The first and next derivatives depend upon the energy parameters U_A, U_B, U_{AA}, U_{BB}, U_{AB}, E_A, E_B and the occupation numbers $\langle n_A\rangle$, $\langle n_B\rangle$ indirectly by

$$\Gamma_A = U_{AA}\langle n_A\rangle - U_{AB}\langle n_B\rangle\,, \tag{4.9a}$$

$$\Gamma_B = U_{BB}\langle n_B\rangle - U_{BA}\langle n_A\rangle\,; \qquad U_{AB} = U_{BA}\,, \tag{4.9b}$$

$$\frac{v_A}{v_B} = \frac{\tilde{\varepsilon}_A - \Sigma}{\tilde{\varepsilon}_B - \Sigma} = V_R - iV_I\,, \tag{4.9c}$$

$$\tilde{\varepsilon}_A = \bar{\varepsilon}_A + (1+\eta)\,U_{AA}\langle n_A\rangle + (1-\eta)\,U_{AB}\langle n_B\rangle\,, \tag{4.9d}$$

$$\tilde{\varepsilon}_B = \bar{\varepsilon}_B + (1+\eta)\,U_{BB}\langle n_B\rangle + (1-\eta)\,U_{BA}\langle n_A\rangle\,, \tag{4.9e}$$

and

$$\bar{\varepsilon}_{A,B} = E_{A,B} + U_{A,B}\langle n_{A,B}\rangle\,. \tag{4.9f}$$

The dependence of the derivatives upon the coherent potential is formal, we can eliminate this dependence by reason of the two equations (4.5) of autocoherence. Treating the second equation (4.5b) as the autocoherence condition of calculations with respect to V_R, V_I (equivalent to autocoherence with respect to Σ_1, Σ_2) and supposing the energy parameters, we determine the occupation numbers $\langle n_A \rangle$, $\langle n_B \rangle$ from the first equation (4.5a).

The above considerations reduce the terms in the expansion of the free energy (4.1) to two,

$$b(T) = T + \left.\frac{\partial^2 \Sigma_1}{\partial \eta^2}\right|_{\eta=0} I(T)\,, \tag{4.10a}$$

$$d(T) = \frac{1}{12}\left[2T + \left.\frac{\partial^4 \Sigma_1}{\partial \eta^4}\right|_{\eta=0} I(T) - 3\left.\left(\frac{\partial^2 \Sigma_1}{\partial \eta^2}\right)^2\right|_{\eta=0} \frac{\partial}{\partial \beta} I(T)\right], \tag{4.10b}$$

where

$$I(T) = \int\limits_{-w}^{w} \varrho_0(\omega) \left[\frac{1}{2} - \frac{1}{\pi} \operatorname{Im} \psi\left(\frac{1}{2} - \frac{\beta \Sigma_2}{2\pi} + i\,\frac{\beta\omega}{2\pi}\right)\right] d\omega\,, \tag{4.10c}$$

$\psi(x)$ denotes the digamma function. The second and forth derivatives are evaluated from (2.8) within the approximation (4.6).

5. Results and Comparisons

The free energy of the binary alloy, in which we observe the second-kind order–disorder phase transition is determined by (4.1) and (10),

$$\mathcal{F}_{\text{tot}} = F_{0s}(T) + b(T)\,\eta^2 + d(T)\,\eta^4 + O(\eta^5)\,. \tag{5.1}$$

We have considered a simple binary alloy with CsCl structure, fixed stoichiometric composition, and non-degenerate 3d orbitals. The mentioned assumptions limit our possibilities to the investigations of the physical character of the mechanisms responsible for the phase transition from disordered to ordered phase. The transition temperature and ordering energy give exhaustive information on the above subject.

Evaluating the transition temperature we have supposed the non-perturbed density of states (3.5a), for this density of states the integral (4.10c) is independent of the temperature (for $T = 1$ to 10^4 K we get the value $I(T) = 0.500065$ to 0.500067). The more exact treatment of the structure of the narrow band permits to modify the critical exponents of Landau's theory. The energy parameters U_{AA}, U_{BB}, U_{AB} measured in units of the halfwidth w of the narrow band have the values 0 to 2. The displacement of the bands in the same units is $\delta = E_A - E_B = 0$ to 5.

The Coulomb interaction of the electrons appertaining to the same atom U_α ($\alpha = $ A, B) determines the value of δ indirectly by (4.9d) to (4.9f), therefore we have $U_A = U_B = 0$.

The equation determining the transition temperature in conformity with the Landau theory is obtained from the condition

$$\left.\frac{\partial^2 \mathcal{F}_{\text{tot}}}{\partial \eta^2}\right|_{\eta=0} = 0$$

which implicates the following relation:

$$b(T) = T + \left.\frac{\partial^2 \Sigma_1}{\partial \eta^2}\right|_{\eta=0} I(T) = 0\,, \qquad I(T) = 0.500067\,. \tag{5.2}$$

The transition temperature as a function of the interactions between the electrons appertaining to two different atoms for fixed values of U_{AA}, U_{BB} is represented on Fig. 1. These dependences demonstrate

(i) that the asymmetry of interactions U_{AA}, U_{BB}, U_{AB} creates the phase transition (if $U_{AA} = U_{BB} = U_{AB}$, then $T_t = 0$ K, curve 1, Fig. 1),

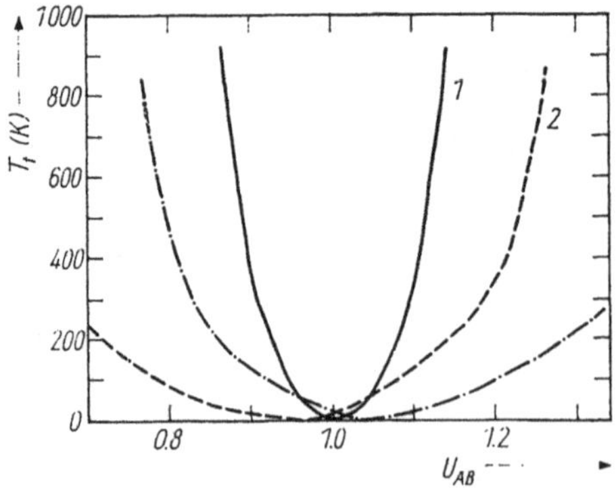

Fig. 1

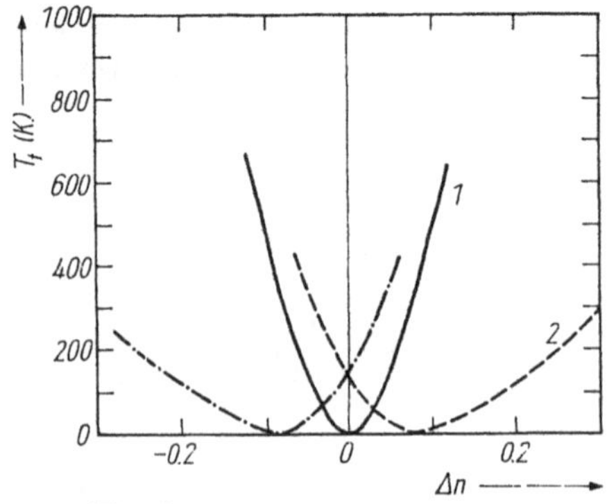

Fig. 2

Fig. 1. Transition temperature T_t as a function of U_{AB} for fixed $\delta = E_A - E_B = 0.5$ and (1) $U_{AA} = 1$, $U_{BB} = 1$; (2) $U_{AA} = 2$, $U_{BB} = 0.5$; the symmetric case $U_{AA} = 0.5$, $U_{BB} = 2$ is represented by the dash-dotted curve

Fig. 2. Ordering energy and transition temperature as functions of the difference of occupation numbers $\Delta n = n_B - n_A$ (the same energy parameters as in Fig. 1 are used)

(ii) more generally, the energy asymmetry between the bands of A and B metals determines transition temperature and ordering energy (if $\Gamma_A + \Gamma_B = 0$, then $T_t = 0$ K, curve 2 of Fig. 1).

The dependence of the ordering energy upon the difference $\Delta n = n_B - n_A$ of the occupation numbers $n_A = \langle n_A \rangle$, $n_B = \langle n_B \rangle$ of the A and B atoms, Fig. 2, confirms the preceding conclusions that the asymmetry of energy inspires the ordering process in a binary alloy. The ordering energy as a function of the displacement of the narrow bands $\delta = E_A - E_B$ for fixed different values of U_{AA}, U_{BB}, U_{AB}, Fig. 3, demonstrates that the second-kind ordering process is more probable in an alloy formed from transition metals with comparable energy levels $E_A \approx E_B$.

The dependence of the order parameter upon temperature is evaluated from the condition

$$\bar{\eta}(T) = \frac{\eta(T)}{\eta(0)} = \left[\frac{1 - T/T_t}{1 + 4T(\partial^4 \Sigma_1/\partial \eta^4|_{\eta=0})^{-1}}\right]^{1/2} \tag{5.3}$$

and given on Fig. 4. The character of the above dependence assured by the autocoherence of the calculations confirms the kind of the phase transitions.

All the graphic results are evaluated for a narrow band of halfwidth $w = 3$ eV and the chosen autocoherence parameters $V_B = -1.2$, $V_A = 0.033$.

The ordering process is caused by the energy asymmetry of the bands of A and B metals. This fact is the main conclusion arising from the results represented on Fig. 1 and 2.

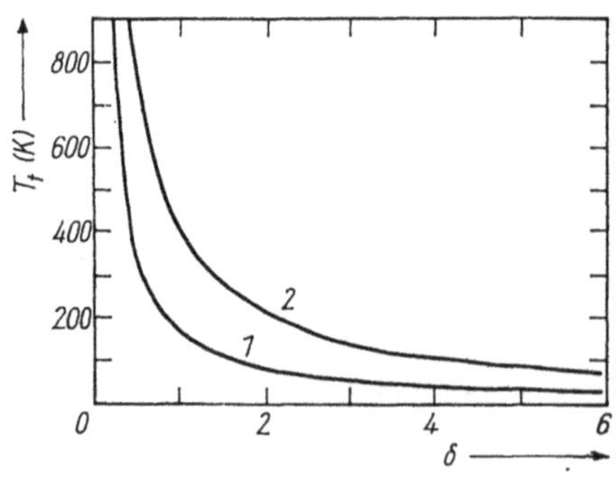

Fig. 3. Transition temperature as a function of δ for fixed (1) $U_{AA} = U_{BB} = 1$, $U_{AB} = 1.094$; (2) $U_{AA} = 2$, $U_{BB} = 0.5$, $U_{AB} = 1.65$

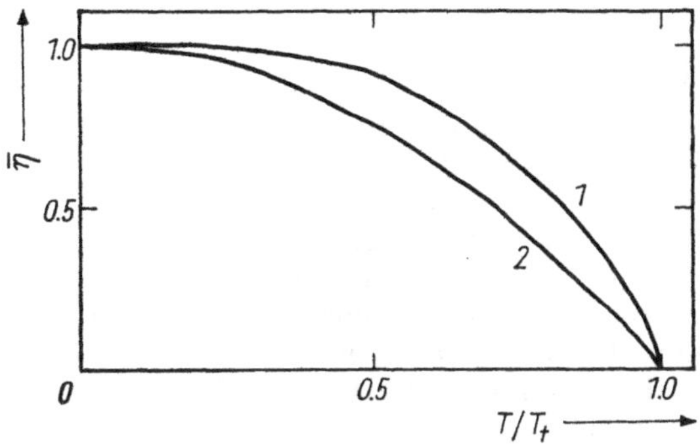

Fig. 4. Dependence of the normalized order parameter $\bar{\eta}$ upon temperature. (1) $U_{AA} = U_{BB} = 1$, $U_{AB} = 1.094$, $\delta = 0.5$, $T_t = 340.8$ K; (2) $U_{AA} = 2$, $U_{BB} = 0.5$, $U_{AB} = 1.65$, $\delta = 0.5$, $T_t = 794.7$ K

In interpretations [9] obtained within the Bragg-Williams approximation connected with the CPA method, the quantity responsible for the phase transitions is the charge transfer. In reality, the charge transfer is the secondary quantity caused by the gradient of energy. Therefore, the energy transfer is the primary cause of the ordering process.

The index of the energy transfer is determined from (4.9a), (4.9b),

$$\Gamma = \Gamma_A + \Gamma_B = U_{AA}\langle n_A\rangle + U_{BB}\langle n_B\rangle - U_{AB}(\langle n_A\rangle + \langle n_B\rangle)\,, \tag{5.4}$$

its dependence upon the interaction and the difference Δn is represented on Fig. 5 and 6, respectively.

The zeros of the energy transfer Γ correspond to the zeros (minima) of the transition temperature as a function of U_{AB} (Fig. 1) and $\Delta n = n_B - n_A$ (Fig. 2).

The considerable decrease (or increase) of the interactions U_{AB} (or U_{AA}, U_{BB}) or of the effective charge transfer Δn below (or above) the point of energy equilibrium,

$$\Gamma = \Gamma_A + \Gamma_B = 0\,,$$

increases the ordering energy, rendering the phase transitions unobservables.

The index of the energy transfer (5.4) in some sense possesses the character of the

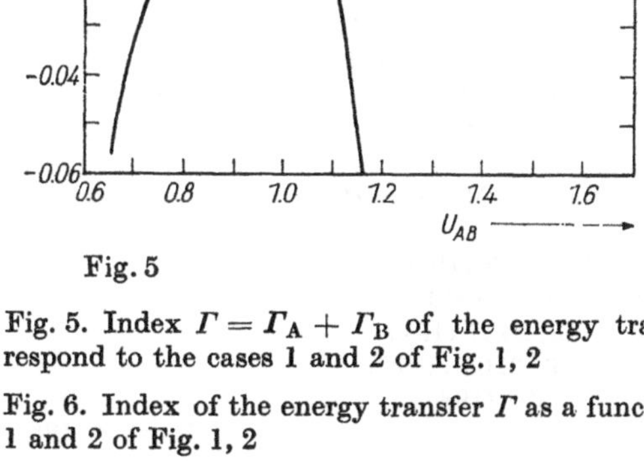

Fig. 5

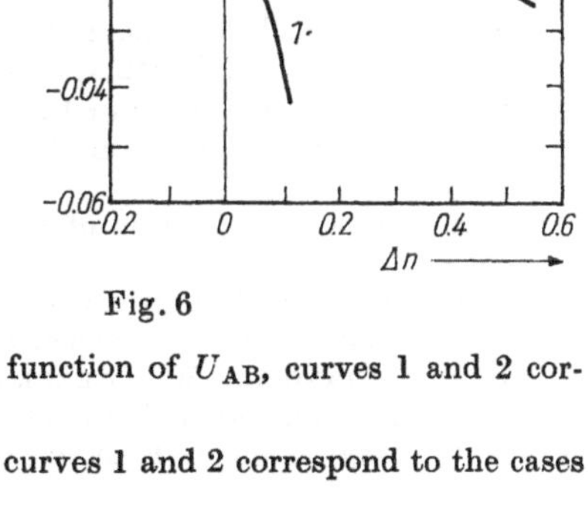

Fig. 6

Fig. 5. Index $\Gamma = \Gamma_A + \Gamma_B$ of the energy transfer as a function of U_{AB}, curves 1 and 2 correspond to the cases 1 and 2 of Fig. 1, 2

Fig. 6. Index of the energy transfer Γ as a function of Δn, curves 1 and 2 correspond to the cases 1 and 2 of Fig. 1, 2

Bragg-Williams ordering energy,

$$E_{\text{order}} = \text{const}\,(U_{AA} + U_{BB} - 2U_{AB})\,.$$

The ordering energy in the Landau theory determined indirectly by the second derivative of the real part of $\Sigma(\omega, \eta)$ from (5.2) is positive for negative and positive values of Γ, but for the characteristic relation of the interactions $U_{AA} \gtrless U_{AB} \gtrless U_{BB}$, except $U_{AA} = U_{BB}$ when $U_{AB} \gtrless U_{AA} = U_{BB}$, curve 1 of Fig. 1.

The transition temperature as a function of the displacement of the narrow bands $\delta = E_A - E_B$ demonstrates that the ordering process in a binary alloy formed from transition metals with comparable energy levels $E_A \approx E_B$ is more probable. This latter fact is confirmed by experimental measurements [10] and theoretical considerations within the Bethe approximation [11].

Acknowledgement

The author would like to express his thanks to Prof. J. Friedel for his valuable consultations and discu ssions.

References

[1] Shang-Keng Ma, Rev. mod. Phys. **45**, 589 (1973).
Shang-Keng Ma, Modern Theory of Phase Transitions, W. A. Benjamin, Inc., New York 1976
G. Toulouse and P. Pfeuty, Introduction au groupe de renormalization et ses applications, Presses Universitaires de Grenoble, Grenoble 1975.
H. E. Stanley, Introduction to Phase Transitions and Critical Phenomena, Clarendon Press, Oxford 1971.
Ya. G. Sinai, Theory of Phase Transitions, Rigorous Results, Akadémiai Kiado, Budadest 1982.

[2] L. Onsager, Phys. Rev. **65**, 117 (1944).

[3] Th. Niemeijer and J. M. J. van Leeuwen, Phase Transitions and Critical Phenomena, Vol. 6, Domb-Green Academic, London 1976.
E. Eggarter and T. P. Eggarter, Phys. Rev. B **15**, 2804 (1977).

[4] N. Boccara, Symétries Brisées (Théories des Transitions avec paramètre d'ordre), Hermann, Paris 1976.
L. Landau and E. Lifshits, Physique Statistique, Chap. 14, Izd. Mir, Moscow 1967.

[5] C. N. Yang and T. D. Lee, Phys. Rev. **87**, 404 (1952).
R. K. Pathria, Statistical Mechanics, Pergamon Press, Oxford/New York/Toronto 1972.

[6] J. Hubbard, Proc. Roy. Soc. **276**, 238 (1963); **277**, 237 (1964).

[7] B. Velický, S. Kirkpatrik, and H. Ehrenreich, Phys. Rev. **175**, 747 (1968).
B. Velický, Phys. Rev. **184**, 614 (1969).
S. Kirkpatrik, B. Velický, and H. Ehrenreich, Phys. Rev. B **1**, 3250 (1970).

[8] D. Ruelle, Ann. Phys. **25**, 109 (1963); Statistical Mechanics, Rigorous Results, W. A. Benjamin Inc., New York 1969.

[9] F. Brouers, F. Ducastelle, F. Gautier, and van der Rest, J. Physique **35**, 64 (1975).
F. Ducastelle, G. Treglia, and F. Gautier, J. Phys. F **8**, 7 (1978).

[10] H. Hansen, Constitution of Binary Alloys, McGraw-Hill Publ. Co., New York 1958.
R. P. Elliott, Constitution of Binary Alloys, 1. Suppl., McGraw-Hill, Publ. Co., New York 1965.
F. A. Shunk, Constitution of Binary Alloys, 2. Suppl., McGraw-Hill, Publ. Co., New York 1969.
M. Shiga and Y. Nakamura, phys. stat. sol. (a) **37**, K89 (1976).
M. S. Seehra and P. Silinsky, Phys. Rev. B **13**, 5183 (1976).
P. A. Montano and M. S. Seehra, Phys. Rev. B **15**, 2437 (1977).
J. A. Oyedele and M. F. Collins, Phys. Rev. B **16**, 3208 (1977).

[11] M. O. Robbins and L. Falicov, Phys. Rev. B **29**, 1333 (1984).

(Received March 4, 1985)

phys. stat. sol. (b) **131**, 605 (1985)

Subject classification: 13.5.1; 22.4.1

Sektion Mathematik/Physik der Pädagogischen Hochschule „Liselotte Herrmann" Güstrow[1])

Many-Body Theory for the Dense Exciton Gas of Direct Semiconductors

III. Response of the Many-Exciton System on an Externally Driven Light Field[2])

By

V. May, F. Boldt, and K. Henneberger

Dedicated to Prof. Dr. Dr. h. c. Dr. E. h. P. Görlich on the occasion of his 80th birthday

The nonlinear response of a dense exciton gas in a platelet on an externally driven field is discussed using non equilibrium Green's function technique. A generalized Boltzmann-equation of excitons containing many-particle corrections in the diffusion-, drift- and collision-term is derived. Using a relaxed distribution (Boltzmann-distribution) and an unrelaxed distribution (macro-occupation) of 1s-excitons a bistability in exciton density as well as in the transmitted light intensity is obtained for a nearly resonant and stationary applied pump-field of intensity 0.1 and 10 kW/cm², respectively. The bistability is obtained as a pure excitonic effect below the Mott-transition.

Für ein Plättchen wird bei Benutzung von Nichtgleichgewichts-Greenfunktionstechnik die nichtlineare Response eines dichten Exzitonengases auf ein externes Feld diskutiert. Es wird eine verallgemeinerte Boltzmanngleichung von Exzitonen abgeleitet, die Vielteilchen-Korrekturen im Diffusions-, Drift- und Stoßterm enthält. Bei Benutzung einer relaxierten Verteilung (Boltzmann-Verteilung) und einer unrelaxierten Verteilung (Makrobesetzung) von 1s-Exzitonen wird in jedem Falle eine Bistabilität sowohl in der Exziton-Dichte als auch in der transmittierten Intensität für ein als nahezu resonant und stationär angenommenes Pump-Feld der Intensität von etwa 0,1 bzw. 10 kW/cm² erhalten. Die Bistabilität wird als ein rein exzitonischer Effekt gerade unterhalb des Mott-Überganges erhalten.

1. Introduction

Connected with the first observation of optical bistability (OB) in the semiconductors InSb [1] and GaAs [2] a lot of mechanisms resulting in optical nonlinearities were investigated. Beside band-filling effects [1] and exciton screening due to free carriers [2] OB was realized in GaAs superlattices [3] and in CuCl connected with the two-photon absorption of excitonic molecules [4, 5]. Optical nonlinearities observed near the exciton levels in CdS [6 to 8] are also explained partly in terms of an OB.

Concerning the theoretical description of OB as well as density bistability (DB) of excited carriers in semiconductors a lot of papers were published in the last years. One group of papers considers band-filling effects [9 to 11]. The optical nonlinearities connected with the excitonic molecule are extensively studied in [12 to 15]. OB in a many-exciton system (MES) was first proposed in [16] as an intrinsic OB [17]. The phenomenological treatment of [16] was further developed in [18, 19]. A microscopic approach for the MES and the resulting OB was already published in [20, 21] but with exciton–exciton interaction treated in a rough approximation. The mean-field approximation is used in [20] whereas exciton–exciton interaction in [21] is replaced by the

[1]) Goldberger Str. 12, DDR-2600 Güstrow, GDR.

[2]) Part I and II see phys. stat. sol. (b) **129**, 717 (1985) and **130**, 675 (1985), respectively.

coupling of the externally excited exciton mode to an effective thermal reservoir formed by all other exciton modes.

We present in the following a microscopic approach for the MES using nonequilibrium many-exciton theory of [22, 23] (called I and II) and its interaction with a semiclassically considered radiation field. This approach allows the description of the kinetics of excitons under the action of the radiation field as well as the feedback of the MES on the propagation properties of the field.

A generalized kinetic equation of excitons was given in I on the basis of nonequilibrium Green's function technique (for a review see [24]). We handled the interaction among the excitons in second Born collision approximation. The corresponding spectral properties of the MES depending on the total exciton density were calculated in Part II. Due to the overcompensation of the repulsive exchange interaction between excitons by an attractive van-der-Waals-like force we obtained with increasing exciton density red shifts of the 1s level as well as of the continuum edge. Here, we study nonlinear response of the MES if the radiation field is resonantly applied to the 1s level. In contrast to [20, 21] the dependence of the exciton self-energy on the actual detuning is taken into account. We derive a Boltzmann-like kinetic equation of excitons with many-particle corrections in the diffusion and drift term as well as in the collisional integral. The application of the semiclassical theory is well established since the field strength is strong enough. However, this approximation results in the neglection of the polariton effect. All resonantly excited particles are distributed in the pure exciton spectrum. But this approximation is justified since our exciton damping becomes greater than the longitudinal–transverse splitting of the exciton level.

Since we are mainly interested in showing the importance of many-exciton effects in connection with nonlinear optical properties we discuss the propagation effects of the field in a platelet in a special case only. Furthermore, the additional resonances just below the 1s exciton level connected with the induced and two-photon-absorption process of biexcitons are not taken into account. First results concerning DB and OB of the MES have already been published in [25].

2. Kinetic Equations of Excitons

In I we derived a generalized nonlocal kinetic equation for the correlation parts $G^{\gtrless}$ of the excitonic Green's function ((29) of Part I) starting with the two-time electron–hole pair propagator. Studying the response of MES on the radiation field we introduce as external perturbation (instead (2a) in I),

$$H_{\text{ext}}(t) = -\frac{1}{\Omega}\sum_{\boldsymbol{k}} \boldsymbol{E}(\boldsymbol{k}, t)\, \boldsymbol{P}^{+}(\boldsymbol{k}, t)\,, \tag{1}$$

$\boldsymbol{E}$ being the electric field strength in the medium coupled to the applied field via boundary conditions. The operator $\boldsymbol{P}$ of the excitonic polarization,

$$\boldsymbol{P}(\boldsymbol{k}) = \sum_{\alpha} (\boldsymbol{p}_{\alpha} b_{\alpha\boldsymbol{k}} + \boldsymbol{p}_{\alpha}^{*} b_{\alpha-\boldsymbol{k}}^{+})\,, \tag{2}$$

contains the dipole moment $\boldsymbol{p}_{\alpha}$ for the transition from the filled valence band to an exciton level labelled by α (at $\boldsymbol{k} = \boldsymbol{0}$) and the exciton operators $b_{\alpha\boldsymbol{k}}$. H_{ext} defines the S-operator contained in all propagation functions (see, e.g., (1) of Part I) and leads to an additional contribution in the equation for $G^{\gtrless}$. Considering, at first, the equation of motion for the electron–hole pair propagator ((6) of Part I) an additional contribution due to H_{ext},

$$-\frac{1}{\Omega}\sum_{\boldsymbol{k}2\text{e}2\text{h}} \boldsymbol{E}(\boldsymbol{k}, t_1)\, e\boldsymbol{r}_{\text{cv}} \langle \hat{F}(1\text{e}1\text{h}, 2\text{e}2\text{h} + \boldsymbol{k}; t_1)\, e_{1\text{e}'}^{+}(t_1')\, h_{1\text{h}'}^{+}(t_1')\rangle\,, \tag{3}$$

occurs, where $e\boldsymbol{r}_{cv}$ is interband matrix element of the dipole operator (for further explanation see Part I). We neglect any correlation between phase-space occupation $\hat{F}$ and the electron–hole pair amplitude e^+h^+ in (3) and use the propagator in a factorized form only. Changing to an exciton representation we obtain the equation for $G^<(1 = \bar{1}t_1, \bar{1} = \alpha_1\boldsymbol{k}_1)$,

$$\int d2 \{g^{(\text{ret})-1}(1, 2)\, G^<(2, 1') - \Sigma_c^<(1, 2)\, G^{(\text{adv})}(2, 1')\} =$$

$$= -\frac{1}{i\hbar} \int d2\, F(1, 2)\, \mathscr{E}(2)\, \Phi^*(1'), \tag{4}$$

with the correlation parts $\Sigma_c^{\gtrless}$ of the self-energy, F as the exciton representation of the phase-space occupation,

$$\mathscr{E}(1) = \frac{1}{\Omega}\, \boldsymbol{p}_{\alpha_1}^*\, \boldsymbol{E}(\boldsymbol{k}_1, t_1), \tag{4a}$$

and

$$g^{(\text{ret})-1}(1, 1') = i\hbar \frac{\partial}{\partial t_1} \delta(1 - 1') - H^{(\text{ret})}(1, 1'). \tag{5}$$

The effective exciton Hamiltonian $H^{(\text{ret})}$ contains the energy $E_{ex}(1)$ of an isolated exciton and all self-energy contributions. We use an approximate form for the self-energy containing exciton–exciton interaction within the second Born approximation ((27b), (27c) of Part I) but no dependence on both the electric field strength and the exciton amplitude

$$\Phi(1) = \langle b_{\alpha_1\boldsymbol{k}_1}(t_1)\rangle. \tag{6}$$

The electric field strength follows from

$$\left(\frac{\partial^2}{\partial t^2} + c^2\boldsymbol{k}^2\right) \boldsymbol{E}(\boldsymbol{k}, t) = -4\pi \frac{\partial^2}{\partial t^2} \boldsymbol{P}_{\text{tot}}(\boldsymbol{k}, t), \tag{7}$$

where the total polarization $\boldsymbol{P}_{\text{tot}}$ contains a background contribution $\chi_\infty \boldsymbol{E}$ and the excitonic contribution

$$\langle \boldsymbol{P}(\boldsymbol{k}, t)\rangle = \sum_\alpha \boldsymbol{p}_\alpha \Phi(\alpha\boldsymbol{k}t) + (\text{c.c.}, \boldsymbol{k} \to -\boldsymbol{k}). \tag{8}$$

Instead a further equation of motion for Φ we consider the variation of Φ with $\boldsymbol{E}$,

$$\frac{\delta\Phi(1)}{\delta\boldsymbol{E}(\boldsymbol{k}_2, t_2)} = -\frac{1}{\Omega} \frac{\theta(t_1 - t_2)}{i\hbar} \langle (b_{\alpha_1\boldsymbol{k}_1}(t_1), \boldsymbol{P}^+(k_2, t_2))_-\rangle. \tag{9}$$

In the general case $\delta\Phi/\delta\boldsymbol{E}$ is a complicated functional of $\boldsymbol{E}$ and (9) cannot be integrated. However, if we assume, as in the case of the self-energy, that any direct dependence of the expectation value in (9) on $\boldsymbol{E}$ can be neglected, the integration becomes trivial. Neglecting additionally the anomal expectation value with respect to the exciton operators we get

$$\Phi(1) = -\int d2\, G^{(\text{ret})}(1, 2)\, \mathscr{E}(2) \tag{10}$$

and the dielectric tensor follows

$$\hat{\varepsilon}(\boldsymbol{k}t, \boldsymbol{k}'t') = \varepsilon_\infty - 4\pi \sum_{\alpha\alpha'} \frac{\boldsymbol{p}_\alpha^* \cdot \boldsymbol{p}_{\alpha'}}{\Omega} \times$$

$$\times \{G^{(\text{ret})}(\alpha\boldsymbol{k}t, \alpha'\boldsymbol{k}'t') + G^{(\text{ret})}(\alpha - \boldsymbol{k}'t', \alpha' - \boldsymbol{k}t)\}, \tag{11}$$

which looks like the linear response expression for a nonstationary, spatial inhomogeneous system depending, however, on the exciton, i.e. excitation density. The system of (4), (7), (10) is completed by the equation for the retarded exciton Green's function (see (30) of Part I, but now with phase-space occupation)

$$\int d2\, g^{(\mathrm{ret})-1}(1, 2)\, G^{(\mathrm{ret})}(2, 1') = F(1, 1') \tag{12}$$

which determines the spectral properties of the MES.[3])

Due to the neglection of any dependence of the self-energy on the exciton amplitude we can use (10) to obtain equations for $G^{<}$, $G^{(\mathrm{ret})}$, and $\boldsymbol{E}$ only. Instead of (4) we have

$$\int d2\, \{g^{(\mathrm{ret})-1}(1, 2)\, G^{<}(2, 1') - \left(\Sigma_c^{<}(1, 2) + \Sigma_{\mathrm{rad}}(1, 2)\right) G^{(\mathrm{adv})}(2, 1')\} = 0 \tag{13}$$

with the radiative self-energy

$$\Sigma_{\mathrm{rad}}(1, 1') = \frac{1}{i\hbar} \int d2\, F(1, 2)\, \mathscr{E}(2)\, \mathscr{E}^*(1')\,. \tag{14}$$

Equation (7) for $\boldsymbol{E}$ becomes

$$\sum_{\boldsymbol{k}'} \int dt' \left\{ c^2 \boldsymbol{k}^2 \delta_{\boldsymbol{k},\boldsymbol{k}'}\, \delta(t - t') + \frac{\partial^2}{\partial t^2}\, \hat{\varepsilon}(\boldsymbol{k}t, \boldsymbol{k}'t') \right\} \boldsymbol{E}(\boldsymbol{k}', t') = 0\,. \tag{15}$$

(12), (13), (15) form the basic equations for further calculations.

2.1 Generalized Boltzmann equation of excitons

If the absorption coefficient and the wavelength of the radiation field are greater than the excitonic Bohr radius, a_{ex}, the radiation field acts as a spatially slowly varying disturbance on the MES. The radiation field is slowly varying also in the time region if the pulse duration is equal or greater than all relaxation times of the MES. In those cases the general nonlocal connection in (12), (13), (15) can be replaced by first derivatives with respect to space and time arguments. (For details we refer to [24]). If the following kinetical equation is integrated over energy argument E, a generalized Boltzmann equation follows ($\lambda = \alpha\boldsymbol{k};\, \boldsymbol{r}t$):

$$\left\{ \frac{\partial}{\partial t} + \left(\boldsymbol{v}(\lambda)\, \nabla_{\boldsymbol{r}}\right) + \left(\boldsymbol{F}(\lambda)\, \frac{1}{\hbar}\, \nabla_{\boldsymbol{k}}\right) \right\} N(\lambda) =$$

$$= P(\lambda) - \left(\frac{\Gamma(\lambda)}{\hbar} + \frac{1}{\tau_{\mathrm{rec}}}\right) N(\lambda) + R^{(\mathrm{into})}(\lambda)\,. \tag{16}$$

This kinetic equation for excitons contains a renormalized group velocity

$$\boldsymbol{v}(\lambda) = \frac{1}{\hbar}\, \nabla_{\boldsymbol{k}} \mathscr{E}_{\mathrm{ex}}(\lambda)\,, \tag{17}$$

and an effective force including exchange-correlation contributions

$$\boldsymbol{F}(\lambda) = -\nabla_{\boldsymbol{r}} \mathscr{E}_{\mathrm{ex}}(\lambda)\,. \tag{18}$$

$\mathscr{E}_{\mathrm{ex}}$ is the renormalized exciton energy (see (13) II), Γ is the damping due to exciton–exciton collisions,

$$\Gamma(\lambda) = -2\, \mathrm{Im}\, \Sigma_c^{(\mathrm{ret})}(\alpha\boldsymbol{k}\mathscr{E}_{\mathrm{ex}}, \boldsymbol{r}t) \tag{19}$$

---

[3]) It is more convenient to use the equations for $G^{<}$ and the equation for $G^{(\mathrm{ret})}$ instead of the equations for $G^{>}$ and $G^{<}$ (see, e.g., [24]).

and

$$R^{(\mathrm{into})}(\lambda) = \frac{1}{i\hbar} \Sigma_{\mathrm{c}}^{<}(\alpha \boldsymbol{k} \mathscr{E}_{\mathrm{ex}}, \boldsymbol{r}t) \tag{20}$$

the scattering rate into the state $(\alpha\boldsymbol{k})$. Furthermore, a phenomenological lifetime constant τ_{rec} of excitons was introduced which results from radiative as well as nonradiative recombination. P is the pump rate of excitons by the light field,

$$P(\lambda) = \frac{i}{\hbar} \left\{ \delta_{\boldsymbol{k}, -\frac{i}{2}(\nabla_{\boldsymbol{r}'} - \nabla_{\boldsymbol{r}''})} \hat{G}\left(\alpha \boldsymbol{k} \frac{i\hbar}{2}\left(\frac{\partial}{\partial t'} - \frac{\partial}{\partial t''}\right); \boldsymbol{r}t\right) \times \right.$$
$$\left. \times\, \tilde{\mathscr{E}}(\alpha \boldsymbol{r}'t')\, \mathscr{E}^*(\alpha \boldsymbol{r}''t'') \right\}_{t=t'=t''}^{\boldsymbol{r}=\boldsymbol{r}'=\boldsymbol{r}''}. \tag{21}$$

The derivatives in the δ-function and in the special function $\hat{G}$ are defined in the sense of Fourier transformation. $\tilde{\mathscr{E}}$ contains the phase-space occupation. In all further calculations we use the approximate form

$$\tilde{\mathscr{E}}(\alpha \boldsymbol{r}t) = [1 - \mathscr{C}(\alpha \boldsymbol{r}t)]\, \boldsymbol{p}_\alpha^* \boldsymbol{E}(\boldsymbol{r}t) \tag{22}$$

with the correction due to the non-bosonic nature of excitons (for explanation see I)

$$\mathscr{C}(\alpha \boldsymbol{r}t) = \frac{1}{\Omega} \sum_{\alpha'\boldsymbol{k}'} C(\alpha 0, \alpha'\boldsymbol{k}')\, N(\alpha'\boldsymbol{k}', \boldsymbol{r}t)\,, \tag{23}$$

$$C(\bar{1}, \bar{2}') = \Omega \sum_{1\mathrm{e}1\mathrm{h},1\mathrm{e}'1\mathrm{h}'} [f_{\bar{1}}^*(1\mathrm{e}'1\mathrm{h})\, f_{\bar{2}}^*(1\mathrm{e}1\mathrm{h}') +$$
$$+ (\bar{1} \leftrightarrow \bar{2})]\, f_{\bar{2}}(1\mathrm{e}'1\mathrm{h}')\, f_{\bar{1}}(1\mathrm{e}1\mathrm{h})\,. \tag{23a}$$

In particular we have

$$C(1\mathrm{s}0, 1\mathrm{s}0) \equiv C = \frac{33\pi}{2}\, a_{\mathrm{ex}}^3\,. \tag{23b}$$

We use $H^{(\mathrm{ret})}$ linearly in the exciton density n_{ex} as calculated in Part II. Neglecting any $\boldsymbol{k}$-dependence of the self-energy contribution in $H^{(\mathrm{ret})}$ we obtained for the 1s level with a background damping Γ_0,

$$H^{(\mathrm{ret})}(1\mathrm{s}0\hbar\omega; \boldsymbol{r}t) = E_{\mathrm{ex}}(1\mathrm{s}) + \bar{\sigma}(\omega)\, n_{\mathrm{ex}}(\boldsymbol{r}t) - i[\Gamma_0 + \bar{\gamma}(\omega)\, n_{\mathrm{ex}}(\boldsymbol{r}t)]/2\,, \tag{24}$$

with e. g. the shift factor $\bar{\sigma}(\omega)$ according to $\bar{\sigma}(\omega) = -\sigma(\omega)\, E_{\mathrm{ex}}^{\mathrm{b}} a_{\mathrm{ex}}^3$ (see Part II).

3. Response of the Many-Exciton System on a Given Light Field

In the present section we study the action of a stationary, linearly polarized, monochromatic light field of given intensity (negletion of any absorption),

$$\boldsymbol{E}(\boldsymbol{r}, t) = \boldsymbol{n}\hat{E}_0\, \mathrm{e}^{i(\boldsymbol{q}\cdot\boldsymbol{r} - \omega t)} + (\mathrm{c.c.})\,. \tag{25}$$

The resulting generation rate of the Boltzmann equation becomes

$$P(\lambda) = \frac{i}{\hbar}\, \delta_{\boldsymbol{k},\boldsymbol{q}} \big(1 - \mathscr{C}(\alpha \boldsymbol{r}t)\big)\, |\boldsymbol{p}_\alpha \boldsymbol{n} \hat{E}_0|^2\, \hat{G}(\alpha \boldsymbol{q} \hbar\omega; \boldsymbol{r}t)\,. \tag{26}$$

Its properties are mainly determined by the spectral function of the MES.

In the following we consider the spectral region around the 1s exciton level only. In such a frequency range $\boldsymbol{q}$ is small in comparison with the characteristic $\boldsymbol{k}$-dependence of $\hat{G}$. Due to the sharp generation rate in $\boldsymbol{k}$-space the exciton distribution will possess a peak at $\boldsymbol{k} = \boldsymbol{q} \approx \boldsymbol{0}$. The occupation of finite $\boldsymbol{k}$-states results from exciton–exciton and exciton–phonon scattering. For solutions of (16) independent of $\boldsymbol{r}t$ we obtain a balance equation for the exciton density due to the balance of generation and recombination.

Neglecting any backscattering in the $\boldsymbol{k} = \boldsymbol{0}$ state and assuming a very quick recombination of excitons in finite $\boldsymbol{k}$-states we use a macro-occupation of excitons in the $\boldsymbol{k} = \boldsymbol{0}$ state in the excitons distribution

$$N(\bar{1}) = \delta_{\alpha_1, 1s}\delta_{k_1, 0}\Omega n_{ex}/4 \,. \tag{27}$$

The balance equation for $\alpha = 1s$ reads

$$n_{ex} = \frac{a(1 - cn_{ex})^2 I_0}{[\Delta\hbar\omega - \bar{\sigma}(\omega)\, n_{ex}]^2 + [\Gamma_0 + \bar{\gamma}(\omega)\, n_{ex}]^2/4}, \tag{28}$$

where $a = 2\Delta E_{LT}/c_m$, $I_0(c_m)$ is the light intensity (velocity) in the medium, ΔE_{LT} the longitudinal–transverse splitting, and $\Delta\hbar\omega = \hbar\omega - E_{ex}(1s)$ the detuning. The shift and damping factors are now functions of the actual frequency. A similar balance equation for the exciton density was obtained in [26] for bose-condensed excitons under the influence of an external field. However, $\bar{\sigma}$ and $\bar{\gamma}$ were not calculated for exciton–exciton scattering.

In calculating (28) from (16) we have assumed for the damping $\Gamma(\hbar\omega) \approx \Gamma(E_{ex}(1s))$ and hence Γ can be cancelled on both sides of (28) (usually $1/\tau_{rec}$ can be neglected in comparison with $\Gamma/\hbar$).

Dropping for a moment cn_{ex} and the damping in (28) a cubic equation for the density is obtained and bistability occurs for negative detuning. This behaviour results from increasing absorption with increasing density due to the energetical lowering of the exciton level. The linear connection between n_{ex} and I_0 becomes hyperlinearly ending up with $n_{ex} \sim I_0^{1/3}$ at high I_0. Fig. 1a shows the situation for parameters appropriate to the $A_{n=1} - \Gamma_5^T$ exciton of CdS.[4]) Curves with negative detuning touch the n_{ex}-axis at $|\Delta\hbar\omega/\bar{\sigma}(\omega)|$. For $\Delta\hbar\omega = -3$ meV this value is about one order of magnitude below the value of the Mott transition given by our theory (see Part II). This n_{ex}-value increases with further lowering of $\Delta\hbar\omega$.

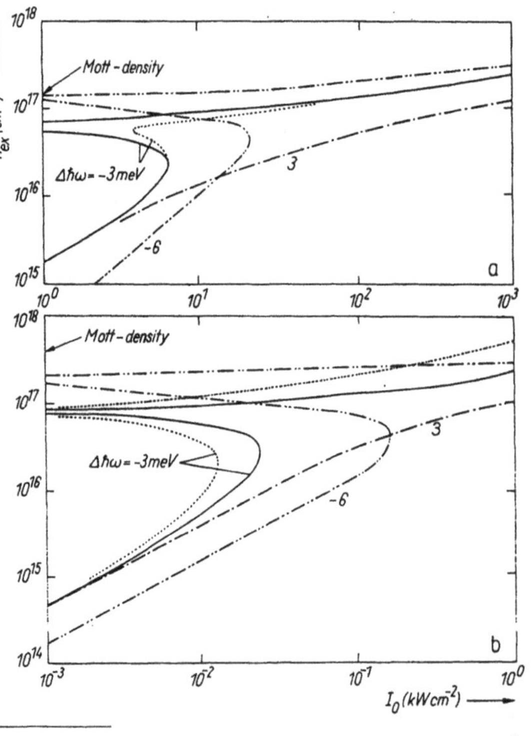

Fig. 1. a) Density of the $A_{n=1}$ excitons in CdS via radiation intensity if a macro-occupation of the $\boldsymbol{k} = \boldsymbol{0}$ state is assumed; dotted curve: full damping; other curves: phonon damping $\Gamma_0 = 0.2$ meV only. b) The same as in a), but for a Boltzmann distribution with $T^* = 30$ K

[4]) We use a geometry ($\boldsymbol{k}, \boldsymbol{E} \perp \boldsymbol{c}$) where only the Γ_5^T-exciton is optically active. Due to the high damping values the energy difference to the Γ_5^L- and Γ_6-excitons can be neglected.

Taking the damping due to exciton–phonon coupling into account the shape of the curves is nearly unchanged for $\Gamma_0 < 1$ meV (in Fig. 1a we have used $\Gamma_0 = 0.2$ meV which correspondes to a lattice temperature of about 10 K [27]). The damping due to exciton–exciton collisions, however, reduces the bistability strongly. The increasing damping with density prevents a remarkable increase of absorption due to the level shift. The influence of the correction (23) due to the non-bosonic nature of excitons is of minor importance in the bistable region.

The case contrary to the macro-occupation is the Boltzmann distribution

$$N(\mathrm{I}) = \delta_{\alpha_1,1s} 2 \left(\frac{\pi\hbar^2}{2M_{\mathrm{ex}}k_{\mathrm{B}}T^*}\right)^{3/2} \mathrm{e}^{-\frac{\hbar^2 k_1^2}{2M_{\mathrm{ex}}k_{\mathrm{B}}T^*}} n_{\mathrm{ex}} \,. \tag{29}$$

Equation (24) must be integrated over $\boldsymbol{k}$ and then the resulting balance equation reads (note the neglection of the $\boldsymbol{k}$-dependence in $\hat{G}$)

$$n_{\mathrm{ex}} = \frac{a[1 - c(T^*)\, n_{\mathrm{ex}}]^2\, I_0}{[\Delta\hbar\omega - \bar{\sigma}(\omega, T^*)\, n_{\mathrm{ex}}]^2 + [\Gamma_0 + \bar{\gamma}(\omega, T^*)\, n_{\mathrm{ex}}]^2/4}\, \frac{\Gamma_0 + \bar{\gamma}(\omega, T^*)\, n_{\mathrm{ex}}}{\hbar/\tau_{\mathrm{rec}}} \,. \tag{30}$$

c, $\bar{\sigma}$, and $\bar{\gamma}$ are now functions of effective temperature. Since annihilation of excitons is produced by interband recombination and not by exciton–exciton collisions as in the case of the macro-occupation, the quotient of intraband and interband damping occurs in (30). Such a balance equation was already introduced in [16] by phenomenological arguments. The field intensity I_0 is scaled by τ_{rec}. In our calculations we use $\tau_{\mathrm{rec}} = 0.5$ ns. Fig. 1b shows n_{ex} as a function of I_0 with and without the damping due to exciton–exciton collisions for the $\mathrm{A}_{n=1} - \Gamma_5^{\mathrm{T}}$ exciton of CdS ($\Gamma_0 = 0.2$ meV, $T^* = 30$ K). In the present case the density bistability is less modified if the damping due to exciton–exciton collision is taken into account. The intensity range of interest is much lower as in the case of the macro-occupation due to the scaling behaviour of τ_{rec}. Increasing I_0 a jump in density occurs somewhat lower as one order of magnitude. But now, decreasing I_0 the high-density value is maintained. The jump to the low-density branch contains more than three orders of magnitude in density.

According to the value of the Mott density obtained in Part II the jump from the low- to the high-density branch is not connected with a Mott transition. Furthermore, the actual value of the exciton binding energy at the high-density branch ($n_{\mathrm{ex}} \approx 10^{17}\ \mathrm{cm}^{-3}$) is about 70% of the value of the unexcited crystal (cf. Fig. 4 of Part II).

4. Response of the Many-Exciton System of a Platelet

We consider the MES extending in a platelet in z-direction from $z = 0$ to $z = d$ under the action of an external field (stationary, linearly polarized, monochromatic, and with normal incidence on the surface).

We neglect any transverse effect of the field (plane-wave ansatz) and assume the diffusion length L to be greater than the platelet thickness d. In this case the exciton distribution is z-independent and, hence, the system is optically homogeneous and the field inside the platelet splits into a forward and a backward plane wave.

To obtain in (28) as well as in (30) the intensity I_i of the incoming field instead of I_0 we follow the procedure of [19] taking the spatial average over the platelet. It results in the replacement of I_0 by $(1 - R - T)\, I_i$ with $R(T)$ as the reflexion (transmission) coefficient of an optically homogeneous platelet with dielectric function according

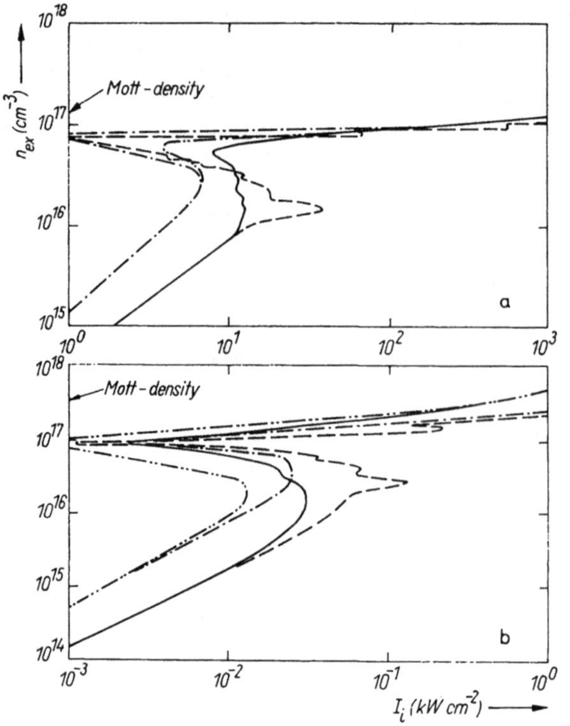

Fig. 2. a) Density of the $A_{n=1}$ excitons in a CdS platelet with thickness $d = 0.5$ μm via incoming intensity I_i if a macro-occupation of the $\boldsymbol{k} = \boldsymbol{0}$ state is assumed ($\Delta\hbar\omega = -3$ meV; $\Gamma_0 = 0.2$ meV; full curve: full damping; dashed curve: phonon damping only; other curves: approximation of the given light field as in Fig. 1; —··— full damping, —·— phonon damping only. b) The same as in a), but for a Boltzmann distribution with $T^* = 30$ K,

to (11),

$$\varepsilon(\omega) = \varepsilon_\infty \left[1 - \frac{\Delta E_{\text{LT}}(1 - cn_{\text{ex}})}{\hbar\omega - E_{\text{ex}}(1\text{s}) - \bar{\sigma}n_{\text{ex}} + i(\Gamma_0 + \bar{\gamma}n_{\text{ex}})/2}\right]. \tag{31}$$

Using (28) as well as (30) we have calculated the respones of n_{ex} and the transmitted intensity I_t on I_i. Fig. 2a and 2b show n_{ex} in dependence on I_i for the macro-occupation and the Boltzmann distribution, respectively. The shape of the curves is somewhat shifted to higher I_i since $1 - R - T$ is smaller than unity. The additional structures in the curves with $\Gamma \approx \Gamma_0$ result from the resonator effect of the platelet. They are smeared out if damping due to exciton–exciton interaction is taken into account.

I_t in dependence on I_i is shown in Fig. 3 and 4. The intensity gap in Fig. 3a results from the low transmission value at the shifted exciton resonance. Again the loops result from the resonator effect of the platelet.

For a pulse excitation with maximum intensity lower than intensity I_c (where I_t is strongly decreased) a hysteresis cycle in I_t over I_i similar to those of e.g. [6, 28] occurs. Our results exhibit also a jump in the absorption coefficient (see Fig. 4). However, the maximum intensities of about 1 MW/cm^2 used in the experiments [6 to 8, 19] are some orders of magnitude greater than our I_c. A similar discrepancy between theory and experiment was also obtained in [19]. Possibly this jump in absorption results from the Mott transition after which strong interband absorption takes place. According to Fig. 2a the Mott transition is expected near 1 MW cm^{-2} for I_i, too. Clearly, a definite description of the nonlinear response of the MES up to the Mott transition becomes only possible if many-exciton theory could be extended to that high-density region. Nevertheless, the present results for the nonlinear effects are already in the

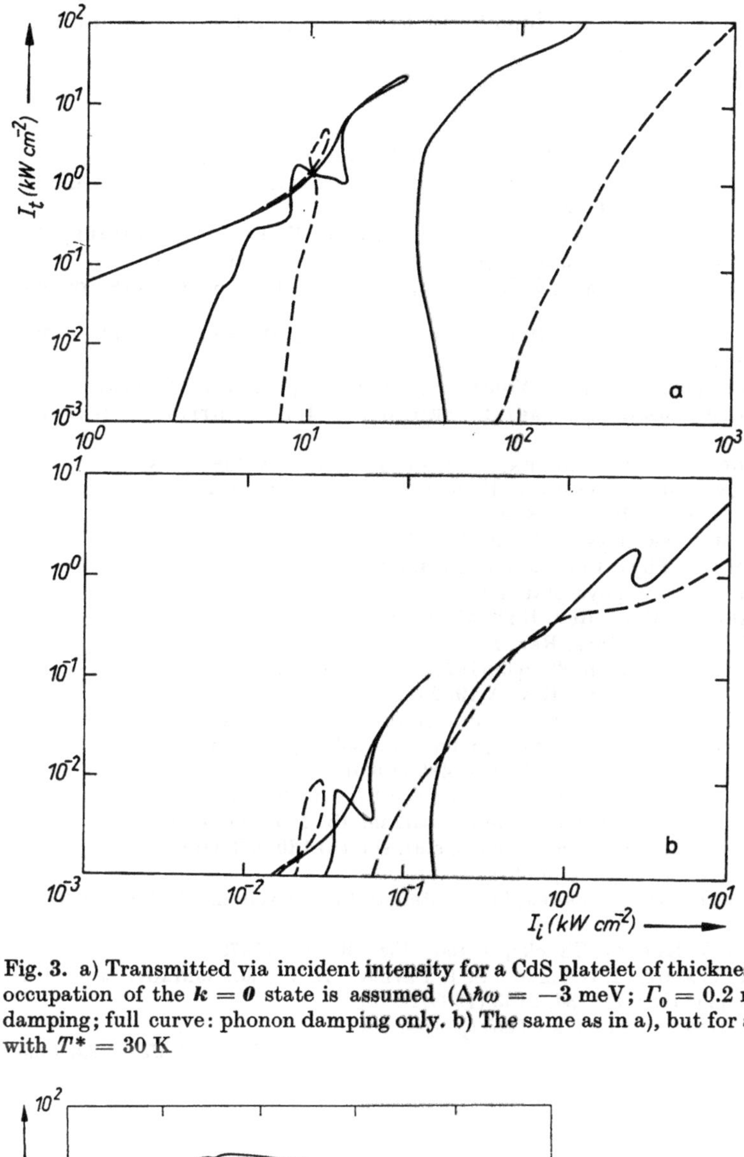

Fig. 3. a) Transmitted via incident intensity for a CdS platelet of thickness $d = 0.5\ \mu m$ if a macro-occupation of the $\boldsymbol{k} = \boldsymbol{0}$ state is assumed ($\Delta\hbar\omega = -3$ meV; $\Gamma_0 = 0.2$ meV); dashed curve: full damping; full curve: phonon damping only. b) The same as in a), but for a Boltzmann distribution with $T^* = 30$ K

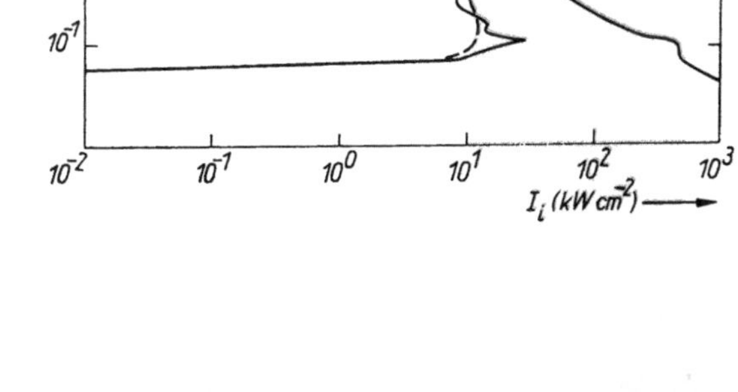

Fig. 4. Absorption coefficient α via incident intensity I_i if a macro-occupation of the $\boldsymbol{k} = \boldsymbol{0}$ state is assumed. ($\Delta\hbar\omega = -3$meV; $\Gamma_0 = 0.2$ meV); dashed curve: full damping; solid curve: phonon damping only

range of validity of our theory. Experiments in the corresponding pump-intensity range could clarify the situation.

References

[1] D. A. B. MILLER, S. D. SMITH, and A. JOHNSTON, Appl. Phys. Letters **35**, 658 (1979).
[2] H. M. GIBBS, S. L. MCCALL, T. N. C. VENKATESAN, A. C. GOSSARD, A. PASSNER, and W. WIEGMANN, Appl. Phys. Letters **35**, 451 (1979).
[3] H. H. GIBBS, S. S. TARUG, J. J. JEWELL, D. A. WEINBERGER, K. TAI, A. C. GOSSARD, S. L. MCCALL, A. PASSNER, and W. WIEGMANN, Appl. Phys. Letters **41**, 221 (1982).
[4] R. LEVY, J. Y. BIGOT, B. HÖNERLAGE, F. TOMASINI, and J. B. GRUN, Solid State Commun. **48**, 705 (1983).
[5] N. PEYGHAMBARIAN, H. M. GIBBS, M. C. RUSHFORD, and D. A. WEINBERGER, Phys. Rev. Letters **51**, 1692 (1983).
[6] H. ROSSMANN, F. HENNEBERGER, and J. VOIGT, phys. stat. sol. (b) **115**, K63 (1983).
[7] N. A. VIDMONT, A. A. MAKSIMOV, I. I. TARTAKOVSKII, and V. M. EDELSHTEIN, Zh. tekh. Fiz., Pisma **9**, 1527 (1983).
[8] K. BOHNERT, H. KALT, and C. KLINGSHIRN, Appl. Phys. Letters **43**, 1088 (1983).
[9] A. ELEI, M. O. SCULLY, and J. M. O'HARE, Phys. Rev. B **27**, 4779 (1983).
[10] J. GOLL and H. HAKEN, Phys. Rev. A **28**, 910 (1983).
[11] TH. HARBICH and G. MAHLER, phys. stat. sol. (b) **117**, 635 (1983).
[12] S. W. KOCH and H. HAUG, Phys. Rev. Letters **46**, 450 (1981).
[13] F. HENNEBERGER and V. MAY, phys. stat. sol. (b) **109**, K139 (1982).
[14] J. ABRAM and A. MARUANI, Phys. Rev. B **26**, 4759 (1982).
[15] C. C. SUNG and C. M. BOWDEN, Phys. Rev. B **29**, 1957 (1984).
[16] Y. TOYOZAWA, Solid State Commun. **28**, 533 (1978); **32**, 13 (1979).
[17] C. M. BOWDEN and C. C. SUNG, Phys. Rev. A **19**, 2392 (1979).
[18] F. HENNEBERGER and H. ROSSMANN, phys. stat. sol. (b) **121**, 685 (1984).
[19] HOANG XUAN NGUYEN and R. ZIMMERMANN, phys. stat. sol. (b) **124**, 191 (1984).
[20] J. GOLL and H. HAKEN, phys. stat. sol. (b) **101**, 489 (1980).
[21] M. L. STEYN-ROSS and C. W. GARDINER, Phys. Rev. A **27**, 310 (1983).
[22] V. MAY, F. BOLDT, and H. HENNEBERGER, phys. stat. sol. (b) **129**, 717 (1985).
[23] F. BOLDT, K. HENNEBERGER, and V. MAY, phys. stat. sol. (b) **130**, 675 (1985).
[24] K. HENNEBERGER and V. MAY, to be published.
[25] V. MAY, F. BOLDT, and K. HENNEBERGER, Proc. Internat. Conf. Excitons 84, Güstrow (GDR) 1984 (p. 119).
[26] V. F. ELESIN, and YU. V. KOPAEV, Zh. eksper. teor. Fiz. **63**, 1447 (1972).
[27] N. N. SINOVEV, L. P. IVANOV, I. G. LANG, S. T. PAVLOV, A. V. PROKASNIKOV, and I. D. YAPOSHEZKII, Zh. eksper. teor. Fiz. **84**, 2153 (1983).
[28] H. ROSSMANN and F. HENNEBERGER, see [25] (p. 186).

(Received March 29, 1985)

phys. stat. sol. (b) 131, 615 (1985)

Subject classification: 13.5.1 and 20.1; 22.5

Institute of Physics, Academy of Sciences of the Ukrainian SSR, Kiev[1])

Secondary Emission Spectra and Energy Relaxation of Polaritons in Layer Polar Semiconductors

By

M. S. BRODIN, I. V. BLONSKII, A. O. GUSHCHA, and V. V. TISHENKO

Dedicated to Prof. Dr. Dr. h. c. Dr. E. h. P. GÖRLICH on the occasion of his 80th birthday

The main processes of energy relaxation of polaritons in dependence on their initial energy and temperature are determined for the layer polar semiconductors 2H–PbI_2 and red HgI_2. It is shown that thermalization of the polaritons is reached due to the "thermal barrier" effect.

Определены основные механизмы энергетической релаксации поляритонов в зависимости от их энергии и температуры образца в слоистых полупроводниках 2H–PbI_2 и красной модификации HgI_2. Показано, что термализация поляритонов достигается благодаря действию эффекта ,,теплового барьера".

1. Introduction

One of the important problems in the physics of polaritons is the study of relaxation processes which are responsible for polariton thermalization. At present two main models are in use to describe how quasi-equilibrium is achieved between excitonic polaritons and phonons. The first approach was introduced by Toyozawa as early as 1955 [1]. Dealing with mixed exciton–photon states within some energy range ΔE_B of the lower polariton branch, he has predicted a so-called "bottleneck" effect which is as follows. Relaxation of polaritons within the range ΔE_B is governed by the emission of acoustic phonons. As the energy of polaritons decreases, the efficiency of the emission drops because of two reasons. First, the density of states becomes smaller when moving from the exciton-like part of the dispersion curve. Second, the energy of acoustic phonons decreases along with the decrease of the polariton wave vector. In terms of lifetime this slowing-down of relaxation can be described as the increase of the polariton lifetime τ in this particular energy range ΔE_B. Under certain experimental conditions (exciton concentration, lattice temperature) the value of τ becomes sufficient for polariton thermalization. With the "bottleneck" model it has become possible to understand the thermalization of polaritons in a great number of II–VI, I–VII, III–V, and some other crystals (see the review article [2] for details). However, certain limitations of the above-discussed approach were revealed later [3]. For instance, one faces some difficulties in trying to explain why the quasi-equilibrium state between polaritons and phonons is achieved at different temperatures in different crystals.

The more comprehensive "thermal barrier" model, proposed by Myasnikov in [4], attributes the slowing-down of polariton relaxation within the energy range ΔE_B to anti-Stokes scattering by equilibrium phonons. The efficiency of this process might be orders of magnitude larger than that of Stokes scattering because the density of final exciton-like anti-Stokes states is up 10^3 times higher than the density photon-

[1]) Prospekt Nauki 144, 252028 Kiev, USSR.

like final states for Stokes processes. Thus, the anti-Stokes scattering prevents further decrease of the polariton energy and gives rise to an accumulation of polaritons within the range ΔE_B of the dispersion curve where both group velocity and probability of transformation into secondary emission are small. For certain parameters of the "thermal barrier" the polariton lifetime turns out to be sufficient for polaritons and phonons to equalize their temperatures. This model provided a good background for the explanation of some peculiarities of the low-temperature polariton kinetics in anthracene crystals [3].

The subject of this paper is the relaxation mechanism for polaritons in layer crystals. In their phonon spectrum some interesting peculiarities exist: low-energy optical phonons (LOP) with $\hbar\omega = 1$ to 4 meV and so-called bending waves (BW) which belong to acoustic vibrations. As we shall see, these two kinds of phonon excitation determine the low-temperature polariton relaxation and give an unambiguous evidence for the existence of a "thermal barrier".

A little more about these phonons. LOP appear in layer crystals with two (or more) structurally identical, yet translationally non-equivalent layers in the unit cell. They result from the Davydov splitting of acoustic modes for layer packets when the vibrations of neighbouring packets are of opposite phase [5].

As for bending waves, first predicted theoretically by Lifshits, they correspond to membrane vibrations of layer packets In a formal way they are obtained when the bending deformation is taken into account in addition to the usual stress and shift deformations. The specific features of bending waves are as follows: a) low (≈ 10 K) of Debye temperature; b) quadratic dispersion law; c) corresponding lattice deformation occurs only in the approximation quadratic in atomic displacements and the contribution of BW to exciton–phonon interaction is described by an approximation which is nonlinear in phonon operators [6].

The contribution of phonons to the interaction process with excitons is essential indeed. This was evidenced in our recent studies of excitonic absorption [7]. Fig. 1 summarizes the results for all the compounds investigated which included PbI_2, HgI_2, GaSe, GaTe, and InSe. Three temperature characteristics of the exciton absorption band are shown: integral intensity (i.e. the area covered by the absorption profile) $S_0(T)$, shift of peak position $E_{exc}(T)$, and halfwidth $H/2$. There are some peculiarities which are specific of layer crystals. First, at the lowest temperatures there exists a temperature shift (short-wavelength in PbI_2 and long-wavelength shift in all other compounds) of the peak position which is not accompanied by a broadening of the

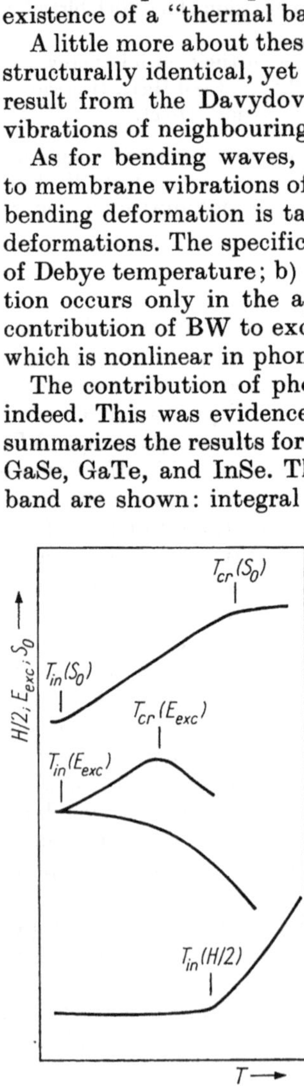

Fig. 1. Generalized temperature dependence of the exciton absorption band parameters in layered crystals. T_{in} and T_{cr} mean the boundaries of variations of the exciton absorption band integral characteristics

absorption band. Second, a rather significant increase of the integral absorption was observed in GaSe, InTe, GaTe within the strikingly broad temperature range from 20 to 150 K. This anomaly can hardly be attributed to the polariton effect because of the small oscillator strength of excitonic transitions in the crystals of interest.

The observed features were explained by the theory of exciton-phonon interaction taking into account the contribution of bending waves and low-energy phonons. In this theory the low-temperature shift of the exciton band is attributed to the nonlinear interaction of excitons with bending waves. A further proof was obtained by independent studies of $E_{exc}(T)$ in intercalated PbI_2 crystals [7]. The temperature dependence of the integral absorption $S(T)$ is understood when one considers the additional "pumping" of direct excitonic transitions by indirect vertical transitions from energy levels of the LOP.

2. Experimental Results and Discussion

Now let us consider how the above-stated concept of polariton relaxation was used to treat the results obtained with HgI_2 and PbI_2 single crystals, where the polaritonic behaviour is quite pronounced. First, let us discuss the polariton kinetics in HgI_2. Some important parameters for this compound, namely, the dispersion curve of polaritons, the longitudinal–transverse splitting Δ_{L-T} which is equal to 5 meV, the above-discussed energy range ΔE_B of strong exciton–photon mixing ($\Delta E_B \geqq 3$ meV) were determined earlier [8 to 10]. The polariton secondary emission spectra have been measured at various temperatures, the energy difference between excitation energy $h\nu_{exc}$ and the energy of the transverse exciton E_T being varied. Both these measurements and the earlier data on the exciton–photon coupling made it possible to determine the main mechanisms of polariton relaxation, including those responsible for thermalization [7].

Fig. 2 shows the polariton emission spectra of HgI_2 at different temperatures. One can see both upper and lower polariton branch emission (bands A_U and A_L, respectively), the emission from bound exciton levels λ, and different phonon replicas of the A_L band.

To determine the main mechanisms of polariton relaxation and the temperature T_e corresponding to the quasi-equilibrium of excitonic polaritons and phonons we have studied the effect of temperature on the excitation spectra of polariton emission bands. The samples were excited by monochromatized Xe-lamp radiation. The measured spectra are shown in Fig. 3. At 4.2 K the excitation spectra exhibit a periodic structure of uniformly spaced bands. The spacing 3.5 meV between them is equal to the energy of the dipole-active phonon of E_{2u} symmetry. A very similar spectrum was obtained in [10] at 4.2 K only. We observed that with the increase of temperature this structure broadens and at 12 K disappears completely. The results obtained give us the possibility to make the following conclusions, based on the well-known ideas of polariton radiation kinetics [2]:

1. At $T < 12$ K polariton emission appears in the stage before quasi-equilibrium between polaritons and phonons is achieved. This view is supported by the dependence of the polariton emission band on $h\nu_{exc}$.

2. The scheme of polariton energy relaxation is as follows: upper branch–lower branch transition followed by relaxation along the lower polariton branch with simultaneous emission of dipole-active phonons E_{2u} ($\hbar\omega_{E_{2u}} = 3.5$ meV).

3. One of the features of HgI_2 crystal is that the range of the lower polariton dispersion curve with the highest probability of transformation into secondary radiation is about the E_{2u} phonon energy. Therefore, the part of acoustic phonons which parti-

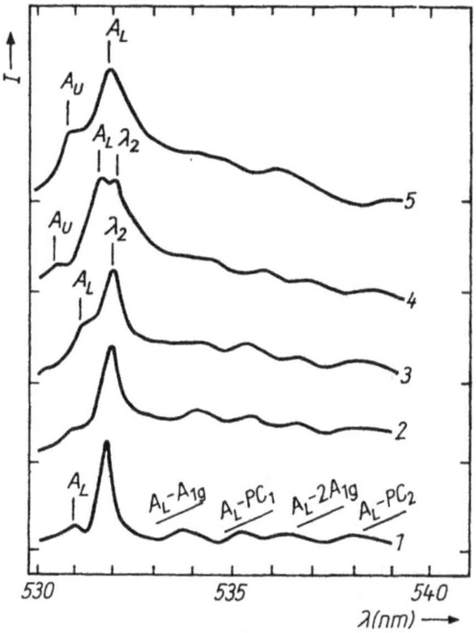

Fig. 2

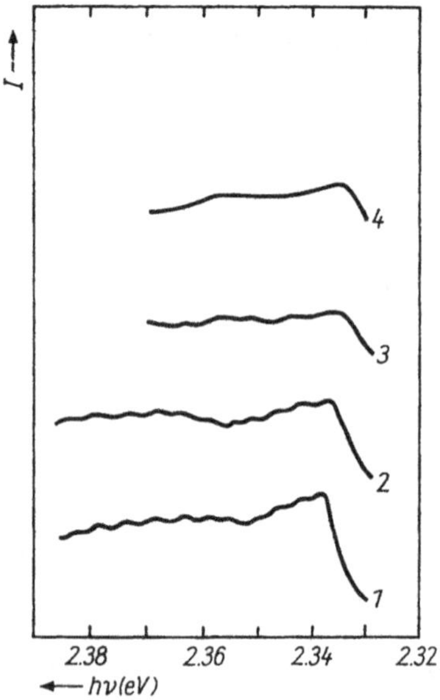

Fig. 3

Fig. 2. Dependence of the HgI_2 resonance photoluminescence spectrum on temperature; T = (1) 4.2, (2) 10, (3) 12, (4) 25, and (5) 35 K. PC_1 and PC_2 mean A_L band phonon replicas

Fig. 3. Temperature dependence of the excitation spectrum of the A_L band for HgI_2. T = (1) 4.2, (2) 10, (3) 12, (4) 35 K

cipate in the energy relaxation of polaritons remains very low even within the energy range ΔE_B and the "bottleneck" effect must be negligible. Nevertheless the thermalization of polaritons in HgI_2 is achieved at a rather low temperature ($\approx$ 12 K). This is indicated by the disappearance of the peaks in the excitation spectra of HgI_2 crystals at $T \geqq 12$ K.

4. Polariton thermalization at $T \geqq 12$ K in HgI_2 crystals can be explained by the influence of the "thermal barrier" effect. Taking into account the value of T_e at which polariton thermalization occurs, such a barrier may be caused by anti-Stokes scattering processes utilizing both rigid-mode (LOP) and piezoacoustic phonons.

The polariton lifetime within the "knee" of the lower polariton branch increases due to the "thermal barrier" and becomes sufficient for polaritons to interact with acoustic phonons and to equalize their temperature with the temperature of the phonon subsystem. Then after multiple attempts polaritons finally penetrate through the barrier and escape from the crystal in the form of thermalized emission.

It is worthwhile to compare the results just described with the data for another layer crystal, the 2H polytype of PbI_2 [11] whose parameters (the value of Δ_{L-T}, exciton–LO-phonon coupling constant, etc.) are quite close to those of HgI_2. However, there are essential differences: in PbI_2 the lowest energy of the TO-phonon is only 6 meV and interaction of polaritons with acoustic phonons, which is described in the deformation potential approximation, weaker than that in HgI_2.

A typical polariton emission spectrum is shown in Fig. 4. Again the upper and lower polariton branch emission bands A_U and A_L along with A_L phonon replicas and bound exciton emission I are present. The excitation spectrum of the A_L branch

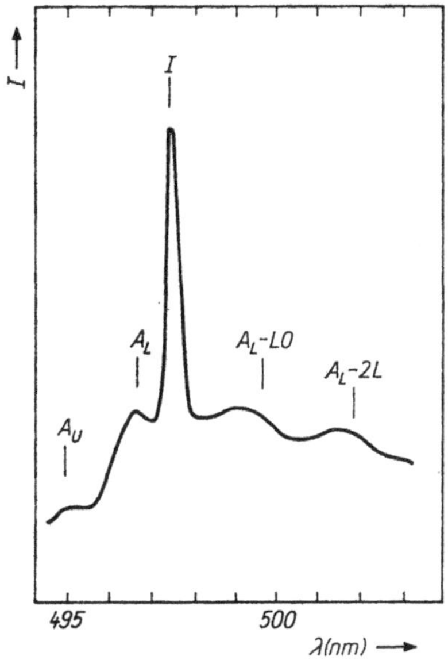

Fig. 4

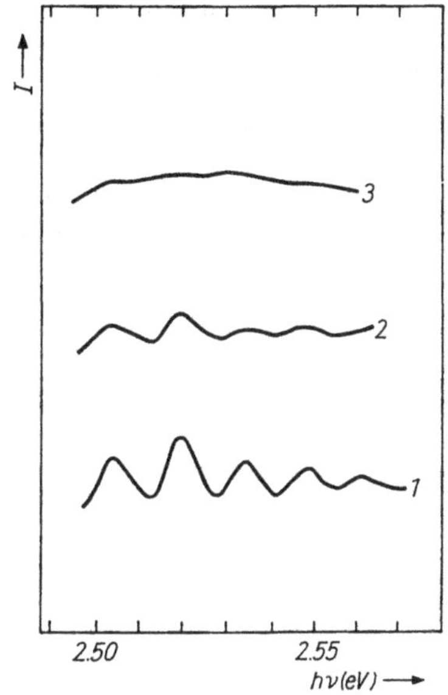

Fig. 5

Fig. 4. Photoluminescence spectrum of PbI_2

Fig. 5. Temperature dependence of the excitation spectrum of the A_L band for PbI_2. T = (1) 4.2, (2) 30, (3) 50 K

is shown in Fig. 5. At 4.2 K it exhibits a structure with equidistant bands separated by 12 meV. This value corresponds to the energy of the LO phonon A_{2u}. With rising temperature this structure gradually broadens and disappears completely at 40 K.

From the results described above one can conclude that though the polariton relaxations in HgI_2 and PbI_2 crystals have much in common there are some evident distinctions between them. Firstly, polariton relaxation in PbI_2 within the energy range ΔE_B is based on acoustic phonons if the initial energy of the polariton $E_p \neq h\nu_{AL} + m\hbar\omega_{LO}$ (here m is an integer). This is a consequence of the fact that the energy of the A_{2u} phonon in PbI_2 exceeds essentially the energy range ΔE_B. Secondly, polaritons become thermalized in PbI_2 samples of such a quality as the samples of HgI_2 crystals at lattice temperatures $T \geqq 45$ K. This is due to the absence of LOP in PbI_2 and the electron–phonon interaction weaker than in HgI_2 (since the space group of the crystal PbI_2 includes inversion among other symmetry elements, the exciton–acoustic phonon interaction is described by the deformation potential approximation). As a result the thermal barrier in the 2H polytype of PbI_2 may be caused by anti-Stokes scattering processes of polaritons on intralayer TO-phonons. This scattering is effective enough to make the process of polariton relaxation slower and to favour the polariton thermalization.

References

[1] Y. Toyozawa, Progr. theor. Phys. (Kyoto) **12**, 112 (1959).
[2] S. Permogorov, phys. stat. sol. (b) **68**, 9 (1975).
[3] M. S. Brodin, S. V. Marisova, and E. M. Myasnikov, Ukr. fiz. Zh. **27**, 801 (1982).
[4] E. N. Myasnikov, Dokl. Akad. Nauk Ukr. SSSR **A8**, 755 (1976).

[5] R. ZALLEN and M. SLADE, Solid State Commun. **17**, 1561 (1975).
[6] I. M. LIFSHITS, Zh. eksper. teor. Fiz. **22**, 471 (1952).
[7] M. S. BRODIN, I. V. BLONSKII, A. S. KROCHUK, B. M. NITSOVICH, and A. V. FRANIV, Zh. eksper. teor. Fiz. **83**, 1052 (1982).
[8] I. AKOPYAN, B. NOVIKOV, S. PERMOGOROV, A. SELKIN, and V. TRAVNIKOV, phys. stat. sol. (b) **70**, 353 (1975).
[9] T. GOTO and Y. NISHINA, Solid State Commun. **31**, 751 (1979).
[10] T. GOTO and Y. NISHINA, Phys. Rev. B **17**, 4565 (1978).
[11] M. S. BRODIN, D. B. GOER, A. O. GUSHCHA, and V. V. TISHENKO, Ukr. fiz. Zh. **27**, 829 (1982).

(Received June 10, 1985)

phys. stat. sol. (b) **131**, 621 (1985)

Subject classification: 13.5.1 and 20.1; 22.5; 22.5.4

D. L. Landau Institute of Theoretical Physics, Academy of Sciences of the USSR, Moscow[1])

The Narrow Bands of Exciton Absorption in Polar Solids and Alloys

Urbach's Rule

By

F. V. Kusmartsev

Dedicated to Prof. Dr. Dr. h. c. Dr. E. h. P. Görlich on the occasion of his 80th birthday

An analytical theory of the Urbach-Martienssen tail for the Wannier-Mott excitons in polar crystals and alloys is developed on the basis of the expansion of all quantities in the small parameter $\Delta \ll E_i(\mu/M)$, where M and μ are the translation and the reduced masses of the exciton; E_i is the exciton Rydberg, and Δ is the photon deficit. The exciton absorption coefficient and the length of the absorption tail are calculated. The latter characterizes the width of the exciton absorption band. The results obtained are compared with experiments on the exciton absorption in layered polar crystals.

Развивается аналитическая теория урбаховского хвоста для экситонов Ванье-Мотта в полярных кристаллах и сплавах, которая основана на разложении всех величин по малому параметру $\Delta \ll E_i(\mu/M)$, где μ и M-приведенная и трансляционная массы экситона; Δ дефицит энергии фотона по отношению к экситонному ридбергу E_i. Вычисляются длина и коэффициент поглощения низкочастотного крыла экситонной полосы. Эта длина характеризует полуширину экситонной полосы поглощения. Полученные результаты сравниваются с экспериментами в слоистых полярных кристаллах.

1. Introduction

Numerous experimental data have convincingly testified the presence of the exciton absorption in layered polar crystals [1 to 8]. Among these substances there exist crystals (e.g. PbI_2 etc.), where quite narrow exciton absorption bands are observed [9]. These exciton states may be described by the Wannier-Mott exciton model (see e.g. [9, 10]). It will be shown in the present paper that the small width of the exciton absorption band is caused by the mutual compensation by electron and hole charges, when the electron and hole masses are close to each other.

Thus, in pure polar crystals the width of the exciton absorption is mainly conditioned by the exciton interaction with polar optical phonons. When the electron mass differs slightly from the hole mass, the fluctuation polarization potential well, in which the exciton is created by the photon, is shallow. This follows from the fact that, if this fluctuation for the hole is a potential well, then it is a hump for the electron and vice versa. On the whole, the exciton interacts with this fluctuation weakly. An analogous situation occurs in solid semiconductor alloys, where the gap changes little with varying composition [11]. The same effect may also occur in polar crystal alloys. In this case we use a model, where the exciton interacts with the composition flutuations via a random polarization: it differs from the model usually applied to solid solutions of semiconductors [12].

[1]) Kosygina 2, 117 940 Moscow V-334, USSR.

If the origin of the very narrow exciton absorption is studied, the light absorption by the Wannier-Mott exciton close to the exciton resonance is considered, so, naturally, the following parameter is introduced:

$$\Delta \ll \frac{\mu}{M} |E_i| , \tag{1}$$

where Δ is the photon energy deficit with respect to the exciton resonance energy $E_i = -\mu e^4/2\varepsilon_\infty^2$ and ε_∞ is the high-frequency dielectric constant, e the electron charge. The calculation is made in the framework of the effective mass method, in which electron and hole bands are considered to be spherically symmetric and non-degenerate with electron and hole masses m_e and m_h.

In the first part of the paper the light absorption coefficient by the Wannier-Mott exciton is calculated in the range of the long-wave Urbach-Martiessen tail caused by the exciton interaction with polar optical phonons. By the tail length we may characterize the width of the exciton absorption band. Here, we have restricted ourselves to the consideration of the two limit cases, namely, quasistatic and dynamic. In both cases the absorption bands are narrow, when the electron and hole masses are close to each other. A more general consideration will be given elsewhere.

In the second part of the paper the narrowing of the exciton band conditioned by the concentration fluctuations of the solid solution components, is studied. The exciton density of the states tail, caused by the composition fluctuations, is calculated. Hence, we have obtained a low-frequency absorption coefficient and the energy characterizing the width of the exciton absorption band.

In the third part of the paper the experimental data are compared with the results obtained. The theory desribes them satisfactorily. The width of the exciton absorption band depends on the mass difference of electron and hole. The less this difference, the narrower is the exciton absorption band.

2. The Width of the Band-Edge Exciton Absorption in Polar Crystals

Let us calculate the low-frequency absorption coefficient of the Wannier-Mott exciton using the formalism suggested by Ioselevich [13]. In this case, the absorption coefficient $K(\Delta)$ is expressed by the following formula:

$$\frac{K(\Delta)}{K(0)} = -\int \mathrm{d}^3r \, \mathrm{e}^{-\Delta\tau} \, \mathrm{Im}\, G(\boldsymbol{r}, \boldsymbol{r}, 0, 0, \tau) , \tag{2}$$

where the two-particle Green function has the form

$$G(\boldsymbol{r}_e, \boldsymbol{r}_h, \boldsymbol{r}'_e, \boldsymbol{r}'_h, \tau) = -i\Psi(\boldsymbol{r}'_e, \boldsymbol{r}'_h, \tau) \, \Psi(\boldsymbol{r}_e, \boldsymbol{r}_h, 0) \, \mathrm{e}^{-S} . \tag{3}$$

Here S is the effective functional equal to

$$S[\Psi] = \int_0^\tau \left\langle \Psi_1^* \left| \frac{\nabla_e^2}{2m_e} - \frac{\nabla_h^2}{2m_h} - \frac{e^2}{\varepsilon_\infty |\boldsymbol{r}_e - \boldsymbol{r}_h|} \right| \Psi_1 \right\rangle \mathrm{d}t_1 + \\ + \frac{e^2\omega}{2\bar{\varepsilon}} \int_0^\tau \int_0^\tau \mathrm{d}t_1 \, \mathrm{d}t_2 \, D(t_1 - t_2) \, \langle\!\langle \Psi_1^* \Psi_2^* | \, |\boldsymbol{r}_{e1} - \boldsymbol{r}_{h2}|^{-1} + \\ + |\boldsymbol{r}_{e2} - \boldsymbol{r}_{h1}|^{-1} - |\boldsymbol{r}_{e1} - \boldsymbol{r}_{e2}|^{-1} - |\boldsymbol{r}_{h1} - \boldsymbol{r}_{h2}|^{-1} \, | \Psi_1 \Psi_2 \rangle\!\rangle \tag{4}$$

and $D(x)$ is the phonon propagator

$$D(x) = N \, \mathrm{e}^{\omega|x|} + (N + 1) \, \mathrm{e}^{-\omega|x|} , \tag{5}$$

where N is the Planck function $N = (\exp(\omega/T) - 1)^{-1}$, T the temperature, ω the frequency of the optical polar phonons, Ψ the normalized wave function of the electron-hole pair; indices 1 and 2 refer to the time moments t_1 and t_2, indices e, h to the electron and the hole, respectively, $\bar{\varepsilon} = (\varepsilon_\infty^{-1} - \varepsilon_0^{-1})^{-1}$ and ε_0 is the static dielectric constant. Here Δ is counted off from the conduction band bottom and τ is the duration of the process in the imaginary time, found of the extremum condition

$$\frac{\partial S}{\partial \tau} = -\Delta\,. \tag{6}$$

Let us introduce the charge density operator

$$\hat{\varrho}\Psi^2(r) = \frac{1}{2\varkappa - 1} \int [\delta(\boldsymbol{r} - \boldsymbol{r}_h) - \delta(\boldsymbol{r} - \boldsymbol{r}_e)]\, \Psi^2(\boldsymbol{r}_e, \boldsymbol{r}_h)\, d^3r_e\, d^3r_h\,, \tag{7}$$

where $\varkappa = m_e/M$, $M = m_e + m_h$.

If we consider the light absorption near the exciton resonance, we use the hard exciton approach [14, 11], where the relative motion of the electron and the hole is just the same as in a free exciton: i.e. the wave function of the electron and the hole in the exciton has the form

$$\Psi(\boldsymbol{r}_e, \boldsymbol{r}_h) = \chi(\boldsymbol{r}_e - \boldsymbol{r}_h)\, \varphi((\boldsymbol{r}_e m_e + \boldsymbol{r}_h m_h)/(m_e + m_h))\,,$$

where $\chi(\boldsymbol{r}) = r_B^{-3/2}\pi^{-1/2}\, e^{-r/r_B}$ is the wave function of the ground state of the relative motion of the electron and the hole in the exciton, $\varphi(r)$ is the wave function of the mass centre. Let us designate $\hat{\varrho}(\Psi^2(\boldsymbol{r}_e, \boldsymbol{r}_h)) = \hat{\varrho}\varphi^2(r)$. Below, we show that this approach is equivalent to the fulfilment of inequality (1).

We restrict ourselves to the exponential accuracy in the calculation $K(\Delta)$. In our approach it is equal to

$$\frac{K(\Delta)}{K(0)} = e^{-(\Delta\tau + S_{ef})} \tag{8}$$

and S_{ef} takes the form

$$S_{ef} = \int d^3r \left(\int_0^\tau \left[\frac{(\nabla\varphi)^2}{2M} - \frac{(2\varkappa - 1)^2\, e^2\omega}{2\bar{\varepsilon}} \int_0^\tau D(t - t')\, \varphi^2(t')\, dt' \right] \hat{\varrho}\hat{K}\hat{\varrho}\varphi^2(t)\, dt \right), \tag{9}$$

where $\hat{K}$ is the Coulomb operator,

$$\hat{K}\varrho(r) = \int d^3r'\, \varrho(\boldsymbol{r}')/|\boldsymbol{r} - \boldsymbol{r}'|\,.$$

In (8) the photon deficit is counted from the exciton resonance energy.

The Schrödinger equation, defining the extrema (9), looks like

$$-\frac{\Delta_r\varphi(t)}{2M} - \frac{e^2(2\varkappa - 1)^2\, \omega\varphi(t)}{\bar{\varepsilon}} \int_0^\tau D(t - t')\, \hat{\varrho}\hat{K}\hat{\varrho}\varphi^2(t')\, dt' = E(t)\, \varphi(t)\,. \tag{10}$$

We can see from (9) and (10) that if the electron mass is equal to the hole mass, then the exciton does not interact with the optical polar phonons.[2]) This fact was firstly marked by Dikman and Pekar [15]. In our case, i.e. when $m_e = m_h$, the broadening of the exciton line does never occur.

---

[2]) Here we excluded the polarization of the exciton, which is a high-order effect.

Let us make the substitution of the variables: $t \to t/\omega$, $r \to rr_B$, $\varphi \to r_B^{-3/2}\varphi$, here r_B is the Bohr radius of the exciton. At first, we study the quasistatic fluctuations, in which the imaginary lifetime satisfies the inequality $\omega\tau \ll 1$.

Then we get from the comparison of (6) and (10) that $E(\tau) = -\Delta$, the action (9) takes the form

$$S_{ef} = S_0\left(\int \left[\frac{(\nabla\varphi)^2}{2\mathcal{M}} - \varphi^2\hat{\varrho}\hat{K}\hat{\varrho}\varphi^2\right] d^3r\right), \tag{11}$$

where

$$S_0 = \frac{(2\varkappa - 1)^2\, e^2(2N + 1)\,\tau^2}{\bar{\varepsilon} r_B \omega} \tag{12}$$

and

$$\mathcal{M} = \frac{M r_B^2 \omega S_0}{\tau}, \tag{13}$$

when $\mathcal{M} \gg 1$ and at large ratios $m_h/m_e \gg 1$ the extrema have been found in [14]. Making the calculation in precisely the same way as in this paper we may show that

$$S_{ef} = \frac{S_0 K_2}{\mathcal{M}^2}. \tag{14}$$

In our case, in contrast to [13] the factor K_2 depends on m_e/m_h ($K_2 = K_2(m_e/m_h)$). Variational calculations show (for an analogous situation cf. [11]) that this dependence is weak, i.e. K_2 changes approximately by 30% when the argument changes from zero up to ∞. Similarily, we get from (10) an expression for E,

$$E = -\frac{S_0 K_3}{\tau \mathcal{M}^3}, \tag{15}$$

where K_3 depends on m_e/m_h analogous to K_2. Having expressed this $\tau = \tau(\Delta)$ and substituted it into (13), we get

$$\mathcal{M} = \sqrt{\frac{2K_3\,|E_i|\,\mu}{\Delta M}}. \tag{16}$$

Having compared this formula with (1), we see that the approach of the hard exciton for quasistatic phonon fluctuations is equivalent to the inequality $\Delta \ll (\mu/M)\,|E_i|$. Having substituted $\tau = \tau(\Delta)$ and $\mathcal{M} = \mathcal{M}(\Delta)$ into (14) and (8) we derive the formula for the absorption coefficient in the low-frequency wing of the exciton band (see e.g. [11]),

$$-\ln\left|\frac{K(\Delta)}{K(0)}\right| = \sqrt{\frac{\Delta}{E_0}} \tag{17}$$

with energy E_0, defining the width of the exciton absorption band, which is equal to

$$E_0 = \frac{M^2\omega^2(2N + 1)^2}{|E_i|}\left(\frac{\varepsilon_0 - \varepsilon_\infty}{\varepsilon_0}\right)^2 \frac{(m_e - m_h)^4}{m_e^3 m_h^3}\, C. \tag{18}$$

Here $C = C(m_e/m_h)$ is a numerical function $\sim K_3^{-1}$. It scarcely depends on m_e/m_h.

On the other hand, in the dynamic limit ($\omega\tau \gg 1$) the exciton absorption coefficient is defined by the functional (for details, see [13])

$$S_{ef} = \tilde{S}_0\left[\int\left(\int_0^{\tau^*} (\nabla\varphi(t))^2\, dt - \frac{L}{2}\int_0^{\tau^*}\int_0^{\tau^*} e^{-t_1-t_2}\varphi^2(t_1)\,\hat{\varrho}\hat{K}\hat{\varrho}\varphi^2(t_2)\, dt_{1,2}\right) d^3r, \right. \tag{19}$$

where $\tilde{S}_0 = (Mr_B^2\omega)^{-1}$ and $L = (2\varkappa - 1)^2 N e^{\tau} 2Mr_B e^2/\bar{\varepsilon}$. The Schrödinger equation has the form

$$-\Delta\varphi(t) - L e^{-t} \varphi(t) \int_0^{\tau^*} e^{-t'} \hat{\varrho}\hat{K}\hat{\varrho}\varphi^2(t')\, dt' = E(t)\, \varphi(t) , \tag{20}$$

where $E(t)$ is the Lagrange factor and τ^* is defined by the condition $E(\tau^*) = 0$. Below we show that the condition $E(t) \ll 1$ is equivalent to the inequality $\Delta \ll \mu|E_i|/M$. In this case (20) has two regions: $r \approx 1$ and $r \gg 1$. [14]. Therefore, the solution of this equation is of the form

$$\varphi(t) = \begin{cases} \dfrac{\sqrt{\alpha(t)}\, e^{-\alpha(t) r}}{\sqrt{2\pi}\, r} & \text{at} \quad r \gg 1 , \\ \sqrt{\alpha(t)}\, \chi_0(r) & \text{at} \quad r \approx 1 , \end{cases} \tag{21}$$

where $E(t) = -\alpha^2(t)$ and $\chi_0(r) \underset{r\to\infty}{\to} r^{-1}/\sqrt{2\pi}$. Let us substitute (21) into (20), then at $\alpha(t) \ll 1$ we obtain

$$\tilde{\alpha}(t) = e^{-t} LBC - A , \tag{22}$$

where

$$A = \int (\nabla\chi_0)^2\, dr , \qquad B = \int \chi_0^2 \hat{\varrho}\hat{K}\hat{\varrho}\chi_0^2\, d^3r ,$$

and

$$C = \int_0^{\tau^*} e^{-t} \alpha(t)\, dt . \tag{23}$$

From (22), (23) and the condition $\alpha(\tau^*) = 0$, we find

$$C = \frac{A}{(\sqrt{LB} - \sqrt{2})\sqrt{LB}} \quad \text{and} \quad \tau^* = \ln\left|\frac{LBC}{A}\right| . \tag{24}$$

If $LB \gg 1$, then

$$S = \tilde{S}_0 \frac{A^2}{BL} \tag{25}$$

and

$$\tau^* = \sqrt{\frac{2}{LB}} \ll 1 , \tag{26}$$

and from (6) we obtain

$$\tau = \ln\left|\frac{A^2\, |E_i|}{B\Delta}\, \frac{\mu}{M}\, \frac{\bar{\varepsilon} e^{-2}}{(2\varkappa - 1)^2 N M r_B}\right| .$$

The value A^2/B is found by the variational method (for details see [14]) with the simplest variation function

$$\chi_0(r) = \frac{C_0}{[r^2 + a^2]^{1/2}} .$$

The inequality $LB \gg 1$ is equivalent to (1). Thus, the absorption coefficient is equal to

$$-\ln\left|\frac{K(\Delta)}{K(0)}\right| = \frac{\Delta}{\omega}\left(1 + \ln\left|\frac{A^2}{BN}\, \frac{|E_i|}{\Delta} \left(\frac{\mu}{M}\right)^2 \frac{\varepsilon_0}{\varepsilon_0 - \varepsilon_\infty}\, \frac{(m_e - m_h)^2}{(m_e + m_h)^2}\right|\right) . \tag{27}$$

This formula may be obtained by another method, the perturbation theory of n-phonon absorption, where n is an integer number. It would be a rougher estimation.

For the exciton absorption band, corresponding to the dipole-allowed transition into the 2S state of the exciton, the numerical factor C or B/A^2 will be by an order greater, i.e. the energy characterizing the exciton broadening, the exciton Rydberg, is replaced by E_i/n_0^2, where n_0 corresponds to the n_0-excited exciton state.

Obviously, if the electron mass is equal to the hole mass, then our theory is not correct. In this case we have to consider higher orders on the parameter (1). If the temperature is low, the quasistaticity condition may be satisfied only at large differences in the electron and hole masses. According to Dikman and Pekar [15], only in this case self-trapping is possible, and the light absorption leads to self-trapping states of the exciton.

3. The Long-Wave Tail of Density of States and the Exciton Absorption in Alloys of Polar Crystals

The solid solution of the layered crystals form two classes. In the first one at a concentration change of the solution its structural composition is not changed. The solid solution $PbCl_{2x}Br_{2-2x}$ is such a system. In the range $0 \leqq x \leqq 1$ the exciton absorption has been investigated in [16]. Solid solution in which with varying x the structural composition is changed, refers to the other type. To this class belongs $PbCl_{2x}I_{2-2x}$, where the exciton absorption, for example, has been investigated in [17], and in these substances the exciton absorption bands have also proved to be narrow [16 to 18] and to depend weakly on composition fluctuations. Below, we show that it is due to the fact that in these substances the difference between electron and hole is small.

Let us consider a simple model, where the exciton interaction with the concentration fluctuation of the solid solution $\xi(r)$ can be written as

$$V_{\text{int}} = V(r_\text{h}) - V(r_\text{e}) \ , \tag{28}$$

where

$$V(r) = \frac{e^2}{\tilde{\varepsilon}} \int \frac{\mathscr{S}(\boldsymbol{r}')}{|\boldsymbol{r} - \boldsymbol{r}'|} \, \mathrm{d}^3r' \, .$$

Here $\tilde{\varepsilon}$ is the effective dielectric constant characterizing the mixed crystal. Particularly, in solutions where the structural composition does not change with x, this value is equal to $\tilde{\varepsilon} = (\varepsilon_{\infty_1}^{-1} - \varepsilon_{\infty_2}^{-1})^{-1}$ where $\varepsilon_{\infty 1}$ and $\varepsilon_{\infty 2}$ are high-frequency dielectric constants of the components of the solid solution. In the general case, $\tilde{\varepsilon}$ is a function of x: $\tilde{\varepsilon} = \tilde{\varepsilon}(x)$.

The probability of the concentration fluctuation depends on the way of sample preparation. Hence, if the sample has been prepared by slow cooling, i.e. if the fluctuation $\mathscr{S}(r) \ll N_0$ (N_0 is the total concentration of the atoms), this probability is equal to

$$W[\mathscr{S}] = \exp\left(-S_{0i}[\mathscr{S}]\right) ; \qquad i = 1, 2 \, ,$$

where the entropy functional is as follows:

$$S_{01} = \int \frac{\mathscr{S}^2(\boldsymbol{r}) \, \mathrm{d}^3r}{2N_0 x(1-x)} . \tag{29}$$

If the solid solution has been prepared by fast cooling, then between different fluctuations Coulomb correlations have been left, characterized by temperature. T_0 is the freezing temperature of diffusion of the solution atoms. Here instead of (29) the following expression should be used:

$$S_{02}[\mathscr{S}] = \frac{e^2}{2\tilde{\varepsilon}} \int \frac{\mathscr{S}(\boldsymbol{r}) \, \mathscr{S}(\boldsymbol{r}') \, \mathrm{d}^3r \, \mathrm{d}^3r'}{|\boldsymbol{r} - \boldsymbol{r}'| \, x(1-x) \, T_0} . \tag{30}$$

Let us calculate the density of states tail of the Wannier-Mott exciton degenerated by the fluctuations. Let us consider the first case — the case of the random fluctuation distribution. We use the method of optimum fluctuations (OF) suggested in

[19 to 21]. Equations for OF have the following form:

$$\left[-\frac{\nabla_e^2}{2m_e}-\frac{\nabla_h^2}{2m_h}-\frac{e^2}{\varepsilon_\infty\,|\boldsymbol{r}_e-\boldsymbol{r}_h|}-\frac{e^2}{\tilde{\varepsilon}}\left(\hat{K}\mathcal{S}(\boldsymbol{r}_e)-\hat{K}\mathcal{S}(\boldsymbol{r}_h)\right)\right]\Psi=\tilde{E}\Psi \tag{31}$$

and

$$\mathcal{S}(\boldsymbol{r})=(2\varkappa-1)\,\beta e^2 N_0 x(1-x)\,\hat{K}\hat{\varrho}\Psi^2(\boldsymbol{r})/\tilde{\varepsilon}\,, \tag{32}$$

where $\beta=\beta(E)$ is the Lagrange multiplier, and the other quantities have just the same meaning as in the previous part. Following [11] we write down an auxiliary functional, the extrema of which are defined by (31) and (32),

$$J=\int\!\!\int\left[\frac{(\nabla_e\Psi)^2}{2m_e}+\frac{(\nabla_h\Psi)^2}{2m_h}-\frac{e^2}{\varepsilon_\infty}\frac{|\Psi|^2}{|r_e-r_h|}\right]\mathrm{d}^3r_e\,\mathrm{d}^3r_h\,-$$

$$-\frac{\delta}{2}\int(\Psi^2\hat{\varrho}\hat{K}^2\hat{\varrho}\Psi^2)\,\mathrm{d}^3r\,, \tag{33}$$

where $\delta=\beta e^4(2x-1)^2\,N_0x(1-x)/\tilde{\varepsilon}^2$. As the density of states tail is studied only close to the exciton resonance, we introduce the approximation of the "hard exciton", i.e. we average over the internal motion of the exciton. This holds true, when $|\tilde{E}-E_i|\ll\ll\mu\,|E_i|/M$ (cf. (1)).

Then the following functional has been derived instead of (33):

$$\tilde{J}=J+|E_i|=\int\left(\frac{(\nabla_{\boldsymbol{R}}\varphi)^2}{2M}-\frac{\delta}{2}\,\varphi^2\hat{\varrho}\hat{K}^2\hat{\varrho}\varphi^2\right)\mathrm{d}^3r\,, \tag{34}$$

where $\boldsymbol{R}$ is the Jacobi coordinate, $\boldsymbol{R}=(m_e\boldsymbol{r}_e+m_h\boldsymbol{r}_h)/(m_e+m_h)$, φ is the wave function of the exciton motion as a whole. It is apparent that the extrema J, and hence, in the approximation of the "hard exciton", $\tilde{J}$, are defined by the equation

$$-\frac{\Delta_{\boldsymbol{R}}\varphi}{2M}-\delta V(\boldsymbol{r})\,\varphi=E\varphi\,, \tag{35}$$

where

$$V(\boldsymbol{r})=\hat{\varrho}\hat{K}\hat{K}\hat{\varrho}\varphi^2(\boldsymbol{r})\,. \tag{36}$$

Let us show that the approach (34) defines the two-scale character of the states of (35). Therefore, we define the asymptotics of (36) at $r\gg r_B$,

$$V(r)=4\pi^2r_B^4\,\varphi^2(r)\,. \tag{37}$$

Suppose that the state radius defined by (35) is $r_E\gg r_B$. Then (37) may be substituted into (35) and E may be found. According to [22] they are equal to

$$E=-\frac{44}{(4\pi^2)^2\,M^3\delta^2r_B^8}\,, \tag{38}$$

$$r_E\sim\sqrt{2ME}\,. \tag{39}$$

Having expressed the function $\beta=\beta(E)$ from (38) and substituting it into (32) and (29), we derive the formula for the density of states tail,

$$-\ln\left|\frac{\varrho(E)}{\varrho(0)}\right|=\sqrt{\frac{E}{E_{01}}}\,, \tag{40}$$

where E_{01} is the characteristic energy, defining the tail length,

$$E_{01}=\frac{2\pi^2}{2}\,\varepsilon_\infty^2\,E_i\left(\frac{M}{\mu}\right)^3\frac{x^2(1-x)^2}{\tilde{\varepsilon}^2}\,(N_0r_B^3)^2\left(\frac{m_e-m_h}{m_e+m_h}\right)^4. \tag{41}$$

It follows from (39) that $r_E \gg r_B$ is equivalent to the inequality

$$|E| \ll \frac{\mu}{M} |E_i| \tag{42}$$

which coincides identically with (1) and characterizes the approximation of the "hard exciton".

Let us consider the second case when in the mixed crystal Coulomb correlations exist between the fluctuations of the composition. Instead of (32) the equation defining the extreme $\mathcal{S}(\boldsymbol{r})$, takes the form

$$\mathcal{S}(\boldsymbol{r}) = \gamma \hat{\varrho} \Psi^2(\boldsymbol{r}) , \tag{43}$$

where

$$\gamma = \beta T_0 x(1 - x)(2\varkappa - 1) .$$

Performing the calculations and analogous to the above-stated, in the approximation of the "hard exciton" (1) we get the formula for the density of states tail,

$$- \ln \left| \frac{\varrho(E)}{\varrho(0)} \right| = \sqrt{\frac{E}{E_{02}}} , \tag{44}$$

where

$$E_{02} = \frac{T_0^2 \varepsilon_\infty^2 x^2 (1 - x)^2}{|E_i| \, \tilde{\varepsilon}^2} \left(\frac{M}{\mu}\right)^3 \left(\frac{m_e - m_h}{m_e + m_h}\right)^4 C . \tag{45}$$

Here the numerical function $C = C(m_e/m_h)$ is just the same as in (18). Let us calculate (see e.g. [24]) the low-frequency absorption coefficient. It is equal to $K(\Delta) \sim e^{-\sqrt{\Delta/E_{01}}}$ and $K(\Delta) \sim e^{-\sqrt{\Delta/E_{02}}}$ for random and correlated fluctuations, respectively.

4. Comparison of the Results Obtained with the Experimental Data

One general conclusion can be drawn from (18) and (27), i.e. the width of the exciton absorption band depends on the mass difference of electron and hole $(m_e - m_h)$. Let us now refer to the experimental results, see Fig. 4 and 5 of [9]. In Fig. 4 the absorption spectra of layered iodide monocrystals are represented. The narrowest band is observed in PbI_2. In this compound the mass difference of electron and hole is minimal. The absorption band next in width is that of BiI_3 characterized by an analogous band structure. We see from this figure that as the difference of the effective masses of electron and hole increases, the width of the exciton absorption band increases, too. A similar situation occurs both for bromides and chlorides, their spectra of the exciton absorption are represented in Fig. 5 of [9]. Again, the narrowest bands are observed for the compounds with Pb. According to [23], for these compounds with Pb the effective masses of electron and hole are approximately of the same order and it is a very important fact accounting for the narrowness of the exciton absorption bands for these compounds. Unfortunately, the absence of experimental values of some quantities used in (18) and (27), hampers a detailed comparison with Fig. 4 and 5 of [9].

It follows from our consideration that (18) and (27) account explicitly for the tendency of the exciton band to widen with increasing mass difference of electron and hole. The physical effect underlying this phenomenon consists in a strong mutual screening or compensation of electron and hole charges, when the masses of electron and hole are close. At low temperatures $T \ll \omega$ only the probability of exciton self-trapping defines the width of the exciton absorption band in pure crystals. In solid solutions the width of such bands is defined by fluctuations in the crystal composition.

According to our calculations, the width of the low-frequency edge of the exciton absorption band is defined by (41) and (45). These formulae account well for the fact that in the solutions $PbCl_{2-2x}Br_{2x}$, $PbCl_{2-2x}I_{2x}$, $PbBr_{2-2x}I_{2x}$ narrow exciton absorption bands [16, 17] are observed at $T = 5$ to 15 K. This is due to the fact that the masses of electron and hole in the compounds with Pb are close, and the width of the exciton absorption band is $(m_e - m_h)^4$. Here, the effective mutual screening or compensation of the electron and the hole charges occurs. It should be noted that though (18), (27), (41), (45) only define the edge width of the exciton absorption, the width of the whole band will be defined by the same formulae, but with large numerical constants (see e.g. [24]). For solid solutions the values E_i, m_e, m_h in (41), (45) should be considered as empirical functions of the compound index x.

Acknowledgements

The author thanks R. A. Kink for his kind report on the experimental data and also E. I. Rashba and I. B. Levinson for useful discussions.

References

[1] R. Hilson and R. W. Pohl, Z. Phys. **48**, 384 (1928).
[2] H. Fesefeld, Z. Phys. **64**, 741 (1930).
[3] G. Bawer, Ann. Phys. (Leipzig) **19**, 434 (1934).
[4] N. M. Kudris and I. N. Schulwa, Optika i Spektroskopiya **2**, 246 (1957).
[5] J. C. Zahner and H. G. Drickmer, J. Phys. Chem. Solids **11**, 92 (1959).
[6] W. P. Doyle and B. D. Pearson, J. chem. Phys. **36**, 1950 (1962).
[7] K. J. Best, Z. Phys. **163**, 309 (1961).
[8] N. L. Kramarenko, W. K. Miloslavskii, and Yu. W. Naboikin, Ukr. fiz. Zh. **15**, 416 (1970).
[9] M. R. Tubbs, phys. stat. sol. (b) **49**, 11 (1972).
[10] G. Harbeke and E. Tosatti, Phys. Rev. Letters **28**, 1967 (1972).
[11] F. V. Kusmartsev, Fiz. Tekh. Poluprov. **8**, (1985) (in the press); Abstract, II. Conf. Electron's Dynamics in Charge-Ordered Structures, Chernogolovka 1984 (p. 23).
[12] S. D. Baranovskii and A. L. Efros, Soviet Phys. — Semicond. **12**, 1328 (1978).
[13] A. S. Ioselevich, Zh. eksper. teor. Fiz **81**, 1508 (1981); 82, 601 (1982); **83**, 743 (1982).
[14] F. V. Kusmartsev and S. V. Meshkov, Soviet Phys. — J. exper. theor. phys. 58, 870 (1983).
[15] I. M. Dikman and S. I. Pekar, Dokl. Akad. Nauk SSSR **83**, 825 (1952).
[16] Ta Awarmara, X. Koppel, G. Lijdij, and A. Haaw, Izv. Akad. Nauk Eston. SSR, Ser. Fiz. Mat. **31**, 59 (1982).
[17] A. I. Ribalka and B. K. Miloslavskii, Fiz. tverd. Tela **18**, 3109 (1976).
[18] A. J. H. Eijkelenkamp and K. Vos, phys. stat. sol. **76**, 769 (1976).
[19] B. I. Halperin and M. Lax, Phys. Rev. **722**, 148 (1966).
[20] J. Zittars and J. S. Langer, Phys. Rev. **148**, 741 (1966).
[21] I. M. Lifshits, Zh. eksper. teor. Fiz. **53**, 743 (1967); Soviet Phys. — J. exper. theor. phys. **26**, 462 (1968).
[22] F. V. Kusmartsev and E. I. Rashba, Zh. eksper. teor. Fiz. **86**, 1142 (1984); Soviet Phys. — J. exper. theor. phys. **59**, 666 (1984); phys. stat. sol. (b) **121**, K87 (1984).
[23] I. M. Tsidilkovskii, Zonnaya struktura poluprovodnikov, Izd. Nauka, Moskva 1978 (p. 126).
[24] N. N. Abliizov, M. E. Raikh, and A. L. Efros, Fiz. tverd. Tela **25**, 353 (1983).

(Received April 26, 1985)

phys. stat. sol. (b) 131, 631 (1985)

Subject classification: 13.1; 22.2

Zentralinstitut für Elektronenphysik der Akademie der Wissenschaften der DDR, Berlin[1]) *(a) and Sektion Physik der Humboldt-Universität zu Berlin*[2]) *(b)*

Charge Self-Consistent Tight-Binding Parameters

Application to III–V Compound Semiconductors

By

R. Strehlow (a), M. Hanke (b), and W. Kühn (a)

Dedicated to Prof. Dr. Dr. h. c. Dr. E. h. P. Görlich on the occasion of his 80th birthday

The description of defects (impurities, surfaces, etc.) in the framework of empirical tight-binding method (ETBM) suffers from the drawback that already the TB parameters of the unperturbed crystal are not uniquely defined when obtained from pure fitting to existing band structures. Such an uncertainty enters the defect model via both, the unperturbed problem itself and the defect potential. It is suggested to overcome these difficulties by determining the diagonal TB parameters and orbital populations always self-consistently. In the present paper and as the first step a method is developed to construct such charge self-consistent TB parameter sets for the pure crystal within the framework of a special fitting procedure including up to second-nearest neighbour interactions. With this additional self-consistency requirement band structures are obtained being even better than those calculated from other parameter sets. The method is applied to the III–V compound semiconductors GaP, GaAs, GaSb, InP, InAs, and InSb. Chemical trends are discussed.

Bei der Beschreibung von Defekten (Störstellen, Oberflächen, usw.) im Rahmen der ETBM erweist sich als nachteilig, daß bereits die TB-Parameter des ungestörten Kristalles nicht eindeutig definiert sind, wenn sie allein durch Anpassung an gegebene Bandstrukturen bestimmt werden. Eine solche Unbestimmtheit überträgt sich auf die Defekteigenschaften sowohl durch das ungestörte Problem als auch durch das Defektpotential. Es wird vorgeschlagen, diese Schwierigkeiten dadurch zu überwinden, daß diagonale TB-Parameter und orbitale Populationen stets selbstkonsistent bestimmt werden. In dieser Arbeit wird als erster Schritt dazu eine Methode zur Konstruktion solcher ladungsselbstkonsistenter TB-Parametersätze für den reinen Kristall im Rahmen einer speziellen Anpassungsprozedur entwickelt, die Wechselwirkungen bis zu zweitnächsten Nachbarn einschließt. Mit dieser zusätzlichen Selbstkonsistenzforderung werden Bandstrukturen erhalten, die sogar besser sind als die, die mit anderen Parametrisierungen erhalten werden. Die Methode wird auf die III–V-Verbindungshalbleiter GaP, GaAs, GaSb, InP, InAs und InSb angewendet. Chemische Trends werden diskutiert.

1. Introduction

The empirical tight-binding method (ETBM) was fully established in the seventies for the study of the properties of solids. The original suggestion by Slater and Koster in 1954 [1] to use the LCAO method as an interpolation scheme by treating the matrix elements of the Hamiltonian as parameters to be fitted to calculated band structure at certain critical points was proved to be efficient for computing energy bands with satisfactory quality. In most cases a Hamiltonian which includes the interactions up to second-nearest (2nn) neighbours is underlying. Alternatively, a nearest neighbour Hamiltonian with an extra excited s-like orbital was introduced recently [17]. More-

[1]) Rudower Chaussee 5, DDR-1199 Berlin, GDR.

[2]) PSF 1297, DDR-1086 Berlin, GDR.

over, one began to use ETBM as an extrapolation scheme to predict the properties of noncrystalline solids and perturbed crystals. A whole variety of calculation schemes such as the cluster-Bethe lattice method, recursion method, slab method, transfer-matrix method, or Koster-Slater scattering-theoretic method were developed and successfully applied to various problems of solid state physics.

Comparing with more sophisticated approaches (pseudopotential calculations) these methods are characterized by modest numerical effort. Their results are of qualitative or even semi-quantitative nature and allow an easy interpretation in physical terms. All that makes ETBM at present one of the most powerful theoretical tools to study the properties of surfaces, interfaces, impurities, amorphous and disordered solids.

Although during the last few years the method was subject to strong criticism in some aspects [2, 3] its popularity has not been affected seriously. On the one hand, such criticism was not always quite well-founded as it was pointed out recently by Lannoo [4], and on the other hand, refinements of the method became available which considerably improve the agreement between results obtained by ETBM and more sophisticated methods [5].

Nevertheless, applying ETBM to perturbed crystals the final result of any calculation is strongly dependent on both the actual values of the bulk TB parameters and the way the perturbation is defined for the problem under consideration. To be more specific, the various results for the T_2 level of the ideal As vacancy in GaAs shall be considered. The reported values range from 0.95 to 1.51 eV [5 to 10]. However, all the different sets of TB parameters used to predict this level via orbital-removal techniques give a fairly accurate description of the valence bands and the band gap, but have their difficulties to reproduce the conduction bands at least approximately. Thus, there seems to be no argument to show that one set is really better than the other [5]. But the ambiguity in determining these sets is reflected in different values for the ionicity (mainly related to the diagonal matrix elements). This seems to be a main reason for the mentioned dispersion [10]. Closely connected with this problem is the description of the perturbation. It is common to nearly all calculations to define the impurity potential by only changing the site-diagonal energies, while all other parameters retain their bulk values. Even in this simplified version the choice of the perturbation is ambiguous. Obviously one can think of taking the site-diagonal perturbation to be the difference between the diagonal TB parameters of the impurity atom and the host atom, however, just this suffers from the uncertainties in defining these parameters and furthermore demands for parameters which have to be defined uniquely with respect to a common energy reference, for instance, universal parameters.

As a semi-empirical scheme ETBM has these intrinsic difficulties which arise from the fact that electronic charge distribution and electronic potential cannot be computed self-consistently from first principles. In the spirit of the method it can be done, however, in a semi-empirical manner and this seems to be the key to further increase the credibility of ETBM.

The TB approximation reflects the self-consistency of charge distribution and potential well-known from one-electron theories such as the Hartree approximation in a self-consistency of orbital population and TB parameters. This idea has already been incorporated in some previous TB calculations. As for the defect problem we know approaches to the charged states of vacancies [11], transition metal impurities [12], and interstitials [13], however, other applications such as to surfaces are found [14]. To achieve self-consistency a linear relation between orbital population and the diagonal matrix elements of the Hamiltonian is usually assumed. A somewhat different method is preferred in recent cluster calculations [5]. Here self-consistency comes into play by requiring a suitable cluster which contains the defect, to be neutral.

As we feel all these former approaches suffer from confining self-consistency only to the defect description, whereas actually it must be fulfilled throughout the whole perturbed crystal. Such a treatment which describes the bulk on equal footing with the defect is what we intend to suggest. For this purpose and as a first step we deal with the construction of charge self-consistent sets of TB parameters for six III–V compound semiconductors in the present paper.

Beside conceptual reasons our parametrization has the following attractive properties:

(i) The resulting orbital eigenenergies are determined from physical requirements instead of being simply obtained from pragmatic fitting. As a consequence the ambiguity in the values for ionicity parameter, polarity, and charge transfer between anion and cation TB orbitals is strongly reduced.

(ii) The self-consistency condition automatically fixes the relative positions of orbital eigenenergies in all cases of macroscopic or microscopic inhomogeneity (impurities, interfaces, alloys) which commonly demands for universal parameter sets. Particularly, different possible charge states of a given defect can be treated.

The plan of this paper is as follows: In Section 2 the method for performing a charge self-consistent parametrization in the framework of a 2nn Hamiltonian is outlined and discussed. The resulting TB sets for six III–V compound semiconductors, their band structures and densities of states are presented in Section 3. Furthermore a comparison with previous parametrization will be given there. A subsequent section contains some concluding remarks.

2. Method of Parametrization

To achieve charge self-consistency a functional dependence of TB eigenenergies on orbital population is needed. In the spirit of the Hartree-Fock method the TB eigenenergies of a given atom are obtained from the free atom values and different terms of Coulomb corrections induced by the charge redistribution of this atom and all other atoms in the solid. The intra-atomic electron–electron repulsion is described by a linear expression so as to yield the correct values of ionization energies of the free atom [13].

The inter-atomic contributions are less easy to estimate. If assuming nonoverlapping spherical orbitals these contributions will be represented by the Madelung energy $-\alpha e^2/d$, where α is the Madelung constant, equal to 1.64 for the zincblende structure, and d the nearest neighbour distance [25]. As a consequence the inter-atomic contributions would nearly completely compensate the intra-atomic ones or even result in negative effective electron–electron interaction parameters indicating an unphysical situation. In fact, as was discussed recently by Harrison [25], the nonvanishing overlap of the atomic orbitals reduces the inter-atomic Coulomb corrections in comparison to the Madelung calculation. Whereas Harrison, however, takes this to mean that Coulomb shifts are to be neglected for the materials under consideration, we have estimated that the influence of orbital overlap can be appreciable and suggest $-e^2/d$ to be a reasonable choice for the inter-atomic term. This yields values $U^{\text{eff}} = U - e^2/d$ between 3 and 7 eV, which not only allow excellent parametrization of the bulk band structure, but also should reproduce charge state splittings of defect levels in the correct order of magnitude [26]. Thus we require the validity of the linear expressions

$$E_s = E_s^A + (N_s - N_s^A)\, U_{ss}^{eff} + (N_p - N_p^A)\, U_{sp}^{eff}\,, \tag{1a}$$

$$E_p = E_p^A + (N_p - N_p^A)\, U_{pp}^{eff} + (N_s - N_s^A)\, U_{sp}^{eff} \tag{1b}$$

to relate the TB eigenenergies E_s, E_p to the orbital population N_s, N_p. Here E_s^A, E_p^A and N_s^A, N_p^A characterize the Hartree-Fock energies and the electron populations of the free neutral atom, respectively. Values for E_s^A, E_p^A and for the intra-atomic electron–electron interaction parameters U_{ss}, U_{pp}, and U_{sp} used in our calculations are taken from [13] (Table 1).

Table 1

Atomic Hartree-Fock energies E_s^A, E_p^A and electron–electron interaction parameters U_{ss}, U_{pp}, and U_{sp} (all in eV) (as given in [13])

	E_s^A	E_p^A	U_{ss}	U_{pp}	U_{sp}
Ga	−11.55	− 5.67	10.13	9.74	9.02
In	−10.14	− 5.37	9.17	10.45	8.72
P	−18.80	−10.12	13.65	9.84	10.85
As	−17.80	− 9.19	12.50	9.27	10.77
Sb	−15.83	− 9.10	11.90	8.33	9.42

Table 2

Charge self-consistent tight-binding parameters (in eV) for GaP, GaAs, GaSb, InP, InAs, and InSb (in the notation of Slater and Koster [1])

Parameter	GaP	GaAs	GaSb	InP	InAs	InSb
$P_1 = E_{ss}(000)_0$	−18.180	−17.312	−15.918	−17.314	−16.720	−15.358
$P_2 \equiv E_{ss}(000)_1$	−13.177	−12.938	−12.971	−11.681	−11.489	−11.590
$P_3 = E_{xx}(000)_0$	− 9.318	− 8.939	− 8.905	− 8.693	− 8.503	− 8.481
$P_4 = E_{xx}(000)_1$	− 6.178	− 5.965	− 6.099	− 5.834	− 5.550	− 5.779
$P_5 = 4E_{ss}(\frac{1}{2}\frac{1}{2}\frac{1}{2})$	− 7.365	− 6.577	− 6.251	− 5.776	− 6.075	− 5.739
$P_6 = 4E_{sx}(\frac{1}{2}\frac{1}{2}\frac{1}{2})_{01}$	2.928	4.040	4.433	2.846	4.750	4.380
$P_7 = 4E_{sx}(\frac{1}{2}\frac{1}{2}\frac{1}{2})_{10}$	4.552	4.259	3.925	4.367	4.606	4.053
$P_8 = 4E_{xx}(\frac{1}{2}\frac{1}{2}\frac{1}{2})$	1.803	1.309	0.450	1.755	1.388	0.394
$P_9 = 4E_{xy}(\frac{1}{2}\frac{1}{2}\frac{1}{2})$	4.722	4.877	4.311	4.313	4.459	4.084
$P_{10} = 4E_{xx}(011)_0$	− 0.196	− 0.202	− 0.282	− 0.386	− 0.501	− 0.506
$P_{11} = 4E_{xx}(011)_1$	− 0.555	− 0.877	− 1.286	− 0.652	− 1.035	− 1.171
$P_{12} = 4E_{xy}(110)_0$	0.934	0.317	− 0.151	0.312	0.392	− 0.143
$P_{13} = 4E_{xy}(110)_1$	− 0.151	0.376	0.493	0.221	0.670	0.746
$P_{14} = 4E_{xx}(110)_0$	0.278	0.212	0.213	0.320	0.461	0.389
$P_{15} = 4E_{xx}(110)_1$	0.788	0.924	0.972	0.541	0.952	0.901
$P_{16} = 4E_{sx}(110)_0$	0.0	0.0	0.0	0.0	0.0	0.0
$P_{17} = 4E_{sx}(110)_1$	0.0	0.0	0.0	0.0	0.0	0.0
$P_{18} = 4E_{ss}(110)_0$	0.165	0.059	0.006	0.073	− 0.137	− 0.103
$P_{19} = 4E_{ss}(110)_1$	0.466	0.257	0.029	0.124	− 0.284	− 0.239
$P_{20} = 4E_{sx}(011)_0$	2.248	1.203	0.689	0.631	0.945	0.602
$P_{21} = 4E_{sx}(011)_1$	− 0.001	0.058	− 0.254	0.261	− 0.597	− 0.381
$P_{22} = 4E_{xy}(011)_0$	0.0	0.0	0.0	0.0	0.0	0.0
$P_{23} = 4E_{xy}(011)_1$	0.0	0.0	0.0	0.0	0.0	0.0
E_Δ	− 9.677	− 9.130	− 8.821	− 9.242	− 8.576	− 8.258

Our method of parametrization works as follows: For a given set of diagonal elements P_1 to P_4 (for the notation see Table 2) the off-diagonal TB parameters P_5 to P_{23} are determined in order to reproduce the energies at critical points of the band structure. The Hartree-Fock values of the neutral atoms are a suitable choice for P_1 to P_4 in the initial step. Then the population of the TB orbitals is calculated and a new set of diagonal elements is found with the help of (1).

Self-consistency is reached by retarded iteration. During each iteration step two problems are to be solved.

2.1 Calculation of the off-diagonal elements

As starting point the following analytic expressions for the energies at the critical points are used:

$$E(\Gamma_{1c/v}) + E_\Delta = (P_1 + 3P_{18} + P_2 + 3P_{19})/2 \pm \pm \left(\tfrac{1}{4}(P_1 + 3P_{18} - P_2 - 3P_{19})^2 + P_5^2\right)^{1/2}, \tag{2a}$$

$$E(\Gamma_{15c/v}) + E_\Delta = (P_3 + P_{10} + 2P_{14} + P_4 + P_{11} + 2P_{15})/2 \pm \pm \left(\tfrac{1}{4}(P_3 + P_{10} + 2P_{14} - P_4 - P_{11} - 2P_{15})^2 + P_8^2\right)^{1/2}, \tag{2b}$$

$$E(X_{1c/v}) + E_\Delta = (P_1 - P_{18} + P_4 + P_{11} - 2P_{15})/2 \pm \pm \left(\tfrac{1}{4}(P_1 - P_{18} - P_4 - P_{11} + 2P_{15})^2 + P_6^2\right)^{1/2}, \tag{2c}$$

$$E(X_{3c/v}) + E_\Delta = (P_2 - P_{19} + P_3 + P_{10} - 2P_{14})/2 \pm \pm \left(\tfrac{1}{4}(P_2 - P_{19} - P_3 - P_{10} + 2P_{14})^2 + P_7^2\right)^{1/2}, \tag{2d}$$

$$E(X_{5c/v}) + E_\Delta = (P_3 - P_{10} + P_4 - P_{11})/2 \pm \pm \left(\tfrac{1}{4}(P_3 - P_{10} - P_4 + P_{11})^2 + P_9^2\right)^{1/2}, \tag{2e}$$

$$E(L_{3c/v}) + E_\Delta = (P_3 + P_{12} + P_4 + P_{13})/2 \pm \pm \tfrac{1}{2}\left((P_3 + P_{12} - P_4 - P_{13})^2 + (P_8 + P_9)^2\right)^{1/2}, \tag{2f}$$

where E_Δ is the energy shift between the calculated band structure and that to be fitted.

With the exception of X_{5c} we require them to be fulfilled exactly. From the trace of the Hamilton matrix (see Appendix) at Γ, X, and L the relation

$$4\sum_{\nu=1}^{8} E_\nu(L) = \sum_{\nu=1}^{8} \left(E_\nu(\Gamma) + 3E_\nu(X)\right) = 4(P_1 + P_2 + 3P_3 + 3P_4 - 8E_\Delta) \tag{3}$$

can be derived, which together with (2) can be used to find an upper and a lower bound to E_Δ. It is common to the III–V compound semiconductors that $E(L_{2c}) > > E(L_{3c})$. This condition, obviously difficult to reproduce with already existing parameter sets [5 to 8, 16] requires

$$\sum_{\nu=1}^{8} E_\nu(L) > E(L_{1v}) + E(L_{2v}) + 2E(L_{3v}) + E(L_{1c}) + 3E(L_{3c}) = S^{L}_{min}\,. \tag{4}$$

From (3) now follows

$$E_\Delta < (P_1 + P_2 + 3P_3 + 3P_4 - S^{L}_{min})/8\,. \tag{5}$$

With decreasing E_Δ and keeping fixed all other energies of Γ and X $E(X_{5c})$ and in consequence of (2e) P_9 increases. On the other hand, (2f) forces P_9 to obey the relation

$$P_8 + P_9 < E(L_{3c}) - E(L_{3v}) \tag{6}$$

which works as a lower bound to E_Δ.

Now we take advantage of the flexibility of E_Δ inside these limits to minimize the absolute values of the 2nn parameters which is in the spirit of a 2nn parametrization.

Assuming a constant ratio between corresponding parameters for cations and anions,

$$P_{11}/P_{10} = P_{15}/P_{14} = P_{19}/P_{18} = A\,, \tag{7}$$

the following system of equations can be deduced from (2):

$$\begin{aligned}
3(A+1)\,P_{18} &= E_1 = E(\Gamma_{1c}) + E(\Gamma_{1v}) + 2E_\Delta - P_1 - P_2\,,\\
(A+1)\,P_{10} + 2(A+1)\,P_{14} &= E_2 = E(\Gamma_{15c}) + E(\Gamma_{15v}) + 2E_\Delta - P_3 - P_4\,,\\
AP_{10} - 2AP_{14} - P_{18} &= E_3 = E(X_{1c}) + E(X_{1v}) + 2E_\Delta - P_1 - P_4\,,\\
P_{10} - 2P_{14} - AP_{18} &= E_4 = E(X_{3c}) + E(X_{3v}) + 2E_\Delta - P_2 - P_3\,.
\end{aligned} \tag{8}$$

The solution

$$\begin{aligned}
A &= (E_1 + 3E_3)/(E_1 + 3E_4)\,,\\
P_{10} &= (E_1 + 3(E_2 + E_3 + E_4))/(6A + 6)\,,\\
P_{14} &= (3(E_2 - E_3 - E_4) - E_1)/(12A + 12)\,,\\
P_{18} &= E_1/(3A + 3)
\end{aligned} \tag{9}$$

determines additionally P_{11}, P_{15}, and P_{19} according to (7). The 1nn parameters P_5 to P_9 now can be deduced by applying (2a) to (2e) to the valence bands. P_{12} and P_{13} follow from (2f) and P_{20} and P_{21} are used to improve the energies at L_{1v}, L_{2v}, and L_{1c}. Finally, P_{16}, P_{17}, P_{22}, and P_{23} are neglected because the energies neither at Γ, X nor at L depend upon them.

2.2 Population of TB orbitals

The population is calculated by weighted summation of squared absolute values of the wave function coefficients of the four valence bands over the irreducible wedge of the Brillouin zone (IBZ). We use a cubic grid with 89$\boldsymbol{k}$ points in the IBZ choosing the weighting factors according to symmetry [15].

3. Results and Discussion

3.1 TB parameters

The values of TB parameters obtained using the procedure explained in the preceding section are presented in Table 2. In our calculations the nonlocal empirical pseudopotential band structures of Chelikowsky and Cohen [19] have been taken as reference for the fit. A general survey shows that the differences of the diagonal elements as well as the order of magnitude of the 1nn parameters fit well into the range expected from existing parameter sets. As for the 2nn parameters it can be noticed that their absolute values as a rule are smaller than in other sets. In some cases the sign of our 2nn parameters differs from that common to other sets and also changes for a given parameter going from one substance to another. This can be observed in the parametrization given by Das Sarma and Madhukar, too [10]. We feel this to be no drawback of a parametrization, for in our opinion there are no physical arguments for a fixed definite sign. What can be required for the sake of internal consistency of a 2nn approximation is that they should be small compared with the 1nn parameters. Just that is generally fulfilled by our sets.

3.2 *Band structure and density of states*

According to our method the available energies at all critical points Γ and X together with L_{3v} and L_{3c} are exactly reproduced with our TB parameters (Table 3). As regards the other energies at the L critical point a special weighting procedure guarantees their energies to be reproduced the better, the closer they are to the fundamental gap. Thus, especially the energies at L_{2v} and L_{1c} are much closer to the exact values than for the parametrizations given in [8, 10]. The energies at L_{1v} generally compare also well with the pseudopotential values (GaP may be viewed as a runaway).

Table 3

Energy levels (in eV) at Γ, X, and L critical points obtained using parameters of Table 2 with diagonal energies shifted by E_Δ to yield $E(\Gamma_{15v}) = 0$. The corresponding values of [19] are given in parentheses. Δq represents the charge transfer from cation to anion (in units of e^-)

	GaP	GaAs	GaSb	InP	InAs	InSb
Γ_{1v}	−12.99 (−12.99)	−12.55 (−12.55)	−12.00 (−12.00)	−11.42 (−11.42)	−12.69 (−12.69)	−11.71 (−11.71)
Γ_{15v}	0.0 (0.0)	0.0 (0.0)	0.0 (0.0)	0.0 (0.0)	0.0 (0.0)	0.0 (0.0)
Γ_{1c}	2.88 (2.88)	1.51 (1.51)	0.86 (0.86)	1.50 (1.50)	0.37 (0.37)	0.25 (0.25)
Γ_{15c}	5.24 (5.24)	4.55 (4.55)	3.44 (3.44)	4.64 (4.64)	4.39 (4.39)	3.16 (3.16)
X_{1v}	− 9.46 (− 9.46)	− 9.83 (− 9.83)	− 9.33 (− 9.33)	− 8.91 (− 8.91)	−10.20 (−10.20)	− 9.20 (− 9.20)
X_{3v}	− 7.07 (− 7.07)	− 6.88 (− 6.88)	− 6.76 (− 6.76)	− 6.01 (− 6.01)	− 6.64 (− 6.64)	− 6.43 (− 6.43)
X_{5v}	− 2.73 (− 2.73)	− 2.99 (− 2.99)	− 2.61 (− 2.61)	− 2.09 (− 2.09)	− 2.47 (− 2.47)	− 2.45 (− 2.45)
X_{1c}	2.16 (2.16)	2.03 (2.03)	1.72 (1.72)	2.44 (2.44)	2.28 (2.28)	1.71 (1.71)
X_{3c}	2.71 (2.71)	2.38 (2.38)	1.79 (1.79)	2.97 (2.97)	2.66 (2.66)	1.83 (1.83)
L_{1v}	−11.62 (−10.60)	−10.91 (−10.60)	−10.18 (−10.17)	− 9.65 (− 9.67)	−11.10 (−10.92)	− 9.97 (− 9.95)
L_{2v}	− 7.09 (− 6.84)	− 6.88 (− 6.83)	− 6.25 (− 6.25)	− 5.82 (− 5.84)	− 6.28 (− 6.23)	− 5.92 (− 5.92)
L_{3v}	− 1.10 (− 1.10)	− 1.42 (− 1.42)	− 1.45 (− 1.45)	− 1.09 (− 1.09)	− 1.26 (− 1.26)	− 1.44 (− 1.44)
L_{1c}	2.75 (2.79)	1.81 (1.82)	1.22 (1.22)	2.20 (2.19)	1.52 (1.53)	1.03 (1.03)
L_{3c}	5.74 (5.74)	5.47 (5.47)	4.43 (4.43)	5.58 (5.58)	5.42 (5.42)	4.30 (4.30)
Δq	0.284	0.203	0.204	0.389	0.314	0.313

We want to stress that our parameter sets yield the correct ordering $E(L_{2c}) > E(L_{3c})$ of the higher conduction bands, which is not reproduced by most of the existing parametrizations [5 to 8, 16]. Furthermore it should be noted that, although not being fitted, the energy of the minimum of the third valence band Σ_{min} comes out fairly well. Summarizing we can say that the overall fit of the band structures for all six III–V compound semiconductors (Fig. 1 a to f) is better than for other param-

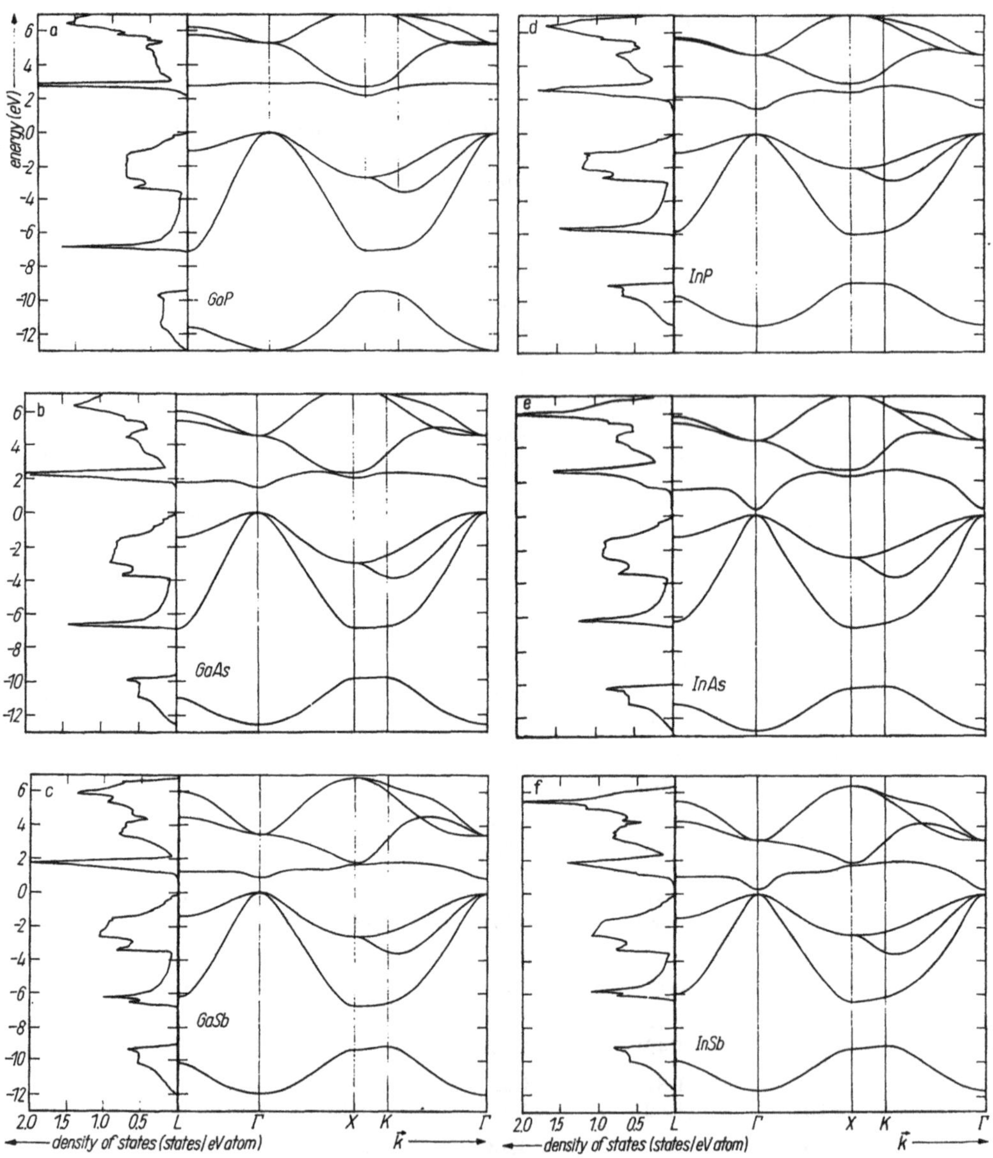

Fig. 1. Band structure and density of states for a) GaP, b) GaAs, c) GaSb, d) InP, e) InAs, and f) InSb calculated with the parameters of Table 2. The energy reference is chosen as explained in the caption of Table 3

etrizations in spite of the constraints required from charge self-consistency. Of course, as an inherent drawback of ETBM the dispersion of the conduction bands is only poorly reproduced. The densities of states which are also shown in Fig. 1 a to f have been obtained with the help of the correctly weighting tetrahedron method recently developped by the authors [18] employing only 89**k** points in the irreducible wedge of the Brillouin zone. They agree fairly well with published data.

3.3 Chemical trends

It is instructive to check whether chemical trends within the group of III–V compound semiconductors are reflected in our sets. Such trends which enter the defect description via both the bulk problem itself and the defect potential should essentially govern corresponding trends in the defect properties.

For the following it is important to note that the atomic Hartree-Fock energies which enter explicitly our procedure turn out to be somewhat uncertain. Thus, the conclusions we are going to draw using the values of Table 1 concerning diagonal energies and heteropolarity, might change slightly for a different choice of atomic energies, but TB diagonal energies are expected to reflect tendencies of the Hartree-Fock energies. Our parameter sets exhibit the following trends:

(i) The anion and cation diagonal energies increase with rising atomic numbers. Similar to Das Sarma and Madhukar [10] the diagonal energies for a given atom are nearly independent of the compound of which the atom is a part. This is, however, better fulfilled for the cations than for the anions because of the smaller values for U^{eff} of the cations.

(ii) For a fixed cation (anion) the sp-splitting of the anions (anions) decreases going down the series P–As–Sb (Ga to In), as the free-atom values do.

On the other hand, the splitting remains nearly constant for a given atom regardless of the other component of the compound.

(iii) With X and Y denoting fixed cation and anion, respectively, the following relations for the charge transfer Δq can be obtained from Table 1:

$$\Delta q(\text{XP}) > \Delta q(\text{XAs}) = \Delta q(\text{XSb})$$

and

$$\Delta q(\text{GaY}) < \Delta q(\text{InY})\,.$$

This is in good correspondence to the relations for the ionicity parameter $\chi = P_2 - P_3$ defined by Mele and Joannopoulos [20],

$$\chi(\text{XP}) \geqq \chi(\text{XAs}) > \chi(\text{XSb})$$

and

$$\chi(\text{GaY}) < \chi(\text{InY})\,.$$

Whereas the ordering of all III–V compounds with respect to heteropolarity is a matter of definition, it is noteworthy that both charge transfer and ionicity parameter as measures often used, exhibit the same chemical trends for our parametrization. It should be emphasized that the same trends in heteropolarity can be obtained from Phillips' ionicity scale [22] and Harrison's bond polarity taken from [21] or recalculated from our parameters.

(iv) Due to different energy references trends in the values of diagonal energies are not comparable with the results of other parametrizations. In [10] experimental values for the photothreshold are used to define the energy reference, and the universal model of Vogl et al. [17] correlates the diagonal energies of the solid with the free-atom values in a quite different manner than we have done. According to their procedure the orbital energies of both anion and cation are always lowered in comparison with the free-atom values, whereas in our approach as a result of charge transfer, cation energies are lowered and anion energies are raised in general.

This statement, of course, is based on our energy scale with the zero at the edge of the continuum part of the atomic spectra. Since the energy zeros of the band structures to be fitted are at the valence band maxima, E_Δ turns out to be the photothreshold in our one-particle description. Taking into account that the difference of 3 to 4 eV can be attributed to many-body correlation effects [24] E_Δ indeed corresponds well with experimental photothreshold data.

(v) The 1nn elements have often been proposed to obey a certain scaling rule with bond length d. In agreement with [17] we also observe a falling tendency for the parameters P_5, P_7, P_8, and P_9 when ordering the materials according to increasing bond length. It is difficult to discriminate between different scaling rules such as $\sim d^{-2}$ [21] or $\sim \exp(-d/d_0)$ [23] as long as we dispose only of the results of six III–V compounds. But tentatively adopting the d^{-2}-rule we find universal parameters close to those of [17]. It should be emphasized, however, that P_6 exhibits a rising tendency and as we have reexamined the same is true for Vogl's parametrization. This should raise some doubt about the validity of a pure scaling with the bond length in general.

4. Concluding remarks

The different existing TB parametrizations for III–V compounds have been able to reproduce the bulk band structures more or less equally well. Problems arose not until the ETBM was extended to defect problems. The defect levels exhibit a strong dispersion due to the nonuniqueness of the parameter sets. As was pointed out recently [10], this ambiguity can be diminished by incorporating some aspects of the charge distribution within the solid. We have taken up this idea by developing a procedure for the charge self-consistent calculation of TB diagonal energies. This approach is especially adapted to defect problems, since a straightforward generalization will allow a unique definition of the on-site perturbation potential for both neutral and charged defects.

It should be emphasized that the additional constraint of charge self-consistency requirement must not imply a loss of quality for the resulting band structures. So our fitting procedure supplies even better fits to the critical point energies than other parameter sets do. In the spirit of a 2nn parametrization we demand the 2nn interaction parameters to be as small as possible. This, however, would have consequences for the treatment of vacancies when simply using the orbital-removal method. The A_1–T_2 splitting would turn out to be much too small. But one knows from [5] that orbital rearrangement around the defect which can drastically alter the parameters is important for improving this splitting.

In a forthcoming paper the parametrization procedure proposed here will be applied to II–VI compounds. Furthermore we are engaged in generalizing our method to the charge self-consistent treatment of point defects using the parameter sets obtained here for the description of the bulk [26, 27].

Acknowledgements

The authors are grateful to Dr. F. Bechstedt for various stimulating discussions. We also wish to thank Dr. J. Pollmann and Dr. P. Vogl for critical and helpful comments.

Appendix

In the following the matrix elements of the 2nn TB Hamiltonian used in our calculations are given, whereby the expressions are corrected for some misprints contained in [8]:

$$\begin{aligned}
\langle s_{0/1} \mid s_{0/1}\rangle &= P_{1/2} + P_{18/19}(C_1C_2 + C_2C_3 + C_3C_1)\,,\\
\langle x_{0/1} \mid x_{0/1}\rangle &= P_{3/4} + P_{10/11}C_2C_3 + P_{14/15}C_1(C_2 + C_3)\,,\\
\langle y_{0/1} \mid y_{0/1}\rangle &= P_{3/4} + P_{10/11}C_1C_3 + P_{14/15}C_2(C_1 + C_3)\,,\\
\langle z_{0/1} \mid z_{0/1}\rangle &= P_{3/4} + P_{10/11}C_1C_2 + P_{14/15}C_3(C_1 + C_2)\,,\\
\langle s_{0/1} \mid x_{0/1}\rangle &= \mp P_{20/21}S_2S_3 + 4iP_{16/17}S_1(C_2 + C_3)\,,\\
\langle s_{0/1} \mid y_{0/1}\rangle &= \mp P_{20/21}S_1S_3 + 4iP_{16/17}S_2(C_1 + C_3)\,,\\
\langle s_{0/1} \mid z_{0/1}\rangle &= \mp P_{20/21}S_1S_2 + 4iP_{16/17}S_3(C_1 + C_2)\,,\\
\langle x_{0/1} \mid y_{0/1}\rangle &= -P_{12/13}S_1S_2 \mp 4iP_{22/23}C_3(S_1 - S_2)\,,\\
\langle y_{0/1} \mid z_{0/1}\rangle &= -P_{12/13}S_2S_3 \mp 4iP_{22/23}C_1(S_2 - S_3)\,,\\
\langle z_{0/1} \mid x_{0/1}\rangle &= -P_{12/13}S_1S_3 \mp 4iP_{22/23}C_2(S_3 - S_1)\,,\\
\langle s_0 \mid s_1\rangle &= P_5g_0\,,\\
\langle x_0 \mid x_1\rangle &= \langle y_0 \mid y_1\rangle = \langle z_0 \mid z_1\rangle = P_8g_0\,,\\
\langle x_0 \mid y_1\rangle &= P_9g_3\,,\\
\langle y_0 \mid z_1\rangle &= P_9g_1\,,\\
\langle z_0 \mid x_1\rangle &= P_9g_2\,,\\
\langle x_0 \mid z_1\rangle &= P_9g_2\,,\\
\langle y_0 \mid x_1\rangle &= P_9g_3\,,\\
\langle z_0 \mid y_1\rangle &= P_9g_1\,,\\
\langle s_0 \mid x_1\rangle &= P_6g_1\,,\\
\langle s_0 \mid y_1\rangle &= P_6g_2\,,\\
\langle s_0 \mid z_1\rangle &= P_6g_3\,,\\
\langle s_1 \mid x_0\rangle &= -P_7g_1^*\,,\\
\langle s_1 \mid y_0\rangle &= -P_7g_2^*\,,\\
\langle s_1 \mid z_0\rangle &= -P_7g_3^*\,.
\end{aligned}$$

The interaction integrals P_1 to P_{23} are defined in Table 2. The trigonometric expressions C_1, C_2, C_3, S_1, S_2, S_3 and g_0, g_1, g_2, g_3 have their usual meaning [8]. The index 0 denotes the anion, whereas 1 denotes the cation. The upper (lower) sign refers to the combination of the first (second) indices.

References

[1] J. C. SLATER and G. F. KOSTER, Phys. Rev. 94, 1498 (1954).
[2] J. B. KRIEGER and P. M. LAUFER, Phys. Rev. B 23, 4063 (1981).
[3] V. A. SINGH, U. LINDEFELT, and A. ZUNGER, Phys. Rev. B 25, 2781 (1982).
[4] M. LANNOO, J. Phys. C 17, 3137 (1984).
[5] J. VAN DER REST and P. PECHEUR, J. Physique 44, 1297 (1983); J. Phys. C 17, 85 (1984).
[6] J. BERNHOLC and S. T. PANTELIDES, Phys. Rev. B 18, 1780 (1978).

[7] M. S. DAW and D. L. SMITH, Phys. Rev. B **20**, 5150 (1979).
[8] D. N. TALWAR and C. S. TING, Phys. Rev. B **25**, 2660 (1982).
[9] P. J. LIN CHUNG and T. L. REINECKE, Phys. Rev. B **27**, 1101 (1983).
[10] S. DAS SARMA and A. MADHUKAR, Phys. Rev. B **24**, 2051 (1981).
[11] M. LANNOO and J. BOURGOIN, Point Defects in Semiconductors I, Springer-Verlag, 1981.
[12] F. D. M. HALDANE and P. W. ANDERSON, Phys. Rev. B **13**, 2553 (1976).
[13] O. F. SANKEY and J. D. DOW, Phys. Rev. B **27**, 7641 (1983).
[14] C. CALANDRA, F. MANGHI, and C. M. BERTONI, J. Phys. C **10**, 1911 (1977).
[15] F. DEL GIALLO, P. MORETTI, and F. PIERALLI, Nuovo Cimento **2** D, 721 (1983).
[16] G. C. OSBOURN and D. L. SMITH, Phys. Rev. B **19**, 2124 (1979).
[17] P. VOGL, H. P. HJALMARSON, and J. D. DOW, J. Phys. Chem. Solids **44**, 365 (1983).
[18] M. HANKE, W. KÜHN, and R. STREHLOW, phys. stat. sol. (b) **123**, K39 (1984).
[19] J. R. CHELIKOWSKY and M. L. COHEN, Phys. Rev. B **14**, 556 (1976).
[20] E. J. MELE and J. D. JOANNOPOULOS, Phys. Rev. B **17**, 1528 (1978).
[21] S. T. PANTELIDES and W. A. HARRISON, Phys. Rev. B **11**, 3006 (1975).
[22] J. C. PHILLIPS, Bonds and Bands in Semiconductors, Academic Press, 1973.
[23] K. C. PANDEY and J. C. PHILLIPS, Phys. Rev. Letters **32**, 1433 (1974).
[24] F. BECHSTEDT, R. ENDERLEIN, and O. HEINRICH, phys. stat. sol. (b) **126**, 575 (1984).
[25] W. A. HARRISON, Phys. Rev. B **31**, 2121 (1985).
[26] R. STREHLOW, M. HANKE, and W. KÜHN, to be published.
[27] M. HANKE, R. STREHLOW, and W. KÜHN, to be published.

(Received July 19, 1985)

phys. stat. sol. (b) **131**, 643 (1985)

Subject classification: 13.1; 4; 22.1; 22.1.1; 22.1.2

Sektion Physik der Humboldt-Universität zu Berlin[1])

Self-Consistent Tight-Binding Method for Total Energy Calculations of Tetrahedral Semiconductors Including Surfaces and Defects

By

F. BECHSTEDT, D. REICHARDT, and R. ENDERLEIN

Dedicated to Prof. Dr. Dr. h. c. Dr. E. h. P. GÖRLICH on the occasion of his 80th birthday

The total energy of macroscopic electron–ion systems as a function of ion positions is an important quantity for studies of atomic and electronic properties of ideal crystals, surfaces, and defects. For the first time a total energy calculation is presented which relies on tight binding theory and takes the electron–electron interaction fully into account. The advantage of the method as compared to the more accurate pseudopotential-density functional approach is its greater simplicity at still existing accuracy. The calculation relates directly to covalently bonded tetrahedrally coordinated semiconductors, but it can also be applied to ionic tetrahedral materials. The intraatomic part of the electron–electron interaction enters the total energy through the characteristic U-parameters whereas the long-range interatomic contribution is approximated by a Madelung-like Coulomb repulsion. Correlation makes that all one-electron energy levels depend on their populations which in turn are determined by the one-electron energies. Hence the total energy has to be calculated self-consistently. Bonding parameters, atomic orbital energies, and parameters of electron–electron interaction are adjusted in such a way that ionization potentials of free atoms and ground-state properties of ideal crystals as cohesive energy and bulk modulus are correctly reflected. The inclusion of the full electron–electron interaction guarantees also the true magnitude of the total energy. The simplicity and accuracy of the presented method makes it suitable for studies of atomic rearrangements near surfaces and defects. This is demonstrated explicitly in the case of relaxed (111) 1 × 1 surfaces of covalent semiconductors.

Die Gesamtenergie eines makroskopischen Elektron–Ion-Systems als Funktion der Ionenlagen ist eine wichtige Größe für Untersuchungen der atomaren und elektronischen Eigenschaften von Kristallen, Oberflächen und Defekten. Erstmalig wird eine Gesamtenergieberechnung durchgeführt, die sich der Methode der starken Bindung bedient und die Elektron–Elektron-Wechselwirkung voll in Rechnung stellt. Der Vorteil dieses Zugangs gegenüber der genaueren Pseudopotential-Dichtefunktional-Methode ist ihre größere Einfachheit bei dennoch vorhandener Genauigkeit. Die Rechnung bezieht sich direkt auf kovalent gebundene tetraedrisch koordinierte Halbleiter, kann aber auch auf ionische tetraedrische Materialien angewandt werden. Der intra-atomare Anteil der Elektron–Elektron-Wechselwirkung wird durch die U-Parameter beschrieben, der langreichweitige inter-atomare Beitrag wird durch eine Madelung-ähnliche Coulomb-Abstoßung ersetzt. Die Korrelation bewirkt, daß alle Einelektronen-Energieniveaus von ihren Besetzungen abhängen, wobei letztere selbst wiederum durch die Einelektronenenergien bestimmt sind. Dadurch muß die Berechnung der Gesamtenergie selbstkonsistent erfolgen. Die Bindungsparameter, atomaren Orbitalenergien und Parameter der Elektron–Elektron-Wechselwirkung werden so angepaßt, daß die Ionisationspotentiale der freien Atome und die Grundzustandseigenschaften der idealen Kristalle wie die Kohäsionsenergie und der Kompressionsmodul korrekt durch den Gesamtenergieausdruck wiedergegeben werden. Die Berücksichtigung der Elektron–Elektron-Wechselwirkung garantiert dann auch die richtige absolute Größe der Gesamtenergie. Wegen ihrer Einfachheit und Genauigkeit ist die präsentierte Methode für Untersuchungen der atomar-geometrischen Struktur von Oberflächen und Defekten mittels Gesamtenergieminimierung geeignet. Dies wird anhand der relaxierten (111) 1 × 1 Oberfläche kovalenter Halbleiter explizit demonstriert.

[1]) Invalidenstr. 110, DDR-1040 Berlin, GDR.

1. Introduction

In recent years, studies of structural properties of ideal semiconducting crystals, surfaces, and defects have made significant progress. It has been demonstrated [1] that structural properties like lattice constants, phonon frequencies, and phase stability criteria of bulk crystals as well as relaxation and reconstruction states of surfaces [2 to 7] can be calculated with high accuracy. The calculations were based on ab initio pseudopotentials and density functionals for the electron–electron interaction [8 to 10]. Because of the large numerical expense, however, such calculations are relatively rare. In order to extend the theoretical predictions of structural properties to a greater variety of materials, simpler methods are needed. For this reason the description of total energies by means of tight binding methods has been developed [11, 12]. An elementary tight-binding theory has been used to calculate dielectric and bonding properties of solids, in particular, chemical trends of such properties [11]. Lattice vibrations [13] and local impurity distortions [14] have been also calculated with sufficient accuracy. The first application of tight binding arguments to surface reconstruction has been presented by Harrison [15]. Subsequently the treatment has been refined [16]. Chemical pseudopotential methods have been used to calculate surface changes of total energies [17]. Chadi [18, 19] has attacked the problem by calculating the bandstructure with high accuracy and by describing the effect of atomic displacements within a Keating-like force constant model.

Despite of their success and their small numerical expense, so far total energy calculations by means of tight-binding methods have an essential disadvantage: The electron–electron interaction is taken into account only in an implicit and incomplete way. Correlation effects and the related self-consistency of the problem which both are extremely important for the derivation of optimal structures by means of total energy minimization [3, 20] are missing in these calculations.

It is the purpose of the present paper to overcome this disadvantage of tight-binding methods. A total energy expression will be given in terms of tight binding parameters which takes the full electron-electron interaction explicitly into account. In Section 2 this will be done for tetrahedral covalent semiconductors forming either bulk crystals, surfaces or defects. The expression approaches the correct total energy of free atoms in the limit of infinite atomic spacing and reflects correctly important solid state properties. In Section 3 this behaviour is used to specify the tight binding and electron–electron interaction parameters in a way that the first four atomic ionization potentials as well as the bulk modulus and the cohesive energy follow from the total energy expression. The theory is tested in Section 4 by total-energy minimization for relaxed (111) 1×1 surfaces of group IV semiconductors.

2. Parametrization of Total Energy

2.1 General expression

Subject to our total energy calculation is a tetrahedrally coordinated crystal with filled valence band and empty conduction band. The crystal is thought to consist of fourfold charged positive ions, located at lattice points $\boldsymbol{R}_i$, and of four valence electrons per ion. The total energy $E_{\text{tot}}(\{\boldsymbol{R}_i\})$ of the crystal, defined originally as the kinetic energy of valence electrons plus the potential energies of electron–electron, electron–ion, and ion–ion interactions [1 to 3], is written in terms of crystal one-electron energies. The latter ones are taken in an approximation which will be specified later. The sum of the one-electron energies upon all occupied states is referred to as band structure energy and denoted by $E_{\text{bs}}(\{\boldsymbol{R}_i\})$. The remaining part of E_{tot}, consisting only of electrostatic energies, is labeled as $E_{es}(\{\boldsymbol{R}_i\})$. The total energy then reads

$$E_{tot}(\{\boldsymbol{R}_i\}) = E_{bs}(\{\boldsymbol{R}_i\}) + E_{es}(\{\boldsymbol{R}_i\}) \,. \tag{1}$$

The electrostatic energy $E_{es}(\{\boldsymbol{R}_i\})$ decomposes in two parts, first the ion–ion interaction energy $E_{ii}(\{\boldsymbol{R}_i\})$, and second the negative electron–electron interaction energy $-E_{ee}(\{\boldsymbol{R}_i\})$. The latter energy is counted twice if total energies are calculated by summing over occupied one-electron energies [8]. It has to be subtracted from $E_{bs}(\{\boldsymbol{R}_i\})$ to get the correct expression for $E_{tot}(\{\boldsymbol{R}_i\})$. Thus the electrostatic energy takes the form

$$E_{es}(\{\boldsymbol{R}_i\}) = E_{ii}(\{\boldsymbol{R}_i\}) - E_{ee}(\{\boldsymbol{R}_i\}) \,. \tag{2}$$

The separation of the total energy into the two contributions E_{bs} and E_{es} offers different possibilities of approximations. The band structure energy E_{bs} can be calculated by means of any theory of crystal one-electron states. In the present paper we use the empirical tight-binding method [11], however, in a version which accounts for electron–electron interaction including correlation. To be consistent, for the calculation of the electrostatic energy E_{es} one has to apply the same approximate treatment of electron–electron interaction as in the case of E_{bs}. We take the Hartree- or Hartree-Fock approximation improved by the inclusion of correlation in a certain sense.

For the electrostatic energy $E_{es}(\{\boldsymbol{R}_i\})$ one can state from general arguments that the $\{\boldsymbol{R}_i\}$-dependent parts of its two contributions, $E_{ii}(\{\boldsymbol{R}_i\})$ and $E_{ee}(\{\boldsymbol{R}_i\})$, will cancel out each other to a large extent. The reason is the nearly complete screening of ion–ion interaction from electrons. The part of electron–electron interaction which does not undergo this cancellation, the interaction of electrons at the same ion, is independent of the atomic structure $\{\boldsymbol{R}_i\}$, but it determines the magnitude of E_{es}. The much smaller structure-dependent part of $E_{es}(\{\boldsymbol{R}_i\})$ is mainly due to the transfer of electronic charge between different ion sites.

2.2 Atomic limit

For the parametrization of the crystal total energy one needs the total energy E_{tot}^{atom} of a free atom with an arbitrary allowed valence electron configuration $s^{n_s}p^{n_p}$. Like in the crystal case one can express $E_{tot}^{atom}(s^{n_s}, p^{n_p})$ as sum over occupied one-electron energy levels $\varepsilon_s(n_s, n_p)$, $\varepsilon_p(n_s, n_p)$, minus the electrostatic energy of electron–electron interaction:

$$E_{tot}^{atom}(s^{n_s}p^{n_p}) = n_s\varepsilon_s(n_s, n_p) + n_p\varepsilon_p(n_s, n_p) - \tfrac{1}{2}\,[n_s^2 U_{ss} + 2n_s n_p U_{sp} + n_p^2 U_{pp}] \,. \tag{3}$$

Here U_{ss}, U_{sp} and U_{pp} are the three intraatomic electron–electron interaction parameters [21]. In the Hartree approximation they represent the Coulomb repulsion between two s electrons, two p electrons, and an s and a p electron, respectively. In the Hartree-Fock approximation the $U_{ll'}$ $(l, l' = \mathrm{s}, \mathrm{p})$ include besides Coulomb integrals also exchange integrals. For practical applications the parameters $U_{ll'}$ are fitted to experimental data. The dependence of one-electron energy levels $\varepsilon_s(n_s, n_p)$, $\varepsilon_p(n_s, n_p)$ on their population exceeds the frame of Hartree or Hartree-Fock approximation, i.e. it accounts for correlation. One can obtain this dependence by means of perturbation theory with respect to the potential change due to the different electronic charge distributions for the actual $n_s - n_p$-configuration and the ground state configuration with $n_s = n_p = 2$. One obtains [22]

$$\varepsilon_l(n_s, n_p) = \varepsilon_l^{atom} + (n_s - 2)\, U_{sl} + (n_p - 2)\, U_{lp} \,, \qquad l = \mathrm{s}, \mathrm{p} \,, \tag{4}$$

with $\varepsilon_l^{\text{atom}}$ the one-electron energies corresponding to the ground state configuration, i.e. in Hartree or Hartree-Fock approximation. Expression (4) is consistent with the electrostatic energy in (3) which, of course, accounts also for correlation. Both expressions (3) and (4) together obey the Janak theorem [23]

$$\varepsilon_l(n_s, n_p) = \frac{\partial E_{\text{tot}}^{\text{atom}}(s^{n_s}, p^{n_p})}{\partial n_l}. \tag{5}$$

The same relation holds within the density-functional theory of total energy [8, 9].

2.3 Band structure energy

To find an analytic expression for the band structure energy we follow Harrison [11, 12] who has presented a reformulation of the old method of linear combinations of atomic orbitals (LCAO). The basic approximation of the LCAO method is the assumption that the one-electron eigenstates can be adequately written in terms of hybrids of atom-like s and p orbitals. For tetrahedrally coordinated crystals four orthogonal and normalized hybrids are needed at a given ion site $\boldsymbol{R}_i$.

Due to the different occupation of orbitals at different ion sites, their energies ε_s^i and ε_p^i as well as their radial wave functions $R_s^i(r)$ and $R_p^i(r)$ will depend on the ion position $\boldsymbol{R}_i$. The j-th hybrid at $\boldsymbol{R}_i$, $h_{ij}(\boldsymbol{x} - \boldsymbol{R}_i)$, is chosen such that it points in the direction of a certain unit vector $\boldsymbol{n}_{ij}$. In general, $\boldsymbol{n}_{ij}$ forms a certain angle $\theta_{ij} \neq 0$ with the direction from $\boldsymbol{R}_i$ to $\boldsymbol{R}_j$ (see Fig. 1). The so defined hybrid h_{ij} is given by the following expression:

$$h_{ij}(\boldsymbol{x} - \boldsymbol{R}_i) = \sqrt{\frac{1}{4\pi}\frac{1}{1 + \lambda_{ij}}}\left[R_s^i(|\boldsymbol{x} - \boldsymbol{R}_i|) + \sqrt{\lambda_{ij}}\,\frac{\boldsymbol{n}_{ij}(\boldsymbol{x} - \boldsymbol{R}_i)}{|\boldsymbol{x} - \boldsymbol{R}_i|}\,R_p^i(|\boldsymbol{x} - \boldsymbol{R}_i|)\right]. \tag{6}$$

The hybridization parameters λ_{ij} characterize the relative s and p portions of a hybrid. Pure atomic s and p states are obtained for $\lambda_{ij} = 0$ and $\lambda_{ij} \to \infty$, respectively. For ion sites $\boldsymbol{R}_i$ at surfaces, less than four next-neighbour positions $\boldsymbol{R}_j$ can occur. Then, if no atom is present at $\boldsymbol{R}_j$, $\boldsymbol{n}_{ij}$ denotes the direction of the dangling hybrid. Note that the ε_h^{ij} of an electron occupying a hybrid of type (6) is given as the on-site matrix element of the crystal Hamiltonian

$$\varepsilon_h^{ij} = \frac{1}{1 + \lambda_{ij}}\,[\varepsilon_s^i + \lambda_{ij}\varepsilon_p^i]. \tag{7}$$

The two hybrids $h_{ij}(\boldsymbol{x} - \boldsymbol{R}_i)$ at $\boldsymbol{R}_i$ and h_{ji} at the nearest neighbour $\boldsymbol{R}_j$ are linearly combined to form bonding states

$$b_{ij}(\boldsymbol{x}) = u_{ij}h_{ij}(\boldsymbol{x} - \boldsymbol{R}_i) + u_{ji}h_{ji}(\boldsymbol{x} - \boldsymbol{R}_j). \tag{8}$$

By means of a variational procedure one obtains

$$u_{ij}^2 = \tfrac{1}{2}\,(1 - \alpha_{ij}) \tag{9}$$

with the corresponding bond energies ε_b^{ij}

$$\varepsilon_b^{ij} = \tfrac{1}{2}\,(\varepsilon_h^{ij} + \varepsilon_h^{ji}) - \sqrt{V_{ij}^2 + W_{ij}^2} + S_{ij}V_{ij}. \tag{10}$$

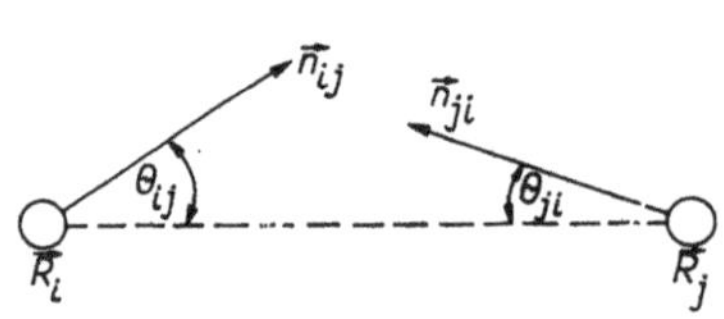

Fig. 1. Binding hybrids located on nearest-neighbour atoms at $\boldsymbol{R}_i$ and $\boldsymbol{R}_j$ and directed along $\boldsymbol{n}_{ij}$ and $\boldsymbol{n}_{ji}$, respectively

The bond polarity α_{ij} is given by

$$\alpha_{ij} = \frac{W_{ij}}{\sqrt{V_{ij}^2 + W_{ij}^2}} \tag{11a}$$

with

$$W_{ij} = \tfrac{1}{2}\,(\varepsilon_{\mathrm{h}}^{ij} - \varepsilon_{\mathrm{h}}^{ji})\,. \tag{11b}$$

The sign of α_{ij} determines the direction of electron transfer along the axis $(\boldsymbol{R}_i - \boldsymbol{R}_j)$. The polarities α_{ij} are zero for bulk samples of covalent materials under consideration. At surfaces, however, the ideal tetrahedron configuration of atoms is destroyed due to relaxation or reconstruction. The ratios of s- and p-parts of the two hybrids forming a chemical bond are generally different. Then the W_{ij} and α_{ij} are no longer zero.

The hybrid covalent energy V_{ij}, defined as the interatomic matrix element of the crystal Hamiltonian between hybrid orbitals of type (6), reads

$$V_{ij} = -\left(\frac{d_0}{|\boldsymbol{R}_i - \boldsymbol{R}_j|}\right)^2 \frac{1}{\sqrt{(1 + \lambda_{ij})\,(1 + \lambda_{ji})}} \times$$
$$\times \{V_{\mathrm{ss}\sigma} - [\sqrt{\lambda_{ij}} \cos\theta_{ij} + \sqrt{\lambda_{ji}} \cos\theta_{ji}]\, V_{\mathrm{sp}\sigma} -$$
$$- \sqrt{\lambda_{ij}\lambda_{ji}} \cos\theta_{ij} \cos\theta_{ji}\, V_{\mathrm{pp}\sigma} + \sqrt{\lambda_{ij}\lambda_{ji}} \sin\theta_{ij} \sin\theta_{ji}\, V_{\mathrm{pp}\pi}\} \tag{12}$$

with $V_{ll'm}$ $(l, l' = \mathrm{s}, \mathrm{p};\ m = \sigma, \pi)$ the corresponding interatomic matrix elements for pure s and p states, taken at the bulk nearest-neighbour distance $d_0 = \sqrt{3}\,a/4$ (a lattice constant). Note that non-zero angles θ_{ij} and θ_{ji} are connected with π-bonding contributions to V_{ij}. Such contributions occur at surfaces and defects [7, 19, 20]. Since we include the latter structures, θ_{ij} and θ_{ji} are allowed to be different from zero. The nearest-neighbour distance d at surfaces differs from d_0 defined in the bulk case. Hence, the scaling of all interatomic matrix elements in (12) with d has to be taken into account. We take the d^{-2} scaling [11], although we are fully aware of its limited applicability [24]. Exponential [13, 25] or more general power laws [26] give sometimes better results. The strength of the d^{-2} scaling is that it explains the d_0^{-2} dependence of certain band gap energies [11] and the d_0^{-5} dependence of the bulk modulus in a quite natural way. The last term in (9) is proportional to the overlap intergral S_{ij} of the two hybrids forming a bond. Owing to the particular form (6) of hybrids, S_{ij} is related to the individual s an p parts, $S_{ll'm}$ $(l, l' = \mathrm{s}, \mathrm{p};\ m = \sigma, \pi)$, by exactly same equation as (12). Only the matrix elements $V_{ll'm}$ must be replaced by $-S_{ll'm}$. The term $S_{ij}V_{ij}$ represents the overlap interaction. It is responsible for forces which prevent the collapse of the crystal against the attractive covalent bonding.

Now the band structure energy can be immediately calculated by using the bond energies $\varepsilon_{\mathrm{b}}^{ij}$ (10). For bulk crystals the bond energies are independent of i, j. The band structure energy is equal to the bond energy times the number of valence electrons. The wave vector dependent part of valence band energies is averaged out. This can can be shown in the bond-orbital or WT-model [11, 12, 27] by writing the band structure energy as the trace of the corresponding secular matrix. One can show that this result can be generalized to crystal surfaces and defects. It reads

$$E_{\mathrm{bs}} = \sum_{i,j} \varepsilon_{\mathrm{b}}^{ij}\,, \tag{13}$$

where the bond energies $\varepsilon_{\mathrm{b}}^{ij}$ have to be replaced by the hybrid energies $\varepsilon_{\mathrm{h}}^{ij}$ if no binding partner exists at a given *site* $\boldsymbol{R}_i$ in a given direction $\boldsymbol{n}_{ij}$.

2.4 Electrostatic energy

Expression (2) for the electrostatic energy can be written in the more explicit form

$$E_{\text{es}} = \frac{e^2}{2} \left\{ \sum_{\substack{i,i' \\ (i \neq i')}} \frac{z_i z_{i'}}{|\boldsymbol{R}_i - \boldsymbol{R}_{i'}|} - \int d^3\boldsymbol{x} \int d^3\boldsymbol{x}' \, \frac{n(\boldsymbol{x}) \, n(\boldsymbol{x}')}{|\boldsymbol{x} - \boldsymbol{x}'|} \right\}. \tag{14}$$

The density of valence electrons $n(\boldsymbol{x})$ is approximated within the hybrid picture, i.e., the bonding states b_{ij} forming the true eigenfunctions of the crystal Hamiltonian are replaced by hybrids h_{ij}. One gets

$$n(\boldsymbol{x}) = \sum_{i,j} (1 - \alpha_{ij}) \, |h_{ij}(\boldsymbol{x} - \boldsymbol{R}_i)|^2 \,. \tag{15}$$

This is sufficient for the calculation of both intraatomic and interatomic electron–electron interaction energies. Indeed, since the localization radii of wave functions $R_s(r)$ and $R_p(r)$ are smaller than the nearest-neighbour distance d, the point-charge approximation can be applied to all interatomic energies. The intraatomic energies can be expressed in terms of atomic Coulomb integrals U_{ss}, U_{sp}, and U_{pp} assuming that their changes due to the formation of the crystal are negligible. In evaluating (14) for E_{es} we take the mean value of the four hybrids located at a given atom by using the relation

$$\sum_{j=1}^{4} \frac{1}{1 + \lambda_{ij}} = 1 \,. \tag{16}$$

This relation indicates that the sum of the four hybrids at each atom includes one s orbital and three p orbitals as in the case of ideal tetrahedra although the ratio of s- and p-parts of an individual hybrid may differ from 1:3. Finally one gets

$$E_{\text{es}} = E^0_{\text{intra}} - E^{Qi}_{\text{intra}} - E_{\text{Mad}} \,, \tag{17}$$

$$E^0_{\text{intra}} = -8 \sum_i U \,, \tag{18a}$$

$$E^{Qi}_{\text{intra}} = \tfrac{1}{2} \sum_i Q_i^2 U \,, \tag{18b}$$

$$E_{\text{Mad}} = \frac{1}{2} \, e^2 \sum_{\substack{i,i \\ (i = i')}} \frac{Q_i Q_{i'}}{|\boldsymbol{R}_i - \boldsymbol{R}_{i'}|} \,, \tag{18c}$$

with the effective intraatomic electron–electron interaction parameter

$$U = \tfrac{1}{16} \left[U_{ss} + 6U_{sp} + 9U_{pp} \right] \tag{19}$$

and the effective ion charges

$$Q_i = \sum_{j=1}^{4} \alpha_{ij} \,. \tag{20}$$

Contributions to (17) linear in the effective ion charge cancel out each other because of the charge neutrality $\sum_i Q_i = 0$ of the crystal.

The first contribution to the electrostatic energy (17), E^0_{intra}, characterizes the intraatomic electron–electron interaction of non-bonded atoms in s^1p^3 configuration. It is independent of the positions of atoms and plays, therefore, no role for the determination of the crystal structure by means of total-energy minimization. It gives, however, a considerable contribution to the absolute values of the total energy. The second term in (17), E^{Qi}_{intra}, describes the change of the intraatomic electron–electron interaction due to electron transfer between atoms as a result of the ionicity of bonding. The last term (17), E_{Mad}, represents the Coulomb interaction of cations and anions with an effective amount of charge equal to eQ_i. For bulk samples of our covalent material it holds $Q_i = 0$ for all i. The Madelung energy vanishes. At surfaces or defects, however, bonds are partially ionic and the corresponding Q_i differ from zero. Hence, the Madelung term gives a non-zero contribution to (17).

2.5 Self-consistency

In Section 2.2 we have seen that one-particle energies of free atoms show a linear dependence on the population numbers of the corresponding orbitals. The same happens for atoms forming a crystal. However, the presence of long-range Coulomb potentials due to the surrounding ions makes this dependence somewhat more complicated. The orbital energies at a given atom depend not only on its own charge state, but also on that of its neighbours. Within the same approximations done in the calculation of the electrostatic energy (17) the occupation dependence of the hybrid energy $\varepsilon_{\text{h}}^{ij}$ from (7) can be written

$$\varepsilon_{\text{h}}^{ij} = \frac{\varepsilon_{\text{s}} + \lambda_{ij}\varepsilon_{\text{p}}}{1 + \lambda_{ij}} - Q_i U - \sum_{\substack{i' \\ (i \neq i')}} \frac{Q_{i'} e^2}{|\boldsymbol{R}_i - \boldsymbol{R}_{i'}|}, \tag{21}$$

where

$$\varepsilon_l = \varepsilon_l^{\text{atom}} - U_{\text{s}l} + U_{l\text{p}}; \qquad l = \text{s, p} \tag{22}$$

denote the one-electron energies of a free atom in s^1p^3 configuration in accordance with (4). Similar expressions for one-particle energies are used in other surface calculations (see [28] and references therein). Expression (21) for hybrid energies indicates two effects of the particular atomic structure. The first, direct one, is due to the hybridization parameters λ_{ij}. The second, indirect one, is caused by the effective ion charges Q_i. The latter effect has an important implication. Since the effective charges Q_i are in turn functions of hybrid energies $\varepsilon_{\text{h}}^{ij}$, due to (10), (12), and (20), the two quantities have to be calculated self-consistently. In practice, self-consistency is reached by means of a retarded iteration. In surface calculations the only problem is the number of atomic layers which are taken into account. Whereas the geometrical relaxation or reconstruction is restricted to a few atomic layers the charge self-consistency influences also deeper layers.

The Q_i dependence of hybrid energies makes also the band structure energy (13) a function of Q_i. It is useful to write this function in the form of two terms, a Q_i-independent term denoted by E_{bs}^0, and a Q_i-dependent rest. By means of (10), (20), and (21) and by using the same approximations as in Section 2.4 one gets

$$E_{\text{bs}} = E_{\text{bs}}^0 + 2E_{\text{intra}}^{Qi} + 2E_{\text{Mad}} + E_{\text{scc}} \tag{23}$$

with E_{intra}^{Qi} and E_{Mad} from (18b) and (18c), respectively, and the self-consistency correction E_{scc} given by

$$E_{\text{scc}} = \sum_i \sum_{j=1}^{4} \left\{ \sqrt{V_{ij}^2 + W_{ij}^{02}} - \frac{V_{ij}^2}{\sqrt{V_{ij}^2 + W_{ij}^2}} \right\}, \tag{24}$$

W_{ij}^0 means W_{ij} from (11b) taken at $Q_i = 0$. Summarizing (1), (17), and (23) the total energy of the crystal can be written as the sum of five well defined energies:

$$E_{\text{tot}} = E_{\text{bs}}^0 + E_{\text{intra}}^0 + E_{\text{intra}}^{Qi} + E_{\text{Mad}} + E_{\text{scc}}\,. \tag{25}$$

Expression (25) shows the expected limiting behaviour for pure covalent and pure ionic bonding, respectively [29]. In the covalent case, i.e. $|W_{ij}| \ll V_{ij}$, E_{tot} is given by E^0_{bs} apart from the structure-independent part of E^0_{intra}. In the ionic case, i.e. $|W_{ij}| \gg V_{ij}$, the structure-dependent part of E_{tot} is given by $E^{Qi}_{\text{intra}} + E_{\text{Mad}}$. The self-consistency condition affects only the last three terms in (25) which are only present for non-vanishing ionicity of bonding. E_{scc} and E^{Qi}_{intra} make ionic charge transfer unfavourable whereas E^0_{bs} and E_{Mad} have the opposite effect.

3. Determination of Parameters

3.1 *Hybridization parameters*

Deviations of the atomic positions from those of an ideal tetrahedron structure cause changes of the hybridization parameters λ_{ij} of an individual hybrid with respect to the bulk value $\lambda_{ij} = 3$. The new values λ_{ij} can be calculated by means of the orthonormalization conditions for the four hybrids located on one atom. With the hybrids from (6) we get the six independent equations

$$1 + \sqrt{\lambda_{ij}\lambda_{ik}}\, \gamma^{i}_{jk} = 0\,, \qquad \gamma^{i}_{jk} = (\boldsymbol{n}_{ij} \cdot \boldsymbol{n}_{ik})\,; \qquad j, k = 1, 2, 3, 4 \quad (j \neq k)\,. \tag{26}$$

If the directions $\boldsymbol{n}_{ij}$ of three of the four hybrids are given (see Fig. 2) one obtains from (26) the four hybridization parameters λ_{ij} and the direction $\boldsymbol{n}_{ij}$ of the fourth hybrid. The solutions are

$$\lambda_{ij} = -\frac{\gamma^{i}_{kl}}{\gamma^{i}_{jk}\gamma^{i}_{lj}}\,; \qquad j, k, l = 1, 2, 3, 4 \quad (j \neq k, l, k \neq l)\,, \tag{27}$$

$$\boldsymbol{n}_{i4} = \sum_{j=1}^{3} c_{ij}\boldsymbol{n}_{ij}\,; \tag{28a}$$

$$\begin{aligned} c_{ij} = {} & \left(1 + \gamma^{i}_{12}\gamma^{i}_{13}\gamma^{i}_{23} - (\gamma^{i2}_{12} + \gamma^{i2}_{13} + \gamma^{i2}_{23})\right)^{-1/2} \left(\gamma^{i2}_{12}\gamma^{i2}_{13} + \gamma^{i2}_{12}\gamma^{i2}_{23} + \gamma^{i2}_{13}\gamma^{i2}_{23} - {}\right. \\ & \left. - 2\gamma^{i}_{12}\gamma^{i}_{13}\gamma^{i}_{23}(\gamma^{i2}_{12} + \gamma^{i2}_{13} + \gamma^{i2}_{23}) + 3(\gamma^{i}_{12}\gamma^{i}_{13}\gamma^{i}_{23})^2\right)^{-1/2} \times \\ & \times (-\gamma^{i}_{kl}) \left[\gamma^{i}_{12}\gamma^{i}_{13}\gamma^{i}_{23} - (\gamma^{i2}_{12}\gamma^{i2}_{13}\gamma^{i2}_{23} - (\gamma^{i2}_{12} + \gamma^{i2}_{23} + \gamma^{i2}_{13})) + \gamma^{i2}_{kl} + \frac{\gamma^{i2}_{jk}\gamma^{i2}_{jl}}{\gamma^{i2}_{kl}}\right]; \\ & j, k, l = 1, 2, 3 \quad (j \neq k \neq l)\,. \end{aligned} \tag{28b}$$

In the bulk case all $\boldsymbol{n}_{ij}$ are directed to the corners of an ideal tetrahedron, i.e. $\gamma^{i}_{jk} = -\frac{1}{3}$ and $\lambda_{ij} = 3$ independent of i, j. All hybrids are equivalent, have the same energy $\varepsilon^{ij}_{\text{h}} = (\varepsilon_{\text{s}} + 3\varepsilon_{\text{p}})/4$, and form therefore only covalent bonds. For the interaction and overlapping matrix elements of two hybrids on neighbouring atoms it holds, respectively, $V_{ij} = V_2(d_0/d)^2$ and $S_{ij} = S(d_0/d)^2$. The corresponding bulk quantities at

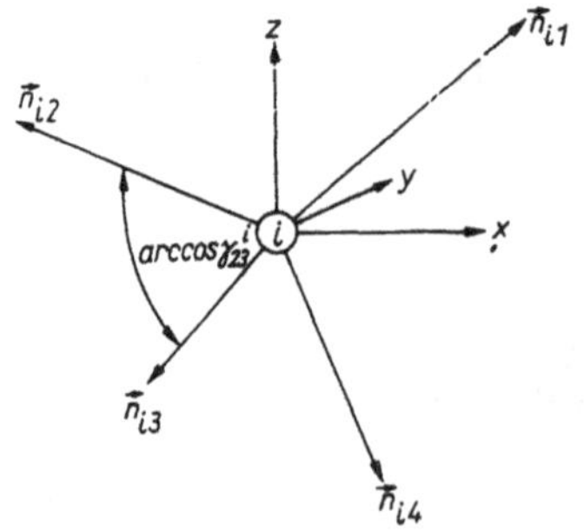

Fig. 2. Directions of the four hybrids on a certain atom i

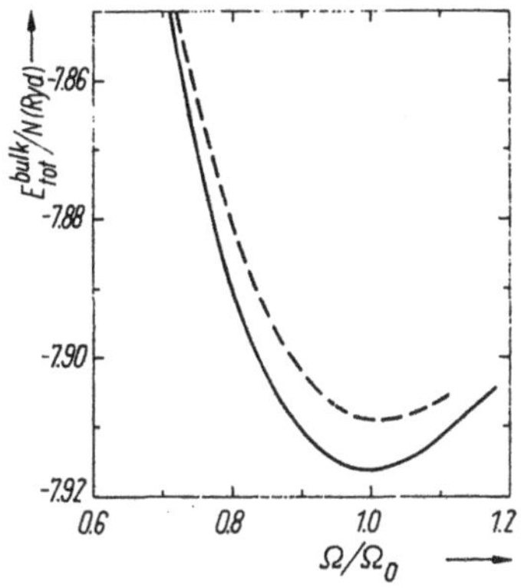

Fig. 3. Bulk total energy per atom, E_{tot}^{bulk}/N, for silicon calculated by means of (29) and parameters listed in Table 2 versus crystal volume (solid line). For comparison results of self-consistent pseudopotential calculations [1] are also shown (dashed line)

the equilibrium nearest-neighbour distance d_0 are defined as $V_2 = -(V_{ss\sigma} - 2\sqrt{3}\,V_{sp\sigma} - 3V_{pp\sigma})/4$ and $S = (S_{ss\sigma} - 2\sqrt{3}\,S_{sp\sigma} - 3S_{pp\sigma})/4$, respectively. For the total energy of a bulk crystal having diamond structure and consisting of N atoms then it follows from (25)

$$E_{tot}^{bulk} = N\left\{\varepsilon_s + 3\varepsilon_p - 4V_2\left(\frac{d_0}{d}\right)^2 + 4SV_2\left(\frac{d_0}{d}\right)^4 - 8U\right\}. \tag{29}$$

The d-dependence of the total energy (29) can be transformed into a dependence on volume Ω by means of $(d_0/d)^3 = \Omega_0/\Omega$. For silicon a plot of the Ω-dependence is shown in Fig. 3. The used parameters are listed in Tables 1 and 2. For comparison the result of a self-consistent pseudopotential calculation [1] is reproduced.

Table 1

Experimental data for covalent semiconductors taken as input for the determination of tight-binding parameters in total-energy calculations: Bulk nearest-neighbour distance d_0 [30], bulk moduli B [29, 30], first four atomic ionization potentials I_1, I_2, I_3, I_4 [31], cohesive energies for bulk materials E_{coh} [29], and bulk total energies per atom (core electrons excluded) E_{tot}^{bulk}/N [29, 31]. Since the valence electron contribution to the total energy is not directly observable we define, according to the usual measurement procedure, $-E_{tot}^{bulk}/N = \sum_{i=1}^{4} I_i + E_{coh}$ as the sum of the first four atomic ionization potentials increased by the cohesive energy

parameter	C	Si	Ge	α-Sn
d_0 (nm)	0.154	0.235	0.244	0.280
B (Mbar)	4.42	0.978	0.772	0.531
I_1 (eV)	11.264	8.149	7.88	7.342
I_2 (eV)	24.376	16.34	15.93	14.628
I_3 (eV)	47.864	33.46	34.21	30.49
I_4 (eV)	64.476	45.13	45.70	40.72
E_{coh} (eV)	7.36	4.64	3.87	3.12
E_{tot}^{bulk}/N (Ryd)	−11.418	− 7.918	− 7.908	− 7.077

Table 2

Tight-binding parameters calculated by means of experimental data listed in Table 1 (all values in eV): Covalent energy V_2 from (31), interatomic matrix elements $V_{ll'm}$ from (32), electron–electron interaction constants $U_{ll'}$ and orbital energies ε_p^{atom}, ε_s^{atom} of free atoms in s^2p^2 configuration from (34), (35), orbital energies ε_p, ε_s of atoms in s^1p^3 configuration from (22), and hybrid–hybrid interaction parameter V_1 from (35)

parameter	C	Si	Ge	α-Sn
V_2	8.73	6.86	6.06	6.30
$V_{ss\sigma}$	− 2.79	− 2.19	− 1.94	− 2.02
$V_{sp\sigma}$	3.67	2.89	2.55	2.65
$V_{pp\sigma}$	6.47	5.08	4.49	4.67
$V_{pp\pi}$	− 1.61	− 1.27	− 1.12	− 1.17
U_{ss}	16.61	11.67	11.49	10.23
U_{sp}	13.20	8.14	8.71	6.98
U_{pp}	13.11	8.19	8.05	7.29
U	13.36	8.39	8.51	7.36
$-\varepsilon_p^{atom}$	4.71	4.05	3.86	3.70
$-\varepsilon_s^{atom}$	13.15	11.34	11.05	11.41
$-\varepsilon_p$	4.80	4.00	4.52	3.39
$-\varepsilon_s$	16.56	14.87	13.67	14.66
V_1	2.11	1.82	1.80	1.93

3.2 Hybrid covalent energy V_2 and overlap integral S

The conditions to be imposed on E_{tot}^{bulk} from (29) at equilibrium are

$$\left.\frac{\partial E_{tot}^{bulk}}{\partial \Omega}\right|_{equ} = 0\,, \qquad \Omega\left.\frac{\partial^2 E_{tot}^{bulk}}{\partial \Omega^2}\right|_{equ} = B\,. \tag{30}$$

Here B is the bulk modulus. Equilibrium means that the nearest-neighbour distance d equals d_0. Applying the simple expression (29) these conditions require the validity of the relations

$$S = \frac{1}{2}\,, \qquad V_2 = \frac{\sqrt{3}}{2}\, d_0^3 B\,, \tag{31}$$

which fix the overlap integral S as well as the interatomic hybrid interaction V_2 for undistorted crystals as far as experimental values of the bulk modulus B are known (see Table 1). In the case of surfaces and defects, however, we need a greater variety of parameters, $S_{ll'm}$ and $V_{llm'}$ with l, l', and m as the appropriate angular-momentum quantum numbers ($l, l' = $ s, p; $m = \sigma, \pi$). The simplest possibility to obtain these quantities is the application of a certain scaling procedure like

$$V_{ll'm} = \varkappa_{ll'm} V_2\,. \tag{32}$$

In the case of $V_{ll'm}$ matrix elements such a scaling has been made by Harrison [11]. He finds $\varkappa_{ss\sigma} = -0.320$, $\varkappa_{sp\sigma} = 0.421$, $\varkappa_{pp\sigma} = 0.741$, and $\varkappa_{pp\pi} = -0.185$. For the overlap integrals $S_{ll'm}$ no analogous scaling is available from the literature. One can apply, however, the relation between the Hamiltonian matrix elements $V_{ll'm}$ and the overlap integrals $S_{ll'm}$ from the extended Hückel theory of quantum chemistry [32]

$$V_{ll'm} = K \frac{\varepsilon_l + \varepsilon_{l'}}{2} S_{ll'm} , \tag{33}$$

where the Wolfsberg-Helmholtz constant K is adjusted so that $S = 0.5$ holds in the case of sp^3 hybrids.

The bulk interatomic Hamiltonian matrix elements V_2 and $V_{ll'm}$ calculated by means of (31) and (32) are listed in Table 2. Their values are generally somewhat larger (10% for Si) than those used in band structure calculations [11, 18]. The magnitude of $V_{ll'm}$ would be somewhat decreased if second-nearest-neighbour interactions, particularly the interaction of each bond with the six surrounding antibonds would be taken into account. This interaction gives rise to additional ppπ contribution to B and, hence, reduces the values of V_2 [14].

3.3 Intraatomic electron–electron interaction parameters and orbital energies

Atomic ionization potentials I_i ($i = 1, \dots 4$) for valence electrons are defined as differences of total energies of free atoms with different valence electron configurations. Applying expression (3) one obtains for the first four ionization potentials

$$\begin{aligned} I_1 &= E_{\text{tot}}^{\text{atom}}(s^2p^1) - E_{\text{tot}}^{\text{atom}}(s^2p^2) = -(\varepsilon_p^{\text{atom}} - U_{pp}) - \tfrac{1}{2} U_{pp} , \\ I_2 &= E_{\text{tot}}^{\text{atom}}(s^2p^0) - E_{\text{tot}}^{\text{atom}}(s^2p^1) = -(\varepsilon_p^{\text{atom}} - 2U_{pp}) - \tfrac{1}{2} U_{pp} , \\ I_3 &= E_{\text{tot}}^{\text{atom}}(s^1p^0) - E_{\text{tot}}^{\text{atom}}(s^2p^0) = -(\varepsilon_s^{\text{atom}} - 2U_{sp} - U_{ss}) - \tfrac{1}{2} U_{ss} , \\ I_4 &= E_{\text{tot}}^{\text{atom}}(s^0p^0) - E_{\text{tot}}^{\text{atom}}(s^1p^0) = -(\varepsilon_s^{\text{atom}} - 2U_{sp} - 2U_{ss}) - \tfrac{1}{2} U_{ss} . \end{aligned} \tag{34}$$

In each case the first term on the right hand side of (34) represents the orbital energy (4) of the electron lifted from the corresponding valence electron shell. The second term describes the Koopmans correction known also as relaxation energy [33]. In a similar way as the potential (34) one obtains the cohesive energy E_{coh} per atom of a given covalent solid. It is defined as the difference of the total energy $E_{\text{tot}}^{\text{atom}}(s^2p^2)$ from (3) of a free atom in the s^2p^2 valence electron configuration and the bulk total energy per atom $E_{\text{tot}}^{\text{bulk}}/N$ from (29) at equilibrium ($d = d_0$, $S = 0.5$). Then, using relation (22) between orbital energies in s^2p^2 and s^1p^3 configuration as well as definition (19) it follows

$$E_{\text{coh}} = E_{\text{tot}}^{\text{atom}}(s^2p^2) - \frac{E_{\text{tot}}^{\text{bulk}}\big|_{\text{equ}}}{N} = 2V_2 - 4V_1 - \frac{1}{2}[U_{ss} - 2U_{sp} + U_{pp}];$$

$$V_1 = \frac{\varepsilon_p^{\text{atom}} - \varepsilon_s^{\text{atom}}}{4} . \tag{35}$$

V_1 characterizes the s–p splitting in the free atom and is usually identified with the Hamiltonian matrix element between two different hybrids located on the same atom. The structure of formula (35) for the cohesive energy per atom is quite obvious. The strongest contribution, $2V_2$, comes from the covalent bonding of the four valence electrons per atom. It is reduced by the s–p promotion energy $4V_1$. The additional reduction of the bonding energy by $(U_{ss} - 2U_{sp} + U_{pp})/2$ is due to changes of electronic correlation within the valence electron shells with s^2p^2 and s^1p^3 configurations. The influence of the electron–electron interaction on the cohesive energy is small ($\leqq V_1$), in contrast to the total energy itself where it gives the main contribution.

With the covalent energy V_2 known from (31), expressions (34) and (35) represent five equations for the five unknown parameters U_{ss}, U_{sp}, U_{pp}, ε_p^{atom}, and ε_s^{atom}. By means of experimental ionization potentials I_i ($i = 1, \dots, 4$) and cohesive energies listed in Table 1, the intraatomic electron–electron interaction constants and the orbital energies may now be evaluated. The results are shown in Table 2. The values for U_{ss} and U_{pp} are exactly the same as found by Sankey and Dow [34]. They are solely determined by the ionization potentials. The resulting U_{sp} are somewhat smaller than those from [34] since they are fitted also to the cohesive energy and not only to the atomic s–p splitting. The average values U from (19) for the s^1p^3 configuration in Table 2 are comparable with the data by Harrison [35] who used the asymptotic form of Coulomb integrals.

For investigations of semiconductor surfaces and defects by means of total energies, absolute positions of orbital energies ε_p^{atom}, ε_s^{atom} and ε_p, ε_s, respectively, are not needed. However, the determination of these energies in a way that expression (29) reproduces experimental values of total energies per atom E_{tot}^{bulk}/N, as listed in Table 1, is an interesting test of our fitting procedure. The extracted values $-\varepsilon_p^{atom}$ or $-\varepsilon_p$ are much smaller than the corresponding energies obtained in atomic Hartree-Fock calculations [36], but they approach the solutions of the Kohn-Sham equations for atoms with s^2p^2 as well as s^1p^3 configuration [1]. The s-orbital energies $-\varepsilon_s^{atom}$ show a similar behaviour whereas the values $-\varepsilon_s$ are closer to the Hartree-Fock energies. The reason is obvious. Since the U-parameters are fitted to experimental ionization potentials, correlation effects are included like in the density-functional theory. Hence, deviations from Hartree-Fock data are to be expected. The s–p splitting for free atoms in s^2p^2 configuration is only weakly influenced by correlation effects, contrary to the case of s^1p^3 configuration. Therefore, the values of V_1 in (35) are nearly the same as used in band structure calculations [11, 18]. Summarizing the results of our adjustment procedure we conclude that total energies of bulk covalent semiconductors can be predicted within a tight-binding theory which takes electron–electron interaction into account in a proper way. By using appropriate parameters ε_l^{atom}, $V_{ll'm}$, $S_{ll'm}$ and $U_{ll'}$ ($l, l' = $ s, p; $m = \sigma, \pi$) we were able to derive equilibrium values of total energies and their variations with crystal volume. Our total energy expression can be also applied to crystal surfaces and defects and should be capable for the optimization of their atomic structure.

4. Application: Relaxed (111) 1 × 1 Surface

The atomic structure of a crystal surface is derived by finding a (local) minimum of the total energy with respect to the configuration $\{\boldsymbol{R}_i\}$. This corresponds to the force theorem due to Ehrenfest [37] which now is often referred to as Hellmann-Feynman theorem: The generalized forces $\boldsymbol{F}_i = -\nabla_{\boldsymbol{R}_i} E_{tot}(\{\boldsymbol{R}_i\})$ on all atoms i have to vanish simultaneously. In real surface calculations one assumes that only a finite number of atomic layers respond to the non-existence of atoms in one half-space whereas atoms in deeper layers remain at their bulk positions. To take advantage of this assumption we consider the difference of total energies of two halfspaces, the first one filled with atoms at relaxed positions $\boldsymbol{R}_i^{bulk} + \Delta\boldsymbol{R}_i$ in surface layers and with ideal bulk positions $\boldsymbol{R}_i^{bulk}$ elsewhere, and the second one filled with atoms at ideal bulk positions throughout.

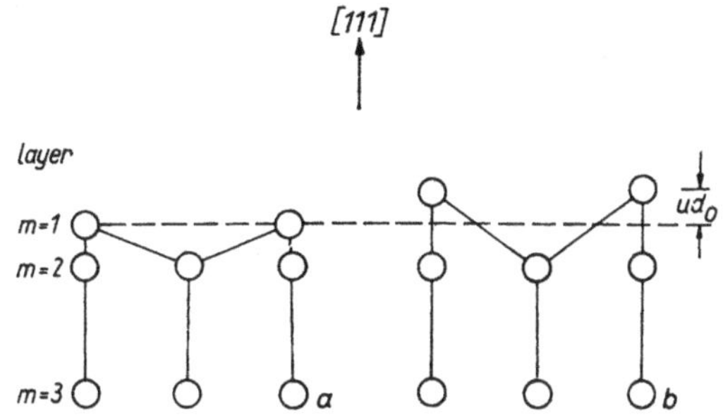

Fig. 4. Side view of the (111) 1 × 1 surface. The first three atomic layers are shown, a) unrelaxed, b) partially relaxed (first layer displaced by ud_0 in the surface normal direction)

Only atoms from surface layers enter this difference. It reads

$$\Delta E_{\text{tot}}(\{\Delta \boldsymbol{R}_i\}) = E_{\text{tot}}(\{\boldsymbol{R}_i^{\text{bulk}} + \Delta \boldsymbol{R}_i\}) - E_{\text{tot}}(\{\boldsymbol{R}_i^{\text{bulk}}\}) \,. \tag{36}$$

The displacements $\Delta \boldsymbol{R}_i$ are varied until $\Delta E_{\text{tot}}(\{\Delta \boldsymbol{R}_i\})$ reaches its minimum at $\Delta \boldsymbol{R}_i = \Delta \boldsymbol{R}_i^0$. Then $-\Delta E_{\text{tot}}(\Delta \boldsymbol{R}_i^0)$ devided by the number of surface atoms is the energy gain per surface atom due to relaxation.

In calculating the total energy of a crystal surface, dangling bonds have to be considered besides of bonds. The unreconstructed, unrelaxed (111) 1 × 1 surface has one dangling bond per first layer (see Fig. 4). One gets

$$E_{\text{tot}}(\{\boldsymbol{R}_i\}) = N_{||}\{\tfrac{1}{4}(\varepsilon_s + 3\varepsilon_p) + 3[\tfrac{1}{4}(\varepsilon_s + 3\varepsilon_p) - \tfrac{1}{2}V_2] - 8U + \\ + 4\sum_{m=1}^{M}[\tfrac{1}{4}(\varepsilon_s + 3\varepsilon_p) - \tfrac{1}{2}V_2 - 2U]\} \,, \tag{37}$$

where $N_{||}$ is the number of surface elementary cells and M the number of atomic layers under the surface which are taken into account.

In the simplest version of surface relaxation only first layer atoms are displaced, and the displacement ud_0 is assumed to be perpendicular to the surface, i.e. in [111] direction. The three hybrids of both first layer and second layer atoms entering the back bonds are dehybridized and twisted in such a way that they are directed to each other within the bonds. Hybrids of atoms in the third layer as well as in deeper layers are assumed to remain bulk-like. Then relaxation influences only the dangling bonds, the back bonds and the bonds between second and third layer atoms, as far as the self-consistent calculation of atomic charges Q_i according to (10), (12), (20), and (21) show, do not exceed the third layer. The dangling hybrids will be s-like for the largest possible value of u, $u = \frac{1}{3}$, and p-like for the smallest one $u = -\frac{1}{3}$. The back bonds remain covalent. Only their covalent energy changes. Bonds between second and third layer atoms become partially ionic. On gets alternatingly charged atomic layers. The effect of this dipole layer on the total energy is, however, small since its Madelung energy vanishes.

In Fig. 5 we show the change of total energy per surface atom, $\Delta E_{\text{tot}}(u)/N_{||}$, as a function of the relaxation displacement u for C, Si, Ge, and α-Sn. Positive values of u correspond to outward relaxation and negative u to inward relaxation. In all cases the total energy change reaches a pronounced minimum at negative atomic displacements u_0. The value of $-u_0$ is nearly constant from C to α-Sn, whereas the energy gain $-\Delta E_{\text{tot}}(u_0)/N_{||}$ decreases in this sequence. The energy gain follows the magnitude of the covalent energy (see Table 3). The atomic displacement u_0, on the other hand,

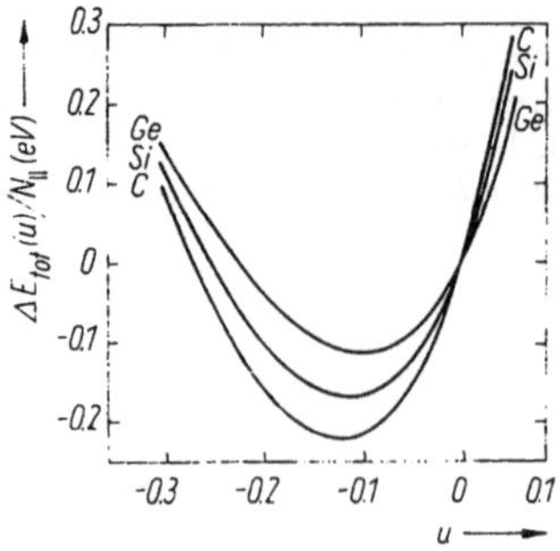

Fig. 5. Total energy change per surface atom for relaxed (111) 1 × 1 surfaces of covalent semiconductors versus first-layer displacement u. The curve for α-Sn coincides with that for Si

Table 3

First-layer atom displacement u_0 and energy gain per surface atom $-\Delta E_{tot}(u_0)/N_{||}$ for relaxed (111) 1 × 1 surfaces

semiconductor	$-u_0$	$-\Delta E_{tot}/N_{\|\|}$ (eV)
C	0.12	0.22
Si	0.11	0.17
Ge	0.10	0.11
α-Sn	0.11	0.17

depends mainly on the balance of covalent energy V_{ij} and overlap interaction $-S_{ij}V_{ij}$ in the back bonds. These energies have an opposite tendency so that u_0 remains nearly constant. Our results for silicon correspond to those of pseudopotential calculations [2, 38, 39] and LEED studies [40, 41] which yield contractions between 0.008 and 0.015, and 0.012 and 0.016 nm, respectively. Of course, our value of 0.024 nm is somewhat larger. On the other hand, other tight-binding calculations [16] predict much smaller relaxations of 0.004 to 0.006 nm. Chadi's total energy minimization procedure [42] indicates that the Si(111) 1 × 1 surface remains unrelaxed. Despite of its simplicity, our version of tight-binding theory with regard of electron–electron interaction seems to be accurate enough to predict surface relaxation and other structural changes. The method should be applicable to a wide range of structural problems. In a forthcoming paper we will use it for a reinvestigation of the various models of (2 × 1) reconstructed (111) and (100) surfaces of covalent semiconductors.

References

[1] M. T. Yin and M. L. Cohen, Phys. Rev. Letters **45**, 1004 (1980); Phys. Rev. B **26**, 5668; 3259 (1982).

[2] J. Ihm and M. L. Cohen, Solid State Commun. **29**, 711 (1979); Phys. Rev. B **21**, 1527 (1980).

[3] J. E. Northrup, J. Ihm, and M. L. Cohen, Phys. Rev. Letters **47**, 1910 (1981).

[4] O. H. Nielsen, R. M. Martin, D. J. Chadi, and K. Kunc, J. Vacuum Sci. Technol. B **1**, 714 (1983).

[5] J. E. Northrup and M. L. Cohen, Phys. Rev. B **29**, 1966 (1984).

[6] D. Vanderbilt and S. G. Louie, Phys. Rev. B **29**, 7099 (1984).

[7] J. E. Northrup, Phys. Rev. Letters **54**, 815 (1985).

[8] P. Hohenberg and W. Kohn, Phys. Rev. **136**, B 864 (1964).

[9] W. Kohn and L. J. Sham, Phys. Rev. **140**, A 1133 (1965).

[10] J. IHM, A. ZUNGER, and M. L. COHEN, J. Phys. C 12, 4409 (1979).
[11] W. A. HARRISON, Electronic Structure and the Properties of Solids, Freeman, San Francisco 1980.
[12] W. A. HARRISON, Phys. Rev. B 23, 5230 (1981).
[13] M. LANNOO, J. Physique 40, 461 (1979).
[14] J. M. BARANOWSKI, J. Phys. C 17, 6287 (1984).
[15] W. A. HARRISON, Surface Sci. 55, 1 (1976).
[16] K. MASUDA, phys. stat. sol. (b) 107, 529 (1981).
[17] K. SUZUKI and T. HOSHINO, J. Phys. Soc. Japan 49, 1055 (1980).
[18] D. J. CHADI, Phys. Rev. Letters 15, 1062 (1978); J. Vacuum Sci. Technol. 16, 1290 (1979); Phys. Rev. B 19, 2074 (1979); 29, 785 (1984).
[19] D. J. CHADI, Phys. Rev. B 26, 4762 (1982).
[20] K. C. PANDEY, Phys. Rev. Letters 47, 1913 (1981); 49, 223 (1982).
[21] P. W. ANDERSON, Phys. Rev. 124, 41 (1961).
[22] J. C. SLATER, The Self-Consistent Field for Molecules and Solids, McGraw-Hill Publ. Co., New York 1974.
[23] J. F. JANAK, Phys. Rev. B 18, 7165 (1978).
[24] S. Y. REN and W. A. HARRISON, Phys. Rev. B 23, 762 (1981).
[25] K. C. PANDEY and J. C. PHILLIPS, Phys. Rev. B 13, 750 (1976).
[26] L. BREY and J. TEJEDOR, Solid State Commun. 48, 403 (1983).
[27] D. WEAIRE and M. F. THORPE, Phys. Rev. B 4, 2508 (1971).
[28] C. CALANDRA, F. MANGHI, and C. M. BERTONI, J. Phys. C 10, 1911 (1977).
[29] C. KITTEL, Introduction to Solid State Physics, Wiley, New York 1966.
[30] K.-H. HELLWEGE (Ed.). Landolt-Börnstein, New Series, Group III, Vol. 17, Subvolume a, Springer-Verlag, Berlin/Heidelberg/New York 1982.
[31] C. E. MOORE, Atomic Energy Levels, Vol. 1 to 3, NBS Circular 467, 1971.
[32] R. HOFFMANN, J. chem. Phys. 39, 1397 (1963).
[33] F. BECHSTEDT, phys. stat. sol. (b) 112, 9 (1982).
[34] O. F. SANKEY and J. D. DOW, Phys. Rev. B 27, 7641 (1983).
[35] W. A. HARRISON, Phys. Rev. B 31, 2121 (1985).
[36] C. F. FISHER, Atomic Data 4, 301 (1972).
[37] P. EHRENFEST, Z. Phys. 45, 455 (1927).
[38] A. REDONDO, W. A. GODDARD, T. C. MCGILL, and G. T. SURRATT, Solid State Commun. 20, 733 (1976).
[39] J. IHM, M. L. COHEN, and D. J. CHADI, Phys. Rev. B 21, 4592 (1980).
[40] H. D. SHIH, F. JONA, D. W. JEPSEN, and P. M. MARCUS, Phys. Rev. Letters 37, 1622 (1976).
[41] J. V. FLORIO and D. W. ROBERTSON, Surface Sci. 24, 173 (1971).
[42] D. J. CHADI, Phys. Rev. Letters 41, 1062 (1978).

(Received August 7, 1985)

phys. stat. sol. (b) **131**, 659 (1983)

Subject classification: 13.1 and 20.1; 21.1; 21.6

Department of Physics, University of California, Los Angeles[1])

Wavelength Modulated Spectra of α-$Ag_{0.7}Zn_{0.3}$ near the Optical Absorption Edge [2])

By

R. STEARNS[3]), R. BRAUNSTEIN, and L. MULDAWER[4])

Dedicated to Prof. Dr. Dr. h. c. Dr. E. h. P. GÖRLICH on the occasion of his 80th birthday

The optical properties of α-phase $Ag_{0.7}Zn_{0.3}$ are determined near the optical absorption edge. This alloy composition dampens the sharp Ag-plasma resonance. The band shifts which are observed are in qualitative agreement with calculations which are performed for the similar alloy, α-$Cu_{0.7}Zn_{0.3}$.

Die optischen Eigenschaften der α-Phase von $Ag_{0.7}Zn_{0.3}$ werden in der Nähe der optischen Absorptionskante bestimmt. Diese Legierungszusammensetzung dämpft die scharfe Ag-Plasma-Resonanz. Die beobachteten Bandverschiebungen sind in qualitativer Übereinstimmung mit Berechnungen, die für die ähnliche Legierung α-$Cu_{0.7}Zn_{0.3}$ durchgeführt wurden.

1. Introduction

An outstanding feature of the optical properties of Ag is the sharp onset of interband transitions near 4.0 eV in the near ultraviolet, a feature which is very near the plasma resonance. This interband absorption edge is due to the overlap of contributions from d-band to Fermi level transitions and Fermi level to conduction band transitions, both in the L (111) direction in the Brillouin zone [1 to 3]. Optical results tend to support this interpretation [4 to 8], although it has been questioned by Christensen [9].

Near the absorption edge, the optical spectrum of Ag should be sensitive to alloying. Since the transitions which are involved in the absorption edge involve free electrons at either the initial or final state, alloying with Zn should shift the Fermi surface enough to

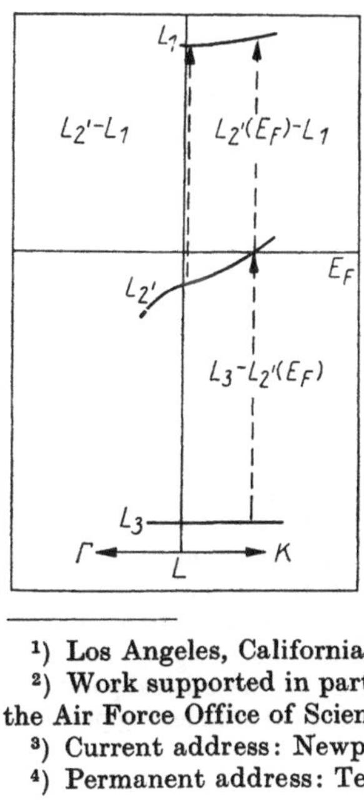

Fig. 1. Energy bands near the L point in pure Ag [6]

1) Los Angeles, California 90024, USA.

2) Work supported in part by the U.S. Army Research Office, Durham, North Carolina, and by the Air Force Office of Scientific Research.

3) Current address: Newport Corporation, Fountain Valley, CA 92708, USA.

4) Permanent address: Temple University, Philadelphia, PA 19122, USA.

separate the two transitions. In noble metals, the L_1 (s-like) level is more sensitive to the crystalline potential than the $L_{2'}$ (p-like) or the L_3 (d-like) levels [6]. (These levels are illustrated in Fig. 1.) Therefore, perturbation of the potential by alloying should result in a separation of the two transitions. The experimental results presented here do show a separation of the transition energies. As expected, the separation appears to be due to a downward shift of the L_1 level with respect to the Fermi level.

2. Sample Preparation

Precisely weighed amounts of 0.9999 pure Ag and 0.99999 pure Zn were sealed in quartz tubes at a pressure of less than 10^{-5} mm Hg. They were melted and thoroughly mixed by being shaken in the liquid states 5000 times and then quenched to room temperature in a water bath to form a homogeneous ingot. This ingot was, in turn, homogenized by heating to 650 °C for 24 hours and then cooling slowly. After cutting and polishing, the sample was given another 24 hour anneal, this time at 225 °C, and then again slow cooled. This was to relieve cold work acquired in the polishing process.

Leavings from the cutting process were assayed for Ag content by the U.S. Mint, which confirmed that the final composition was within 0.1% of $Ag_{0.7}Zn_{0.3}$.

3. Experimental Method

Wavelength-modulated spectroscopy [10, 11] was used because of the unambiguous lineshapes obtained by this technique and the resulting ease of interpretation of these lineshapes. The theory of wavelength-modulation spectroscopy is well detailed elsewhere [12, 13]. The theory of lineshapes near a three-dimensional critical point was given by Batz [14, 15]. It is sufficient here to give the theoretical lineshapes for the sake of reference. All three dimensional critical point lineshapes may be expressed

Table 1

Derivative of critical points with broadening included in terms of $F(W)$
$F(W) = (W^2 + 1)^{1/2} + W]^{1/2}/(W^2 + 1)^{1/2}$, $W = (\omega - \omega_g)/\eta$

critical point	$2\eta^{1/2} \frac{d\varepsilon_1}{d\omega}$	$2\eta^{1/2} \frac{d\varepsilon_2}{d\omega}$
M_0	$F(-W)$	$F(W)$
M_1	$-F(W)$	$F(-W)$
M_2	$-F(-W)$	$-F(W)$
M_3	$F(W)$	$-F(-W)$

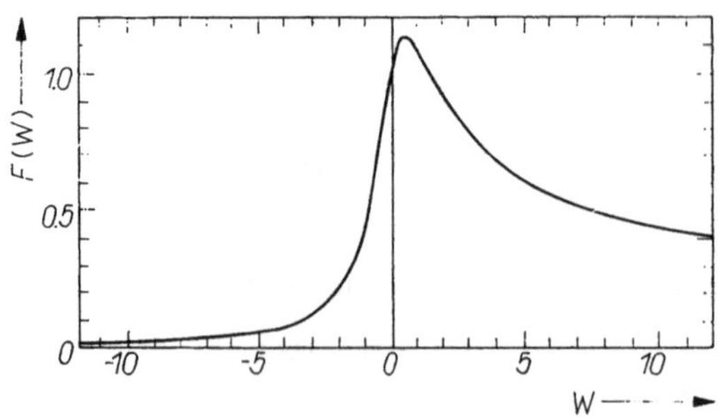

Fig. 2. Universal function, $F(W)$, derived by Batz [14, 15] for lineshapes near critical points as seen in wavelength modulation spectroscopy. $F(W) = [W^2 + 1]^{-1/2} \times [(W^2 + 1)^{1/2} + W]^{1/2}$, where W is the reduced frequency t$\omega - \omega_g)/\eta$. $\hbar\omega_g$ is the interband energy (a the critical point and η is a phenomenological broadening parameter

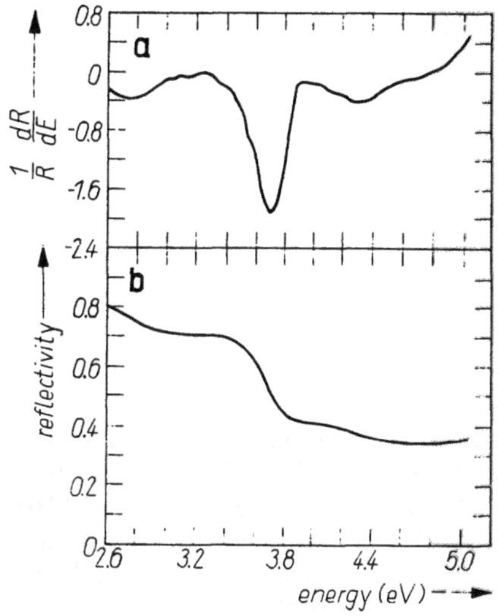

Fig. 3. a) Wavelength-modulated logarithmic derivative of the reflectivity of α-$Ag_{0.7}Zn_{0.3}$. b) Reflectivity of α-$Ag_{0.7}Zn_{0.3}$, found by integrating the data from Fig. 3a

in terms of a single function, $F(W)$, which is reproduced in Fig. 2. The behavior of critical points in terms of $F(W)$ is listed in Table 1. Derivative data with a spacing of 0.02 eV between data points were taken between 2.60 and 5.10 eV.

4. Experimental Results

The wavelength-modulated logarithmic derivative spectrum of the reflectivity, which is the direct result of the experiment, is displayed in Fig. 3a. This is integrated to give the reflectivity of the sample, as shown in Fig. 3b. In order to perform a Kramers-

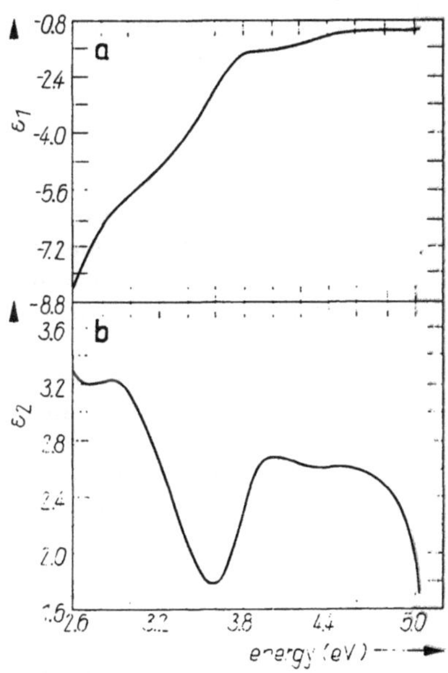

Fig.4

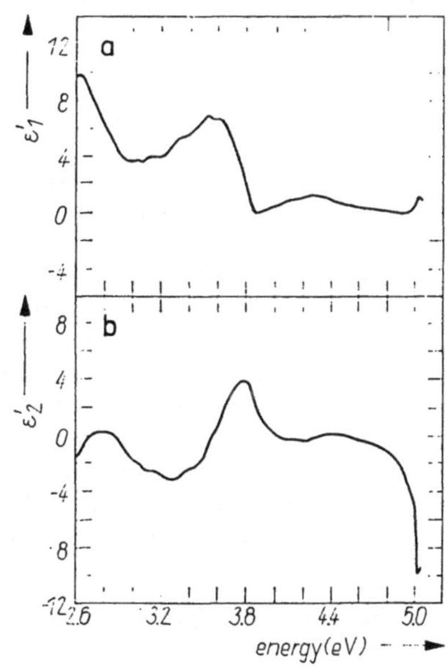

Fig.5

Fig. 4. Dielectric function of α-$Ag_{0.7}Zn_{0.3}$ determined by a Kramers-Kronig analysis of the reflectivity data. a) Real part, ε_1; b) imaginary part, ε_2

Fig. 5. Derivative with respect to energy of the dielectric function of α-$Ag_{0.7}Zn_{0.3}$. a) Real part, ε_1'; b) imaginary part, ε_2'

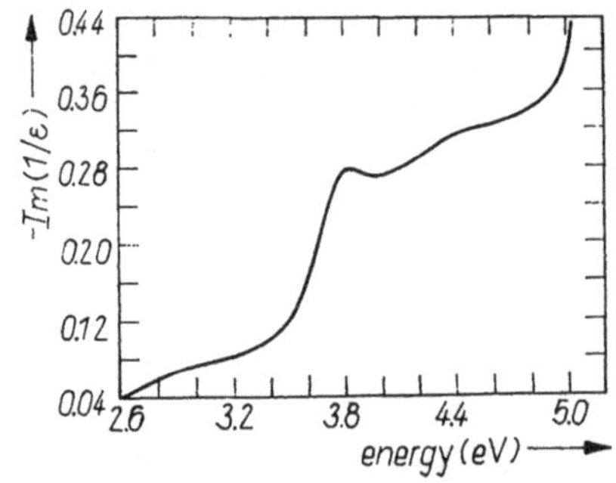

Fig. 6. Energy loss function of α-$Ag_{0.7}Zn_{0.3}$

Kronig analysis of the data and separate the real and imaginary parts of the dielectric function, extrapolations over all energy must be done. For the infrared region, a Hagen-Rubens extrapolation to a reflectivity of 100% at zero energy was used [16]. For the ultraviolet extrapolation, modified silver data [17] were used. Using this, the real and imaginary parts of the dielectric function, ε_1 and ε_2, are calculated. These are given in Fig. 4a and b. The dielectric function is then differentiated to give ε_1' and ε_2' (Fig. 5).

The important point to note concerning the data is that the plasma resonance which dominates the experimental spectrum of Ag [17] is greatly damped in this alloy, as may be seen from the fact that ε_1 does not go through zero at any point in the experimental energy range, and from the strongly damped peak in the energy loss function at 3.85 eV (Fig. 6). The ε_1 shown in Fig. 4 is in excellent agreement with that of previous results [5]. Note also that the principal dip in the reflectivity is at a considerably lower energy than in pure Ag. This occurs between about 3.6 and 3.8 eV in the alloy, compared with a minimum of about 3.85 eV in pure Ag.

5. Free-Electron Effects

The free and bound parts of the dielectric function have been separated by a de-convolution method which has been described elsewhere [18, 19]. The Drude effective mass was found to be 1.23 free electron masses and the Drude relaxation time was found to be 1.87×10^{-14} second. These results compare well with the values found for Ag. Previous results for the effective mass of silver are $m^*/m = 1.03$ [1] and 0.97 [20]. The relaxation time in Ag was found to be 1.6×10^{-14} s at 3 eV [17].

We thus see that the Drude part of the dielectric function predicts some flattening of the bands at the Fermi level. Although alloying will be seen to have a considerable damping effect on the interband properties of Ag, the long relaxation time found here indicates that the behavior of free electrons is not significantly changed. We may conclude that this amount of alloying has only small consequences for those electronic properties described by the Drude theory.

6. Interband Transitions

Once the Drude effective mass and relaxation time are known, they may be used to subtract the intraband contributions to the dielectric function from the total experimental function. The resulting real and imaginary parts of the bound-electron contribution to the dielectric function are shown in Fig. 7. Their derivatives are given in Fig. 8. It is these derivatives of the bound part of the dielectric function which are used to assign interband transitions.

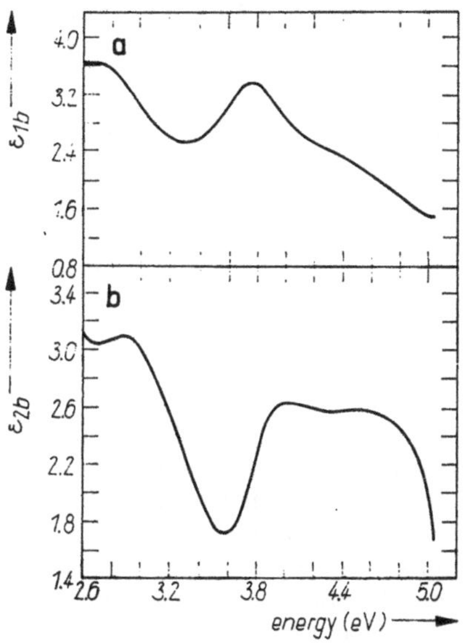

Fig. 7

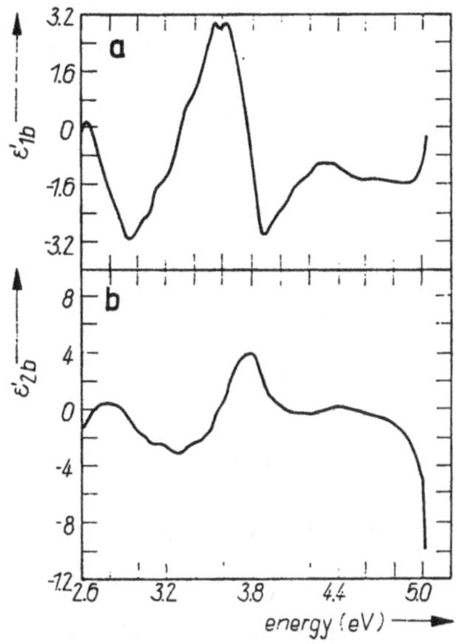

Fig. 8

Fig. 7. Bound part of dielectric function α-$Ag_{0,7}Zn_{0,3}$ determined by subtracting the Drude contribution from the data of Fig. 5. a) Real part, ε_{1b}; b) imaginary part, ε_{2b}

Fig. 8. Derivative with respect to energy of the bound part of the dielectric function of α-$Ag_{0.7}Zn_{0.3}$. a) Real part, ε'_{1b}; b) imaginary part, ε'_{2b}. With reference to the energy bands in Fig. 1, the following transitions are identified: $L_{2'}(E_F) \rightarrow L_1$ (3.55 eV), $L_{2'} \rightarrow L_1$ (3.83 eV), $L_3 \rightarrow L_{2'}(E_F)$ (4.37 eV)

It is interesting to compare ε_{2b} of Fig. 7 with the ε_2 results shown in Fig. 9e of [5]. The earlier work shows maxima in ε_2 at 3.1 and 4.9 eV with a minimum at 4.1 eV. The current work has a maximum at 2.8 eV, a minimum at 3.6 eV, and maxima at 4.0 eV, which are resolved here where they were not previously. We thus see that the features found in this study are at lower energies than in [5].

The current literature does not firmly fix the exact value of the Ag interband transitions of interest in this study. While the energy gap for the $L_{2'} \rightarrow L_1$ transitions ranges from 3.8 eV [7] to 4.1 eV [5], the values quoted for the $L_3 \rightarrow L_{2'}(E_F)$ transition range from 4.1 eV [7] to 4.4 eV [5]. Clearly, there is enough discrepancy in the experimental values for these energies that it will be difficult to accurately assess the effects of alloying on them.

No band structure calculation has been reported for α-$Ag_{0.7}Zn_{0.3}$ but calculations do exist for α-$Cu_{0.7}Zn_{0.3}$ [21, 22]. In the work of Bansil et al. [22], predictions are made for the shifts in the interband transitions of interest due to alloying. The $L_3 \rightarrow L_{2'}(E_F)$ transition is predicted to move upward by about 0.3 eV and the $L_{2'} \rightarrow L_1$ transition is to move downward in energy by 0.1 to 0.8 eV depending on the amount of lattice dilation due to alloying.

Since both transitions would be at saddle points in the Brillouin zone, a dip in ε'_{1b} would be expected near the critical point for each of them [14, 15]. The minimum in ε'_{1b} at 3.88 eV is the expected structure due to $L_{2'} \rightarrow L_1$. The associated structure in ε'_{2b} is the peak at 3.78 eV. This places the critical point at 3.83 eV, coinciding with the peak in the energy loss function. The peak in ε'_{1b} at 3.55 eV is due to the onset of transitions at $L_{2'}(E_F) \rightarrow L_1$. We may then conclude that the $L_{2'}$ critical point lies 0.28 eV below the Fermi level.

Ehrenreich and Phillipp [17] assigned the mean ε_2 Ag peak at 4.4 eV to the direct transition from the d-band to the Fermi surface $L_3 \rightarrow L_{2'}(E_F)$. The $L_{2'}$ band is filled at the critical point, so the actual critical point transition structure for the $L_3 \rightarrow L_{2'}$ transition will not be seen. A break in the derivatives of ε_b should be seen at the energy where transitions to the Fermi level begin to occur. The $L_3 \rightarrow L_{2'}(E_F)$ transition will produce a downward break in ε'_{1b} with an associated break in ε'_{2b} at the same energy, which is the energy of the onset of interband transitions. Such breaks are seen at 4.37 eV in Fig. 8. Thus, our data show that there is very little shift of the transition in the α-$Ag_{0.7}Zn_{0.3}$ alloy. This places the $L_3 \rightarrow L_{2'}$ critical point gap at 4.09 eV when the results from the data for the $L_{2'} \rightarrow L_1$ transition are used.

7. Discussion

The results of this study indicate that the effects of alloying in the $Ag_{0.7}Zn_{0.3}$ are actually fairly small. The shifts in the interband transitions must be on the lower end of those predicted for the α-$Cu_{0.7}Zn_{0.3}$ system by Bansil et al. [22]. If the $L_3 \rightarrow L_{2'}(E_F)$ transition with an energy of 4.37 eV has been shifted by +0.3 eV, as might be expected from the α-$Cu_{0.7}Zn_{0.3}$ study, then this transition should have an onset at about 4.1 eV in pure Ag. This value is just within the range of experimental values which have been quoted. However, there is also good evidence that this transition is essentially unshifted from that in pure Ag [5, 17].

The actual shift in α-$Cu_{0.7}Zn_{0.3}$ for the $L_{2'} \rightarrow L_1$ transition is about −0.2 eV [23, 24]. Using this value would place the $L_{2'} \rightarrow L_1$ transition in pure Ag at 4.0 eV which is about the middle of the experimental values for this transition. Previous work which shows that $L_3 \rightarrow L_2(E_F)$ is unshifted closely agrees with this assignment for $L_2 \rightarrow L_1$ in pure Ag [5, 17].

The following may be cited as the important results of this study:

(i) In this alloy composition, ε_1 is dampened such that the main Ag plasma resonance is greatly reduced.

(ii) The Drude effective mass predicts a slight flattening of the bands. The fact that the energy difference $L_3 \rightarrow L_{2'}(E_F)$ is 0.28 eV is not any help in confirming the flattening since values for this difference for Ag are reported to lie between 0.2 and 0.3 eV.

(iii) The fact that the Drude relaxation time for this alloy is approximately the same for pure Ag indicates that this amount of alloying has not greatly affected the mean free path of conduction electrons.

(iv) The shifts found in the interband transitions between pure Ag and the alloy indicate that the transition usually more sensitive to perturbation of the crystal potential ($L_{2'} \rightarrow L_1$) has a shift of about −0.2 eV, while the less sensitive transition ($L_3 \rightarrow L_{2'}$) may be essentially unshifted. This indicates that, at least in this portion of the Brillouin zone, the rigid-band model may be used to describe the interband transitions of this alloy, with only a shift of the s-like (L_1) band.

References

[1] D. BEAGLEHOLE, Proc. Phys. Soc. London. 85, 1007 (1965); 87, 461 (1966).
[2] F. M. MUELLER and J. C. PHILLIPS, Phys. Rev. 157, 600 (1967).
[3] R. M. MORGAN and D. W. LYNCH, Phys. Rev. 182, 719 (1969).
[4] R. M. MORGAN and D. W. LYNCH, Phys. Rev. 172, 628 (1968).
[5] E. L. GREEN and L. MULDAWER, Phys. Rev. B2, 330 (1970).
[6] H. G. LILJENVALL and A. G. MATHEWSON, J. Phys. C. 3, 341 (1970).
[7] P. O. NILSSON and B. SANDELL, Solid State Commun. 8, 721 (1970).
[8] B. F. SCHMIDT and D. W. LYNCH, Phys. Rev. B3, 4015 (1971).

[9] N. E. Christensen, phys. stat. sol. (b) **54**, 551 (1972).
[10] M. Welkowsky and R. Braunstein, Rev. Sci. Instrum. **43**, 399 (1972).
[11] R. Stearns, J. Steele, and R. Braunstein, Rev. sci. Instrum. **54**, 984 (1983).
[12] M. Cardona, Modulation Spectroscopy, Academic Press, 1969.
[13] M. Welkowsky, Ph. D. Thesis, University of California at Los Angeles, 1971.
[14] B. Batz, Ph. D. Thesis, Univ. Libre de Bruxelles, 1967.
[15] B. Batz, Solid State Commun. **4**, 241 (1965).
[16] N. F. Mott and H. Jones, The Theory of the Properties of Metals and Alloys, Dover Publ., Inc., New York 1951.
[17] H. Ehrenreich and J. R. Phillipp, Phys. Rev. **128**, 1622 (1962).
[18] M. Burd, R. Stearns, and R. Braunstein, phys. stat. sol. (b) **117**, 101 (1983).
[19] R. Stearns, Ph. D. Thesis, University of California at Los Angeles, 1982.
[20] L. G. Schultz, Suppl. Phil. Mag. **6**, 102 (1957).
[21] H. Amar and K. Johnson, in: Optical Properties and Electronic Structure of Metals and Alloys, Ed. F. Abeles, North Holland Publ. Co., 1966 (p. 586).
[22] A. Bansil, H. Ehrenreich, L. Schwartz, and R. E. Watson, Phys. Rev. B **9**, 445 (1974).
[23] M. A. Biondi and J. A. Rayne, Phys. Rev. **115**, 1522 (1959).
[24] G. P. Pells and H. Montgomery, J. Phys. C **3**, 5330 (1970).

(Received July 3, 1985)

phys. stat. sol. (b) 131, 667 (1985)

Subject classification: 13.5

Institute of Physics, Charles University, Prague[1])

Stochastic Liouville Equation Model for Nonperiodic Systems and Finite Temperatures

By

V. Čápek and V. Szöcs

Dedicated to Prof. Dr. Dr. h. c. Dr. E. h. P. Görlich on the occasion of his 80th birthday

Theoretical background leading to the recently proposed generalization of the stochastic Liouville equation model of the excitation transfer to nonperiodic systems and finite temperatures is reformulated here using the time-convolutionless generalized master equations. In the lowest order and for linear coupling to phonons, the Haken-Strobl-Reineker parametrization becomes acceptable after a short initial time-interval following the creation of the excitation, but with time-dependent parametrization, the theory applies to arbitrarily short time-intervals. For separable initial conditions, the time development of the excitation is calculated numerically for one-dimensional models describing the dynamics of the excitation near traps, trap-luminescence etc. provided that the time-development in the initial time interval is negligible. The equations correctly provide non-equal asymptotic site-occupation probabilities for non-equal sites if the excitation lifetime is infinite.

Der theoretische Hintergrund, der zu der kürzlich vorgeschlagenen Verallgemeinerung des Modells stochastischer Liouville-Gleichungen des Anregungstransfers in nichtperiodischen Systemen und bei endlichen Temperaturen führt, wird erneut mit den zeit-konvolutionslosen verallgemeinerten Mastergleichungen formuliert. In der niedrigsten Ordnung und für eine lineare Kopplung der Phononen, wird die Haken-Strobl-Reineker-Parametrisierung akzeptabel nach einem kurzen anfänglichen Zeitintervall, das der Bildung der Anregung folgt, jedoch mit zeitabhängiger Parametrisierung läßt sich die Theorie auf beliebig kurze Zeitintervalle anwenden. Für separable Anfangsbedingungen wird die zeitliche Entwicklung der Anregung numerisch für eindimensionale Modelle, die die Dynamik der Anregung in der Nähe von Haftstellen, Haftstellenlumineszenz usw. beschreiben, berechnet, vorausgesetzt, daß die zeitliche Entwicklung im Anfangszeitintervall vernachlässigbar ist. Die Gleichungen ergeben korrekt die asymptotischen Nichtgleichgewichts-Platzbesetzungswahrscheinlichkeiten für Nichtgleichgewichtsplätze, wenn die Anregungslebensdauer unendlich ist.

1. Introduction

The Haken-Strobl-Reineker model or more generally, the stochastic Liouville equation approach [1 to 6] provides probably the most developed and simultaneously the best founded method of investigation and modelling of the excitation dynamics in molecular condensates. Its history, as far as the idea of modelling the influence of the reservoir on the dynamics of the system in question by a stochastic Gaussian Markovian δ-correlated potential is concerned, goes back to the early fifties [7, 8]. This fruitful idea is, on the other hand, the very source of severe limitations of the model to periodic systems and infinite temperatures. As regards the latter limitation, it is partly shifted assuming a finite decay time of the potential correlation functions [9, 3]. While this assumption makes the approach more realistic at times $t \leqq \tau^{P}$, the phonon dephasing time ($\gtrsim$ 1 ps), from the point of view of the short-time asympto-

[1]) Ke Karlovu 5, 12116 Prague 2, Czechoslovakia.

tics of the site-occupation probabilities $P_n(t)$, it does not help in the long-time domain ($t \gtrsim \tau^P$) since the model still gives equal $P_n(t)$ for non-equal sites when the system approaches equilibrium. In this respect, from the point of view of the detailed balance condition, the theory describes the situation with infinite temperature even for finite decay time of the potential correlations.

In order to improve the situation, one must take into account the quantum character of the potential field (phonons). As the quantum field admits spontaneous processes (with respect to the field), the asymmetry of hopping rates between non-equivalent sites (required by the detailed balance condition) is readily established. Technically, however, it is not so easy to bring the problem to a manageable form. In [10], we presented a method of doing that based on the time-convolution generalized master equations. Since the Born-Markov approximation used may not be very reliable in connection with the (generally non-Markovian) excitation dynamics at very short as well as very long times, we present here a new formal approach to the problem based on the time-convolution-less generalized master equations as well as results of extensive numerical calculations made hitherto for the simplest models. In our opinion, the method opens a possibility of direct and reliable computer modelling of the excitation dynamics in non-periodic systems (vicinity of traps in crystals, photosynthetic units, etc.) in nano-, pico-, as well as subpicosecond time intervals.

2. Theory

We start with the Hashitsume et al. and Shibata et al. [11, 12] identity

$$i \frac{\partial}{\partial t} D\hat{\varrho}(t) = D\hat{\mathscr{L}}(t) \left[1 + i \int_{t_0}^{t} \mathscr{G}(t,\tau)\,(1-D)\,\hat{\mathscr{L}}(\tau)\,DG(t,\tau)\right]^{-1} D\hat{\varrho}(t) +$$

$$+ D\hat{\mathscr{L}}(t) \left[1 + i \int_{t_0}^{t} \mathscr{G}(t,\tau)\,(1-D)\,\hat{\mathscr{L}}(\tau)\,DG(t,\tau)\right]^{-1} \mathscr{G}(t,t_0)\,(1-D)\,\hat{\varrho}(t_0) \qquad (1)$$

with

$$G(t,\tau) = \exp_{\rightarrow}\left[i \int_{\tau}^{t} ds\, \hat{\mathscr{L}}(s)\right] =$$

$$= 1 + i \int_{\tau}^{t} ds\, \hat{\mathscr{L}}(s) + (i)^2 \int_{\tau}^{t} ds_1 \int_{\tau}^{s_1} ds_2\, \hat{\mathscr{L}}(s_2)\,\hat{\mathscr{L}}(s_1) + \dots \qquad (2)$$

and

$$\mathscr{G}(t,\tau) = \exp_{\leftarrow}\left[-i \int_{\tau}^{t} ds (1-D)\,\hat{\mathscr{L}}(s)\right] =$$

$$= 1 - i \int_{\tau}^{t} ds\,(1-D)\,\hat{\mathscr{L}}(s) +$$

$$+ (-i)^2 \int_{\tau}^{t} ds_1 \int_{\tau}^{s_1} ds_2\,(1-D)\,\hat{\mathscr{L}}(s_1)\,(1-D)\,\hat{\mathscr{L}}(s_2) + \dots\,. \qquad (3)$$

In (1) to (3)

$$D = D^2 \qquad (4)$$

is an arbitrary projection superoperator to be specified below. For $D = 1$, (1) reduces to the Liouville equation

$$i \frac{\partial}{\partial t}\,\hat{\varrho}(t) = \frac{1}{\hbar}\,[\mathscr{H}(t), \hat{\varrho}(t)] \equiv \hat{\mathscr{L}}(t)\,\hat{\varrho}(t) \qquad (5)$$

from which it was originally derived. As seen from (4), we use the interaction picture (designated by sign "$\hat{...}$"), i.e.

$$\hat{A}(t) = \exp\left(\frac{i}{\hbar} H_0 t\right) A \exp\left(-\frac{i}{\hbar} H_0 t\right) = \exp(iL_0 t)\, A\,, \tag{6a}$$

$$\hat{B}(t) = \exp(iL_0 t)\, B \exp(-\, iL_0 t) \tag{6b}$$

for an arbitrary operator A and superoperator B. The Liouvillians L_0 and $\mathcal{L}$ read

$$L_0 = \frac{1}{\hbar}[H_0, \ldots] \tag{7}$$

and

$$\mathcal{L} = \frac{1}{\hbar}[\mathcal{H}, \ldots]\,, \tag{8}$$

where

$$H = H_0 + \mathcal{H} \tag{9}$$

is the total Hamiltonian of the problem.

We assume the standard model Hamiltonan

$$H = H_e + H_{ph} + H_{e\text{-}ph} = \sum_{mn} [\delta_{mn}\varepsilon_n + (1 - \delta_{mn})J_{mn}]\, a_m^+ a_n + \sum_k \hbar\omega_k b_k^+ b_k + \\ + \frac{1}{\sqrt{N}} \sum_k \sum_{mn} g_k^{mn} \hbar\omega_k a_m^+ a_n [b_k + b_{-k}^+]\,; \quad g_k^{mn} = (g_{-k}^{nm})^* \tag{10}$$

and work with just a single excitation in the system. In what follows, we assume it to be the exciton; the theory (generalized to include the spin indices), however, applies to the electron case, too. The meaning of ε_n, J_{mn}, ω_k, g_k^{mn}, $a_m^+(a_m)$, and $b_k^+(b_k)$ is then the site energy, hopping integral, harmonic phonon frequency, exciton–phonon coupling constant, creation (annihilation) operator of the exciton residing at site m and that of the phonon, respectively.

Two possibilities of the choice of H_0 and $\mathcal{H}$ are at hand. Usually,

$$H_0 = H_e + H_{ph}\,, \\ \mathcal{H} = H_{e\text{-}ph} \tag{11a}$$

so that $H_0|_{J=0}$ is diagonalized by

$$|m\mu\rangle \equiv |m, \{\mu_k\}\rangle = a_m^+ \prod_k \left\{\frac{1}{\sqrt{\mu_k!}} (b_k^+)^{\mu_k}\right\} |0\rangle \tag{11b}$$

with the corresponding energies

$$E_{m\mu} = \varepsilon_m + \sum_k \hbar\omega_k \mu_k \tag{11c}$$

but we may also choose the small-polaron basis

$$|m\mu\rangle \equiv |m, \{\mu_k\}\rangle = a_m^+ \prod_k \left\{\frac{1}{\sqrt{\mu_k!}} \left(b_k^+ + \frac{1}{\sqrt{N}} g_k^{mm}\right)^{\mu_k}\right\} \times \\ \times \exp\left\{\frac{1}{\sqrt{N}} \sum_k [g_k^{mm} b_k - g_k^{mm*} b_k^+]\right\} |0\rangle \tag{12a}$$

which corresponds to the choice

$$H_0 = H_e + H_{ph} + \\ + \frac{1}{\sqrt{N}} \sum_m \sum_k g_k^{mm} \hbar\omega_k a_m^+ a_m (b_k + b_{-k}^+) . \tag{12b}$$

Then $H_0|_{J=0}$ is diagonalized by (12a) with the corresponding energies

$$E_{m\mu} = \varepsilon_m + \sum_k \hbar\omega_k \mu_k - \frac{1}{N} \sum_k |g_k^{mm}|^2 \hbar\omega_k . \tag{12c}$$

Then

$$\mathcal{H} = \frac{1}{\sqrt{N}} \sum_{m \neq n} \sum_k g_k^{mn} \hbar\omega_k a_m^+ a_n (b_k + b_{-k}^+) . \tag{12d}$$

Advantages of the choice (11) or (12) will be seen below in connection with the initial condition and the choice of D. We take the Argyres and Kelley [13] projector

$$(DA)_{m\mu, n\nu} \equiv \sum_{r\varrho, s\sigma} D_{m\mu, n\nu, r\varrho, s\sigma} A_{r\varrho, s\sigma} = \delta_{\mu\nu} p_\nu \sum_\lambda A_{m\lambda, n\lambda} , \tag{13a}$$

$$\sum_\mu p_\mu = 1 \tag{13b}$$

(where the first equality in (13) introduces matrix elements of an arbitrary linear superoperator). It is easy to verify that (4) is well fulfilled as long as (13b) applies which may be achieved taking

$$p_\mu = \prod_k [e^{-\beta\hbar\omega_k\mu_k} (1 - e^{-\beta\hbar\omega_k})] . \tag{14}$$

$T = (k_B\beta)^{-1}$ then receives the meaning of the initial phonon temperature once we assume the initial condition

$$\hat{\varrho}(t = t_0) = \hat{\varrho}^{ph}(t = t_0) \otimes \hat{\varrho}^{el}(t = t_0) , \tag{15a}$$

$$[\hat{\varrho}^{ph}(t = t_0)]_{\mu\nu} = \delta_{\mu\nu} p_\nu , \tag{15b}$$

since (14) then becomes the Boltzmann weight of the initial phonon distribution. The initial state (15a) therefore corresponds to statistically independent electrons and phonons, with the latter being in thermal equilibrium in the unrelaxed (choice (11a) to (11c)) or relaxed (choice (12a) to (12d)) state.

From (15) and (13), it follows that in either case

$$(1 - D)\, \hat{\varrho}(t_0) = 0 , \tag{16}$$

so that the last term in (1) disappears. Since

$$\big(D\hat{\varrho}(t)\big)_{m\mu, n\nu} = p_\mu \delta_{\mu\nu} \sum_\lambda \hat{\varrho}_{m\lambda, n\lambda}(t) \equiv \\ \equiv p_\mu \delta_{\mu\nu} \hat{\varrho}_{mn}^{el}(t) , \tag{17}$$

(1) then yields the closed identity for the electron density matrix

$$\frac{\partial}{\partial t} \hat{\varrho}_{mn}^{el}(t) = \sum_{rs} \hat{\Lambda}_{mnrs}(t, t_0)\, \hat{\varrho}_{rs}^{el}(t) ,$$

$$\hat{\Lambda}_{mnrs}(t, t_0) = -i \sum_{\mu\nu} \{\hat{\mathcal{L}}(t) [1 + i \int_{t_0}^{t} \mathcal{G}(t, \tau)\, (1 - D)\, \hat{\mathcal{L}}(\tau)\, DG(t, \tau)]^{-1}\}_{m\mu, m\mu, r\nu, s\nu}\, p_\nu = \\ = - \sum_{\mu\nu} \{\hat{\mathcal{L}}(t) \int_{t_0}^{t} (1 - D)\, \hat{\mathcal{L}}(\tau)\}_{m\mu, m\mu, r\nu, s\nu}\, p_\nu + O(g^2) \tag{18}$$

(see e.g. [14, 15]). Here, we have omitted or just partially summed up to infinity the terms beyond second order in g in case of choice (11a) to (11c) or (12a) to (12d), respectively. Returning then to the Schrödinger picture

$$\varrho^{\mathrm{el}}_{mn}(t) = \sum_{\lambda} \varrho_{m\lambda,n\lambda}(t) = \sum_{\lambda} [\mathrm{e}^{-iL_0 t}\,\hat{\varrho}^{\mathrm{el}}(t)]_{mn} = [\mathrm{e}^{-iL_e t}\,\hat{\varrho}^{\mathrm{el}}(t)]_{mn}\,, \tag{19a}$$

$$L_{\mathrm{e}} = \frac{1}{\hbar}\,[H_{\mathrm{e}}, \ldots] \tag{19b}$$

we have

$$\frac{\partial}{\partial t}\,\varrho^{\mathrm{el}}_{mn}(t) = -\frac{i}{\hbar}\,([H_{\mathrm{e}}, \varrho^{\mathrm{el}}(t)])_{mn} + \sum_{rs} \Lambda_{mnrs}(t, t_0)\,\varrho^{\mathrm{el}}_{rs}(t) \tag{20}$$

with

$$\Lambda_{mnrs}(t, t_0) = \{\mathrm{e}^{-iL_e t}\,\hat{\Lambda}(t, t_0)\,\mathrm{e}^{iL_e t}\}_{mnrs} = \\ = -\int_0^{t-t_0} \mathrm{d}\tau \sum_{\mu\nu} \{\hat{\mathcal{L}}(\tau)\,\mathcal{L}\}_{m\mu,n\mu,r\nu,s\nu}\,p_\nu + O(g^2)\,. \tag{21}$$

Here, we have used the fact that because of the structure of (10), $D\hat{\mathcal{L}}(\tau)\,D = 0$. Writing down $\Lambda_{mnrs}(t, t_0)$ in (21) explicitly in terms of the matrix elements of $\mathcal{H}$ yields

$$\Lambda_{mnrs}(t, t_0) = \frac{1}{\hbar^2} \int_0^{t-t_0} \sum_{\mu\nu} [\hat{\mathcal{H}}(\tau)_{m\mu,r\nu}\,\mathcal{H}_{s\nu,n\mu}p_\nu + \\ + \hat{\mathcal{H}}(\tau)_{s\nu,n\mu}\,\mathcal{H}_{m\mu,r\nu}p_\nu - \delta_{ns} \sum_a \hat{\mathcal{H}}(\tau)_{m\nu,a\mu}\,\mathcal{H}_{a\mu,r\nu}p_\nu - \\ - \delta_{mr} \sum_a \hat{\mathcal{H}}_{a\mu,n\nu}(\tau)\,\mathcal{H}_{s\nu,a\mu}p_\nu]\,\mathrm{d}\tau\,. \tag{22}$$

One should turn attention to the fact that in case of choice (12), the matrix elements $\hat{\mathcal{H}}(\tau)_{m\mu,r\mu}$ (diagonal in the phonon indices) are nonzero, contributing in general by the phonon-less terms to the integrand in (22). These contributions are not damped as $\tau \to \infty$ even after taking the sum $\sum_{\mu\nu}$ in (22) [16], yielding oscillating terms in Λ_{mnrs}. As the latter contributions are, however, of the order g^4 at least [16], we shall disregard them for our purposes here. The remaining terms in (22) get damped (in case of choice (11), the integrand gets always damped) so that Λ_{mnrs} attain constant values once $|t - t_0| \gtrsim \tau^{\mathrm{P}}$ where τ^{P} is some phonon-dephasing time. Then (20) may be solved by standard methods.

3. The Haken-Strobl-Reineker Parametrization

First, there is the identity

$$\sum_m \Lambda_{mmpq}(t, t_0) = 0 \tag{23}$$

following from e.g. (22) which connects different coefficients. As a consequence, we have, for example,

$$\Lambda_{mmmm}(t, t_0) = -\sum_{n(\neq m)} \Lambda_{nnmm}(t, t_0)\,. \tag{24}$$

Let us therefore take into account the coefficients Λ_{mmnn}. First we introduce

$$\begin{aligned}\gamma_{mn}(t, t_0) &= \gamma_{mn}^*(t, t_0) = \\ &= \frac{1}{2\hbar^2} \int_0^{t-t_0} d\tau \sum_{\mu\nu} [\hat{\mathcal{H}}(\tau)_{m\mu, n\nu} \mathcal{H}_{n\nu, m\mu} + \hat{\mathcal{H}}(\tau)_{n\nu, m\mu} \mathcal{H}_{m\mu, n\nu}] p_\nu \approx \\ &\approx \frac{\pi\hbar}{N} \sum_k \omega_k^2 |g_k^{mn}|^2 G_k^{mn} \{\Delta_{t-t_0}(\varepsilon_m - \varepsilon_n + \hbar\omega_k) [n_B(\hbar\omega_k) + 1] + \\ &+ \Delta_{t-t_0}(\varepsilon_m - \varepsilon_n - \hbar\omega_k) n_B(\hbar\omega_k)\}\,, \end{aligned} \tag{25a}$$

$$n_B(z) = [e^{\beta z} - 1]^{-1}\,, \tag{25b}$$

$$\Delta_t(\varepsilon) = \frac{1}{\pi} \frac{\sin(\varepsilon t/\hbar)}{\varepsilon} \underset{t\to\infty}{\to} \delta(\varepsilon)\,, \tag{25c}$$

$$G_k^{mn} = \begin{cases} 1 & \text{in case of choice (11)}\,, \\ \exp\left[-\frac{1}{N} \sum_k (g_k^{mm} - g_k^{nn})^2 [1 + 2n_B(\hbar\omega_k)]\right] & \text{in case of choice (12)}\,. \end{cases} \tag{25d}$$

In second equality on the right-hand side of (25a), we have omitted higher (than zero -) order terms in the hopping integrals and in case of choice (12), we omitted all but dominating single-phonon terms. Consequently, for $m \neq n$, we have

$$\Lambda_{mmnn}(t, t_0) = 2\gamma_{mn}(t, t_0)\,. \tag{26}$$

Because of (25b, c),

$$\lim_{t\to\infty} \gamma_{mn}(t, t_0) = e^{-\beta(\varepsilon_m - \varepsilon_n)} \lim_{t\to\infty} \gamma_{nm}(t, t_0) < \lim_{t\to\infty} \gamma_{nm}(t, t_0) \tag{27}$$

for $\varepsilon_m > \varepsilon_n$ as long as the terms omitted on the right-hand side of (25a) become negligible. One can then easily reveal that (27) is the lowest-order detailed balance condition. In higher orders, the equivalence of $\gamma_{mn}(t, t_0)$ and $\gamma_{nm}(t, t_0)$ in the long-time domain is also not restored for non-equal sites with $\varepsilon_m \neq \varepsilon_n$. This distinguishes the present treatment from the usual formulation of the problem [3, 4].

Now, let us take the elements $\Lambda_{mnmn}(t, t_0)$, $m \neq n$. It is

$$\begin{aligned}\Lambda_{mnmn}(t, t_0) = &-\frac{1}{2\hbar^2} \sum_a \int_0^{t-t_0} d\tau \sum_{\mu\nu} [\hat{\mathcal{H}}(\tau)_{a\mu, m\nu} \mathcal{H}_{m\nu, a\mu} + \hat{\mathcal{H}}(\tau)_{m\nu, a\mu} \mathcal{H}_{a\mu, m\nu}] p_\nu - \\ &-\frac{1}{2\hbar^2} \sum_a \int_0^{t-t_0} d\tau \sum_{\mu\nu} [\hat{\mathcal{H}}(\tau)_{a\mu, n\nu} \mathcal{H}_{n\nu, a\mu} + \hat{\mathcal{H}}(\tau)_{n\nu, a\mu} \mathcal{H}_{a\mu, n\nu}] p_\nu + \\ &+\frac{1}{2\hbar^2} \sum_a \int_0^{t-t_0} d\tau \sum_{\mu\nu} (\hat{\mathcal{H}}(\tau)_{a\mu, m\nu} \mathcal{H}_{m\nu, a\mu} - \hat{\mathcal{H}}(\tau)_{m\nu, a\mu} \mathcal{H}_{a\mu, m\nu}] p_\nu + \\ &+\frac{1}{2\hbar^2} \sum_a \int_0^{t-t_0} d\tau \sum_{\mu\nu} [\hat{\mathcal{H}}(\tau)_{n\nu, a\mu} \mathcal{H}_{a\mu, n\nu} - \hat{\mathcal{H}}(\tau)_{a\mu, n\nu} \mathcal{H}_{n\nu, a\mu}] p_\nu + \\ &+\frac{1}{\hbar^2} \int_0^{t-t_0} d\tau \sum_{\mu\nu} [\hat{\mathcal{H}}(\tau)_{n\nu, n\mu} \mathcal{H}_{m\mu, m\nu} + \hat{\mathcal{H}}(\tau)_{m\mu, m\nu} \mathcal{H}_{n\nu, n\mu}] p_\nu\,. \end{aligned} \tag{28}$$

The first and second terms on the right-hand side of (28) equal $-\sum_a \gamma_{am}(t, t_0)$ and $-\sum_a \gamma_{an}(t, t_0)$, respectively. Designating

$$\Delta\varepsilon_m = -\frac{i}{2\hbar} \sum_a \int_0^\infty \mathrm{d}\tau \sum_{\mu\nu} [\hat{\mathscr{H}}(\tau)_{m\nu, a\mu} \mathscr{H}_{a\mu, m\nu} - \hat{\mathscr{H}}(\tau)_{a\mu, m\nu} \mathscr{H}_{m\nu, a\mu}] p_\nu \approx$$
$$\approx \sum_a \sum_{\mu\nu} |\mathscr{H}_{m\nu, a\mu}|^2 \frac{1}{E_{m\nu} - E_{a\mu}} p_\nu + O(g^2) , \tag{29}$$

the sum of the third and fourth terms on the right-hand side of (28) turns to $-\frac{i}{\hbar}(\Delta\varepsilon_m - \Delta s_n)$ when $t \to \infty$ and may be interpreted as a temperature-dependent second-order renormalization of the diagonal part of H_e in the first term on the right-hand side of (20). We shall drop this term henceforth. The last term on the right-hand side of (28) has no single-phonon contribution in the lowest order in J and the long-time limit and will be ignored below as a higher-order term.

At last, we take the term $\Lambda_{mnnm} (m \neq n)$ which reads

$$\Lambda_{mnnm}(t, t_0) = 2\bar{\gamma}_{mn}(t, t_0) =$$
$$= \frac{1}{\hbar^2} \int_0^{t-t_0} \mathrm{d}\tau \sum_{\mu\nu} [\hat{\mathscr{H}}(\tau)_{m\mu, n\nu} \mathscr{H}_{m\nu, n\mu} + \hat{\mathscr{H}}(\tau)_{m\nu, n\mu} \mathscr{H}_{m\mu, n\nu}] p_\nu \approx$$
$$\overset{t-t_0\to\infty}{\approx} \frac{1}{N} \sum_k (g_k^{mn})^2 G_k^{mn} \omega_k^2(-i\hbar) [2n_\mathrm{B}(\hbar\omega_k) + 1] \times$$
$$\times \left[\frac{1}{\varepsilon_n - \varepsilon_m + \hbar\omega_k - i\delta} + \frac{1}{\varepsilon_n - \varepsilon_m + \hbar\omega_k - i\delta}\right] \tag{30}$$

in the same approximation and notations as on the right-hand side of (25a). It is then always

$$\bar{\gamma}_{mn}(t, t_0) = \bar{\gamma}_{nm}(t, t_0)^* , \qquad t - t_0 \to \infty . \tag{31}$$

For equivalent sites ($\varepsilon_m = \varepsilon_n$), the lowest-order contribution to $\bar{\gamma}_{mn}$ (see (30)) as well as that to γ_{mn} (see (25a)) disappear. Keeping only the above-mentioned coefficients $\Lambda_{mnpq}(t, t_0)$ (which are just the same as those in the original Haken-Strobl-Reineker model) and omitting the renormalization corrections as well as the last term on the right-hand side of (28), we have

$$\frac{\partial}{\partial t} \varrho_{mn}^{\mathrm{el}}(t) = -\frac{i}{\hbar} [H_\mathrm{e}, \varrho^{\mathrm{el}}(t)]_{mn} +$$
$$+ \delta_{mn} 2 \sum_{r(\neq m)} [\gamma_{mr}(t, t_0) \varrho_{rr}^{\mathrm{el}}(t) - \gamma_{rm}(t, t_0) \varrho_{mm}^{\mathrm{el}}(t)] -$$
$$- (1 - \delta_{mn}) \sum_r [\gamma_{rm}(t, t_0) + \gamma_{rn}(t, t_0)] \varrho_{mn}^{\mathrm{el}}(t) +$$
$$+ 2(1 - \delta_{mn}) \bar{\gamma}_{mn}(t, t_0) \varrho_{nm}^{\mathrm{el}}(t) +$$
$$+ 0.5 \sum_s \Gamma_s[\delta_{ms} + \delta_{ns}] \varrho_{mn}^{\mathrm{el}}(t) . \tag{32}$$

Here, we have complemented the right-hand side of (32) by adding a term describing the effect of the finite lifetime (Γ_s^{-1} at the site s) of the excitation [17]. One can easily find that for $|t - t_0| \geqq \tau^\mathrm{P}$, when $\gamma_{mn}(t, t_0)$ and $\gamma_{mn}(t, t_0)$ attain constant values, (32) is a simple generalization of the Haken-Strobl-Reineker model to a non-periodic system, with the asymmetric hopping rates $2\gamma_{mn}$.

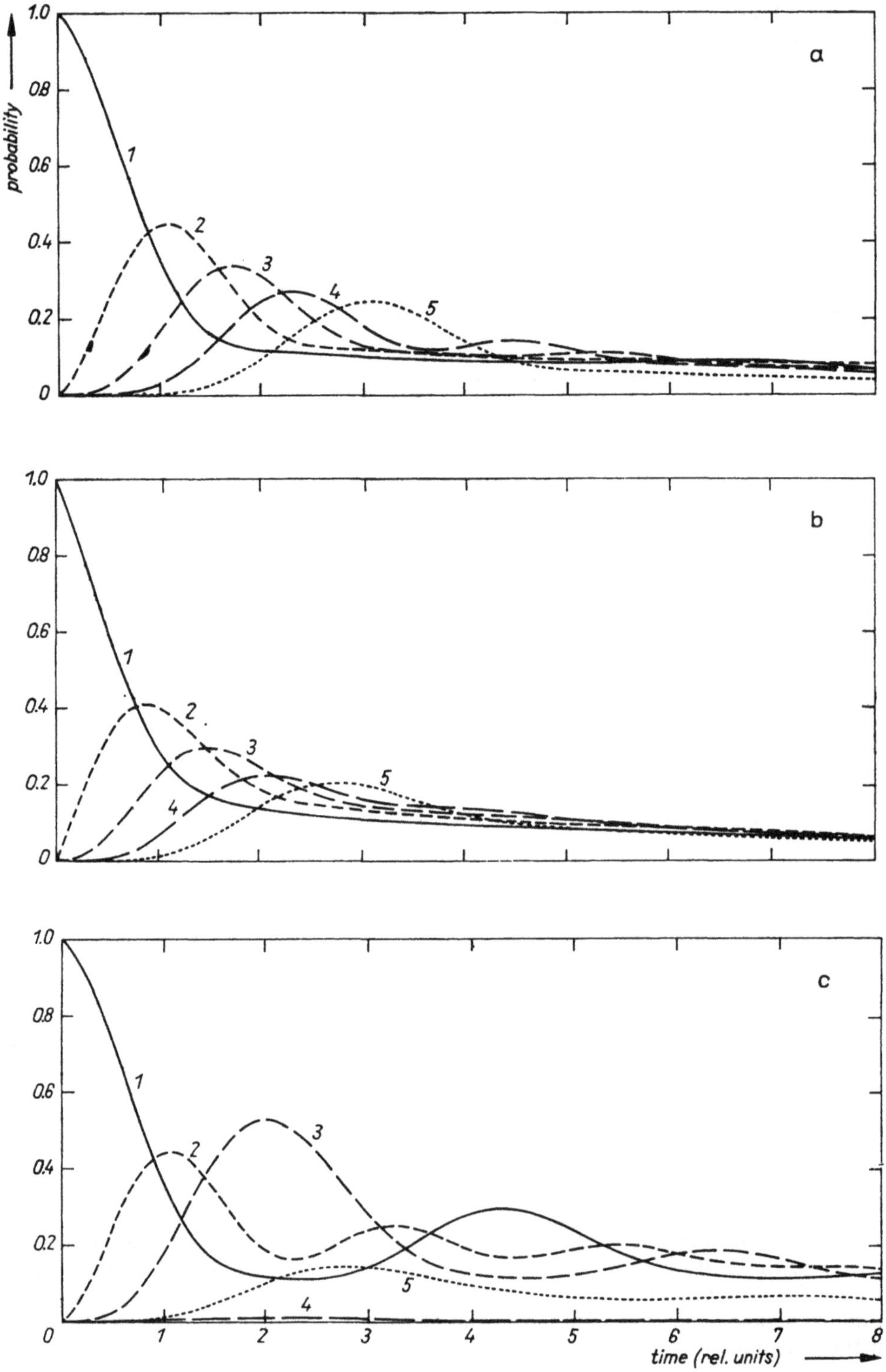
probability
1.0
0.8
0.6
0.4
0.2
0
a
b
c
1
2
3
4
5
0
1
2
3
4
5
6
7
8
time (rel. units)

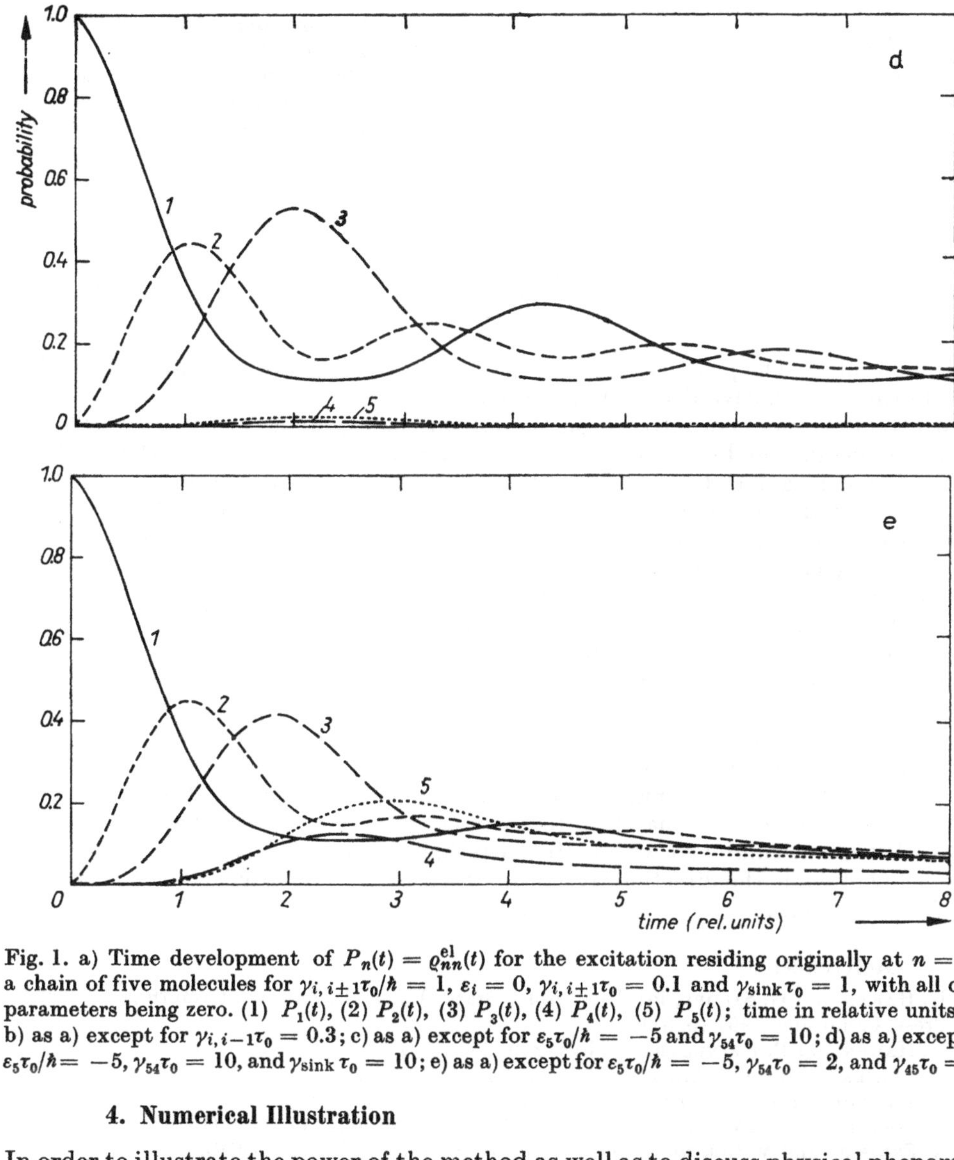

Fig. 1. a) Time development of $P_n(t) = \varrho^{el}_{nn}(t)$ for the excitation residing originally at $n = 1$ in a chain of five molecules for $\gamma_{i,i\pm 1}\tau_0/\hbar = 1$, $\varepsilon_i = 0$, $\gamma_{i,i\pm 1}\tau_0 = 0.1$ and $\gamma_{\text{sink}}\tau_0 = 1$, with all other parameters being zero. (1) $P_1(t)$, (2) $P_2(t)$, (3) $P_3(t)$, (4) $P_4(t)$, (5) $P_5(t)$; time in relative units t/τ_0; b) as a) except for $\gamma_{i,i-1}\tau_0 = 0.3$; c) as a) except for $\varepsilon_5\tau_0/\hbar = -5$ and $\gamma_{54}\tau_0 = 10$; d) as a) except for $\varepsilon_5\tau_0/\hbar = -5$, $\gamma_{54}\tau_0 = 10$, and $\gamma_{\text{sink}}\tau_0 = 10$; e) as a) except for $\varepsilon_5\tau_0/\hbar = -5$, $\gamma_{54}\tau_0 = 2$, and $\gamma_{45}\tau_0 = 0.5$

4. Numerical Illustration

In order to illustrate the power of the method as well as to discuss physical phenomena which get now within the reach of a quantitative discussion, we solved (32) for a number of one-dimensional chains of five molecules, with the initial condition at $t_0 = 0$

$$\varrho^{el}_{mn}(0) = \delta_{mn}\delta_{m1} \tag{33}$$

(i.e. with the excitation residing solely on the site 1 at $t = 0$) assuming that all the coefficients in (32) attain constant values within a time interval which is negligible from the point of view of the time development of the excitation. In all these cases, we chose $\bar{\gamma}_{mn} = 0$ and assumed that $\Gamma_s = \delta_{s5}\gamma_{\text{sink}}$, i.e. that the decay of the excitation takes place just at the fifth molecule, either due to natural radiative or non-radiative processes or due to a sink system (e.g. reaction centre) draining the chain and appended

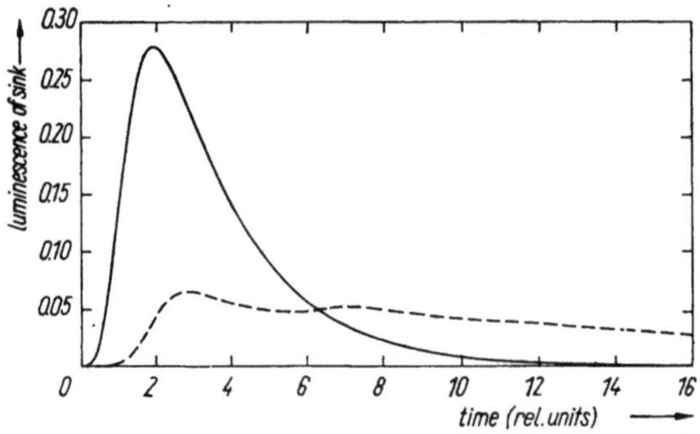

Fig. 2. Time dependence of the sink luminescence $\gamma_{\text{sink}}\tau_0 P_5(t)$ for staircase-like site energies; $(\varepsilon_i - \varepsilon_{i+1})\, \tau_0/\hbar = 1$, $\gamma_{i,i+1}\tau_0/\hbar = 1$, $\gamma_{\text{sink}}\tau_0 = 1$ and $\gamma_{i,i\pm 1}\tau_0 = 0.1$ (dashed line) or $\gamma_{i,i+1}\tau_0 = 0.1$, $\gamma_{i,i-1}\tau_0 = 1$ (full line)

to the molecule 5. The product $\gamma_{\text{sink}} P_5(t)$ (with $P_n(t) = \varrho^{\text{el}}_{nn}$) is now the rate of luminescence in case of radiative decay at the site 5.

Our attention was concentrated on the role of the asymmetry of γ_{mn} and changes of the site energies and γ_{sink} on the time development of $P_n(t)$ in a partly coherent regime. Comparing Fig. 1a with Fig. 1b, c, d, and e shows a typical change of the form of $P_n(t)$ with the asymmetry of all nonzero γ_{mn} along the chain, with the change of the sink-site energy as well as down-to-sink rate, with the same changes combined with the exaggerated value of γ_{sink} and with a modest value of γ_{sink} and modest asymmetry of γ_{mn} at the sink, respectively. Our attention should be concentrated on novel features, namely the depletion of the values of $P_4(t)$ in Fig. 1c and the suppression of the transition rate due to the exaggerated value of γ_{sink} in Fig. 1d. The asymmetry of the hopping rate parameters γ_{mn} near the sink is shown to influence the transition rate even in the neighbourhood of the sustained region near the sink (Fig. 1c as compared to Fig. 1a) and the depletion of $P_4(t)$ in Fig. 1c is not directly due to the change of the site energy ε_5 but due to the high asymmetry of $\gamma_{45} \neq \gamma_{54}$ (compare Fig. 1d). Finally, Fig. 2 shows that the step-like form of the site energies in the direction to the sink may lead to a partial reflection off the sink (luminescence centre) if it is not combined with sufficiently high values of $\gamma_{i,i\pm 1}$ (dashed line); the asymmetry of γ_{ij} in the direction to the sink then positively influences the rate of the transfer to the sink.

References

[1] H. Haken and G. Strobl, in: The Triplet State, Ed. A. B. Zahlan, Cambridge University Press, Cambridge 1967 (p. 311).
[2] H. Haken and P. Reineker, Z. Phys. **249**, 253 (1972).
[3] P. Reineker, Springer Tracts mod. phys. **94**, 111 (1982).
[4] C. Aslangul and Ph. Kottis, Adv. chem. Phys. **41**, 321 (1980).
[5] V. Gorini and A. Kossakowski, J. Math. Phys. (Cambridge) **17**, 1298 (1976).
[6] A. Frigerio and V. Gorini, J. Math. Phys. (Cambridge) **17**, 2123 (1976).
[7] H. Primas, Helv. phys. Acta **34**, 36 (1961).
[8] M. Lax, Rev. mod. Phys. **32**, 25 (1960).
[9] H. Sumi, J. Phys. Soc. Japan **36**, 770 (1974).
[10] V. Čápek, Z. Phys. B **60**, 101 (1985).
[11] N. Hashitsume, F. Shibata, and M. Shingu, J. statist. Phys. **17**, 155 (1977).
[12] F. Shibata, Y. Takahashi, and N. Hashitsume, J. statist. Phys. **17**, 171 (1977).
[13] P. N. Argyres and P. L. Kelley, Phys. Rev. **134**, A98 (1964).
[14] S. Chaturvedi and F. Shibata, Z. Phys. B **35**, 297 (1979).
[15] M. Saeki, Progr. theor. Phys. (Kyoto) **67**, 1313 (1982).
[16] V. Čápek, J. Physique **44**, 767 (1983).
[17] V. Čápek and V. Szöcs, phys. stat. sol. (b) **125**, K137 (1984).

(Received June 19, 1985)

phys. stat. sol. (b) 131, 677 (1985)

Subject classification: 13.4; 13.5.1; 20.3; 22

Institute of Applied Physics, Academy of Sciences of the Moldavian SSR, Kishinev[1])

Shake-Effect in Radiative Auger Recombination of Carriers on Two-Electron Impurity Centres in Semiconductors

By

V. A. Kovarskii, L. V. Chernysh, and M. K. Sheinkman

Dedicated to Prof. Dr. Dr. h. c. Dr. E. h. P. Görlich on the occasion of his 80th birthday

A calculation of the Auger-capture cross-section of a hole on two-electron impurity centres is performed. For the first time in a semiconductor (and probably in solid state, in general) the "shake-effect" accompanying this process is taken into account. Consideration of this effect gives the possibility of truthfull explanation of selection rules and magnitude of "two-electron" luminescence in crystals.

Проведен расчет поперечного сечения Оже-захвата дырки двух-электронным примесным центром. Впервые в полупроводниках (по-видимому, и в твердом теле, вообще), учтен эффект ,,встряски", сопровожающий такой процесс. Учет этого эффекта позволяет правильно объяснить правила отбора и величину ,,двухэлекэронной" люминесценции в кристаллах.

1. Introduction

The Auger recombination mechanism is one of the effective mechanisms of non-radiative recombination in semiconductors. During the Auger recombination process the recombination energy may be transferred to a free carrier [1, 2], as well as, a bound one [3 to 6]. The latter type processes were considered in detail in [5, 6] and were widely investigated in different semiconductors. The Auger processes which occur with participation of carriers bound to the neighbouring centre were found to be rather important. These processes take place both in case of strong doping and in presence of impurity complexes. In the latter case the main interaction mechanism of the Auger process is the dipole–dipole interaction between two centres when one of them captures a free carrier.

The Auger processes under consideration may be very effective in multicharged centres [3 to 5]. Calculations of such processes [3, 7 to 10] were made by the methods commonly used for the Auger processes, but the "shake-effect" was not taken into account [11]. Such "shake-effects" were studied in the radiative Auger transitions in many-electron atoms (see [12]).

This paper is devoted to the investigation of the "shake-effect" in semiconductors with multicharged centres. In semiconductors the "shake-effect" may be observed, for example, in luminescence spectra due to the presence of both free electrons (holes) and free (or bound) excitons.

It is important that the calculations given below take into account the electron-phonon interaction (EPI) in impurity centres. Calculations of the probabilities for Auger processes in impurity centres were given in many papers, but the effect of EPI was underlined only in [13, 14]. In [13] the Auger-process coefficients calculated with consideration of EPI and omitting it were compared. While in [14] the cross-section σ_{ph} of the Auger process involving phonons with energy transfer to a carrier

---

[1]) Ul. Grosul 5, 277028 Kishinev, USSR.

bound on the neighbouring centre was compared with a similar cross-section σ_{phl} of the phononless Auger process. The substantial value of the ratio σ_{ph}/σ_{phl} obtained in [14], shows the significant role of the EPI for the Auger-processes with the participation of localized carriers.²)

Here we consider the shake-effect on the radiative Auger capture of a hole by multicharged centre with account for the interaction of localized carriers with phonons (for simplicity the Einstein model of a crystal is used).

The "shake-effect" is based on the theory of abrupt perturbation [15]. In this case the "shake-approximation" corresponds to the quick change of the mean field for one of the bound electrons (the "shaked out" one) during the quantum transition of the second electron (for example, with the emission of a photon $\hbar\Omega$) (see [12]).

Let us consider a semiconductor with two-electron centres. Such centres have a special feature: after quantum transition of one of the two impurity electrons the second impurity electron has both the ground and excited states. The atoms of transition elements Ni, Co, Fe, Cr with partially filled 3d-shells in semiconductors A_2B_6 (CdS [16], ZnSe etc.), A_3B_5 (GaP, GaAs etc.), A_4 (Ge, Si) serve as examples of such two-electron centres (see [17]). The "shake-effect" can be observed in case of radiative recombination of the hole on the two-electron centre with simultaneous "shaking off" of the second electron into the conduction band, or its "shaking up" into the excited discrete s-levels of the impurity centre (Fig. 1). The shake-effect gives rise to special selection rules for quantum numbers which characterize the states of "shaking up" electrons, i.e. the orbital momentum and its projection must be conserved, $\Delta l = \Delta m = 0$ [12]. In the latter case the radiation which accompanies such a process may contain in the "red" part (long-wavelength part)³) of the spectrum some lines analogous to the atomic Auger lines (frequencies of these lines are determined by the energies of the final quantum states of the "shaken up" electron) and simultaneously the "blue" lines due to radiative recombination of the excited ("shaken up") electron with the hole (it is expected that optical transitions of the "shaken up" electron from the excited s-states to the ground state of the impurity centre are dipole-forbidden).

For the "shaken up" electron we use the mean field approximation in its simplest form, i.e. we introduce the effective charge of the centre Z^* due to the change of the screening field for the first electron after a fast transition of the second electron.

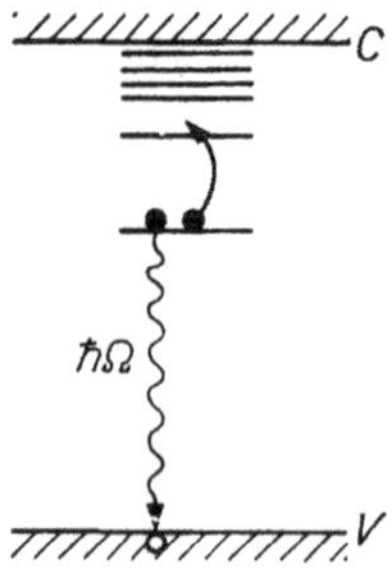

Fig. 1. Radiative Auger process on a two-electron impurity centre. The first electron radiatively recombines with a hole and simultineously "shake up" the second electron into the excited discrete levels of the impurity centre

²) If the coefficient of the Auger process C_1, calculated with the consideration of EPI (in a crystal with fixed magnitude of EPI) is compared with the coefficient of the Auger process C_{10}, calculated without taking into account EPI (i.e., for a crystal with the magnitude of EPI equal to zero), the ratio of C_1/C_{10} is equal to ≈ 4 [14]. If such a problem is solved accounting for EPI, the probability of the phononless Auger process is 10 to 10^3 times less than the probability of the Auger process involving phonons (in the given crystal with fixed magnitude of EPI).

³) With respect to the frequency of the radiative electron transition from the ground impurity level to the valence band (see Fig. 1).

2. Calculation of the Cross-Section of Radiative Recombination with Taking into Account "Shake-Effect"

The expression for the radiative multiphonon Auger capture cross-section of a hole by a centre $\sigma^{\text{opt}}_{\text{Auger}}$ may be obtained with the usual methods of the multiphonon process theory [18, 19] in second order of the dipole–dipole interaction for an optical transition. For the sake of simplicity the electrons are considered to be recognizable. The change of the Coulomb field for a "shaken up" electron is accounted for by the introduction of the shaking parameter $\Delta Z^*/Z^* = (Z - Z)^*/Z^* < 1$ (see [12]). The wave function of the ground electron state $\Psi_\text{i}(\boldsymbol{r}_1, \boldsymbol{r}_2, q)$ of a two-electron centre and the final one $\Psi_\text{f}(\boldsymbol{r}_1, \boldsymbol{r}_2, q)$ may be written in the form

$$\Psi_\text{i}(\boldsymbol{r}_1, \boldsymbol{r}_2, q) = U_\text{c}(\boldsymbol{r}_1)\, U_\text{c}(\boldsymbol{r}_2)\, \psi_\text{s}(\boldsymbol{r}_1)\, \psi_\text{s}(\boldsymbol{r}_2) \prod_f \Phi_{n_f}(q_f - q_f^{(\text{z})})\,, \tag{1}$$

$$\Psi_\text{f}(\boldsymbol{r}_1, \boldsymbol{r}_2, q) = U_\text{v}(\boldsymbol{r}_1)\, U_\text{c}(\boldsymbol{r}_2)\, \psi_{k_f}(\boldsymbol{r}_1)\, \psi_b(\boldsymbol{r}_2) \prod_f \Phi_{n_f}(q_f - q_f^{(\text{b})})\,. \tag{2}$$

Here $\Phi_{n_f}(q_f)$ is the oscillator wave function of the phonon subsystem; $q_f^{(\text{z})}$ is the shift of the equilibrium lattice configuration caused by two localized s-electrons; $q_f^{(\text{b})}$ the analogous quantity determined by the "shaken up" electron in the excited state of the impurity centre; $U_{\text{c(v)}}(\boldsymbol{r})$ the Bloch periodic function of the conduction band (valence band); $\psi_\text{s}(\boldsymbol{r}_1)$, $\psi_\text{s}(\boldsymbol{r}_2)$, $\psi_{k_f}(\boldsymbol{r}_1)$ are the wave functions of the initial two-electron state and of the hole in the valence band, respectively; $\psi_b(\boldsymbol{r})$ is the wave function of the "shaken up" electron (for charge Z^*). Note that the wave function of the two-electron centre may be adopted as a helium-like one, as in [7, 20]. The use of the helium-like wave function to describe two-electron impurity centres (and deep ones), as was shown in [7, 20], gives reasonable agreement with experimental data. It is possible to use the H^--like wave function (H^- is the negative hydrogen ion). The stability of H^- complexes was investigated in [21]. H^- complexes were observed experimentally in Ge and Si [22].

For the cross-section $\sigma^{\text{opt}}_{\text{Auger}}$ we found

$$\sigma^{\text{opt}}_{\text{Auger}} = \sigma_{1\text{s}} + \sum_{b \neq 1\text{s}} \sigma_b\,, \tag{3}$$

where $\sigma_{1\text{s}}$ is the cross-section of radiative multiphonon capture of a hole by an impurity centre, which is not accompanied by the "shake up" of the impurity 1s-electron, and σ_b is the cross-section of the radiative multiphonon Auger transition, which caused the "shake up" of the localized carrier into the b-excited impurity centre state of the s-type; $\sum_{b \neq 1\text{s}}$ is the sum over all excited states of the impurity centre (including the continuous spectrum);

$$\sigma_{1\text{s}} = \frac{4}{3\hbar c^3} \sum_{p=-\infty}^{p_0} \varkappa(\Omega_{p_0})\, \Omega^3_{p_0} R_p(S_0)\, |d_{\text{cv}}|^2\,, \tag{4}$$

$$\sigma_b = \frac{4}{3\hbar c^3} \sum_{p=-\infty}^{p_b} \varkappa(\Omega_{p_b})\, \Omega^3_{p_b} R_p(S_b)\, |d_{\text{cv}}|^2\, |T_b|^2\,. \tag{5}$$

Here

$$d_{\text{cv}} = \frac{1}{V_0} \int_{V_0} \text{d}^3r\, U^*_\text{c}(\boldsymbol{r})\, (e\boldsymbol{r})\, U_\text{v}(\boldsymbol{r}) \int \text{d}^3\varrho\, \psi^*_s(\varrho)\, \psi_{k+}(\varrho)\,, \tag{6}$$

$$R_p(S) = I_p(S) \exp\left[-a\left(\bar{n} + \frac{1}{2}\right) + \frac{p\hbar\omega}{2k_0T}\right], \qquad S \equiv S_0, S_b\,, \tag{7}$$

$$S_0 = a_0 \sqrt{\bar{n}(\bar{n}+1)}; \qquad S_b = a_b \sqrt{\bar{n}(\bar{n}+1)}; \qquad \bar{n} = (e^{\frac{\hbar\omega}{k_0 T}} - 1)^{-1}, \tag{8}$$

$$a_0 \equiv \sum_f (q_f^{(r)} - q_f^{(1)})^2; \qquad a_b \equiv \sum_f (q_f^{(2)} - q_f^{(b)})^2, \tag{9}$$

$$\Omega_{p_0} = (p_0 - p)\,\omega; \qquad \Omega_{P_b} = (p_b - p)\,\omega, \tag{10}$$

$$p_0 = \left\langle \frac{E_g - E_1}{\hbar\omega} \right\rangle; \qquad p_b = \left\langle \frac{E_g - E_1 - E_b}{\hbar\omega} \right\rangle, \langle A \rangle \equiv \text{h.p.}\, A, \tag{11}$$

$$T_b = \int d^3r\, \psi_s^*(\boldsymbol{r})\, \psi_b(\boldsymbol{r}). \tag{12}$$

Here $\varkappa(\Omega)$ is the crystal refraction coefficient; c the light velocity; $I_p(S)$ the modified Bessel function; ω optical phonon frequency; V_0 the unit cell volume; T_b the overlapping integral of the initial state wave function $\psi_s(\boldsymbol{r})$ (with charge Z) with the final state wave function $\psi_b(\boldsymbol{r})$ (with charge Z^*) (an electron is "shaken out" into the b-state); E_g is the band gap value; E_1 and E_2 are the impurity electron ionization energies; E_b is the electron "shaking up" energy (for a helium-like impurity $E_b = -(2e^2/a_0^2 n^2)$, e the electron charge; a_0 the first Bohr radius).

If the wave function of the initial state of the neutral impurity centre is helium-like: $\Psi_i(\boldsymbol{r}_1, \boldsymbol{r}_2) = \psi_s(\boldsymbol{r}_1)\, \psi_s(\boldsymbol{r}_2) = Z/\pi a_0^3\, e^{-(zr_1/a_0) - (zr_2/a_0)}$; $Z = 2 - 5/16$ [23], and the final state wave function is H_e^+-like (with charge $Z^* = 2$, main quantum number n, and orbital quantum number $l = 0$), T_b has the form

$$T_b = \frac{\Delta Z^*}{Z^*} \left(\frac{Z}{Z^*}\right)^{3/2} \frac{2^3}{n^{3/2}} \frac{\left(1 - \frac{1}{n}\right)^{n-2}}{\left(1 + \frac{1}{n}\right)^{n+2}}. \tag{13}$$

(In the case of shaking up in the 2s-state $T_{2s} \approx 0.56(\Delta Z^*/Z^*)\,(Z/Z^*)^{3/2}$, in 3s-state $T_{3s} \approx 0.43(\Delta Z^*/Z^*)\,(Z/Z^*)^{3/2}$, and so on). The case of "shaking out" in the spectrum of continuous states may be calculated similarly.

3. "Shake-Effect" in Crystals with Weak Electron–Phonon Coupling ($a \ll 1$)

In this case in the "red" (long-wavelength) and "blue" (short-wavelength) parts of the luminescence spectrum pairs of phononless lines (PLL) with frequencies Ω_{P_b} and Ω_c may be observed. Their frequencies satisfy the relation

$$2\Omega_{p_0} = \Omega_{p_b} + \Omega_c|_{p=0}, \tag{14}$$

where

$$\Omega_c = \Omega_{p_0} + \frac{E_b}{\hbar}. \tag{15}$$

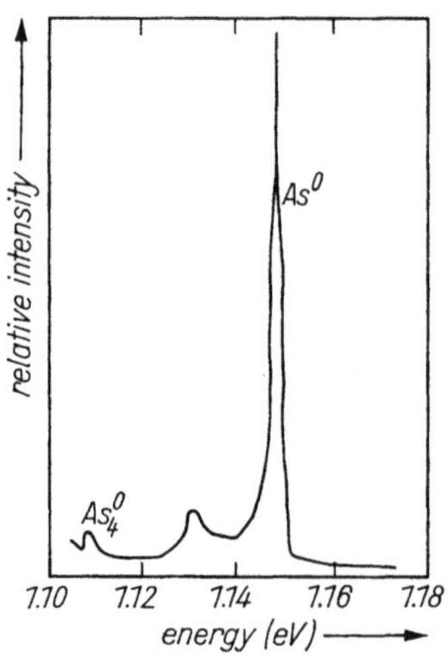

Fig. 2. Photoluminescence spectrum of an exciton bound at a neutral donor impurity centre in Si ([22]). The As^0 line corresponds to the radiative exciton decay only, while the As$_4^0$ line is identified with the "two-electron" transition (when the impurity is excited from 1s- to 2s-state)

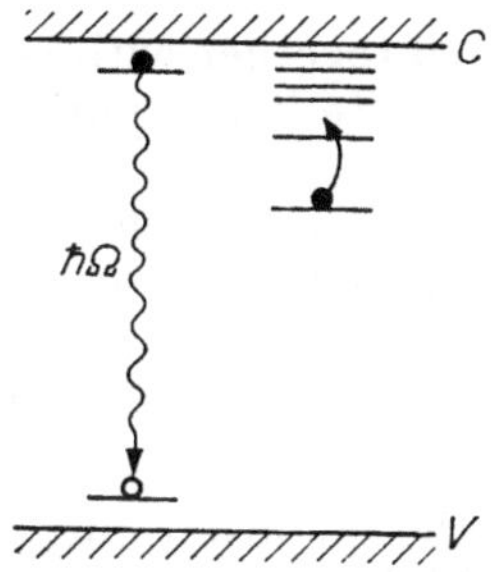

Fig. 3. Schematic drawing of radiative recombination of an exciton bound at the neutral donor centre with simultaneous excitation of a donor electron in a discrete impurity level

The number of such bound PLL pairs will be determined by the number of discrete excited states of the impurity centre. The cross-section of the phononless optical Auger process may be easily obtained,

$$\sigma^{(0)}_{\text{Auger}} = \sigma^{(0)}_{1s} + \sum_{b \neq 1s} \sigma^{(0)}_b \,, \tag{16}$$

where

$$\sigma^{(0)}_{1s} = \frac{4}{3\hbar c^3} |d_{cv}|^2 \varkappa(\Omega_{p_0})\, \Omega^3_{p_0} R_0(S_0) \,, \tag{17}$$

$$\sigma^{(0)}_b = \frac{4}{3\hbar c^3} |d_{cv}|^2 \varkappa(\Omega_{p_b})\, \Omega^3_{p_0} \, |T_b|^2 \,. \tag{18}$$

For helium-like centres in silicon, for example, the ratio $\sigma^{(0)}_{2s}/\sigma^{(9)}_{1s}$ is equal to $\approx 10^{-2}$.

Note that "two-electron" luminescence in silicon, caused by radiative recombination of an exciton bound at the neutral donor centre was experimentally observed in [22]. There, the additional radiation maxima (see Fig. 2) corresponding to the excitation of the donor electron onto discrete s-levels appear in the radiation band "red" (long-wavelength) wing (see [2], too). (Fig. 3 is taken from [2], where this effect is clearly shown). The intensity of the additional lines was two orders less than the main radiation line intensity. Most likely that consideration of the "shake-effect" gives the possibility of trustworthy explanation of selection rules and magnitude of such "two-electron" luminescence in silicon (under the assumption of weak correlation of electron and hole motion). The same explanation may be also used for the interpretation of the selection rules for the donor two-electron transitions in other materials (CdS [24], CdSe [25]).

4. "Shake-Effect" in Crystals with Strong Electron–Phonon Coupling ($a \gg 1$)

In order to observe the "shake-up" effect in this case we must compare σ_b and σ_{1s} for the frequency of the "shake-up" luminescence band maximum: $\Omega^{\max}_{p_b} = (p_b - a_b/2)\,\omega$, this frequency is situated in the "red" wing of the "unshaken" radiation band. For the logarithm of the ratio (if $\beta = \hbar\omega/k_0T < 1$)

$$\ln \frac{\sigma_b}{\sigma_{1s}}\bigg|_{\Omega^{\max}_{p_b}} \equiv \xi = \frac{(\Omega^{\max}_{p_0} - \Omega^{\max}_{p_b})^2}{2a_0\omega^2}\, \beta + 2 \ln |T_b| + \frac{1}{2} \ln \frac{a_0}{a_b} , \tag{19}$$

where

$$\Omega^{\max}_{p_0} = \left(p_0 - \frac{a_0}{2}\right) \omega \,. \tag{20}$$

If $\xi > 0$, the "shaken" "red" luminescence bands will be observed together with the unshaken recombination luminescence band. The estimations for the He-like impurites in silicon show the reality of the $\xi > 0$ condition.

For $a = 3$, the ratio $\sigma_{2s}/\sigma_{1s} \leqq 10^{-2}$. At high temperature ($\beta < 1$) the phonon-involving process has the activation-dependent Auger-recombination cross-section

$$\sigma \sim \exp\left(-\frac{\Delta}{k_0 T}\right); \qquad \Delta = \frac{(|p| - a_b/2)^2}{2a_b} \hbar\omega - \text{activation energy}\,. \tag{21}$$

If the non-radiative Auger-recombination input is considered, the quantum output η of long-wavelength and short-wavelength luminescence should fall down according to the law

$$\eta(T) = \left(1 + A\,\mathrm{e}^{-\frac{\Delta}{k_0 T}}\right)^{-1}, \tag{22}$$

where A is a constant, Δ the activation energy.

In conclusion we want to underline that the investigation of frequency and temperature dependences of the "shake-up" spectra gives an additional way to gather information on the energy structure of excited states of a multicharge impurity centre and complicate complexes of different type.

References

[1] A. R. Beattie and P. T. Landsberg, Proc. Roy. Soc. **A249**, 16 (1959).
[2] P. T. Landsberg, phys. stat. sol. (b) **41**, 457 (1970).
[3] L. Bess, Phys. Rev. **111**, 129 (1958).
[4] M. K. Sheinkman, Fiz. tverd. Tela **5**, 2780 (1963).
[5] M. K. Sheinkman, Fiz. tverd. Tela **7**, 28 (1965).
[6] E. I. Tolpygo, K. B. Tolpygo, and M. K. Sheinkman, Fiz. tverd. Tela **7**, 1790 (1965).
[7] G. Jones and A. R. Beattie, phys. stat. sol. (a) **54**, 193 (1971).
[8] M. Jaros, Solid-State Commun. **25**, 1071 (1978).
[9] F. A. Riddoch and M. Jaros, J. Phys. C **13**, 6181 (1980).
[10] G. P. Neumark, Phys. Rev. B **7**, 3802 (1973).
[11] A. M. Dyhne and G. L. Yudin, Uspekhi fiz. Nauk **125**, 377 (1978).
[12] V. I. Matveev and E. S. Parilis, ibid. **138**, 573 (1982).
[13] J.-T. Rebsch, Solid-State Commun. **31**, 377 (1979).
[14] L. V. Chernysh, M. K. Sheinkman, E. P. Sinyavskii, and V. A. Kovarskii, phys. stat. sol. (b) **100**, K149 (1980).
[15] L. D. Landau and E. M. Lifshits, Quantovaya mechanika, Izd. Nauka, Moscow 1974.
[16] R. Pappalardo and R. E. Dietz, Phys. Rev. **123**, 1188 (1961).
[17] E. M. Omelyanovskii and V. I. Fistul, Primesi prekhodnych metallov v poluprovodnikakh, Metallurgiya, Moscow 1983.
[18] Yu. E. Perlin, Uspekhi fiz. Nauk **50**, 553 (1963).
[19] V. A. Kovarskii, Mnogokvantovie perekhody, Shtiintsa, Kishinev 1974.
[20] A. Glodeanu, phys. stat. sol. (b) **19**, K43 (1967).
[21] M. A. Lampert, Phys. Rev. Letters **1**, 450 (1958).
[22] P. J. Dean, J. R. Haynes, and W. F. Flood, Phys. Rev. **161**, 711 (1967).
[23] H. A. Bethe and E. E. Salpeter, Quantum Mechanics of One- and Two-Electron Atoms, New York, 1957.
[24] C. H. Henry and K. Nassau, Phys. Rev. B **2**, 997 (1970).
[25] C. H. Henry, K. Nassau, and J. W. Shilver, Phys. Rev. B **5**, 458 (1972).

(Received April 25, 1985)

phys. stat. sol. (b) 131, 683 (1985)

Subject classification: 4; 10.2; 14.2; 22.9

Department of Physics, University of Lancaster[1])

Critical Lengths for Formation of Kink Structures in Finite Trans-Polyacetylene Chains

By

R. J. WATTS-TOBIN

Dedicated to Prof. Dr. Dr. h. c. Dr. E. h. P. GÖRLICH on the occasion of his 80th birthday

Polyacetylene chains can have defects such as kinks in their bonding patterns. As these defects are long compared with the carbon–carbon spacing they can be described by a continuum model. In this paper the continuum model is applied to finite chains of polyacetylene, which are of greater practical interest than very long chains. A perturbative calculation shows that there are critical lengths of polymer chain for the appearance of structures containing increasing numbers of kinks. The energy of formation of these structures is also estimated.

Die Polyazetylenketten können Defekte, z. B. Kinks in ihrem Bindungsmuster besitzen. Da diese Defekte lang sind im Vergleich zum Kohlenstoff–Kohlenstoff-Abstand können sie durch ein Kontinuummodell beschrieben werden. In der Arbeit wird das Kontinuummodell auf endliche Ketten von Polyazethylen angewendet, die von größerem praktischem Interesse als sehr lange Ketten sind. Eine Störungsrechnung zeigt, daß es eine kritische Länge der Polymerkette für das Auftreten von Strukturen mit steigender Zahl von Kinks gibt. Die Bildungsenergie dieser Strukturen wird ebenfalls berechnet.

1. Introduction

It has been known for a few years that the theory of bond alternation patterns in polyenes can be turned into a version of the BCS theory of superconductivity. The relation between the systems is mathematical rather than physical. The order parameter in the polyene is just the difference between adjacent bond lengths. The pairing is between electronic states which can be scattered into each other by a lattice distortion of wavelength $2a$. a is the bond length of the undistorted, conjugated, polymer (more accurately the projection of the bond length along the polyene chain) so that the distorted lattice is periodic with period $2a$. The polymer theory is one-dimensional and it has to take into account end groups on the polymer chain and other molecules or groups on or near the chain. These are the main differences from superconductivity theory.

Su et al. [1] derived the form of BCS theory relevant to polyenes from an atomic model and applied it to calculating the ground state of an alternating trans-polyacetylene polymer, and to a numerical estimate of the structure of a "kink" in the bonding pattern, when the polarity of the bond alternation is reversed. Since the order parameter changes sign at a kink it is analogous to a vortex in a type II superconductor.

Takayama et al. [2] pointed out that structures in the bond pattern of polyenes change only over a coherence length of about seven bond lengths. It therefore makes sense to take a continuum limit so as to remove the structure on an atomic scale.

[1]) Lancaster LA1 4YB, England.

The resulting field theory is greatly simplified and becomes even more similar to standard BCS theory. Being in one dimension the theory is much easier, however. Takayama et al. [2] calculated the structure of a kink exactly, and Horovitz [3] and Nakahara and Maki [4] calculated the structure of a periodic array of kinks and antikinks. This is analogous to calculating the structure of a superconductor in the mixed state.

In the avalanche of theoretical papers on polyenes most of the interest has been in very long polymers. Practical samples tend not to be very long: trans-polyacetylene is polymerised from acetylene in the *cis*-form by catalysis and allowed to isomerise by annealing. The resulting polymer has chains of 20 to 70 carbon atoms which are cross-linked to a considerable extent.

2. Structures and Boundary Conditions in Finite Polymers

The length of the polymer will certainly affect the possible bond structures because of their coherence length ξ. The bond pattern cannot change significantly in a distance less than ξ. Short chains will also be greatly affected by the terminating groups at their ends. Chance et al. [5] have made numerical calculations of the structure of kinks in finite polyene chains.

In this paper I show how the simpler models of Su et al. [1] and Takayama et al. [2] can be applied to finite molecules. It becomes possible to describe the structures on them systematically. The results of the theory consist of simple formulae giving critical lengths for the appearance of distorted structure in the bonding pattern of polyenes, expressions for the amplitude of the distortions when the length is greater than critical, and estimates of the energy of formation of the structures. These estimates are important in assessing the stability of the structures.

To consider a finite polyene chain along the lines of Takayama et al. [2] it is first necessary to take account of the fact that the π-electron energy levels are discrete. This has a profound effect on the possibility of distorting the molecule, as was pointed out by Longuet-Higgins and Salem [6]. Fig. 1 illustrates this by showing the Hückel approximations to the π-electron energy levels of the closed-ring polyenes benzene (C_6H_6) and cyclobutadiene (C_4H_4). In benzene the lowest three (bonding) orbitals are well separated in energy from the uppermost three (antibonding)orbitals. Any small distortion of the molecule would have to do work against the σ-bonds without gaining any appreciable energy from mixing the π-orbitals. Benzene is known to be conjugated and undistorted in its ground state. In cyclobutadiene, however, there are two degenerate nonbonding orbitals in the undistorted structure. In the ground state these would be singly occupied. A distortion lowering the symmetry from D_{4h} to D_{2h} would recombine these levels, giving a doubly occupied bonding orbital and an unoccupied antibonding orbital. In such a case the Jahn-Teller distortion always pays energetically. As the number of carbon atoms in the ring is increased, the separation between π-electron energy levels is decreased. The distorted structure with alternating bonds

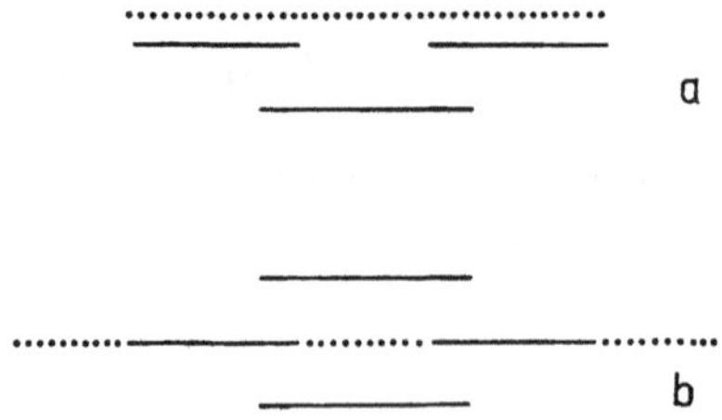

Fig. 1. Energy level diagrams for the π-electrons in small conjugated rings. The Fermi energy is marked as a dotted line. a) Benzene, b) cyclobutadiene

has a gap in its π-electron energy spectrum. When the separation of the single-particle energy levels becomes comparable with the energy gap in an infinite polymer, a polymer with $(4n + 2)$ carbon atoms can distort, as was shown by Longuet-Higgins and Salem [6]. As the ring size is further increased the amount of distortion increases to a definite limit. Correspondingly in a polyene ring with $4n$ carbon atoms the distortion decreases to the same limit as n increases. These effects are described by Salem [7].

Much the easiest way of describing the finite length of a polymer is to use electronic wave functions which are periodic over the length of the polymer. This puts in the discreteness of the energy levels, which is the most important effect of the finite length. It is well to remember that (except for geometrically perfect cyclic polyenes) periodic boundary conditions only give information about structures in which the end groups of the polymers do not play a significant rôle.

3. The Peierls Distortion as a Perturbation

I confine attention to a polymer with $(4n + 2)$ carbon atoms. This makes it possible to describe the nucleation of kink–antikink pairs. The theory is a perturbation expansion about the undistorted, conjugated state of the polymer. The small parameter in the perturbation theory is the amount of distortion in the polymer. As might be expected, one of the approximations to the BCS theory of superconductivity should apply. However, this is not the simple Ginzburg-Landau (GL) theory. The GL theory applies at a second-order phase transition in the superconductor, where the coherence length (the distance over which paired states are correlated) tends to infinity. This can never happen in a finite polymer. As a result one has to use a "non-local" theory: the order parameter becomes small but its gradient does not become small. Helfand and Werthamer [8] have derived the theory of the onset of superconductivity in a type II superconductor in a decreasing magnetic field. The analogy with the onset of distortion in a polymer with increasing magnetic field is quite straightforward. In the superconductor the magnetic flux density penetrating it prescribes the number of vortices per unit area because the flux is quantised. The upper critical field H_{c2} is the highest field the superconductor can support without going into the normal state. H_{c2} is merely the highest of a sequence of critical fields, below each of which a more complicated structure in the order parameter can appear. In the model of a polyene discussed here, the critical length for the appearance of a Peierls distortion found by Longuet-Higgins and Salem [6] appears as just the shortest of a sequence of critical lengths, above each of which a more complicated distortion pattern can appear.

4. Green's Functions for the π-Electrons

The perturbation theory in the distortion $\Delta(x)$ developed directly from the Bogolyubov equations and self-consistency condition in the form given by Takayama et al. [2]. Because it is a perturbation theory it is much easier to use a version of the theory adapted to iterative solution of the equations. I therefore start by rewriting the theory of Takayama et al. [2] in the language of Green's functions.

There is no need to consider finite-temperature effects on the π-electrons. In analogy with the zero-temperature formalism of Abrikosov et al. [9] I start with a spinor-valued annihilation operator $\psi_\sigma(xt)$ in the Heisenberg picture at time t. The top component of ψ acts on valence band electronic states and the bottom component on conduction band states. The single-particle Green's function for a possibly distorted polymer with conjugated electrons is

$$G_\sigma(xt, x't') = -i\langle T[\psi_\sigma(xt)\, \psi_\sigma^+(x't')]\rangle\,, \tag{4.1}$$

where G is a 2×2 matrix.[2])

The Hamiltonian used by Takayama et al. is

$$H = \frac{1}{\lambda} \int \Delta^2(x)\, \mathrm{d}x + \sum_\sigma \int \mathrm{d}x\, \psi_\sigma^+(x) \left[-iv_\mathrm{F}\sigma_3 \frac{\mathrm{d}}{\mathrm{d}x} + \sigma_1 \Delta(x)\right] \psi_\sigma(x)\,. \tag{4.2}$$

Here

$$\frac{1}{\lambda} = \frac{\omega_Q^2}{2g^2} \tag{4.3}$$

is a reciprocal electron–phonon coupling constant, v_F is the Fermi velocity, and σ_1, σ_3 are Pauli spin matrices.

Before writing down the equation of motion for G, I discuss the admissible wave functions, because the periodic boundary conditions are put into the theory by restricting them. Kivelson et al. [10] have shown that for a ring with $(4n + 2)$ atoms, the wave functions are antiperiodic over the ring. This is easy to see. Takayama et al. [2] moved the zero of momentum to the Fermi momentum, which can only be done in one dimension. In their formalism a wave function $\varphi(x) = \mathrm{e}^{ikx}$ represents a state whose true wave function is $\tilde{\varphi}(x) = \mathrm{e}^{i(k+k_\mathrm{F})x}$. The periodic boundary condition applies to $\tilde{\varphi}$, i.e.

$$\tilde{\varphi}(x + L) = \tilde{\varphi}(x)\,, \tag{4.4}$$

where L is the circumference of the ring. To go further one has to look at the original atomic model of Su et al. [1]. One finds that the Fermi wavenumber is

$$k_\mathrm{F} = \frac{\pi}{2a}\,. \tag{4.5}$$

Thus

$$\varphi(x + L) = \mathrm{e}^{-iL\pi/2a}\, \varphi(x) \tag{4.6}$$

and

$$\frac{L}{a} = 4n + 2 \tag{4.7}$$

so

$$\varphi(x + L) = -\varphi(x)\,. \tag{4.8}$$

The equation of motion for the Green's function is obtained from equations (4.1) and (4.2) in the form

$$\left[i \frac{\partial}{\partial t} + iv_\mathrm{F}\sigma_3 \frac{\partial}{\partial x} - \sigma_1 \Delta(x)\right] G_\sigma(xt, x't') = \delta(x - x')\, \delta(t - t')\,. \tag{4.9}$$

G_σ is antiperiodic with period L in both x and x', and so is $\delta(x - x')$. $\Delta(x)$ must be periodic with period L, as follows from the self-consistency equation arising from $\delta H/\delta\Delta(x) = 0$,

$$\Delta(x) = i\lambda\, \mathrm{Tr}\, \sigma_1 G(x, t, x, t + \delta)\,. \tag{4.10}$$

The limit as $\delta \to 0+$ is independent of t because $G(tt')$ depends only on $t - t'$.

The perturbation theory developed here is in an obvious sense the simplest possible because the unperturbed state is the undistorted polymer. In the undistorted state the

[2]) It is important to perform the matrix multiplication of ψ and ψ^+ before ordering the matrix elements with the time-ordering operator T.

Green's function satisfies

$$\left[i\frac{\partial}{\partial t} + iv_F\sigma_3\frac{\partial}{\partial x}\right] G_\sigma^0(x - x', t - t') = \delta(x - x')\,\delta(t - t')\,. \tag{4.11}$$

From now on I will generally leave out the spin indices, but they must be taken into account in calculating physical properties by summing over states. It is easy to calculate G^0, and it follows from (4.9) and (4.11) that the derived perturbation theory comes from iterating the equation

$$G(xx', t - t') = G^0(x - x', t - t') + \\ + \int_0^L \mathrm{d}x'' \int_{-\infty}^{\infty} \mathrm{d}t''\, G^0(x - x'', t - t'')\, \sigma_1 \Delta(x'')\, G(x''x', t'' - t')\,. \tag{4.12}$$

Assuming the polymer is undoped so that the Fermi energy is zero in the formalism of Takayama et al. [2],

$$G^0(xt) = -i\theta(t)\frac{1}{L}\sum_{k_m>0} \exp\,(i\sigma_3 k_m x - iv_F kt) + \\ + i\theta(-t)\frac{1}{L}\sum_{k_m<0} \exp\,(i\sigma_3 k_m x - iv_F kt), \tag{4.13}$$

where

$$k_m = \frac{(2m + 1)\,\pi}{L} \tag{4.14}$$

is a "Fermi" wave number; this makes $G^0(x)$ antiperiodic. In fact, with the usual inner product

$$\langle f \mid g\rangle = \int_0^L f^*(x)\, g(x)\,\mathrm{d}x \tag{4.15}$$

the set of functions $L^{-1/2} \exp\,(ikx)$ for $m = ..., -2, -1, 0, 1, 2, ...$ forms a complete orthonormal set of antiperiodic functions on $0 \leqq x \leqq L$.

It is convenient to take the Fourier transform of G^0,

$$G^0(k_n\Omega) = \int_{-\infty}^{\infty} \mathrm{d}t \int_0^L \mathrm{d}x\, \mathrm{e}^{-ikx+i\Omega t}\, G^0(xt) = (\Omega - v_F k_n \sigma_3)^{-1}\,. \tag{4.16}$$

This form is valid whenever $\mathrm{Im}\,(\Omega) \neq 0$.

For real values of Ω one must take limits

$$\mathrm{Im}\,(\Omega) \to 0- \quad \text{when} \quad \mathrm{Re}\,(\Omega) < \mu$$

and

$$\mathrm{Im}\,(\Omega) \to 0+ \quad \text{when} \quad \mathrm{Re}\,(\Omega) > \mu\,.$$

As $\Delta(x)$ is periodic it must be transformed using "Bose" wave numbers,

$$q_m = \frac{2m\pi}{L}, \tag{4.17}$$

so that

$$\Delta_q = \int_0^L \Delta(x)\, \mathrm{e}^{-iqx}\,\mathrm{d}x\,. \tag{4.18}$$

The basic integral equation (4.12) becomes

$$G(kk'\Omega) = G^0(k\Omega)\, L\delta_{kk'} + \frac{1}{L} \sum_q G_0(k\Omega)\, \sigma_1 \Delta_q G(k-q, k'; \Omega)\,. \tag{4.19}$$

The perturbation series for G in powers of Δ is obtained by iterating this equation.

Many physical quantities can be worked out from G. A compact derivation of the forms and energies of formation of kink structures proceeds by first working out the energy. Using the definition of G in (4.1), the Hamiltonian in (4.2), and the equation of motion (4.9) the expectation value of the energy is

$$H = \frac{1}{\lambda} \int_0^L \Delta^2(x)\, dx + \lim_{t' \to t+} \sum_\sigma \int_0^L dx\, \mathrm{Tr}\, \frac{\partial G_\sigma(xx, t-t')}{\partial t}\,. \tag{4.20}$$

On taking the Fourier transform with respect to x and t, the expression

$$G(t-t') = \int_{-\infty}^{\infty} \frac{d\Omega}{2\pi}\, e^{-\Omega(t-t')}\, G(\Omega) \tag{4.21}$$

enables us to move the upper limit of the Ω-integration round the upper half-plane because $t - t' < 0$. We are left with a contour which starts at $-\infty$, goes just below the real Ω-axis to $\Omega = 0$, and returns to $-\infty$ just above the real axis.[3]) The resulting contour integral will be denoted by $\oint d\Omega$. In this way our general formula for the energy becomes

$$H = \frac{1}{\lambda L} \sum_q |\Delta_q|^2 + \frac{1}{L} \sum_k \oint \frac{\Omega\, d\Omega}{\pi i}\, \mathrm{Tr}\, G(kk\Omega)\,. \tag{4.22}$$

When the perturbation series for G derived from (4.19) is substituted into this formula, only the even-order terms contribute because they have an even number of σ_1-matrices under the trace operation.

From (4.20) we may proceed directly to the "self-consistency equation" for the order parameter. This is just the equilibrium condition

$$\frac{\partial H}{\partial \Delta_q} = 0\,. \tag{4.23}$$

However, it is often useful to work out the energy as a functional of the Δ_qs before finding the Δ_qs which minimises it.

5. Example: A Uniformly Distorted Polymer

I first solve the equation in the case of a uniform order parameter. The results are not new: they merely rewrite the work of Longuet-Higgins and Salem in the continuum approximation of Takayama et al. [2]. However, this calculation gives a simple illustration of the Green's function method. It also helps in interpreting the calculation of more complicated structures which are the main results reported here.

In the case of a polymer with alternating carbon–carbon bond lengths, Δ is independent of x. We obtain

$$G(k\Omega) = \frac{\Omega + v_F k\sigma_3 + \Delta\sigma_1}{\Omega^2 - E^2}\,, \tag{5.1}$$

---

[3]) This deformation is only necessary in the lowest order of perturbation theory.

where

$$E = \sqrt{v_F^2 k^2 + \Delta^2} . \tag{5.2}$$

The self-consistency equation follows from (4.22),

$$\Delta = -\frac{\lambda}{L} \sum_{v_F|k| < \Lambda} \frac{\partial}{\partial \Delta} \oint \frac{\Omega \, d\Omega}{\pi i} \operatorname{Tr} G(k\Omega) . \tag{5.3}$$

Λ is a cut-off because bonding patterns on a scale smaller than the interatomic separation cannot occur.

In this way we obtain the equation determining Δ on a finite chain,

$$\frac{1}{\lambda} = \frac{1}{L} \sum_{v_F|k| < \Lambda} \frac{1}{\sqrt{v_F^2 k^2 + \Delta^2}} = \frac{1}{\pi v_F} \int_0^{\Lambda} \tanh \frac{L \sqrt{\varepsilon^2 + \Delta^2}}{2 v_F} \frac{d\varepsilon}{\sqrt{\varepsilon^2 + \Delta^2}} . \tag{5.4}$$

Letting $L \to \infty$ we find the order parameter in an infinite polymer chain; in the weak coupling limit it is

$$\Delta_\infty = 2\Lambda \exp - \frac{\pi v_F}{\lambda} . \tag{5.5}$$

The integral in (5.4) is a monotonically increasing function of L, and a decreasing function of Δ. It follows that Δ is an increasing function of L, and (putting $\Delta = 0$) that if $L < L_c$, where the critical length is

$$L_c = \frac{\pi \, e^{-\gamma} \, v_F}{\Delta_\infty} , \tag{5.6}$$

the uniform bond alternation cannot exist. This is the result of Longuet-Higgins and Salem [6]. The separation between π-electron energy levels of the undistorted polymer is

$$\Delta E = \frac{2\pi v_F}{L} . \tag{5.7}$$

So another way of expressing the result is that

$$\Delta E < 2 \, e^{\gamma} \, \Delta_\infty \tag{5.8}$$

is necessary for distortion to occur.

6. Critical Lengths

Just as there is a critical length L_c for the onset of a uniform alternation of the carbon–carbon bond lengths, so there is a sequence of critical lengths for the nucleation of more complicated structures. The obvious way of calculating these critical lengths is by linearising the theory. If one wants the critical length for the appearance of the Fourier component Δ_q of the order parameter one can linearise the theory with respect to Δ_q and Δ_{-q} because they are small by definition. In practice one gets simple answers only if the theory is linearised with respect to all Fourier components of Δ. This is entirely correct for the components with shorter wavelengths than Δ_q as these components must vanish at the critical length for Δ_q. However, the Fourier components of Δ with longer wavelengths than Δ_q may not be small, so that linearising the theory with respect to their amplitudes may not be accurate.

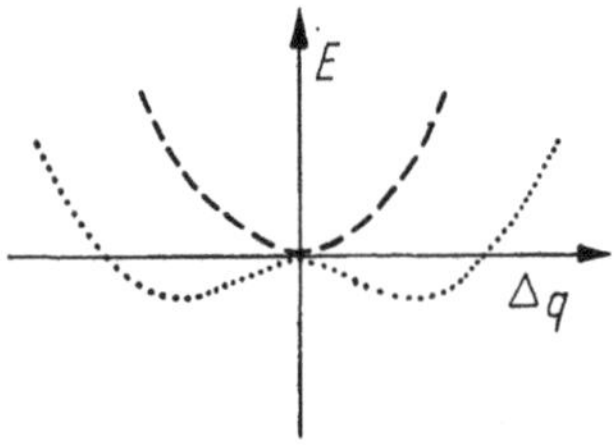

Fig. 2. Graphs of energy E distortion Δ_q in distorted polyacetylene structure with wave number q, for which the critical length is L_q. The dashed line is for a polymer whose length L is less than L_q, and the dotted line for L greater than L_q

The calculation starts from (4.19) expanded to quadratic order in Δ,

$$G(kk'\Omega) = G^0(k\Omega)\,\delta_{kk'}L + G^0(k\Omega)\,\sigma_1\Delta_{k-k'}G^0(k'\Omega) + \\ + \frac{1}{L}\sum_q G^0(k\Omega)\,\sigma_1\Delta_q G^0(k-q)\,\sigma_1\Delta_{k-k'-q}\sigma_1 G^0(k'\Omega)\,. \tag{6.1}$$

It is shown in Appendix A that the quadratic term in Δ leads to a contribution to the energy of the form

$$\Phi_2 = \frac{1}{\pi v_F L}\sum_q |\Delta_q|^2 \ln\frac{L_q}{L}, \tag{6.2}$$

where the critical length L_q is defined in (A9). Thus the perturbation expansion of G up to quadratic order shows that the order parameter $\Delta(x)$ can have a Fourier component Δ_q if $L < L_q$. It will be shown in the next section that the fourth-order term in the energy always has a positive coefficient, so that the energy versus distortion curve has the shape sketched in Fig. 2.

It is interesting to compare these results with the corresponding ones found for an infinite polyacetylene chain by Su et al. [1]. They find that the π-electron energy has a singularity as a function of distortion. It grows faster than quadratically, and always dominates the σ-energy. The π-energy may be identified with the second term in (A1) and the σ-energy with the first term. Our result for a finite chain is that both terms are quadratic for small distortion, the coefficients of the σ- and π-terms being positive and negative, respectively. For a given mode q, the negative π-energy outweighs the positive σ-energy when $L > L_q$. The singularity in the π-energy of an infinite chain ensures that for a finite chain the critical length always exists however weak the coupling constant λ may be.

Table 1 shows critical lengths L_q (measured as the number of carbon atoms in the polymer) calculated from (6.2) using the parameters for trans-polyacetylene given by

Table 1

Critical lengths for trans-polyacetylene

mode number m	critical length L/a
0	12.6
1	50.4
2	93.1
4	134.0
5	181.3

The wave vector of the order parameter in mode m is $q = 2\pi m/L$, and the critical length is given as the number of atoms in the chain.

Su et al. [1]. It is remarkable that L_c ($m = 0$) is much smaller than the others. As m increases, L_q rapidly approaches its asymptotic form

$$L_q \sim \frac{2\pi v_F |m|}{\Delta_\infty}. \tag{6.3}$$

We have a clear prediction that in the absence of doping, kink–antikink pairs cannot be formed on polyacetylene chains of less than about 50 carbon atoms, even as excited states.

7. Amplitudes and Energies of Excited States

The "self-consistency" equation which determines the order parameters Δ from the electronic Green's function (or the Bogolyubov amplitudes) was derived in (4.10). The energy of the system, including the elastic energy of the σ-bonds, was made stationary with respect to variations in $\Delta(x)$. Thus it is assumed that the state is an equilibrium one, though it may be unstable. Bearing in mind this restriction to equilibrium states, I discuss the onset of successively more complex patterns in the order parameter.

The most useful quantity to calculate is the energy, given by (4.22). The calculation in the previous section was equivalent to expanding H in a perturbation series in each Fourier component Δ_q up to $O(\Delta_q^2)$. To describe the growth of the q-mode when the length of the polymer exceeds the critical length L_q it is necessary to carry out the expansion up to $O(\Delta_q^4)$. This is also necessary to calculate the energy of the structure.

In this quartic-order expansion one obtains terms coupling modes of different q as well as those referring to the same mode $\pm q$. This is to be expected: In the kink lattice found by Horovitz, the order parameter is given by a Jacobian elliptic function and not by a pure sine wave. The structure of the off-diagonal terms in perturbation theory is complicated, and I will confine attention to structures consisting of a single mode. In appendix B it is shown that the energy is

$$H - H_0 = -\frac{1}{L}\left[\frac{|\Delta_q|^2}{\pi v_F}\ln\frac{L}{L_q} + \frac{|\Delta_q|^4}{32\pi^3 v_F^3}\,\psi^{(2)}\left(\frac{1}{2} + \frac{|q|\,L}{2\pi}\right)\right]. \tag{7.1}$$

The critical length L_q is defined in (A9), H_0 is the energy of the undistorted polymer, and $\psi^{(2)}$ is the tetragamma function.

In using (7.1) to calculate the energy of a structure it must be remembered that there is a difference between the uniform structure and the other modes. The uniform structure is described just by the $q = 0$ term of the expansion. In the higher modes the q- and $-q$-terms must be considered together: as $\Delta(x)$ is a real function of position its Fourier components satisfy $\Delta_{-q} = \Delta_q^*$. The result is that the energy of the $q = 0$ mode is given up to quartic order by (7.1), but the energy for a higher mode q is got by adding the equal contributions of q and $-q$.

I next use (7.1) to calculate the amplitude of the order parameter and the displacements of the carbon atoms in a definite mode. When $L > L_q$ there is a relative minimum in the energy when

$$|\Delta_q|^2 = -16\pi^2 v_F^2 \ln\frac{L}{L_q}\Big/\psi^{(2)}\left(\frac{1}{2} + \frac{|m|}{2}\right). \tag{7.2}$$

This formula for the equilibrium value of Δ_q is true for all q including $q = 0$. From (7.2) we can calculate immediately the amplitude of the Peierls distortion in the m-th mode.

In the notation of Su et al. [1] the displacement of the n-th atom is[4])

$$u_n = \frac{(-1)^n}{4\alpha L} \sum_p \Delta_p \, e^{ipna} . \tag{7.3}$$

Writing

$$\Delta_p = |\Delta_p| \, e^{ix} = \Delta^*_{-p} \tag{7.4}$$

we have

$$u_n = \frac{(-1)^n \Delta_0}{4\alpha L} + \frac{(-1)^n}{2\alpha L} \sum_p |\Delta_p| \cos (pna + \beta) . \tag{7.5}$$

So although the prefactor in (7.2) for Δ_0 is the same as that for Δ_q' when $q \neq 0$, a given value of Δ_0 gives a distortion only half as great as the same value of $|\Delta_q|$ when $q \neq 0$. As in the case of the energy, this effect is due to the linking of Δ_q and Δ_{-q} when $q = 0$. With only one mode excited, the amplitude of the Peierls distortion is

$$d = \frac{\Delta_0}{4\alpha L} \tag{7.6}$$

for $q = 0$, and

$$d = \frac{|\Delta_q|}{2\alpha L} \tag{7.7}$$

if $q \neq 0$.

For comparison, the homogeneous distortion amplitude in an infinite chain is

$$u = \frac{\Delta_\infty}{4\alpha} .$$

The perturbation formulas are only valid if $0 < L - L_q \ll L_q$. So in (7.2) we may write

$$\ln (L/L_q) = \frac{L - L_q}{L_q}$$

and put $L = L_q$ in the prefactor.

Using (7.6) and (7.7) we find the ratio of the distortion to that in an infinite chain as

$$\frac{d}{u} = \frac{\sqrt{(L - L_q)/L_q}}{\exp \psi(\tfrac{1}{2}) \sqrt{-\psi^{(2)} (\tfrac{1}{2})}} \tag{7.8}$$

when $m = 0$, and

$$\frac{d}{u} = 2 \sqrt{\frac{L - L_q}{L_q}} \Bigg/ \left[\exp \psi \left(\frac{1}{2} + \frac{m}{2}\right) \sqrt{-\psi^{(2)} \left(\frac{1}{2} + \frac{m}{2}\right)} \right] \tag{7.9}$$

when $m \neq 0$. These formulae are tabulated for a fixed ratio $(L - L_q)/L_q = 0.1$ in Table 2. Thus L has been varied as q is. This keeps the length of the chain just above the critical length L_q, so that the perturbation theory is valid.

Again, the energy may be calculated by substituting (7.2) into(7.1).Making the same approximations we have

$$\frac{\langle H - H_0 \rangle}{\Delta_\infty} = 2 \left(\frac{L - L_c}{L_c}\right)^2 \Bigg/ \left[\psi^{(2)} \left(\frac{1}{2}\right) \exp \psi \left(\frac{1}{2}\right) \right] \tag{7.10}$$

[4]) The Δ_p in this formula does not have the same meaning as Δ_p in the paper of Su et al. [1], where the formalism refers to a homogeneous distortion.

Table 2
Amplitudes of the atomic displacements and energies. The polymer length is varied so as to be 10% greater than the critical length of the mode

mode number m	atomic displacement	energy
0	0.549	−0.0085
1	0.726	−0.0296
2	0.670	−0.0465
3	0.652	−0.0649
4	0.644	−0.0838
5	0.640	−0.1031

The amplitude of the atomic displacement is measured in units of the displacement in a long polymer in the uniform mode $m = 0$. The energy is given in units of the band gap Δ_∞ in a long polymer. Su et al. [1] give $u = 0.004$ nm and $\Delta_\infty = 1.4$ eV.

for the uniform mode $m = 0$, and

$$\frac{\langle H - H_0 \rangle}{\Delta_\infty} = 4 \left(\frac{L - L_q}{L_q} \right)^2 \Bigg/ \left[\psi^{(2)} \left(\frac{1}{2} + \frac{m}{2} \right) \exp \psi \left(\frac{1}{2} + \frac{m}{2} \right) \right] \tag{7.11}$$

for $m = 0$.

The most noticeable feature of these results is that the finite length of the polymer has a considerable effect on the energy and the atomic displacement in a homogeneous polymer ($m = 0$) and in a polymer carrying one kink–antikink pair ($m = 1$). For the longer values of m the polymer is necessarily longer and the energy and atomic displacements rapidly take their asymptotic forms as $m \to \infty$. In fact the energy per kink–antikink pair is

$$\frac{H - H_0}{m} \to 2 \left(\frac{L - L_q}{L_q} \right)^2 \Delta_\infty \tag{7.12}$$

and the displacement is

$$d \to 2u \sqrt{\frac{L - L_q}{L_q}} \,. \tag{7.13}$$

The energies of the states calculated here are all negative and decrease with mode number. This is because they are calculated relative to the undistorted state, and the length is varied to keep within the regime of validity of perturbation theory. For a polymer of fixed length L, the lowest energy state is always the uniformly distorted state $m = 0$, provided $L > L_c$. The energy of this state is only given accurately by perturbation theory when the length of the polymer is slightly greater than the critical length L_c. States with kinks, i.e. $m > 0$, are excited states of the polymer unless they are stabilised by doping.

8. Conclusion

In this paper I have shown that the model of kinks in trans-polyacetylene proposed by Takayama et al. [2] may be extended to describe a finite polymer. By applying periodic boundary conditions to describe the length of the polymer, I have shown that there is a sequence of critical lengths L_q below which the number of kink–antikink pairs in the bonding pattern is less than $m = Lq/2\pi$.

Above $L = L_q$, I have derived formulae for the structure and energy of the state with m kinks. The structure and energy are considerably affected by the finite length of the polymer when $m = 0$ (uniform distortion) and $m = 1$ (one kink–antikink pair).

The cyclic boundary conditions can provide a guide to the behaviour of real polymers. Applied to a polymer with $(4n + 2)$ carbon atoms, these boundary conditions make it possible to describe the transition from the conjugated (undistorted) state to the uniformly (Peierls-) distorted state, as well as the more complicated states carrying kinks. These boundary conditions have the advantage that they lead to a perturbation expansion about the undistorted state of the polymer. One could also develop the theory of a finite polymer with $4n$ atoms subjected to periodic boundary conditions. The perturbation expansion would be about the Peierls-distorted state, and would be considerably more complicated. Such a perturbation theory could not describe kinks. At a kink the order parameter vanishes and has therefore changed by a large amount from its value in the Peierls-distorted state. However, the theory should be able to describe polarons. These are localised dips in the order parameters.

The field theory of Takayama et al. [2] should be a good approximation for the purposes of this paper, because it only considers polymer structures which vary slowly with position on the scale of the coherence length. However, an atomic model would be necessary to describe a polymer with more realistic boundary conditions than the cyclic ones I have used. On the face of it, cyclic boundary conditions apply to cyclic polyenes. However, cyclic polyenes cannot be made with more than about 20 carbon atoms in the ring, and this is too few to support kink structures. Also they have many flexing modes which would interfere with the Peierls distortion.

The effect of the cyclic boundary conditions I have applied to a ring containing $(4n + 2)$ carbon atoms is to force the ring into the undistorted state, as benzene is, unless the ring is longer than the critical length. It is important to see what information this model can give about polyenes with terminated chains. Most end groups that might be attached to a polyene chain would tend to force it to distort to some extent, though it is possible to have end groups which would encourage the undistorted structure. By contrast, the $4N$ boundary condition could be used to study biologically interesting chain compounds like the carotenes in which a Peierls distortion is forced by the end groups. An interesting question is whether the distortion is uniform along the chain, or whether polarons can form.

Cyclic boundary conditions should not be regarded as a literal description of finite polyene chains, but under favourable circumstances they can be used to model more realistic boundary conditions.

One central topic in the theory of these electroactive polymers must be mentioned. It is the effect of doping the polymers. This plays a very important rôle both experimentally and theoretically. For example, the electrical conductivity of polyacetylene increases dramatically at a very high level of doping. Again, in the calculation of the structure of a lattice of kinks in polyacetylene by Horovitz [3] and by Nakahara and Maki [4] it was assumed that the polymer was doped. The calculations of these authors can be adapted to an undoped polymer, leading to the conclusion that the undoped kink lattice would be unstable, and the kinks and antikinks would annihilate each other. The doped structure is stable, however.

In the same approximation as used by these authors, the effect of doping can be incorporated into the formalism of this paper. The approximation consists in regarding the dopant molecules as simply altering the Fermi energy of the electrons, and neglecting the interaction between the dopant molecules and the polymers. The results of extending the formalism of this paper in this way are reasonably simple. I will illustrate them by describing the effect on the critical lengths of the polymers. Raising the Fermi level μ has no effect on the critical lengths L_q until M (say) of the levels $v_F k$ of the undistorted system have been brought below μ. Then the critical lengths for structures involving more than $2M$ kink–antikink pairs are reduced below their values for $\mu = 0$. The critical lengths for structures involving fewer than $2M$ kink–antikink pairs actually become shorter than the critical length for the uniform distortion at the same value of μ. The kink structures are stabilised by doping.

The formula for the critical length L_q is

$$\ln \frac{L_q \Delta_\infty}{4\pi v_F} = \frac{1}{2}\left[\psi\left(\frac{1}{2} + \left|M + \frac{m}{2}\right|\right) + \psi\left(\frac{1}{2} + \left|M - \frac{m}{2}\right|\right)\right]. \tag{8.1}$$

Here as usual $q = 2\pi m/L_q$ and M, the number of levels pulled below the Fermi energy, is the integer in the range

$$\frac{L\mu}{2\pi v_F} - \frac{1}{2} < M < \frac{L\mu}{2\pi v_F} + \frac{1}{2}. \tag{8.2}$$

The effect of doping will almost certainly be neeced to describe biologically interesting conjugated polymers. In the interest of clarity I have described the theory of finite polymers in detail only when $\mu = 0$. In this case, states of the polymer carrying kinks are excited states. They are local minima of the energy, but the global minimum is the uniformly distorted state $m = 0$.

Appendix A

Critical lengths and the linearised self-consistency condition

As explained in Section 6, to work out the critical lengths we need the self-consistency condition linearised in each Δ_q, or the energy up to quadratic order. Substituting (6.1) for the Green's function into (4.22) for the energy, the second-order correction to the energy becomes

$$\Phi_2 = \frac{1}{\lambda L}\sum_q |\Delta_q|^2 + \frac{1}{L}\sum_q |\lambda_q|^2 \frac{1}{L}\sum_k \frac{d\Omega}{2\pi i} \operatorname{Tr} G^0(k\Omega)\, \sigma_1 G^0(k - q, \Omega)\, \sigma_1 . \tag{A.1}$$

Here we have used the cyclic invariance of the trace and the identity

$$G^0(k\Omega)\, G^0(k\Omega) = -\frac{\partial}{\partial\Omega} G^0(k\Omega) \tag{A.2}$$

to integrate the expression by parts. Writing $\Omega = iv_F z$ we find

$$\oint \frac{d\Omega}{2\pi i} \operatorname{Tr} G^0(k\Omega)\, \sigma_1 G^0(k - q, \Omega)\, \sigma_1 =$$

$$= -\frac{1}{\pi v_F}\int_{-\infty}^{\infty} \frac{dz}{(z + ik)\,(z - ik + iq)} = I \text{ (say)}. \tag{A.3}$$

I vanishes unless $q - k < 0 < k$ when it is $-2/v_F(2k - q)$, or $k < 0 < q - k$ when it is $+2/v_F(2k - q)$. So for $q \geqq 0$

$$I = \begin{cases} -\dfrac{2}{v_F(2k - q)} & (k > q)\,, \\[2ex] \dfrac{2}{v_F(2k - q)} & (k < 0) \end{cases} \tag{A.4}$$

and I vanishes for $0 < k < q$. Thus

$$\frac{1}{L}\sum_k I = -\frac{4}{v_F L}\sum_{k>0}\frac{1}{2k+q}. \tag{A.5}$$

Similarly, when $q < 0$

$$\frac{1}{L}\sum_k I = \frac{4}{v_F L}\sum_{k>0}\frac{1}{2k-q}. \tag{A.6}$$

The k-summation must be restricted by $v_F\,|k| < \Lambda$, and we obtain

$$\frac{1}{L}\sum I = -\frac{1}{\pi v_F}\sum_{n=0}^{\infty}\left\{\frac{1}{n+\frac{1}{2}+\frac{|m|}{2}} - \frac{1}{n+\frac{1}{2}+\frac{|m|}{2}+[L\Lambda/(2\pi v_F)]}\right\}$$

$$= -\frac{1}{\pi v_F}\left[\ln\frac{L\Lambda}{2\pi v_F} - \psi\left(\frac{1}{2}+\frac{|m|}{2}\right)\right], \tag{A.7}$$

where ψ is the digamma function. We are assuming that the polymer is sufficiently long that the band width Λ is much greater than the separation of single-particle states $2\pi v_F/L$. Using (5.5) for the order parameter on an infinite chain, (A7) may be substituted into (A1) to give the second-order contribution in the form

$$\Phi_2 = \frac{1}{L}\sum_q |\Delta_q|^2\,\frac{1}{\pi v_F}\left[\psi\left(\frac{1}{2}+\frac{|m|}{2}\right) - \ln\frac{L\Delta_\infty}{4\pi v_F}\right]. \tag{A.8}$$

The zero-order energy Φ_0 is independent of each Δ_q. The undistorted polymer will be unstable for the distortion described by Δ_q provided the coefficient of $|\Delta_q|^2$ in (A8) is negative. So for each mode there is a *critical length* L_q given by

$$\ln\frac{L_q\Delta_\infty}{4\pi v_F} = \psi\left(\frac{1}{2}+\frac{m}{2}\right), \tag{A 9}$$

where $q = 2\pi m/L$. The mode can only exist for $L > L_q$. For $m = 0$, L agrees with the critical length for uniform distortion given in (5.6).

Appendix B

Quartic contributions to the energy

From (4.19) and (4.21), expanded to quartic order in all Δ_q, it follows that the quartic contribution to the energy is

$$\Phi_4 = \frac{1}{L^3}\sum_{pqr} C_{pqr}\,\Delta_p\Delta_q\Delta_r\Delta_{-p-q-r}\,, \tag{B.1}$$

where

$$C_{pqr} = \int_{-i\infty}^{i\infty}\frac{d\Omega}{4\pi i}\,\frac{1}{L}\sum_k \mathrm{Tr}\;[G^0(k\Omega)\,\sigma_1 G^0(k-p,\,\Omega)\,\sigma_1\,\times$$

$$\times\, G^0(k-p-q,\,\Omega)\,\sigma_1 G^0(k-p-q-r,\,\Omega)\,\sigma_1]\,. \tag{B.2}$$

In writing down (B2) I have used (A2) and integrated (4.22) by parts, as in Appendix A.

The expression for C_{pqr} may be written

$$C_{pqr} = \frac{1}{2\pi v_F^3} \int_{-\infty}^{\infty} dz \frac{1}{L} \sum_k T \,, \tag{B.3}$$

where

$$T = \frac{1}{(k - iz)(k - p + iz)(k - p - q - iz)(k - p - q - r + iz)} . \tag{B.4}$$

We are interested in the terms in Φ_4 containing only Δ_p and Δ_{-p}. The possibilities are (i) $r = -q = p$, (ii) $-r = q = p$, (iii) $r = q = -p$. In case (iii),

$$\begin{aligned} \frac{1}{L} \sum_k T &= \frac{1}{L} \sum_k \frac{1}{(k - iz)^2 (k - p + iz)(k + p + iz)} \\ &= \frac{1}{2} \frac{\partial}{\partial z} \frac{\tanh (Lz/2)}{(2iz - p)(2iz + p)} . \end{aligned} \tag{B.5}$$

We may assume $p \neq 0$ ($p = 0$ comes under case (i)), so the expression vanishes on integrating with respect to z. Case (ii) gives the same expression replacing $k - p$ by k and z by $-z$. So both contributions to Φ_4 vanish. In case (i),

$$T = \frac{1}{(2 + ik)^2 (z - ik + ip)^2} . \tag{B.6}$$

It is convenient to integrate with respect to z before summing over k. If $p \geqq 0$,

$$\int_{-\infty}^{\infty} T \, dz = \begin{cases} \dfrac{4\pi}{(2k - p)^3} & (k > p) \,, \\ 0 & (0 < k < p) \,, \\ -\dfrac{4\pi}{(2k - p)^3} & (k < 0) \,, \end{cases} \tag{B.7}$$

$$C_{p,-p,p} = \frac{4}{v_F^3 L} \sum_{k>0} \frac{1}{(2k + p)^3} . \tag{B.8}$$

Similarly, if $p < 0$,

$$C_{p,-p,p} = \frac{4}{v_F^3 L} \sum_{k>0} \frac{1}{(2k - p)^3} . \tag{B.9}$$

These terms may be combined to give

$$\Phi_4 = -\frac{1}{32 v_F^3 L} \sum_p |\Delta_p|^4 \, \psi^{(2)} \left(\frac{1}{2} + \frac{L|p|}{4\pi}\right) . \tag{B.10}$$

References

[1] W. P. Su, J. R. Schrieffer, and A. J. Heeger, Phys. Rev.Letters42, 1698 (1979); Phys. Rev. B 22, 2099 (1980).
[2] H. Takayama, Y. R. Lin-Liu, and K. Maki, Phys. Rev. B 21, 2388 (1980).

[3] B. HOROVITZ, Phys. Rev. Letters **46**, 742 (1981).
[4] M. NAKAHARA and K. MAKI, Phys. Rev. B **24**, 1045 (1981).
[5] R. R. CHANCE, D. S. BOURDREAUX, J. L. BREDAS, and R. SILBEY, Phys. Rev. B **28**, 6927 (1983).
[6] H. C. LONGUET-HIGGINS and L. SALEM, Proc. Roy Soc. **A251**, 172 (1959).
[7] L. SALEM, Molecular Orbital Theory of Conjugated Systems, W. A. Benjamin, Inc., New York 1966 (p. 503).
[8] E. HELFAND and N. R. WERTHAMER, Phys. Rev. Letters **13**, 686 (1964); Phys. Rev. **147**, 288 (1966).
[9] A. A. ABRIKOSOV, L. P. GORKOV, and I. E. DZHYALOSHINSKI, Methods of Quantum Field Theory in Statistical Physics, Chap. 2, Prentice-Hall, Inc., Englewood Cliffs (N.J.) 1964.
[10] S. KIVELSON, T.-K. LEE, Y. R. LIN-LIU, I. PESCHEL, and LU YU, Phys. Rev. B **25**, 4173 (1982).

(Received January 2, 1985)

phys. stat. sol. (b) 131, 699 (1985)

Subject classification: 13.1 and 13.5; 21.4

Physikalisches Institut der Technischen Universität Clausthal[1])
and Sonderforschungsbereich 126 der Deutschen Forschungsgemeinschaft, Göttingen/Clausthal

Low Energy Electron Exchange Excitation of 4f Transitions and Location of 4f Levels in Rare Earths

By

E. BAUER and J. KOŁACZKIEWICZ[2])

Dedicated to Prof. Dr. Dr. h. c. Dr. E. h. P. GÖRLICH on the occasion of his 80th birthday

Eu, Gd, and Tb single crystal films and atomically dispersed layers adsorbed on W(110) surfaces are studied by low energy electron loss spectroscopy. A very sharp dominant loss peak at about 4 eV is analyzed in detail and attributed to spin exchange excitation of the 4f level. This interpretation agrees very well with self-consistent spin-polarized free atom and band structure calculations. The results are compared with XPS and other experimentel data which leads to a re-consideration of the interpretation of the 4f XPS binding energies.

Eu-, Gd- und Tb-Einkristallschichten und atomar disperse Adsorptionsschichten auf W(110)-Oberflächen werden mit Energieverlustspektroskopie mit langsamen Elektronen untersucht. Ein sehr scharfes dominierendes Verlustmaximum bei etwa 4 eV wird im Detail analysiert und einer Spinaustauschanregung im 4f-Band zugeschrieben. Diese Interpretation ist in sehr guter Übereinstimmung mit selbstkonsistenten spin-polarisierten Rechnungen für freie Atome und mit Bandstrukturrechnungen. Die Ergebnisse werden mit XPS und anderen experimentellen Daten verglichen, was zu einer kritischen Prüfung der Interpretation von 4f-XPS-Bindungsenergien führt.

1. Introduction

Recently Sticht and Kübler [1] reported results of an accurate band structure calculation of Gd and pointed out the lack of *experimental* data with which the results could be compared. It is the purpose of this paper to present such data which were obtained a number of years ago but not published because of their apparent contradiction to other experimental data [2 to 6] and because of the lack of *theoretical* data which would have allowed a trustworthy interpretation. The specific problem which we want to address here is the spin–orbit splitting of the 4f states of Eu, Gd, and Tb metals in which the spin-down states are close to the Fermi level and either unoccupied (Eu, Gd) or partially occupied (Tb).

It has been accepted for a long time that the binding energies of the 4f levels of Eu and Gd are located about 2 and 8 eV, respectively, below the Fermi level. These values were obtained already in the early XPS studies [2, 3] and confirmed by later more refined XPS experiments [4, 5] and synchrotron radiation UPS experiments [6]. They agree well with renormalized atom calculations involving 5d screened neutral final states [7, 8] and thermodynamical calculations of the binding energies [9]. The assumption of complete screening of the 4f hole which was made in these calculations

---

[1]) D-3392 Clausthal-Zellerfeld, FRG.

[2]) Present address: Institute of Experimental Physics, University of Wrocław, ul. Cybulskiego 36, 50-205 Wrocław, Poland.

was questioned, however [10], and other theoretical considerations have led to much *larger* 4f binding energies (25 to 26 eV) for Eu, Gd, and Tb [11]. Much *smaller* values than the generally accepted ones were suggested by other experimental evidence: (i) From soft X-ray spectroscopy a maximum distance between occupied 4f states and Fermi level of 1.2 eV for Eu and 5.5 eV for Gd was deduced [12]. (ii) An optical absorption threshold in Gd at 6.1 eV was unambiguously associated with a transition from a 4f level, presumably into a d-state [13]. This puts an upper limit of 6.1 eV to the 4f binding energy because the final state of the transition must be above the Fermi level. Also in disagreement with the XPS-derived, accepted values are more recent spin-polarized scalar-relativistic band structure calculations for Gd [1, 14] which both give a binding energy of 4.5 to 4.6 eV, a value which is also obtained with different forms of the exchange correlation potential [15]. The exchange splitting between 4f spin-up and spin-down levels is also the same (about 5 eV) in both calculations [1, 14].

2. Low Energy Electron Loss Spectroscopy on Rare Earth Single Crystals

In order to clarify the discrepancy between theory and experiment obviously other experimental techniques have to be used. Low energy electron loss spectroscopy (LEELS) is uniquely suited for the present problem: (i) At low energies spin exchange scattering is very probable which allows a direct measurement of the exchange splitting. (ii) Due to the high density of states and the small width of the 4f levels — even taking spin–orbit coupling into account — sharp features are obtained in the loss spectrum. (iii) Although non-direct transitions occur at low energies [16] the flatness of the 4f bands ensures that no significant broadening of the loss peak occurs.

We have, therefore, applied LEELS to the study of the 4f levels of Eu, Gd, and Tb, using a cylindrical mirror analyzer with a co-axial gun, normal incidence, and primary energies down to 20 eV. The resolution as determined by the inflection width of the elastic peak was as low as 0.6 eV at the lower energies. The samples were thick epitaxial layers grown on a W(110) surface and adsorption layers on this surface in the coverage range from 0 to about 1.5 monolayers. In the submonolayer range the rare earths at room temperature form a two-dimensional gas (below 0.2 monolayers) and with increasing coverage a series of ordered phases [17, 18]. Thus it is possible to study energy losses from atomically dispersed to single crystalline samples. The pressure in the system was in the 10^{-11} Torr range after extended outgassing of the evaporators and the cleanness of the sample surface was monitored by Auger electron spectroscopy (AES) which was also used to determine the coverage. In addition to AES and LEED work function change ($\Delta\varphi$) measurements were made. The loss spectra were taken in the derivative mode in which the zeros of the signal (or the midpoint between maximum positive and negative amplitudes) determine the locations of loss peak maxima and the amplitude is a measure of the peak intensity for comparable inflection widths.

Fig. 1 shows spectra from thick epitaxial layers typical for low primary energies before excitations, which are excited at higher energies, dominate the spectra. Sharp loss peaks are seen at 3.90, 4.45, and 3.79 eV (± 0.04 eV) for Eu, Gd, and Tb, respectively. Intense sharp energy losses between 3 and 5 eV have been repeatedly observed in low energy electron scattering and attributed to various causes. A 4.74 eV loss in Gd and a 5.2 eV loss in Dy which increase with decreasing energy (from 600 to 150 eV) were assigned to surface plasmon excitation at the metal–oxide interface [19]. This interpretation is excluded by later studies on Lu [20], Er [21], Yb [22], Tm [23], and Gd [24] which showed that the sharp 3 to 5 eV loss peak was eliminated by oxidation.

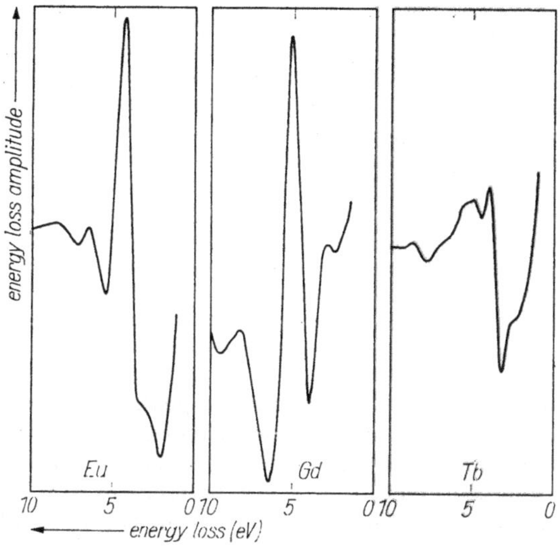

Fig. 1. Low energy electron loss spectra from Eu, Gd, and Tb single crystal layers on W(110) in the energy range from about 1.5 to 10 eV. First derivative mode, modulation voltage 0.5 V_{pp}, primary energy 50 eV

The loss was, therefore, attributed to an unspecified valence band–conduction band transition but a 4f state as initial state was excluded on the basis of the accepted large values of the 4f binding energy [20 to 23]. In a recent systematic LEELS study this loss was found in the complete rare earth series [24] and tentatively attributed to an one-electron surface transition because it increases in intensity with decreasing incident energy and is eliminated by oxidation.

In our study the 4 eV loss peak at low primary energy had the same inflection width as the elastic peak in the case of Gd and Tb, only in the case of Eu in which this peak is less separated from the 2.4 eV loss the inflection width is larger. This suggests that the convolution of initial and final state widths is at or below the energy resolution of the system which requires very narrow initial and final states, a condition which is fulfilled by 4f states. Although optical transitions between spin-up and spin-down states are forbidden, such transitions can occur easily at low energies by spin exchange scattering. This interpretation is supported by the following observations: (i) The loss is approximately equally strong in Eu and Gd which both have the 4f configuration $4f^7$, but significantly weaker in Tb with the ground state configuration $4f^7\,4f^2$. (ii) The loss intensity increases strongly with energy decreasing from 1000 to 250 [24] to 175 eV [23]. It should be noted, however, that this increase does not continue down to energies as which one would expect the maximum of the spin exchange excitation cross-section (about twice the threshold energy). Our excitation functions (Fig. 2) show a maximum at about 100 eV. There are at least two reasons for this unexpected result:

(i) In our experimental configuration inelastically scattered electrons can be observed only when excitation is preceded or followed by diffraction, because the differential cross-section for inelastic scattering decreases strongly with increasing scattering angle [25]. The analyzer accepts only electrons scattered into cone angles between 132° and 143°, the single crystal layer produces at low energies only a few diffracted beams which deviate increasingly from the acceptance cone with decreasing energy. As a consequence the measured inelastic intensity decreases in spite of an increasing cross-section.

(ii) In all heavy rare earth metals the optical properties of which have been studied [26, 27], $\varepsilon_1 = -1$ at (4.2 ± 0.3) eV [26, 28]. Thus a surface plasmon loss peak occurs

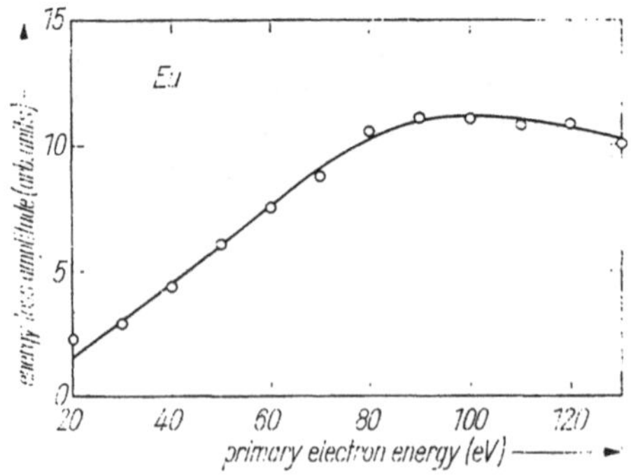

Fig. 2. Excitation function of the 3.9 eV loss from an Eu single crystal layer on W(110). Similar data are obtained also for Gd and Tb. The sharp loss amplitude shown in Fig. 1 is normalized to constant incident current and plotted

at about the same energy as the postulated $4f_\uparrow \rightarrow 4f_\downarrow$ excitation peak. Our data indeed indicate peak shape changes (broadening and shifting) with increasing energy which suggest a composite peak as do the data of [24] (Fig. 1d), at least for Gd and Tb. $f_\uparrow \rightarrow f_\downarrow$ transition and surface plasmon creation have certainly different excitation functions so that a complicated average excitation function results. On the basis of our data up to about 500 eV we suggest that at 250 eV the 4 eV loss is mainly due to surface plasmon excitation. This explains the observed disappearance upon oxidation [20 to 24] and its slow decrease with increasing filling of the $4_\downarrow$ shell [26]. In Yb and Lu it would be solely due to surface plasmons.

3. Low Energy Electron Loss Spectroscopy on Rare Earth Adsorption Layers

More light is shed on the nature of the 4 eV loss by comparing the spectra from rare earth single crystals with those of atomically dispersed rare earths. If the loss is due to the excitation of (6s5d) electrons it should be strongly influenced by the different environment. If it is an intrinsic surface transition on the crystalline rare earth, it should disappear. But if it involves only 4f electrons as is the case in a $4f_\uparrow \rightarrow 4f_\downarrow$ excitation, then it should be rather insensitive to the environment.

The LEELS experiments on Eu, Gd, and Tb adsorbed on W(110) and on Eu and Tb adsorbed on W(112) very clearly show the latter phenomenon. In all cases loss peaks near 4 eV loss are seen which are as sharp as those from rare earth single crystals. Also the peak energies are nearly the same as on single crystals. Shifts up to 0.3 eV were observed but it is not certain whether these are caused by the superposition of adsorbate and substrate losses or whether they are true environment-induced shifts of the energy levels involved. This question is not very relevant here because the unchanged sharpness of the 4 eV peak alone excludes (6s5d) electrons and its large intensity excludes the surface transition explanation. The coverage dependence of the 4 eV loss amplitude is shown in Fig. 3 for three different primary energies. Up to about 0.2 monolayers (ML) the curves nearly coincide, but at higher coverages they differ strongly. This can be explained by differences in the diffraction patterns. Below about 0.2 ML the layer is disordered at room temperature, the diffraction is diffuse at all energies and the chance for inelastically scattered electrons to enter the analyzer is approximately independent of energy. Above 0.2 ML sharp superlattice diffraction patterns with extra beams develop which allow more or fewer inelastically scattered electrons to enter the analyzer depending upon their direction with respect to the acceptance cone. Above 0.6 ML a hexagonal pattern forms whose spot distribution does not change significantly with further increasing coverage. Only some intensity changes occur at lower energies. The loss amplitudes are closely related to these changes.

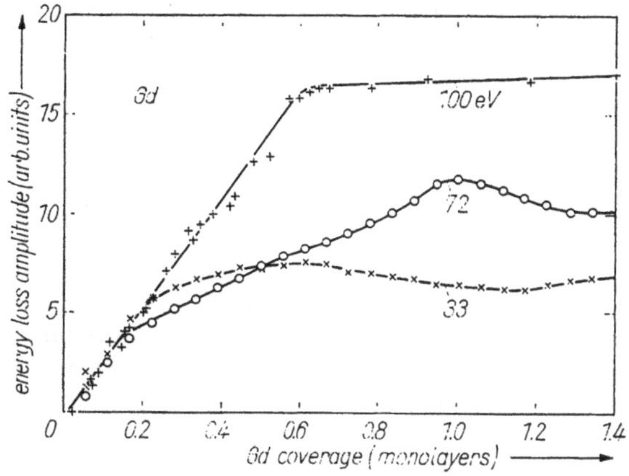

Fig. 3. Coverage dependence of the 4.45 eV loss from Gd adsorption layers on W(110) for three different primary energies (33, 72, and 100 eV) normalized to constant incident current

It is obvious that excitation functions as the one shown in Fig. 2 are strongly influenced by diffraction and that they are not simply related to the true excitation function. A better approach to the true function is to limit the analyzer acceptance to a constant angular region around *one* diffracted beam or to use unoriented polycrystalline samples but even in this case the extraction of the true excitation function is complicated by multiple scattering [29].

Although the spin exchange excitation mechanism cannot be proved via excitation functions there is strong experimental evidence supporting it, mainly the sharpness of the loss peaks, their appearance both in single crystals and in atomically dispersed adsorbed layers, and their absence in optical spectra.

4. Comparison with Theory

Additional support for our interpretation of the 4 eV loss is provided by a comparison with theory. In view of the insensitivity of the 4 eV loss to the environment — crystal or adsorbed layer — a feature noted long ago also in the 4f XPS spectra of the rare earths, the comparison can be made both with band structure calculations and with free atom calculations. As mentioned in the Introduction, reliable spin-polarized band structure data are available for Gd [1, 14]. They give an exchange splitting — measured between $f_\uparrow$ and $f_\downarrow$ band centers — of 5.1 [1] and 4.8 eV [14] and a $4f_\uparrow$ binding energy — measured from the Fermi level — of 4.6 [1] and 4.5 eV [14]. If spin–orbit coupling is introduced as a perturbation, the $4f_\uparrow$ band splits into seven narrow subbands with a total width of about 0.72 eV and the $4f_\downarrow$ band assumes a similar width [1]. The calculated $f_\uparrow$–$f_\downarrow$ splitting of 4.8 to 5.1 eV agrees well with our energy loss value for Gd, 4.45 eV, but so does also the $4f_\uparrow$–$4f_\downarrow$ splitting obtained with the same local exchange-correlation potential for the free Gd atom, 5.16 eV [30]. It is interesting to note in passing that the binding energy of the $4f_\uparrow$ electron in the free atom is 10.9 eV while that in the crystal is 7.6 eV referred to the vacuum level (assuming a work function of 3.1 eV [31]) in contrast to the value of 11.3 eV (with the same reference level) obtained by Herbst et al. [8] in agreement with the XPS data.

In view of the environment independence of the 4f features mentioned before — which is also evident in calculations using the same approximations — and because we are not aware of any self-consistent spin-polarized band structure calculations for Eu and Tb, we will make our comparisons in the following only with the free atom calculations of Gollisch [30]. He used also a non-local correlation potential [32] for Gd and Eu which gave similar results for the $4f_\uparrow$–$f_\downarrow$ splitting $\Delta E_{\uparrow\downarrow}$ as the standard local potential, 4.87 versus 5.17 eV for Gd and 3.87 versus 4.48 eV for Eu, so that the comparison will be limited to the local potential calculations. In these he obtained for Tb $\Delta E_{\uparrow\downarrow} = 3.40$ and 4.51 eV in the lowest energy configuration of the free atom, f^7f^2, and of the atom in the solid, $f^7f^1d^1$, respectively. The three $\Delta E_{\uparrow\downarrow}$ values 4.48, 5.17, and 3.40 eV compare favorably with the observed losses of 3.90, 4.45, and 3.65 eV for Eu, Gd, and Tb, respectively.

It should be kept in mind, however, that the experiment does not measure $\Delta E_{\downarrow\uparrow}$ but more precisely the energy difference between the configurations $4f^nf^{n'}$ and $4f^{n-1}f^{n'+1}$: the incident electron with spin-down "drops" into the $4f_\downarrow$ level and "kicks" one $4f_\uparrow$ electron out. The difference between the total energies of these configurations is 3.875, 4.458, and 3.788 eV for Eu, Gd, and Tb ($f^7f^1d^1$ configuration) in excellent agreement with the measured data for Eu and Gd. It appears unlikely that the agreement in all three cases is accidental so that the best presently available theory strongly supports our interpretation.

5. Consequences for XPS Analysis

If our interpretation of the 4 eV loss in Eu, Gd, and Tb is correct — an assumption strongly supported by experimental and theoretical evidence — then the interpretation of 4f XPS spectra, at least those of the rare earths, must be revised. The basis for the good agreement between the measured 4f XPS energies and the widely accepted theoretical models was the assumption that the 4f hole created by the photoionization process is completely "screened" by charge instantaneously rushing in form the environment of the ion, occupying a 5d level at the Fermi level and thus neutralizing the ion before the photoelectron escapes. It is difficult to understand in this model the environment insensitivity of 4f features. In particular, it is difficult to visualize the formation of a d^{m+1} configuration in trifluorides which has led already earlier to a questioning of the model [10].

The model predicts for Gd a $\Delta E_{\uparrow\downarrow}$ value in utter disagreement with the self-consistent spin-polarized calculations discussed above: 11.5 [8] instead of 5.16 eV [30]. $\Delta E_{\uparrow\downarrow}$ should, however, be independent of the screening model because screening is expected to be about the same for spin-up and spin-down states [8]. Thus the cause of the discrepancy may not only be in the model but also in the computational procedure. In [8] $\Delta E_{\uparrow\downarrow}$ is obtained from the total energies E_{n-1}, E_n, E_{n+1} per unit cell of the metal of the configurations $4f^{n-1}(6s5d)^{m+1}$, $4f^n(6s5d)^m$, and $4f^{n+1}(6s5d)^{m-1}$ as $\Delta E'_{\uparrow\downarrow} = |(E_{n-1} - E_n) + (E_{n+1} - E_n)|$. Therefore, we consider in our comparison for Gd not only the final (ionic) configurations $4f^66s^25d$ and $4f^76s^25d^0$ but also the screened configurations $4f^66s^25d^2$ and $4f^86s^25d^0$ which are possible only when the atom is imbedded in an electron-donating environment. Gollisch [30] obtained the following differences between the total energies of the relevant configurations and the ground state configurations: $E_{622} - E_{721} = 10.25$, $E_{820} - E_{721} = -0.58$, $E_{621} - E_{721} = 18.61$, and $E_{720} - E_{721} = 6.90$ eV for Gd and for the corresponding (unscreened) ionic configurations of Eu and Tb $E_{620} - E_{720} = 11.60$ and $E_{820} - E_{920} = 8.22$ eV, respectively.

With these numbers at hand we can now resolve the discrepancies noted before. The energy splitting $\Delta E'_{\uparrow\downarrow}$ for Gd obtained here, $\Delta E'_{\uparrow\downarrow} = |(E_{622} - E_{721}) + (E_{820} - E_{721})| = 10.83$ eV is comparable to that calculated by Herbst et al. [8], 11.5 eV. Thus, neither computational differences nor screening are responsible for the discrepancy between $\Delta E'_{\uparrow\downarrow}$ and $\Delta E'_{\uparrow\downarrow}$ but rather the definition of the 4f level splitting used in the two quantities: in $\Delta E'_{\uparrow\downarrow}$ two electrons are transferred from the 5d level to the exchange-split 4f levels, in $\Delta E'_{\uparrow\downarrow}$ one electron is transferred from the 4f spin-up level to the 4f spin-down level which is the correct definition of the 4f splitting. Thus the quantity $\Delta E'_{\uparrow\downarrow}$ calculated by Herbst et al. [8] is *not* the energy difference between occupied and unoccupied 4f levels and no significant discrepancy exists between our measurements and Gollisch's calulations, on the one hand, and previous calculations, on the other.

The absolute energy values differ more, however, as a comparison of the 4f binding energies of the free Eu, Gd, and Tb atoms shows. For Eu an experimental value for the minimum 4f binding energy is available from photoemission experiments (9.6 eV [33]) and for all three atoms estimates of this energy based on spectroscopic data: 9.8, 17.2, and 6.5 eV for Eu, Gd, and Tb, respectively [8]. Herbst et al. [8] report 7.5, 14.8, and 3.0 eV, respectively, while Gollisch [30] obtains 11.6, 18.6, and 8.2 eV, respectively, somewhat closer to experiment. Absolute energies have, therefore, to be taken with some caution and the good agreement with previous theory [7 to 9] and 4f XPS binding energies may well be accidental.

6. Summary and Conclusions

In this paper we have reported an experimental study on Eu, Gd, and Tb single crystals and adsorbed layers of the 4 eV energy loss of electrons in the energy range below about 100 eV. The experimental evidence leaves little doubt that this loss is at these low energies predominantly due to a spin exchange excitation of a $4f_{\uparrow}$ electron into the $4f_{\downarrow}$ level. As a consequence of this interpretation the $4f_{\uparrow}$ band in Gd can lie at most about 4.5 eV below the Fermi level in agreement with spin-polarized band structure calculations [1, 14] but in strong disagreement with the widely accepted XPS-derived location, 8 eV. This location is strongly supported by calculations requiring complete screening of the 4f hole by electrons drawn from the environment of the hole [7 to 10].

The screening model predicts for Gd also a much larger $4f_{\uparrow}$–$4f_{\downarrow}$ exchange splitting, 11.5 eV [8], than deduced here, 4.5 eV, in agreement with the spin-polarized band structure calculations, 5 eV [1, 14]. This discrepancy is shown to be due mainly to an inappropriate definition of the exchange splitting energy in [8].

Our data do not give absolute values for the 4f binding energies $E_{4f\uparrow}$ in metals but rather upper limits for them because the $4f_{\downarrow}$ band must be at least partially above the Fermi level in order to be partially unoccupied, a prerequisite for a spin flip transition. They confirm previous suggestions of upper $E_{4f\uparrow}$ limits for Gd of 5.5 [12] and 6.1 eV [13]. They also strongly support spin-polarized band structure calculations [1, 14] and call for a re-evaluation of 4f excitation energies by extending the spin-polarized calculations to other rare earth metals.

Acknowledgements

One of the authors (E.B.) wishes to thank H. Gollisch for making his data available before publication, the other (J.K.) acknowledges the 1980 Welch Foundation fellowship during the initial phase of the work and support by the DFG via SFB 126 for the remaining period.

References

[1] J. Sticht and J. Kübler, Solid State Commun. **53**, 529 (1985).
[2] P. O. Hedén, H. Löfgren, and S. B. M. Hagström, Phys. Rev. Letters **26**, 432 (1971); phys. stat. sol. (b) **49**, 721 (1972).
[3] F. R. McFeely, S. P. Kowalczyk, L. Ley, and D. A. Shirley, Phys. Letters A **45**, 227 (1973).

[4] M. CAMPAGNA, G. K. WERTHEIM, and Y. BAER, in: Photoemission in Solids, Vol. 2, Ed. L. LEY and M. CARDONA, Springer-Verlag, Berlin 1979 (p. 217).
[5] P. A. COX, J. K. LANG, and Y. BAER, J. Phys. F 11, 113 (1981).
[6] R. KAMMERER, J. BARTH, F. GERKEN, A. FLODSTRØM, and L. J. JOHANSSON, Solid State Commun. 41, 435 (1982).
[7] J. F. HERBST, D. N. LOWY, and R. E. WATSON, Phys. Rev. B 6, 1913 (1972).
[8] J. F. HERBST, R. E. WATSON, and J. W. WILKINS, Phys. Rev. B 13, 1439 (1976); 17, 3089 (1978).
[9] B. JOHANSSON, J. Phys. F 4, L169 (1974).
[10] E. I. ZABOLOTSKII, YU. P. IRKHIN, and L. D. FINKELSHTEIN, Soviet Phys. – Solid State 16, 733 (1974).
[11] YA. M. KSENDOV, Soviet Phys. – Solid State 22, 995 (1980).
[12] J.-M. MARIOT, R. C. KARNATAK, and C. BONNELLE, J. Phys. Chem. Solids 35, 657 (1974).
[13] J. L. ERSKINE and C. P. FLYNN, Phys. Rev. B 14, 2197 (1976).
[14] B. N. HARMON, J. Physique 40, C5-G5 (1979).
[15] L. A. COLE and B. N. HARMON, J. appl. Phys. 52, 2149 (1981).
[16] E. BAUER, Z. Phys. 224, 19 (1969).
[17] E. BAUER and J. KOŁACZKIEWICZ, Proc. IX. Internat. Vacuum Congr. and V. Internat. Congr. Solid Surfaces, Madrid 1983 (p. 363).
[18] E. BAUER and J. KOŁACZKIEWICZ, to be published.
[19] E. V. TSVEIMAN, V. S. REDKIN, V. V. ZASHKVARA, and M. I. KORSUNSKII, Soviet Phys. – Solid State 13, 2339 (1972).
[20] J. ONSGAARD, S. TOUGAARD, P. MORGAN, and F. RYBORG, J. Electron Spectroscopy and related Phenomena 18, 29 (1980).
[21] E. BERTEL, F. P. NETZER, and J. A. D. MATTHEW, Surface Sci. 103, 1 (1981).
[22] E. BERTEL, G. STRASSER, F. P. NETZER, and J. A. D. MATTHEW, Surface Sci. 118, 387 (1982).
[23] J. ONSGAARD, I. CHORKENDORFF, and O. SØRENSEN, Physica Scripta T4, 169 (1983).
[24] F. P. NETZER, G. STRASSER, G. ROSINA, and J. A. D. MATTHEW, Surface Sci. 152/153, 765 (1985).
[25] E. BAUER, J. Vacuum Sci. Technol. 7, 3 (1970); Vacuum 22, 539 (1972).
[26] J. G. ENDRIZ and W. E. SPICER, Phys. Rev. B 2, 1466 (1970).
[27] J. H. WEAVER and D. W. LYNCH, Phys. Rev. Letters 34, 1324 (1975).
[28] J. H. WEAVER, C. KRAFKA, D. W. LYNCH, and E. E. KOCH, Physics Data 18-2, Fachinformationszentrum Karlsruhe, 1981.
[29] W. DIETZEL, G. MEISTER, and E. BAUER, Z. Phys. B47, 189 (1982).
[30] H. GOLLISCH, to be published.
[31] H. B. MICHAELSON, J. appl. Phys. 48, 4729 (1977).
[32] L. FRITSCHE and H. GOLLISCH, Z. Phys. B48, 209 (1982).
[33] S.-T. LEE, S. SÜZER, E. MATTHIAS, R. A. ROSENBERG, and D. A. SHIRLEY, J. chem. Phys. 66, 2496 (1977).

(Received July 1, 1985)

phys. stat. sol. (b) **131**, 707 (1985)

Subject classification: 13.4; 16; 19; 20.1; 22.2.1

Zentralinstitut für Elektronenphysik der Akademie der Wissenschaften der DDR, Berlin[1])

Photoinduced Optical Absorption of Cr^{4+} in Semi-Insulating GaAs:Cr

By

W. ULRICI

Dedicated to Prof. Dr. Dr. h. c. Dr. E. h. P. GÖRLICH on the occasion of his 80th birthday

The optical absorption of Cr^{4+} created during illumination with $h\nu > 0.75$ eV in semi-insulating GaAs:Cr is investigated at low temperatures (< 150 K). Analyzing the concentrations of Cr^{2+}, Cr^{3+}, and Cr^{4+} during illumination and measuring the excitation spectrum of Cr^{4+}, the creation mechanism of photoinduced Cr^{4+} conclusively proved to be the two step process $Cr^{3+} + h\nu \rightarrow Cr^{2+} + hole_{vb}$ and $hole_{vb} + Cr^{3+} \rightarrow Cr^{4+}$. The concentration of photoinduced Cr^{4+} is measured in dependence on the contents of Cr^{2+} and Cr^{3+} present in the samples before illumination, on temperature, and on illumination intensity. Its time dependence of rise after switching on and of decay after switching off the illumination is measured under various conditions. The results are interpreted using dynamic equations for the concentration of Cr^{2+}, Cr^{3+}, and Cr^{4+} during illumination basing on the assumption that only excitation and capture processes via the valence band contribute to the formation of photoinduced Cr^{4+} and that the tunneling process $Cr^{4+} + Cr^{2+} \rightarrow 2\,Cr^{3+}$ is responsible for its decay.

Die durch Bestrahlung mit Licht $h\nu > 0{,}75$ eV bei tiefen Temperaturen (< 150 K) in semiisolierendem GaAs:Cr erzeugte optische Absorption des Cr^{4+} wird untersucht. Aus der Analyse der Konzentrationen von Cr^{2+}, Cr^{3+} und Cr^{4+} während der Bestrahlung und aus der Messung des Anregungsspektrums des Cr^{4+} wird eindeutig bewiesen, daß der Erzeugungsmechanismus des photoinduzierten Cr^{4+} der Zwei-Schritt-Prozeß $Cr^{3+} + h\nu \rightarrow Cr^{2+} + hole_{vb}$ und $hole_{vb} + Cr^{3+} \rightarrow Cr^{4+}$ ist. Die Konzentration des photoinduzierten Cr^{4+} wird in Abhängigkeit vom Gehalt an Cr^{2+} und Cr^{3+} in den Proben vor der Bestrahlung, von der Temperatur und von der Bestrahlungsintensität gemessen. Außerdem werden deren zeitlicher Verlauf des Anwachsens nach Einschalten und des Abfallens nach Abschalten der Bestrahlung unter verschiedenen Bedingungen gemessen. Die Ergebnisse werden durch dynamische Gleichungen für die Konzentrationen des Cr^{2+}, Cr^{3+} und des Cr^{4+} während Bestrahlung interpretiert, deren Gültigkeit auf der Voraussetzung beruht, daß nur Anregungs- und Einfangprozesse über das Valenzband zur Bildung des photoinduzierten Cr^{4+} beitragen und darauf, daß der Tunnelprozeß $Cr^{4+} + Cr^{2+} \rightarrow 2\,Cr^{3+}$ für dessen Zerfall verantwortlich ist.

1. Introduction

Semi-insulating (si) GaAs:Cr is characterized by the presence of both Cr^{2+} and Cr^{3+} on Ga lattice sites. Besides these stable charge states, chromium can also be converted into the transient charge state Cr^{4+} by illumination with sub-band-gap light at low temperatures. This has been shown unequivocally by EPR experiments providing the following main results [1 to 4]. (i) During illumination with light in the energy range 0.75 to 1.5 eV a new isotropic EPR spectrum due to Cr^{4+} appears and the Cr^{2+} signal existing before increases in connection with a simultaneous decrease of the Cr^{3+} signal. (ii) After switching on the light, the Cr^{2+} and Cr^{4+} signals rise to their saturation values with a time constant depending on light intensity. After

---

[1]) Hausvogteiplatz 5–7, DDR-1086 Berlin, GDR.

switching off the light, these signals decay non-exponentially with a characteristic time scale of seconds to minutes. (iii) Illumination with light in the energy range 0.45 to 0.75 eV during the decay of the signals accelerates this decay.

In an earlier note [5] on a photoinduced optical absorption of si GaAs:Cr at 78 K was reported occurring in the energy region 0.4 to 0.8 eV. From the behaviour of this absorption being completely analogous to that of the EPR signal of Cr^{4+} ((i) to (iii)) its assignment also to Cr^{4+} was deduced. This interpretation was strongly supported by the results of photo-Hall and photoconductivity experiments on si GaAs:Cr [6, 7]. From recent experiments on p-conducting GaAs:Cr:Zn, in which all chromium is in the charge state Cr^{4+}, the spectral shape of the optical absorption due to the presence of Cr^{4+} in the whole energy region from its onset at 0.4 eV up to the absorption edge of GaAs at 1.52 eV has been derived and analyzed in terms of photoionization transitions [8]. This knowledge allows now to make a more sophisticated analysis of the photoinduced optical absorption of Cr^{4+}.

In the present paper the properties of the photoinduced optical absorption of Cr^{4+} (furtheron labelled Cr^{4+}_{pi}) will be reported in more detail. It is aimed at clarifying the creation mechanism of Cr^{4+}_{pi}, the dependence of the maximum attainable amount of Cr^{4+}_{pi} on the chromium content, on temperature, and on the illuminating light intensity. Furtheron, the transient behaviour of Cr^{4+}_{pi}, i.e. the time dependence of Cr^{4+}_{pi} after switching on or off the exciting illumination, is investigated. The results of all of these experiments will be discussed on the basis of dynamic equations for the concentration of the three charge states Cr^{2+}, Cr^{3+}, and Cr^{4+} under illumination taking into account only excitation and capture processes via the valence band as well as the destruction of Cr^{4+}_{pi} by tunneling.

2. Experimental

All samples investigated were grown by the liquid encapsulated Czochralski (LEC)-method doped with chromium added to the melt in the range 0.1 to 1 mg Cr/g GaAs. The boules were obtained from two different pullers, a normal pressure one and a high pressure one. Additionally some samples from other suppliers (Czechoslovakia, Soviet Union) were investigated.

The samples (about $7 \times 12\ mm^2$, thickness 2 to 5 mm) were polished mirror-like on the large as well as the small long faces and mounted in a variable temperature (6 to 500 K) metal cryostat with LiF windows. For the illumination during the absorption measurements (perpendicular to the measuring light beam) a tungsten halogen lamp together with metal interference filters were used. All optical absorption experiments described were carried out after bleaching the absorption due to the presence of $EL2^0$ centres (see [9, 10]) at $T = 78$ K. The chromium concentrations in the samples ($[Cr^{2+}]$, $[Cr^{3+}]$) were evaluated from optical absorption at $T = 78$ K as described in detail in [10].

3. Results

3.1 The creation mechanism of Cr^{4+}_{pi}

In Fig. 1 the basic experimental result is shown. Curve 1 is the optical absorption spectrum of a si GaAs:Cr sample after removing the superimposed absorption due to $EL2^0$. Curve 2 is measured during illumination of the sample with $h\nu_{exc} = 1.24$ eV light. The most remarkable feature seen in curve 2 is the appearance of a photoinduced absorption in the energy region 0.4 to 0.8 eV. As has been inferred in [5] from the close correlation between the behaviour of this photoinduced absorption

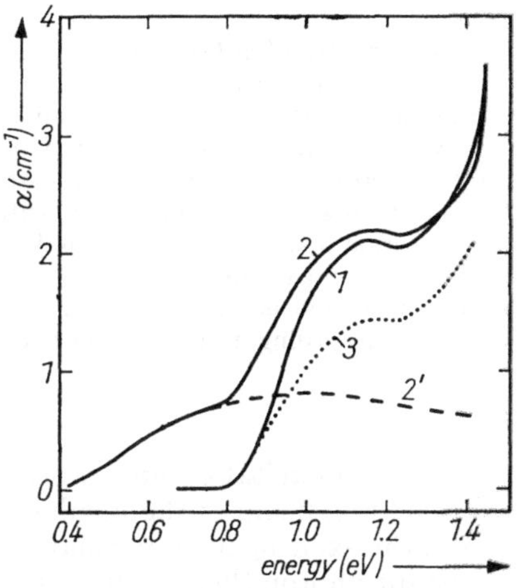

Fig. 1. Optical absorption spectra of a semi-insulating (si) GaAs:Cr sample at $T = 78$ K; (1) after removing the absorption due to $EL2^0$ as described in [10], (2) during illumination with 1.24 eV light, (2') extrapolated Cr^{4+} absorption using the spectral shape given in [8], (3) difference spectrum between (2) and (2')

and that of the Cr^{4+} EPR signal this absorption is due to Cr^{4+} formed during illumination, what is to be demonstrated in more detail in the following sections. According to [5] the formation of Cr^{4+}_{pi} is a two step process: by illumination into the photoionization absorption of Cr^{3+} an electron is excited from the valence band forming Cr^{2+}

$$Cr^{3+} + h\nu_{>0.75\,eV} \rightarrow Cr^{2+} + hole_{vb} \tag{1}$$

and the hole remaining in the valence band can be trapped by another Cr^{3+} forming Cr^{4+}

$$hole_{vb} + Cr^{3+} \rightarrow Cr^{4+}\,. \tag{2}$$

As a result of these excitation and capture processes holds

$$2\,Cr^{3+} \xrightarrow[h\nu_{exc}]{} Cr^{2+} + Cr^{4+}\,. \tag{3}$$

That the creation of Cr^{4+} actually takes place via this mechanism can be tested by the following procedure. The content of chromium in the charge states Cr^{2+} and Cr^{3+} before illumination ($[Cr^{2+}]^0$, $[Cr^{3+}]^0$) can be evaluated from spectrum 1 as described in [10]. During illumination the absorption spectra of Cr^{4+}, Cr^{3+}, and Cr^{2+} are superimposed in the energy region 0.8 to 1.5 eV. However, the spectral shape of the absorption due to the presence of Cr^{4+} in the energy region 0.4 to 1.5 eV is known from experimentals on p-conducting GaAs:Cr:Zn where all chromium is in the Cr^{4+} charge state [8]. Therefore, the Cr^{4+}_{pi} absorption can be extrapolated into the energy region 0.8 to 1.5 eV as the dashed curve 2' in Fig. 1 shows. The difference between curve 2 and the Cr^{4+}_{pi} absorption spectrum in the region 0.8 to 1.5 eV (dotted curve 3 in Fig. 1) represents the absorption due to Cr^{3+} and Cr^{2+} present in the sample during illumination and their concentrations $[Cr^{2+}]^i$ and $[Cr^{3+}]^i$ can be evaluated. The concentration of Cr^{4+}_{pi} is obtained from the absorption coefficient of curve 2 at 0.75 eV and the absorption cross section of Cr^{4+} given in [8]. Carrying out these measurements on more than 20 si GaAs:Cr samples in each case the following relations were found:

$$[Cr^{3+}]^0 - [Cr^{3+}]^i = 2[Cr^{4+}_{pi}]\,,$$

$$[Cr^{2+}]^i - [Cr^{2+}]^0 = [Cr^{4+}_{pi}]\,,$$

confirming convincingly relation (3).

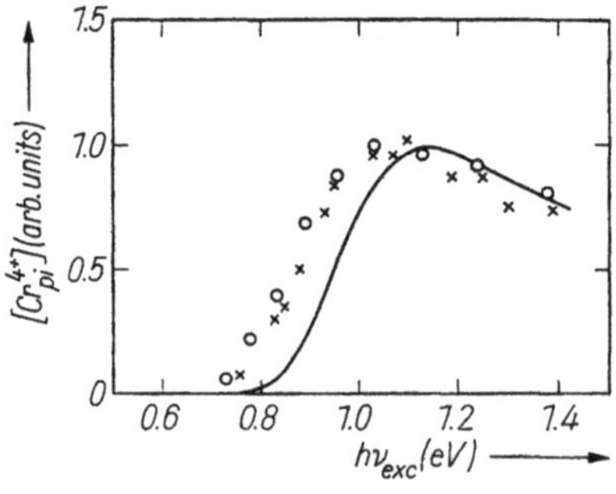

Fig. 2. Cr^{4+} excitation spectra of si GaAs:Cr samples. ○ concentration of Cr^{4+}_{pi} taken from optical absorption at 0.7 eV (the data points have been normalized to the maximum value at 1.1 eV and corrected for constant photon flux), after each point the Cr^{4+}_{pi} absorption was bleached to zero by illumination with $h\nu_b = 0.65$ eV, see Section 3.4. × Cr^{4+} EPR signal intensity normalized to the maximum value at 1.1 eV taken from [3]. The solid line is the photoionization cross section $\sigma_{Cr^{3+}}(h\nu)$ according to process (1) taken from [8]

In order to test if (1) and (2) really describe the underlying excitation and capture processes, the excitation spectrum of Cr^{4+}_{pi} was measured. Fig. 2 shows the results of the measurement of $[Cr^{4+}_{pi}]$ in dependence on the energy of the exciting light, normalized to the maximum value at 1.1 eV and corrected for constant photon flux. Additionally the data points of the corresponding EPR experiments taken from [3] are included. The agreement[2]) between both series of data points and the spectral shape $\sigma_{Cr^{3+}}(E)$ of the photoionization transition (1) derived from a fit of the Cr^{3+} absorption spectrum [8] proves unequivocally that process (1) together with the capture process (2) is the dominant mechanism.

A possible competing mechanism could be

$$Cr^{3+} + h\nu \rightarrow Cr^{4+} + e^-_{cb} , \tag{4}$$

$$e^-_{cb} + Cr^{3+} \rightarrow Cr^{2+} \tag{5}$$

eading also to relation (3). That process (4) actually occurs in si GaAs:Cr was shown by the analysis of the absorption spectrum of GaAs:Cr^{3+}, from which the photoionization spectrum connected with (4) was separated in the region 1.2 to 1.5 eV [8]. However, from the measured excitation spectra (Fig. 2) of Cr^{4+}_{pi} no contribution from process (4) could be detected. That means the probabilities of processes (4) and (5) are much lower than that of (1) and (2).

3.2 Dependence of Cr^{4+}_{pi} on chromium content

For a large number of samples the attainable concentration of Cr^{4+}_{pi} in dependence on the chromium content, i.e. on $[Cr^{2+}]^0$ and $[Cr^{3+}]^0$ was investigated. In all experiments a constant illumination intensity $I_0 = 0.27$ W/cm² and the illumination energy $h\nu_{exc} = 1.24$ eV were applied.

A simple correlation could be established for all samples taken from the seed end of LEC GaAs:Cr boules, i.e. for $g \leq 0.3$[3]) being characterized by a small content of Cr^{2+} ($[Cr^{2+}]^0 \leq 5$ to 8×10^{15} cm^{-3}). Fig. 3 shows that for such samples $[Cr^{4+}_{pi}] = C [Cr^{3+}]^0$ holds with $C = 0.2 \pm 0.04$ meaning the number of Cr^{4+}_{pi} being produced by illumination is proportional to the Cr^{3+} content in the sample. The saturation value of $[Cr^{4+}_{pi}]$ is according to (3) $[Cr^{4+}_{pi}]^s = 0.5 [Cr^{3+}]^0$ with all Cr^{3+} converted to Cr^{4+} and Cr^{2+}.

---

[2]) The difference between the data points of both experiments and the $\sigma_{Cr^{3+}}(E)$ curve in the region of the steep wing is caused probably by the large half-width of the metal interference filters used for selection of the illumination energies.

[3]) g is the fraction of melt solidified at a certain distance from the seed.

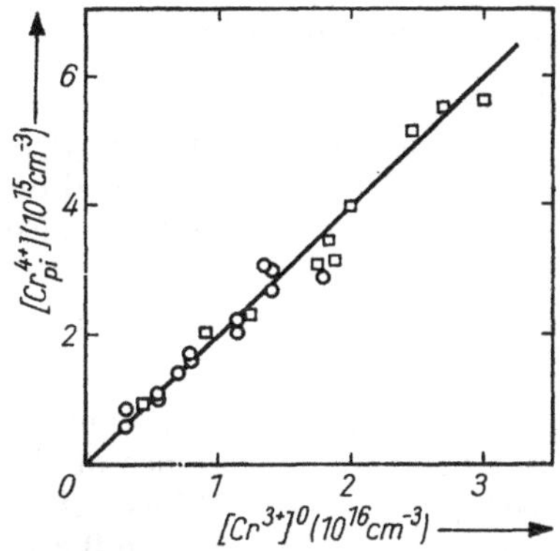

Fig. 3

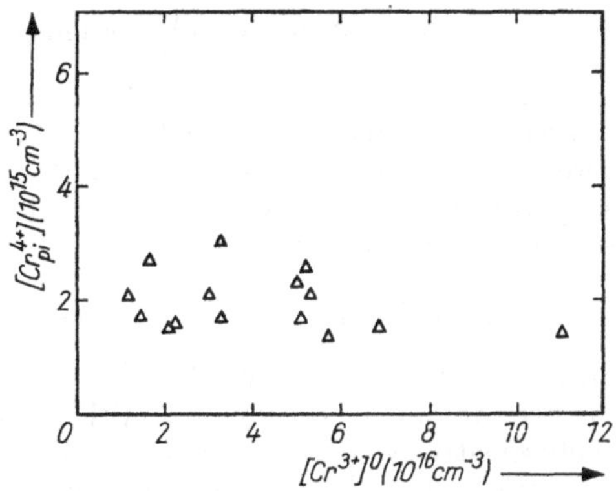

Fig. 4

Fig. 3. Dependence of $[Cr_{pi}^{4+}]$ created under illumination with $h\nu_{exc} = 1.24$ eV and $I = I_0$ on the concentration of Cr^{3+} present in the samples without illumination; $T = 78$ K. The circles and squares differentiate samples from different pullers. All samples were taken from the seed end ($g < 0.3$) of LEC-GaAs:Cr boules

Fig. 4. Dependence of $[Cr_{pi}^{4+}]$ created under illumination with $h\nu_{exc} = 1.24$ eV and $I = I_0$ on the concentration of Cr^{3+} present in the samples without illumination; $T = 78$ K. All samples were taken from the tail end ($g > 0.7$) of LEC-GaAs:Cr boules

On the other hand, in samples taken from the tail end of LEC boules ($g \geqq 0.7$) $[Cr_{pi}^{4+}]$ does not depend on $[Cr^{3+}]^0$ as it can be seen from Fig. 4. For this kind of samples $[Cr_{pi}^{4+}] = (2 \pm 1) \times 10^{15}\ cm^{-3}$ was measured in the concentration region $1 \times 10^{16}\ cm^{-3} \leqq \leqq [Cr^{3+}]^0 \leqq 12 \times 10^{16}\ cm^{-3}$. In order to illustrate the strong dependence of $[Cr_{pi}^{4+}]$ on the position in the boule along the growth direction, i.e. on the content of $[Cr^{2+}]^0$ and may be on impurities until now unidentified with distribution coefficients $k \ll 1$, in Fig. 5 the concentrations $[Cr_{pi}^{4+}]$, $[Cr^{3+}]^0$, and $[Cr^{2+}]^0$ determined at five different positions along a 250 mm long crystal are shown. Although it appears to be evident that there is a correlation between $[Cr_{pi}^{4+}]$ and $[Cr^{2+}]^0$ no clear analytical dependence could be established until now.

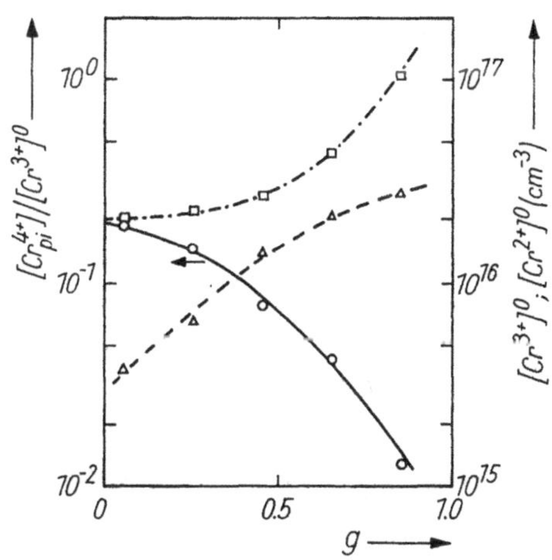

Fig. 5. Distribution of $[Cr^{3+}]^0$ (□), $[Cr^{2+}]^0$ (△), and $[Cr_{pi}^{4+}]/[Cr^{3+}]^0$ (○) along the growth direction of a large LEC-GaAs:Cr boule. Illumination conditions as in Fig. 3 and 4

3.3 Dependence of Cr_{pi}^{4+} on temperature and illumination intensity

Investigating the amount of Cr_{pi}^{4+} measured during illumination with I_0 and $h\nu_{exc} = 1.24$ eV in dependence on temperature two clearly distinct temperature regions were found as can be seen from Fig. 6: (i) for $T \leqq 110$ K $[Cr_{pi}^{4+}]$ does not depend on temperature (measured down to 27 K), (ii) above about 130 K an exponential decrease is measured according to

$$[Cr_{pi}^{4+}](T)/[Cr_{pi}^{4+}]\ (78\ \mathrm{K}) \sim \exp\ (\Delta E/kT) \tag{6}$$

with an activation energy $\Delta E = 80 \pm 10$ meV. This is exactly the same behaviour as has been found by Ridley et al. [6] investigating the temperature dependence of the photosensitivity (chopped-light photoconductivity per photon) at 0.47 eV of si GaAs:Cr under additional illumination with light containing the energy region 0.8 to 1.2 eV. This photosensitivity was interpreted by the authors as being proportional to the number of Cr^{4+} centres created by the additional illumination, the activation energy was found to be $\Delta E = 85 \pm 5$ meV.

The dependence of $[Cr_{pi}^{4+}]$ on illumination intensity I was measured in relation to the highest intensity available I_0 (creating the steady state concentration $[Cr_{pi}^{4+}]_0$ by using wire grid filters for diminishing the intensity. The results do not depend on temperature below $T = 100$ K as is shown for 78 and 27 K in Fig. 7, in which two intensity regions can be distinguished:

(i) up to about two orders of magnitude below I_0 $[Cr_{pi}^{4+}]$ decreases only very weakly with decreasing intensity (at $I/I_0 = 5 \times 10^{-2}$ $[Cr_{pi}^{4+}]/[Cr_{pi}^{4+}]_0 = 0.9$);

(ii) at even lower intensities $[Cr_{pi}^{4+}]$ decreases logarithmically with I/I_0 (at $I/I_0 \approx 10^{-4}$ $[Cr_{pi}^{4+}]/[Cr_{pi}^{4+}]_0 \approx 0.1$ was found, a value near the detection limit of this experiment)

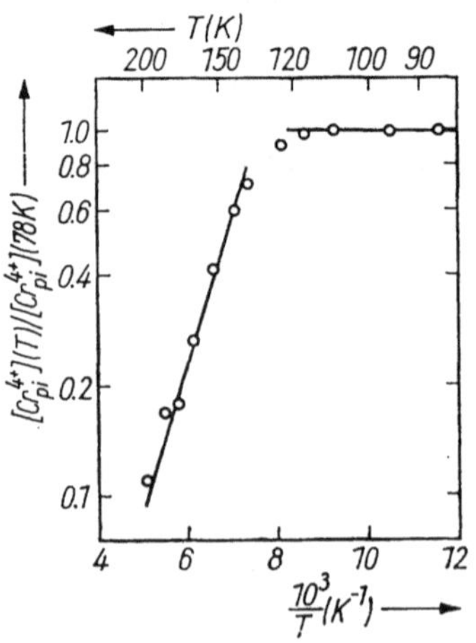

Fig. 6

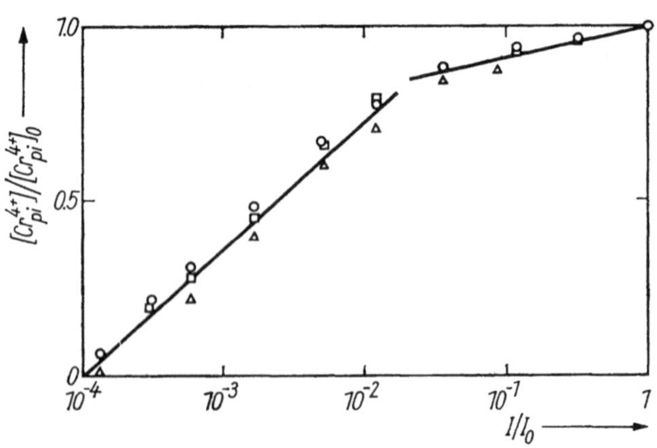

Fig. 7

Fig. 6. Temperature dependence of $[Cr_{pi}^{4+}]$. The illumination conditions are those of Fig. 3 and 4

Fig. 7. Dependence of $[Cr_{pi}^{4+}]$ on illumination intensity I ($h\nu_{exc} = 1.24$ eV) at two different temperatures (○, □ $T = 78$ K, △ $T = 27$ K). $[Cr_{pi}^{4+}]_0$ is the steady state concentration at $I = I_0$ with $I_0 = 0.27$ W/cm²

3.4 Transient behaviour of Cr^{4+}_{pi}

In the foregoing sections the steady state $[Cr^{4+}_{pi}]$ was investigated whereas now the dynamics of its rise and decay is of interest. Fig. 8 illustrates the general time behaviour found on all si GaAs:Cr samples.

(i) After switching on the illumination, $[Cr^{4+}_{pi}]$ rises up to a steady state value $[Cr^{4+}_{pi}]_{ss}$ depending on $h\nu_{exc}$, T and I as discussed in Sections 3.1 to 3.3. The time evolution up to reaching $[Cr^{4+}_{pi}]_{ss}$ also depends strongly on the illumination intensity I. As can be seen from Fig. 9, for a certain I the time dependence of the rise of $[Cr^{4+}_{pi}]$ can be approximately fitted by an exponential law

$$[Cr^{4+}_{pi}](t)/[Cr^{4+}_{pi}]_{ss} = 1 - \exp\left(-t/\tau_r\right), \tag{7}$$

where τ_r is a time constant characterizing the rise of $[Cr^{4+}_{pi}]$ depends approximately on illumination intensity according to $\tau_r \sim (I/I_0)^{-1}$ in the intensity region investigated (see caption of Fig. 9).

(ii) After switching off the illumination a non-exponential decay is observed, in the beginning very fast and than more and more slowly. For large t ($\geqq 10^2$ s) and samples with small $[Cr^{2+}]^0$ (seed end samples) a time dependence of the decay

$$[Cr^{4+}_{pi}](t)/[Cr^{4+}_{pi}]_{ss} \sim \frac{1}{1 + t/\tau_d} \tag{8}$$

was found, with τ_d a time constant characterizing the decay of $[Cr^{4+}_{pi}]$. Decay experiments on 15 seed end samples gave τ_d values in the region 200 to 400 s. The most important result of the decay experiments is that for a certain sample the decay transient does not depend on temperature (measured at 100, 78, and 27 K). This is consistent with the finding of White et al. [1] that the decay transient of the Cr^{4+} EPR signal does not depend on temperature in the region 4 to 10 K where the time

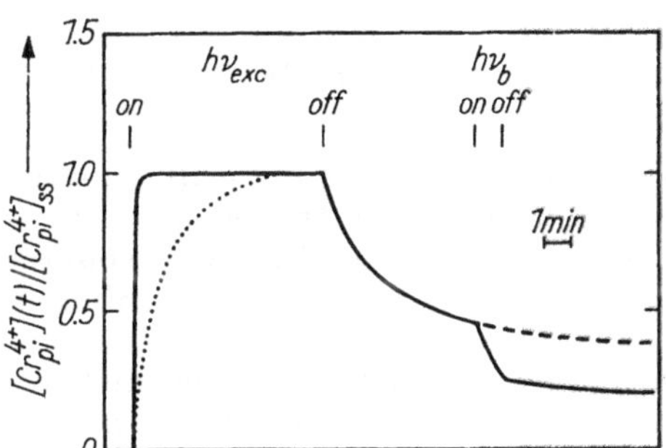

Fig. 8

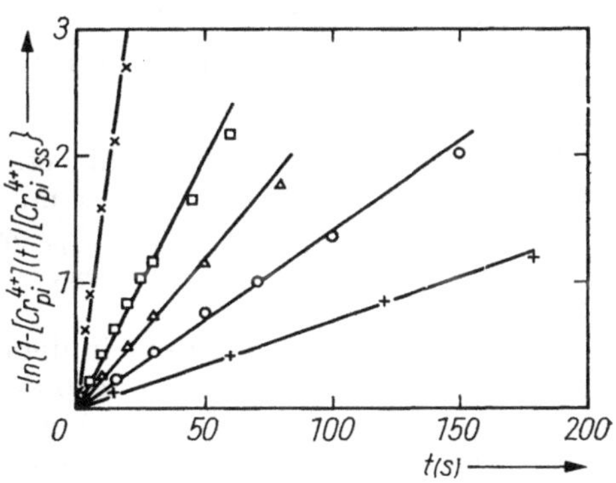

Fig. 9

Fig. 8. Transient behaviour of $[Cr^{4+}_{pi}]$ in rising up to the steady state concentration $[Cr^{4+}_{pi}]_{ss}$ after switching on $h\nu_{exc}$ and the decay from this after switching off $h\nu_{exc}$. Solid line $I = I_0$, dotted line $I/I_0 = 6 \times 10^{-3}$, dashed line decay without bleaching illumination with $h\nu_b$

Fig. 9. Time dependence of the $[Cr^{4+}_{pi}]$ rise up to the steady state concentration $[Cr^{4+}_{pi}]_{ss}$ after switching on $h\nu_{exc} = 1.24$ eV measured at different illumination intensities I/I_0. × $I/I_0 = 5 \times 10^{-2}$, $\tau_r = 8$ s; □ 2×10^{-2}, 26; △ 1.2×10^{-2}, 43; ○ 6×10^{-3}, 72; + 2×10^{-3}, 150

scale is the same as found here. Because all thermally caused releasing and capture processes are more or less temperature dependent, obviously the destruction of Cr_{pi}^{4+} after switching off the exciting illumination (free decay) is dominated by the tunneling process [1]

$$Cr^{4+} + Cr^{2+} \rightarrow 2\ Cr^{3+} \tag{9}$$

a process characterized by a tunneling constant K_i being different for each Cr^{4+}–Cr^{2+} pair with different distance in the sample, but independent of temperature.

(iii) If the sample during the free decay of $[Cr_{pi}^{4+}]$ after switching off $h\nu_{exc}$ will be illuminated with light $0.45\ eV \leqq h\nu_b \leqq 0.75\ eV$, the decay of $[Cr_{pi}^{4+}]$ is accelerated dramatically (see Fig. 8). This bleaching effect occurs because additionally to the tunneling process (9) governing the free decay, the destruction of Cr_{pi}^{4+} centres takes place by the processes

$$Cr^{4+} + h\nu_b \rightarrow Cr^{3+} + hole_{vb}\ , \tag{10}$$

$$hole_{vb} + Cr^{2+} \rightarrow Cr^{3+}\ . \tag{11}$$

Process (10) is the photoionization transition responsible for the absorption both the photoinduced Cr^{4+} in si GaAs:Cr and the stable one in p-GaAs:Cr:Zn [8].

3.5 Persistent Cr_{pi}^{4+}

It has been found that in some si GaAs:Cr samples the Cr_{pi}^{4+} deviates in its behaviour in some respect from that described until now. This will be called persistent Cr_{pi}^{4+}. The term persistent Cr_{pi}^{4+} covers the following experimental facts:

(i) after switching off $h\nu_{exc}$ $[Cr_{pi}^{4+}]$ does not decay to zero also after very long time (measured up to several hours) but reaches a constant value $0 \leqq [Cr_{pi}^{4+}]\ (t = \infty) \leqq$ $\leqq [Cr_{pi}^{4+}]\ (t = 0)$ depending on sample;

(ii) the persistent $[Cr_{pi}^{4+}]\ (t = \infty)$ cannot be destroyed optically by illumination with $0.45\ eV \leqq h\nu_b \leqq 0.75\ eV$;

(iii) persistent Cr_{pi}^{4+} can be destroyed thermally by warming up the sample to $T \approx 140\ K$.

The general explanation of this effect has to go as follows. During illumination and creation of Cr_{pi}^{4+}, Cr^{2+} centres must be annihilated by any trapping process so that after switching off $h\nu_{exc}$ the Cr_{pi}^{4+} or a part of them do not find their Cr^{2+} counterpart for recombining by free decay according to (9) or by bleaching according to (10) and (11).

In [11] it will be pointed out in detail that the Cr^{2+} annihilation is connected with the presence of unfilled $EL2^+$ deep donor centres besides the filled $EL2^0$ centres in the samples. Supposing this, during the illumination the $El2^+$ will be converted to metastable $(El2^0)^*$ (or quenched configuration according to the model proposed by Levinson [12]) and the hole left behind in the valence band is captured by Cr^{2+} forming Cr^{3+} according to the processes

$$EL2^+ + h\nu_{exc} \rightarrow EL2^0 + hole_{vb}\ , \tag{12}$$

$$EL2^0 + hole_{vb} + h\nu_{exc} \rightarrow (EL2^0)^* + hole_{vb}\ , \tag{13}$$

and (11).

If $[Cr^{2+}]^0 < [EL2^+]$ is valid in the sample, that part of $[Cr_{pi}^{4+}]$ being equal $[EL2^+] -$ $- [Cr^{2+}]^0$ cannot find the $[Cr^{2+}]$ created simultaneously with $[Cr_{pi}^{4+}]$ and needed now for its recombination. These Cr^{2+} have been converted to Cr^{3+} and $(EL2^0)^*$ due to processes (11) to (13) and therefore this part $[Cr_{pi}^{4+}]$ becomes persistent.

Actually, persistent Cr_{pi}^{4+} has been found only in samples with small $[Cr^{2+}]^0$ and in which the broad absorption band at 0.9 eV due to the photoionization transition (12) was detectable in the absorption spectrum before illumination (see [11]). The given interpretation is also supported by the fact that the two temperature regions in which the persistent Cr_{pi}^{4+} vanish and in which the metastable state $(EL2^0)^*$ will be thermally depopulated (110 to 140 K, see [13, 14]) are exactly the same.

4. Interpretation of Some Cr_{pi}^{4+} Properties

The population of the different charge states of Cr in si GaAs: Cr depending on temperature and external illumination has been described in general by Blakemore [15] including all possible transition processes (thermal as well as optical ones) via both the valence band and the conduction band. The experimental results presented here suggest that discussing the creation of Cr_{pi}^{4+}, all processes via the conduction band can be neglected. Likewise the thermal release of holes from Cr^{4+} and Cr^{3+} (labelled usually p_3 and p_2) will be neglected because nearly all experiments were carried out at 78 K or below, whereas the destruction of Cr_{pi}^{4+} due to the tunneling process (9) has to be included. Then the set of dynamic equations describing the time dependence of the concentration of Cr^{2+}, Cr^{3+}, and Cr^{4+} during illumination are

$$\frac{d[Cr^{2+}]}{dt} = e_2[Cr^{3+}] - pc_2[Cr^{2+}] - K[Cr^{2+}][Cr^{4+}], \tag{14}$$

$$\frac{d[Cr^{3+}]}{dt} = pc_2[Cr^{2+}] - e_2[Cr^{3+}] - pc_3[Cr^{3+}] + e_3[Cr^{4+}] + 2K[Cr^{2+}][Cr^{4+}], \tag{15}$$

$$\frac{d[Cr^{4+}]}{dt} = pc_3[Cr^{3+}] - e_3[Cr^{4+}] - K[Cr^{2+}][Cr^{4+}], \tag{16}$$

where the symbols have the following meaning: p is the concentration of free holes in the valence band, $e_2 = \sigma_2 I$ the optical hole emission rate from Cr^{3+} (process (1)), $e_3 = \sigma_3 I$ the optical hole emission rate from Cr^{4+} (process (10)), c_2 the hole capture coefficient of Cr^{2+} (process (11)), and c_3 the hole capture coefficient of Cr^{3+} (process (2)).

Additionally the condition for the total chromium content

$$[Cr] = [Cr^{2+}]^0 + [Cr^{3+}]^0 = [Cr^{2+}] + [Cr^{3+}] + [Cr^{4+}] \tag{17}$$

holds, and according to the proved creation mechanism of Cr_{pi}^{4+},

$$[Cr^{3+}] = [Cr^{3+}]^0 - 2[Cr^{4+}], \tag{18}$$

$$[Cr^{2+}] = [Cr^{2+}]^0 + [Cr^{4+}]. \tag{19}$$

At first we consider the steady state concentrations during illumination, i.e. if $d[Cr^{n+}]/dt = 0$. Then (14) and (16) lead to a complicated cubic equation for $[Cr^{4+}]$, in which besides c_2, c_3, e_2, e_3, also $[Cr^{2+}]^0$ and $[Cr^{3+}]^0$ are parameters. For simplification we neglect the tunneling terms. This seems to be justified for high illumination intensities I because the destruction of Cr_{pi}^{4+} by tunneling acts much more slowly than optical excitation and capture of holes. Using this approximation the following equation for $[Cr^{4+}]$ results from (14) and (16)

$$[Cr^{4+}]^2 - [Cr^{4+}]\,\frac{4[Cr^{3+}]^0 + \alpha[Cr^{2+}]^0}{4 - \alpha} + \frac{([Cr^{3+}]^0)^2}{4 - \alpha} = 0 \tag{20}$$

with $\alpha = e_3c_2/e_2c_3 = \sigma_3c_2/\sigma_2c_3$. Evidently, $[Cr^{4+}]$ does not depend on I, a result being practically confirmed by the experiments showing that for large I $[Cr^{4+}]$ changes only very weakly with I. Assuming small $[Cr^{2+}]^0/[Cr^{3+}]^0$, the following approximate solution of (20) will be obtained

$$[Cr^{4+}] = \frac{[Cr^{3+}]^0}{2 - \alpha/2} \left\{1 - \left(\frac{\alpha}{4}\right)^{1/2} + \frac{[Cr^{2+}]^0}{[Cr^{3+}]^0} \left(\frac{\alpha}{4} - \left(\frac{\alpha}{4}\right)^{1/2}\right)\right\}. \tag{21}$$

This solution gives for small $[Cr^{2+}]^0/[Cr^{3+}]^0$ (realistic for seed end samples) indeed the dependence $[Cr^{4+}] = C\,[Cr^{3+}]^0$ measured on seed end samples. From the measured $C = 0.2$ the value $\alpha = 9$ is estimated. As $\sigma_3/\sigma_2 \approx 2$ at $h\nu_{exc} = 1.24$ eV [8, 10], $c_2/c_3 \approx 4$ can be derived from this, i.e. the hole capture coefficients of Cr^{2+} and Cr^{3+} prove to be of the same order of magnitude.

Considering the time dependence of the decay of $[Cr^{4+}_{pi}]$ we have to start from (16) with $I = 0$ and $[Cr^{4+}] = [Cr^{4+}]_{ss}$ at $t = 0$ leading to

$$\frac{d[Cr^{4+}]}{dt} = -K[Cr^{4+}]\,([Cr^{2+}]^0 + [Cr^{4+}])\,. \tag{22}$$

Neglecting $[Cr^{2+}]^0$ (22) has the solution

$$[Cr^{4+}]/[Cr^{4+}]_{ss} = \frac{1}{1 + [Cr^{4+}]_{ss}\,Kt}. \tag{23}$$

Equation (23) describes the time behaviour measured on seed end samples for large t. The exact solution of (22) is given by

$$\frac{[Cr^{4+}](t)}{[Cr^{4+}]_{ss}} = \frac{[Cr^{2+}]^0}{[Cr^{4+}]_{ss}} \sum_i \frac{1}{(1 + [Cr^{2+}]^0/[Cr^{4+}]_{ss}) \exp(K_i[Cr^{2+}]^0\,t) - 1}, \tag{24}$$

where the summation over i affects the different tunneling constants K_i belonging to the Cr^{4+}–Cr^{2+} pairs with different distances. Equation (24) explains the measured time dependence of the decay as a superposition of many exponential decays with different K_i, where immediately after switching off $h\nu_{exc}$ the decay is determined by the destruction of Cr^{4+}–Cr^{2+} pairs with small distances and therefore large K_i, and later on by the destruction of Cr^{4+}–Cr^{2+} pairs having larger distances and smaller K_i.

The order of magnitude of K_i can be estimated in the following way. The lower limit K_l (long time after switching off) can be obtained from (8) with the measured τ_d being $K_l = ([Cr^{4+}]_{ss}\,\tau_d)^{-1} \approx 10^{-18}$ cm³ s⁻¹. The upper limit K_s can be derived from the slope of $[Cr^{4+}](t)/[Cr^{4+}]_{ss}$ at $t = 0$ (from the measured decay curves

$$d\{[Cr^{4+}](t)/[Cr^{4+}]_{ss}\}/dt = 1 \text{ to } 4 \times 10^{-2}\,s^{-1} \quad \text{at} \quad t = 0$$

are obtained). From (24) then follows $K_s \approx 10^{-17}\,cm^3\,s^{-1}$ if $[Cr^{2+}]^0 \approx 5 \times 10^{15}\,cm^{-3}$ is supposed.

The value $c_2 \approx 10^{-10}\,cm^3\,s^{-1}$ at 100 K estimated in [6] and the above obtained result $c_2/c_3 \approx 4$ demonstrate that c_2, c_2 are many orders of magnitude larger than the K_i, so that the justification of the approximation leading to (20) appears to be confirmed.

5. Conclusions

The experiments presented here provide a clear quantitative evidence that the formation of photoinduced optical absorption of Cr^{4+} in si GaAs:Cr is caused by the two processes (1) and (2), a result being in accordance with the suggestions derived from

EPR experiments [1, 2]. Furtheron it is shown that the free decay of Cr^{4+}_{pi} after switching off the exciting light is governed by the tunneling process (9) and the tunneling constants were found to be in the range $K_t \approx 10^{-17}$ to 10^{-18} cm³ s⁻¹. The appearance of persistent Cr^{4+}_{pi} demonstrates impressively how the existence of the two configurations ($EL2^0$) and $(EL2^0)^*$ affects the transient behaviour also of other recharging processes in si GaAs.

A set of dynamic equations for Cr^{n+} ($n = 2, 3, 4$) taking into account merely excitation and capture processes via the valence band proves to be adequate to explain the dependence of Cr^{4+}_{pi} on chromium content in seed end samples and on illumination intensity in the region of comparatively high intensities as well as to discuss the free decay of Cr^{4+}_{pi}.

Nevertheless, some questions connected with Cr^{4+}_{pi} remain open. The proposed model fails in explaining the dependence of Cr^{4+}_{pi} on chromium content in tail end samples because on the one hand the presupposition $[Cr^{2+}]^0/[Cr^{3+}]^0 \ll 1$ is not valid but possibly also the presence of other yet unidentified traps can affect the recharging mechanisms under illumination. The considerable variations from sample to sample of the τ-values characterizing the free decay and depending also on sample history found especially in tail end samples support such suggestions.

The obtained results appear to be valuable in presenting the possibility to monitor recharging processes in si GaAs:Cr by measuring the Cr^{4+}_{pi} absorption. This can be very helpful in interpreting photoconductivity experiments which obviously requires to include not only processes (1) and $Cr^{2+} + h\nu \rightarrow Cr^{3+} + e^-_{cb}$ but also the Cr^{4+} related processes (4) and (10).

Acknowledgements

The author wishes to thank M. Seifert and H. Witte for supplying samples and A. Riedel for technical assistance.

References

[1] A. M. White, J. J. Krebs, and G. H. Stauss, J. appl. Phys. **51**, 419 (1980).
[2] G. H. Stauss, J. J. Krebs, S. H. Lee, and E. M. Swiggard, Phys. Rev. B **22**, 3141 (1980).
[3] U. Kaufmann and J. Schneider, Solid State Commun. **20**, 143 (1976).
[4] J. J. Krebs and G. H. Stauss, Phys. Rev. B **26**, 2296 (1982).
[5] W. Ulrici, phys. stat. sol. (b) **114**, K87 (1982).
[6] B. K. Ridley, M. C. Arikan, P. J. Bishop, M. F. M. Hassan, and W. V. Machado, J. Phys. C **15**, 6865 (1982).
[7] J. S. Blakemore, S. G. Johnson, and S. Rahimi, in: Semi-Insulating III-V Materials, Evian 1982 Shiva, Nantwich, 1982 (p. 172).
[8] W. Ulrici and P. Kleinert, phys. stat. sol. (b) **129**, 339 (1985).
[9] G. M. Martin, Appl. Phys. Letters **39**, 747 (1981).
[10] W. Ulrici, phys. stat. sol. (a) **84**, 243 (1984).
[11] W. Ulrici, to be published.
[12] M. Levinson, Phys. Rev. B **28**, 3660 (1983).
[13] A. Mittoneau and A. Mircea, Solid State Commun. **30**, 157 (1979).
[14] G. Vincent, D. Bois, and A. Chantre, J. appl. Phys. **53**, 3643 (1982).
[15] J. S. Blakemore, in: Semi-Insulating III-V Materials, Nottingham 1980, Shiva, Orpington, 1980 (p. 29).

(Received March 21, 1985)

phys. stat. sol. (b) 131, 719 (1985)

Subject classification: 13.4; 14.3; 20.1; 22.2.1

Zentralinstitut für Elektronenphysik der Akademie der Wissenschaften der DDR, Berlin[1] *(a) and Department of Physics, University of Nottingham*[2] *(b)*

Optical and Electrical Properties of Vanadium-Doped GaAs

By

W. ULRICI (a), K. FRIEDLAND (a), L. EAVES (b), and D. P. HALLIDAY (b)

Dedicated to Prof. Dr. Dr. h. c. Dr. E. h. P. GÖRLICH on the occasion of his 80th birthday

The results of optical absorption, luminescence, and temperature-dependent Hall-effect measurements on GaAs:V, p-type GaAs:V:Zn, and n-type GaAs:V:Si grown by LEC-technique are reported. It is found that the V^{3+}_{Ga} ground state is located very close to or within the valence band and that the V^{2+}_{Ga} acceptor level is at $E_c - 0.14$ eV. The near mid-gap level in GaAs:V responsible for its semi-insulating behaviour is proposed to be due to the complex (V^{2+}_{Ga}–X), where the correlation between the photoionization absorption connected with this centre and the EL2 concentration suggests that X is an arsenic vacancy. From this level scheme conditions for growing GaAs:V which is semi-insulating at room temperature are derived. The optical experiments on GaAs:V with different Fermi-level positions confirm the assignment of the triple-peaked absorption band at 1.1 eV to the $^3A_2 \rightarrow {}^3T_1(F)$ transition of V^{3+}_{Ga}. The weak absorption due to the $^3A_2 \rightarrow$ $\rightarrow {}^3T_2(F)$ transition is found at 0.79 eV. The analysis of the temperature dependence of the line shape of the $\rightarrow {}^3T_1(F)$ absorption band as due to strong dynamical $T \otimes \tau_2$ Jahn-Teller effect provides the JT-parameters $E_{JT} = 780\ cm^{-1}$, $\hbar\omega_\tau = 75\ cm^{-1}$, and $c = 4.7$ eV/nm. The temperature dependent asymmetry of the line shape is explained on the basis of earlier theoretical predictions of the simultaneous action of strong linear and quadratic JT-coupling.

Die Ergebnisse der Messungen der optischen Absorption, der Lumineszenz und des temperaturabhängigen Hall-Effekts an GaAs:V, p-leitendem GaAs:V:Zn und n-leitendem GaAs:V:Si werden mitgeteilt. Es wird gefunden, daß der Grundzustand des V^{3+}_{Ga} sehr nahe oder innerhalb des Valenzbandes liegt und daß das V^{2+}_{Ga}-Akzeptorniveau sich bei $E_c - 0,14$ eV befindet. Das für das semiisolierende Verhalten verantwortliche Niveau in der Mitte des Gaps wird durch den Komplex (V^{2+}_{Ga}–X) verursacht, wobei die Korrelation zwischen der mit diesem Zentrum verbundenen Photoionisationsabsorption und der EL2-Konzentration nahelegt, daß X eine Arsenlücke ist. Daraus werden die Bedingungen für die Züchtung von GaAs:V abgeleitet, das auch bei Raumtemperatur semiisolierend ist. Die optischen Experimente an GaAs:V mit verschiedenen Lagen des Fermi-Niveaus bestätigen die Zuordnung der Absorptionsbande bei 1,1 eV zum $^3A_2 \rightarrow {}^3T_1(F)$-Übergang des V^{3+}_{Ga}. Die schwache Absorption durch den $^3A_2 \rightarrow {}^3T_2(F)$-Übergang wird bei 0,79 eV gefunden. Die Analyse der Temperaturabhängigkeit der Linienform der $\rightarrow {}^3T_1(F)$-Absorption, die durch einen starken dynamischen $T \otimes \tau_2$-Jahn-Teller-Effekt verursacht wird, liefert die JT-Parameter $E_{JT} = 780\ cm^{-1}$, $\hbar\omega_\tau = 75\ cm^{-1}$ und $c = 4,7$ eV/nm. Die temperaturabhängige Asymmetrie der Linienform wird auf der Grundlage früherer theoretischer Voraussagen über das gleichzeitige Wirken einer starken linearen und quadratischen JT-Kopplung erklärt.

1. Introduction

Vanadium as a dopant in GaAs has been gaining attention since it was shown that GaAs:V can be semi-insulating (si) in the same manner as GaAs:Cr ($\varrho \approx 10^8\ \Omega$cm at room temperature) but with possibly superior substrate properties due to the one

---

[1]) Hausvogteiplatz 5—7, DDR-1086 Berlin, GDR.

[2]) Nottingham NG7 2RD, Great Britain.

order of magnitude smaller diffusion coefficient of vanadium in GaAs [1]. Vasiliev et al. [2] and Kaufman et al. [3] have reported previously on semi-insulating GaAs:V and both groups have found a V-related luminescence with zero-phonon line (zpl) at 0.739 eV with some phonon replica ($\hbar\omega_{ph} = 9$ meV) and a triple peaked absorption band centred at 1.1 eV. From temperature-dependent Hall effect (TDH) measurements the location of the V-related acceptor level at $E_c - 0.77$ eV was deduced [2], and the isotropic EPR signal at $g = 1.957$ was assigned to isolated V_{Ga}^{3+} ($^3A_2$ ground state) [3]. Very recently, an anisotropic spectrum ascribed to V^{2+} was found on si GaAs:V by thermally detected EPR experiments [4] with the detected V^{2+} being probably not an isolated impurity.

On the other hand, Mircea-Roussel et al. [5] and Litty et al. [6] obtained n-conducting material by doping with vanadium and the optical absorption and the luminescence proved to be identical with those found on si GaAs:V. DLTS measurements revealed two electron traps connected with V at $E_c - 0.14$ eV and $E_c - 0.23$ eV [6]. The assignment of the photoluminescence spectrum with zpl at 0.739 eV was contradictory for a certain period but was unequivoacclly clarified very recently to be due to the $^3T_2 \rightarrow {}^3A_2$ transition of V_{Ga}^{3+} by Zeeman [7] and uniaxial-stress [8] experiments. This assignment is especially convincing as it is in nearly complete analogy to that for the corresponding luminescence with zpl at 0.705 eV of InP:V [9 to 11].

Although there has been some progress during the last two years in understanding the properties of GaAs:V, it has to be stated that nearly all problems connected with V in GaAs remained open. Among these the most important ones are:

(i) The origin of the near-mid-gap level responsible for the si state of GaAs:V is unknown.

(ii) The growth conditions leading to si GaAs:V are not established. Also the growth of V-doped si epitaxial layers by $AsCl_3$-VPE [12] and MOCVD [13] gave no insight into this problem.

(iii) The variety, the level positions within the gap, and the electronic as well as the microscopic structure of vanadium-related centres in GaAs are not yet clarified.

The present paper aims at contributing to the solution of these problems by investigating optical absorption, luminescence, and TDH of GaAs:V as well as of n-conducting GaAs:V:Si and p-conducting GaAs:V:Zn. By these experiments carried out on V-doped samples with different Fermi-level positions arguments will be accumulated leading to a proposal for the nature of several V-centres and their spectroscopic and electrical properties.

2. Experimental

The samples investigated here were cut from different positions (seed end to tail end) of boules grown by the liquid encapsulated Czochralski (LEC) technique. The doping was made by adding metallic vanadium to the melt (1 to 2 mg V/g GaAs). The boules were grown using two different pullers, a normal pressure one and a high pressure one, under stoichiometric conditions. The double doped boules were grown with the normal pressure puller adding to the melt, together with V, either silicon or zinc in order to get either n-conducting GaAs:V:Si crystals or p-conducting GaAs:V:Zn crystals.

The samples for measuring optical absorption and luminescence (area 6×7 mm^2, thickness 2 to 18 mm) were mirror-like polished and mounted in a variable temperature (6 to 500 K) metal cryostat or in cryostats in which they are immersed in liquid He. For measuring the temperature-dependent Hall (TDH) effect (80 to 500 K), samples with dimensions of about $7 \times 3 \times 0.5$ mm^3 were used with ohmic Au contacts in a standard Hall configuration.

3. Results

3.1 GaAs:V

All vanadium-doped GaAs boules grown by LEC technique under stoichiometric conditions are n-conducting at room temperature with a free electron concentration $n \approx 10^{14}$ to 10^{15} cm^{-3}. Fig. 1 shows the results of the measurement of the temperature-dependent Hall effect which are typical for all investigated samples. Qualitatively, it clearly demonstrates that the n-conductivity cannot be due to residual shallow donors because n is declining drastically with decreasing temperature and the samples become semi-insulating ($\varrho \approx 10^8\,\Omega$cm) at $T = 80$ K. From the analysis of the $\ln(n/T^{3/2}) = f(1/T)$ plot (see [14]) follows that the n-conductivity is caused by a V-related trap at $E_c - 0.14 \pm 0.01$ eV.

Fig. 2 shows the optical absorption spectra at $T = 78$ K of GaAs:V grown in a high-pressure puller and Fig. 3 the corresponding spectra of GaAs:V grown in a normal-pressure puller. All spectra are characterized by the following features. A dominant triple-peaked absorption band centred at 1.1 eV occurs and additionally three different photoionization-like absorptions exist called here α_I, α_{II}, and α_{III} with the following onsets:

$$\alpha_I: \quad E_I = 0.58 \pm 0.03 \text{ eV},$$
$$\alpha_{II}: \quad E_{II} = 1.0 \pm 0.10 \text{ eV},$$
$$\alpha_{III}: \quad E_{III} = 1.38 \pm 0.02 \text{ eV}.$$

The absorption α_{III} can be seen more clearly from Fig. 6 in Section 3.2.

All spectra in Fig. 2 and 3 are measured after removing the superimposed absorption of EL2^0 by illumination with light of $h\nu = 1.24$ eV [15]. For illustration in Fig. 2 the spectrum of the seed-end sample is shown measured after cooling down in the dark before removing the EL2^0 absorption. From the difference between the absorption coefficients at 1.24 eV before and after removing the superimposed absorption due to EL2^0, the concentration [EL2^0] in the samples can be evaluated [15]. The fact that by illumination with $h\nu = 1.24$ eV the EL2^0 can be transferred to (EL2^0)* in the GaAs:V samples is another proof for its semi-insulating behaviour at low temperatures (cf. [16, 17]). Fig. 4 shows that [EL2^0] can be roughly correlated to the intensity of the photoionization absorption α_{II} (measured at 1.3 eV) meaning that the larger [EL2^0] the lower is α_{II} (1.3 eV), i.e. $\alpha_{II} \sim 1/[\text{EL2}^0]$.

Using the absorption coefficient at the energy of the central peak (1.1 eV) of the absorption band (which is assigned to V_{Ga}^{3+}, see Section 4) as a measure of the concentration of V_{Ga}^{3+}, it can be seen from Fig. 5 that the photoionization absorption α_{III} (measured at 1.425 eV) is proportional to $[V_{Ga}^{3+}]$.

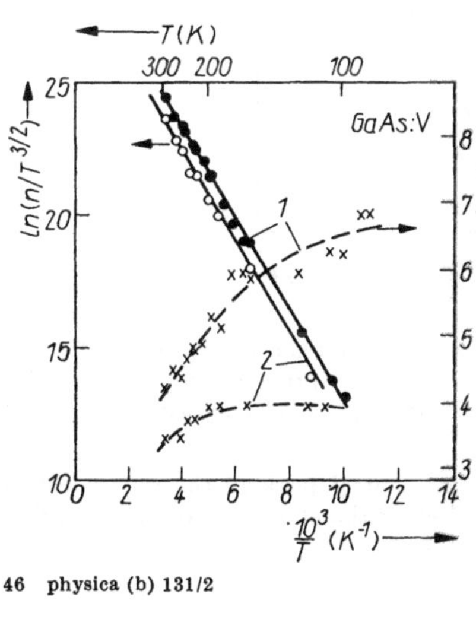

Fig. 1. Temperature dependence of carrier concentration n and mobility μ_H from Hall effect measurements. (1) Sample from the seed end and (2) sample from the tail end of a 250 mm long LEC-grown GaAs:V boule

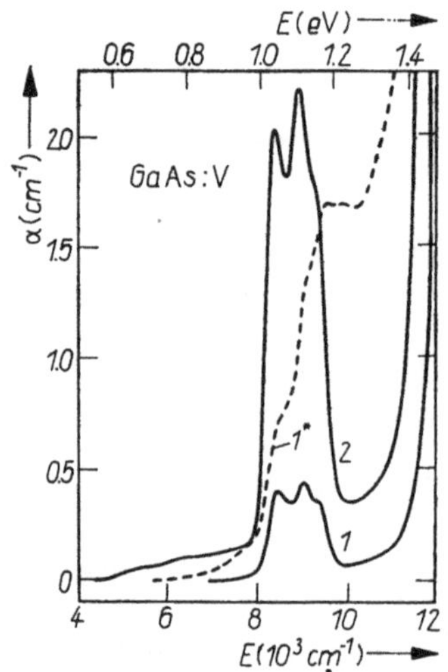

Fig. 2

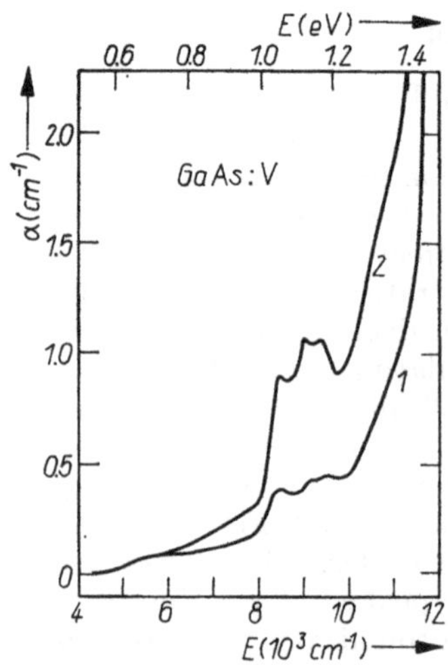

Fig. 3

Fig. 2. Optical absorption spectra of GaAs:V samples from the seed end (curve 1) and the tail end (curve 2) of a boule grown in a high-pressure puller; $T = 78$ K. Curve 1* is the absorption spectrum of the seed end sample after cooling down in the dark and before removing the absorption due to $EL2^0$ by illumination with 1.24 eV light. The samples for the TDH measurements (see Fig. 1) were taken very close to those measured here

Fig. 3. Optical absorption spectra of GaAs:V samples from the seed end (curve 1) and the middle part (curve 2) of boule grown in a normal-pressure puller; $T = 78$ K

In case of α_I and α_{II} being large, a weak absorption band at 0.69 ± 0.02 eV superimposed to the α_I spectrum can be observed. As can be seen from Fig. 2 in samples with very intensive absorption band at 1.1 eV a second weak absorption band at 0.79 eV can be detected, the oscillator strength of which is smaller than that of the 1.1 eV band by a factor of about 2×10^2.

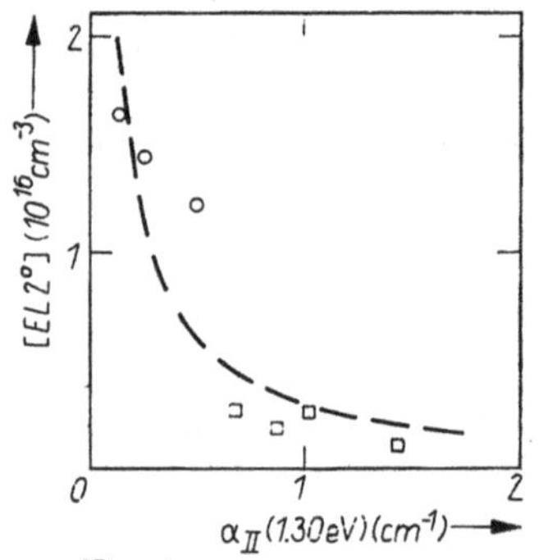

Fig. 4

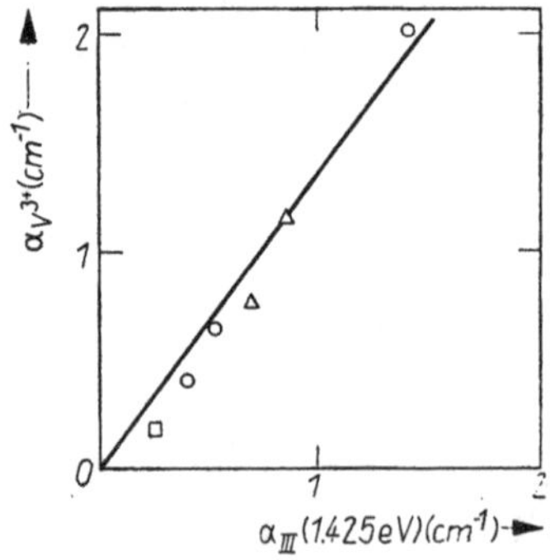

Fig. 5

Fig. 4. Correlation between the content of $EL2^0$ in GaAs:V samples and the absorption coefficient at 1.30 eV of the photoionization absorption α_{II}. ○ samples from high-pressure puller, □ samples from normal-pressure puller. The dashed line is calculated from $\alpha_{II} = \gamma/[EL2^0]$ with $\gamma = 0.3 \times 10^{16}\ cm^{-4}$

Fig. 5. Correlation between the absorption coefficient of the central peak of the triple-peaked absorption band ($\alpha_{V^{3+}}$) and the absorption coefficient at 1.425 eV of the photoionization absorption α_{III}. Different signs label samples from different boules

The luminescence was investigated at 4.2 K (excitation with Kr and Nd:YAG laser) and for all GaAs:V samples the luminescence spectrum with zpl at 0.7387 eV and a hot line at 0.7400 eV (confirmed by measuring at $T = 27$ K) followed by phonon replica with $\hbar\omega_{ph} = 9.0 \pm 0.3$ meV was found in complete accordance with the results of other authors [3, 5 to 8].

3.2 GaAs:V:Zn

The optical absorption spectra of a tail-end sample of double doped GaAs:V:Zn are shown in Fig. 6. The additional doping with zinc causes the whole boule to be p-conducting with $p = 4$ to 7×10^{16} cm^{-3} due to the shallow Zn acceptors. The absorption spectrum is characterized by the presence of the absorption band at 1.1 eV and the very intense absorption α_{III} (for comparison the spectrum of an undoped sample is shown), and by the absence of the photoionization-like absorptions α_I and α_{II}. However, for the tail-end (intensive 1.1 eV band) the weak absorption band at 0.79 eV could be also detected. The photoluminescence spectrum with zpl at 0.739 eV was found in GaAs:V:Zn samples, too.

Fig. 7 shows the triple-peaked absorption band at $T = 4.2$ K and a zpl at 8132 $\pm$ $\pm$ 1 cm^{-1} (1.008 eV) detected at this temperature. The zpl is followed by two weak structures at zpl + 65 cm^{-1} and zpl + 198 cm^{-1} which can be interpreted as phonon replica ($\hbar\omega_{ph} = 65$ cm^{-1} = 8.1 meV), and by a somewhat more pronounced structure zpl + 105 cm^{-1}. Measurements with high spectral resolution revealed a half-width of the zpl of 5 cm^{-1} and no further fine structure could be resolved. The half-width of the line at 8237 cm^{-1} amounts to about 10 cm^{-1}. Both the zpl and the following structures are unequivocally connected with the broad triple-peaked absorption band because they have been measured also on the GaAs:V samples described in Section 3.1.

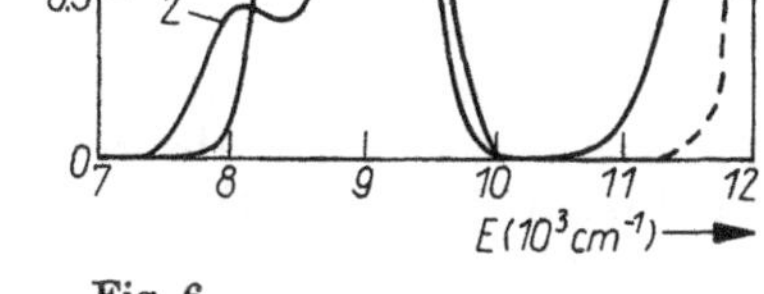

Fig. 6

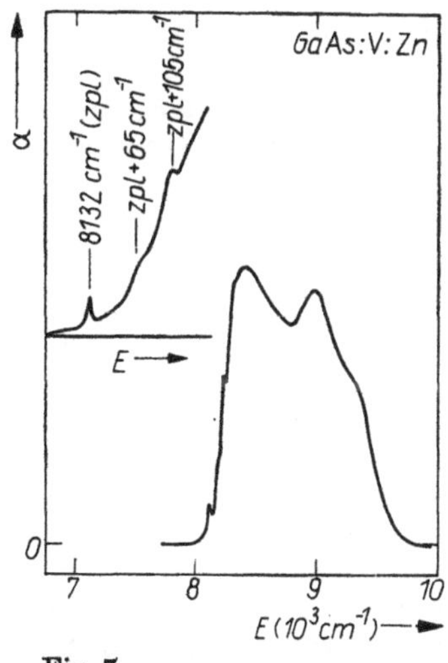

Fig. 7

Fig. 6. Optical absorption spectra of a p-conducting GaAs:V:Zn sample ($p = 4 \times 10^{16}$ cm^{-3} at 300 K). The underlying absorption due to free holes has been subtracted. (1) $T = 78$ and (2) 300 K. The dashed curve is measured on undoped GaAs at 78 K

Fig. 7. Triple-peaked optical absorption band measured on a GaAs:V:Zn sample at $T = 4.2$ K. The inset shows in more detail the zero-phonon line (zpl) and the following structures

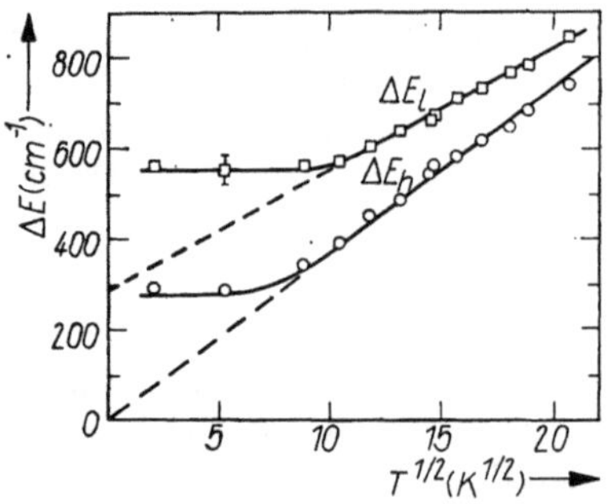

Fig. 8. Temperature dependence of the distances between the central peak and the high-energy peak (ΔE_h) and between the central peak and the low-energy peak (ΔE_l) of the triple-peaked absorption band

As the triple-peaked absorption band in GaAs:V:Zn samples is not superimposed by α_I and α_{II}, a more accurate analysis of the lineshape of this band and its temperature dependence can be carried out. The overall lineshape is seen to be triple-peaked in the whole investigated temperature range 4.2 to 430 K. However, the skewness of the asymmetric lineshape of the band changes drastically. At high temperatures a well-revolved triplet structure is measured with the high-energy sub-band being the most intense one, whereas at low temperatures the low-energy sub-band becomes the most intense one. Both the distance between the low-energy and the central peak (ΔE_l) and between the high-energy and the central peak (ΔE_h) increase with increasing temperature and can be fitted with the relations

$$\Delta E_h(T) = \Delta E_h(0) \left[\coth\left(\frac{\hbar\omega_\tau}{2kT}\right)\right]^{1/2}, \tag{1}$$

$$\Delta E_l(T) = \Delta E_1 + \Delta E_l(0) \left[\coth\left(\frac{\hbar\omega_\tau}{2kT}\right)\right]^{1/2}, \tag{2}$$

where the fit yields the parameters $\Delta E_h(0) = \Delta E_l(0) = (270 \pm 10)\ cm^{-1}$, $\Delta E_1 = (280 \pm 10)\ cm^{-1}$, and $\hbar\omega_\tau = (75 \pm 5)\ cm^{-1} = (9.3 \pm 0.6)$ meV. In Fig. 8 the temperature dependence of ΔE_h and ΔE_l is shown in a $\Delta E = f(T^{1/2})$ plot providing the evidence that the data points are on a straight line at high temperatures, as is required by (1) and (2).

3.3 GaAs:V:Si

The double doping of GaAs with V and Si resulted in a n-conducting boule ($n = 3$ to $7 \times 10^{16}\ cm^{-3}$) on which the n-conductivity has been checked by TDH measurements as due to the shallow Si-donors. Fig. 9 shows the optical absorption spectra of such GaAs:V:Si samples. The most striking feature in the spectra is the absence of the triple-peaked absorption band. Also the luminescence with zpl at 0.739 eV could not been detected on the samples using Nd:YAG laser excitation[3]). The photoionization-like absorptions α_I and α_{II} occur as well as the absorption band at 0.69 eV as they were found also in GaAs:V samples described in Section 3.1. A new absorption band measured only in GaAs:V:Si samples is located at 1.03 eV. However, the intensity of the 1.03 eV band appears not to be correlated to the intensity of neither α_I nor α_{II}, whereas the 0.69 eV band is probably connected with α_I.

The absorption spectra of GaAs:V:Si in the energy region $E > 1.0$ eV are difficult to analyze because of the superposition with the absorption of $EL2^0$ which can be

[3]) However, a weak luminescence was observed with intense Kr-laser excitation. On the other hand, in GaAs:V and GaAs:V:Zn samples the luminescence with zpl at 0.739 eV is much stronger with Nd:YAG laser excitation compared with that using Kr-laser excitation.

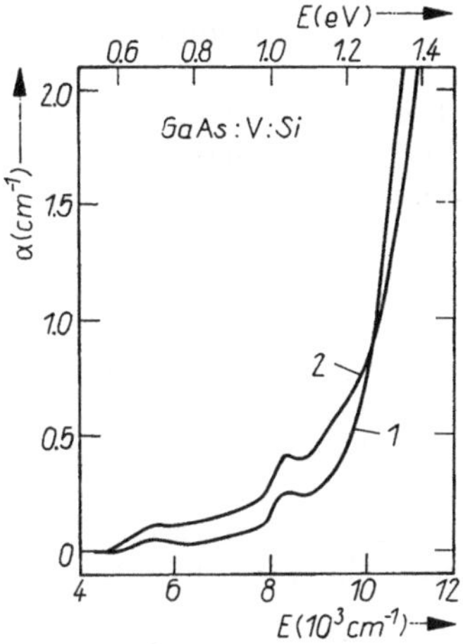

Fig. 9. Optical absorption spectra of n-conducting GaAs:V:Si samples. (1) Seed end sample ($n = 4 \times 10^{16}$ cm^{-3} at 300 K), (2) tail end sample ($n = 7 \times 10^{16}$ cm^{-3} at 300 K)

present in samples with $n \leqq 10^{17}$ cm^{-3} [18] but cannot be removed by illumination [16, 17].

4. Charge States and Level Positions of Vanadium-Related Centres

Considering the effects of shifting the Fermi level through the gap of GaAs:V as described in Section 3 the following conclusions concerning the V-centres can be drawn.

(i) The disappearance of both the luminescence with zpl at 0.739 eV and the absorption band at 1.1 eV if the Fermi level is located close to the conduction band (GaAs:V:Si) strongly supports their assignment to inner-shell transitions of V_{Ga}^{3+} as already proposed by other authors [9, 20], i.e. in GaAs:V:Si samples in which the n-conductivity is caused by shallow Si_{Ga} donors, V_{Ga}^{3+} does not exist.

(ii) TDH and DLTS [6] results give evidence of a V-related electron trap at $E_c - 0.14$ eV assigned to the acceptor level of V_{Ga}^{2+} originating from the capture of a free electron by the neutral V_{Ga}^{3+}. Supposing this, the photoionization transition

$$V_{Ga}^{3+} + h\nu_{>1.38\,eV} \rightarrow V_{Ga}^{2+} + \text{hole}_{vb} \quad (3)$$

can be made responsible for the absorption α_{III}. The correlation $[V_{Ga}^{3+}] \sim \alpha_{III}$ (see Fig. 5) confirms this suggestion.

(iii) In GaAs:V:Zn, p-conducting by the dominance of shallow Zn-acceptors, only the spectroscopic features connected with V_{Ga}^{3+} have been found. Therefore, the ground state $^3A_2$ of V_{Ga}^{3+} is located very close to or within the valence band and a V^{3+}/V^{4+} hole trapping obviously does not occur in contrast to chromium in GaAs.

(iv) The fact that GaAs:V can become semi-insulating also at room temperature ($E_{th} = 0.77$ eV [2]) together with the existence of the V-related absorptions α_I and α_{II} found here, require that one or more V-acceptor centres have to exist with structures necessarily more complicated than the isolated V_{Ga}^{2+} mentioned in (ii). We propose the complex (V_{Ga}^{2+}–X) to be responsible for the near mid-gap acceptor level, where X is thought to be an intrinsic defect rather than any residual impurity atom. Following this, the photoionization transition

$$(V_{Ga}^{+2} - X) + h\nu_{>1.0\,eV} \rightarrow (V_{Ga}^{3+} - X) + e_{cb}^{-} \quad (4)$$

can be regarded as producing the absorption α_{II}. Therefore, a Franck-Condon shift of about 0.23 eV is connected with this transition. The correlation $\alpha_{II} \sim 1/[EL2^0]$ suggests that the concentration of (V-X) centres increases with increasing Ga-excess in the melt during growth so that X can possibly be identified as being an arsenic vacancy. This appears to be reasonable because the proposed assignment requires a very strong down shift of the V_{Ga}^{2+} acceptor level within the gap (about 0.6 eV) caused by the association with the intrinsic defect X. A shift of such remarkable amount can only be generated by a nearest neighbour defect.[4])

(v) In order to explain absorption $\alpha_{\rm I}$ together with the V-related electron trap at $E_c - 0.23$ eV found by DLTS [6], we put forward the hypothesis that the (V_{Ga}^{2+}-X) centre is able to capture another electron forming (V_{Ga}^{+}-X), and this centre is responsible for absorption $\alpha_{\rm I}$ owing to the photoionization transition

$$(V_{Ga}^{+}\text{-X}) + h\nu_{>0.58\,eV} \rightarrow (V_{Ga}^{2+}\text{-X}) + e_{cb}^{-} . \tag{5}$$

The weak absorption band at 0.69 eV always connected with $\alpha_{\rm I}$ could than be interpreted as being the $^5T_2 \rightarrow {}^5E$ transition of ($V_{Ga}^{+}(3d^4)$-X). For an even tentative assignment of the absorption band at 1.03 eV measured in GaAs:V:Si additional investigations are necessarily to be made. However, from a comparison of the triple-peaked absorption band lineshapes measured on GaAs:V and GaAs:V:Zn it seems to be clear that the 1.03 eV band is not hidden detectable under the 1.04 eV low-energy side peak in GaAs:V.

(vi) Obviously the (V_{Ga}-X) centres are absent in GaAs:V:Zn samples because in the region of the $^3T_2 \rightarrow {}^3A_2$ luminescence and $^3A_2 \rightarrow {}^3T_1(F)$ absorption of V_{Ga}^{3+} no changes or new structures could be detected indicating the presence of (V_{Ga}^{3+}-X) centres. Furtheron if (V_{Ga}^{3+}-X) centres would exist, the photoionization transition (V_{Ga}^{3+}-X) + $h\nu$ → (V_{Ga}^{2+}-X) + hole$_{vb}$ should occur with an optical threshold near 0.8 to 0.9 eV, but no such absorption was found.

Summarizing the foregoing conclusions, we have to state that in our LEC-grown GaAs:V samples the concentration of the mid-gap acceptor centres, [(V_{Ga}^{2+}-X)], is smaller than the concentration of free electrons $n = N_D - N_A$ originating from residual impurities. After filling the (V_{Ga}-X) centres (at first forming (V_{Ga}^{2+}-X), then possibly also (V_{Ga}^{-}-X)), the isolated V_{Ga}^{3+} capture the remaining free electrons forming V_{Ga}^{2+} and this acceptor level at $E_c - 0.14$ eV determines mainly the electrical properties of GaAs:V (n-conducting at room temperature, semi-insulating at low temperatures). Therefore, for producing GaAs:V which is semi-insulating also at room temperature the condition [(V_{Ga}-X)] $> N_D - N_A$ must be fulfilled. This can be done either by increasing the concentration of (V_{Ga}-X) centres through changed growth conditions or by lowering the concentration of residual donors.

Fig. 10 summarizes in an energy level diagram the interpretation of the optical and electrical measurements carried out till now on vanadium-doped GaAs. This diagram brings all known data into a consistent picture. However, much work is necessary to prove or improve it in detail.

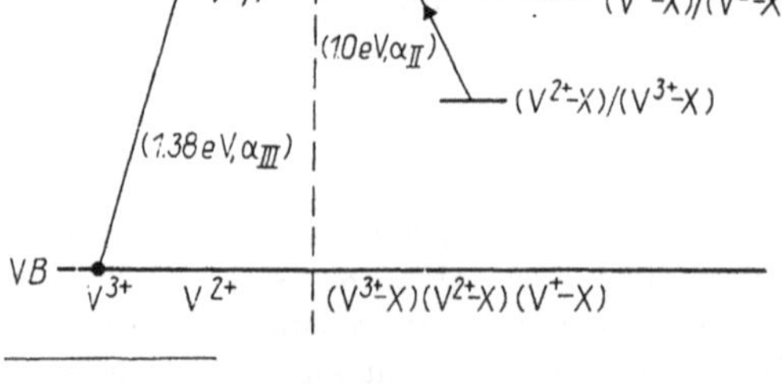

Fig. 10. Energy level diagram showing thermal energies of vanadium-related levels in GaAs. Optical threshold energies are shown in parentheses

[4]) It is worth noting that the analogous centre (Cr_{Ga}^{2+}–X) with X being an arsenic vacancy is investigated and debated in si GaAs:Cr (trigonal Cr^{2+}(II) centre). However, as far as the authors are aware, the position of the ground state of this centre within the gap is unknown until now.

5. Analysis of the $^3A_2 \to {}^3T_1(F)$ Absorption Band of V_{Ga}^{3+}

The experiments described in Section 2 confirm undoubtedly that the triple-peaked absorption band at 1.1 eV has to be assigned to the $^3A_2 \to {}^3T_1(F)$ transition of V_{Ga}^{3+}. The position of the zpl of the $^3T_2(F) \to {}^3A_2$ luminescence transition provides the value of the crystal-field strength $10\,Dq = 0.739$ eV. The absorption transition $^3A_2 \to$ $\to {}^3T_2(F)$ is parity-forbidden in T_d symmetry and its oscillator strength is expected to be two orders of magnitude smaller than that of the $\to {}^3T_1(F)$ transition. Therefore, the weak absorption band at 0.79 eV, which is always observed if the 1.1 eV band is very intense, can be assigned to the $^3A_2 \to {}^3T_2(F)$ transition of V_{Ga}^{3+}.

The increase of the distances between the side peaks and the central peak of the $\to {}^3T_1(F)$ absorption band with increasing temperature clearly indicates that this triplet structure is due to a strong $T \otimes \tau_2$ dynamical Jahn-Teller effect (for a comprehensive survey on the theoretical approaches and results as well as on the experimental evidence of this kind of JT-interaction see chapters 7 and 10 in [19]). From the slope of the $\Delta E_h = f(T^{1/2})$ dependence at high temperatures the JT-energy $E_{JT} = 780$ cm^{-1} can be derived according to (1) and using the relations $\Delta E(0) = 0.91c\,(\hbar\omega_\tau/K_\tau)^{1/2}$ and $E_{JT} = 2c^2/3K_\tau$ with $K_\tau = M\omega_\tau^2$, where c is the $T \otimes \tau_2$ — JT coupling constant, ω_τ the frequency of the interacting vibrational modes of symmetry τ_2, and M the mass of an arsenic atom (cluster model of the electron–phonon coupling). From $\Delta E_h(0)$ it follows that $c = 0.47$ eV/Å and the effective phonon energy $\hbar\omega_\tau = 75$ cm^{-1} in accordance with the value derived from the fit with (1). The resulting parameter $S_\tau =$ $= E_{JT}/\hbar\omega_\tau = 10$ indicates the presence of a strong JT-coupling justifying the application of the used relations following from the semi-classical approach.

Assuming this strong $T \otimes \tau_2$ coupling, we expect a nearly complete quenching of the spin–orbit splitting within the vibronic ground state of the $^3T_1(F)$ state (Ham reduction factor $K(T_1) < 10^{-2}$ for $S_\tau > 5$). The vibronic ground state triplet and the first excited vibronic singlet converge, the tunneling splitting between them amounts to $\delta = 1.2E_{JT}\exp(-1.21E_{JT}/\hbar\omega_\tau) \approx 3 \times 10^{-3}$ cm^{-1}. Therefore, it is not surprising that only one zpl[5]) is found originating from the transition to this fourfold degenerated vibronic ground state. The following phonon replica provide a phonon energy (65 cm^{-1}) close to that derived from the $\Delta E(T)$ dependence, both falling into the TA-phonon density -of-states maximum [21] (the replica at zpl + $2\hbar\omega_{ph}$ is probably masked by the line at 8237 cm^{-1} [6])). At present it cannot be decided whether the half-width of the zpl is due to the remaining non-quenched spin–orbit splitting within the vibronic ground state or due to strains within the sample. The weakness of the zpl compared with the broad band (their intensity ratio amounts to $\approx 10^{-4}$) is consistent with the derived large coupling strength $S_\tau = 10$.

The $\Delta E_l(T)$ dependence fitted with (2) indicates the presence of another interaction besides the JT-coupling leading to a splitting $\Delta E_l = 280$ cm^{-1} of the $^3T_1(F)$ state comparable with the splitting $\Delta E(0)$ due to JT-coupling. The value of ΔE_l suggests that the most probable origin of this additional distortion is the spin–orbit coupling (overall splitting of the $^3T_1(F)$ state $\Delta E_{so} = 3\lambda_0 = 310$ cm^{-1}, λ_0 is the free-ion spin–orbit coupling parameter), which is unquenched in highly excited vibronic states. Such an effect of the spin–orbit coupling in the absorption lineshape of transitions to excited states with strong $T \otimes \tau_2$ JT coupling has been calculated by Cho [22] for a $\to {}^2T_1$ transition and has been found experimentally already in transition metal impurities if $E_{JT}/\Delta E_{so} < 5$ (see Chapter 10 in [19]).

[5]) This zpl was found also by Clerjaud et al. [20] at 8131 cm^{-1} with a half-width of 3 cm^{-1}.

[6]) The origin of the line at 8237 cm^{-1} remains unexplained at present and needs to be clarified by a complete crystal-field analysis of V_{Ga}^{3+} including as many spin-forbidden transitions as found experimentally (cf. also [20]). Its absence in GaAs:V:Si samples confirms the assignment to V_{Ga}^{3+}

The pronounced changes of the skewness of the triple-peaked lineshape with temperature can be regarded as an example for the truth of the predictions made by Nasu and Kojima [23] using the independent-ordering approximation (IOA) for calculating the absorption lineshape. If only the linear JT-interaction is considered, the calculations predict at low temperatures a strong positive skewness (weak high-energy side peak, lineshape nearly "doublet like") as observed here, and for high temperatures a symmetrical triple-peaked lineshape. However, as it was emphasized by the authors, a strong linear JT-interaction is often connected with a strong quadratic JT-interaction so that at high temperatures the lineshape becomes more and more asymmetric with increasing negative skewness just as is measured here for the $^3A_2 \rightarrow$ $\rightarrow {}^3T_1(F)$ transition.

Acknowledgements

The authors wish to thank Prof. C. A. Bates and Dr. V. W. Rampton for valuable discussions, M. Seifert for supplying samples, and A. Riedel for technical assistance.

References

[1] W. KÜTT, D. BIMBERG, M. MAIER, H. KRÄUTLE, F. KÖHL, and E. BAUER, Appl. Phys. Letters **44**, 1078 (1984).
[2] A. V. VASILIEV, G. K. IPPOLITOVA, E. M. OMELYANOVSKII, and A. I. RYSKIN, Fiz. Tekh. Poluprov. **10**, 571 (1976).
[3] U. KAUFMANN, H. ENNEN, J. SCHNEIDER, R. WÖRNER, J. WEBER, and F. KÖHL, Phys. Rev. B **25**, 5598 (1982).
[4] A. M. VASSON, A. VASSON, C. A. BATES, and F. LABADZ, J. Phys. C **17**, L 837 (1984).
[5] A. MIRCEA-ROUSSEL, G. M. MARTIN, and J. E. LOWTHER, Solid State Commun. **36**, 171 (1980).
[6] F. LITTY, P. LEYRAL, S. LOUALICHE, A. NOUAILHAT, G. GUILLOT, and M. LANNOO, Physica (Utrecht) **117B/118B**, 182 (1983).
[7] G. ARMELLES, J. BARRAU, D. THEBAULT, and M. BROUSSEAU, J. Physique **45**, 1795 (1984).
[8] G. ARMELLES, J. BARRAU, and D. THEBAULT, J. Phys. C **17**, 6883 (1984).
[9] M. S. SKOLNICK, P. J. DEAN, M. J. KANE, CH. UIHLEIN, D. J. ROBBINS, W. HAYES, B. COCKAYNE, and W. R. MACEWAN, J. Phys. C **16**, L 767 (1983).
[10] K. J. NASH, M. S. SKOLNICK, B. COCKAYNE, and W. R. MACEWAN, J. Phys. C **17**, 6199 (1984).
[11] M. J. KANE, M. S. SKOLNICK P. J. DEAN, W. HAYES, B. COCKAYNE, and W. R. MACEWAN, J. Phys. C **17**, 6455 (1984).
[12] H. TERAO, H. SUNAKAWA, K. OHABA, and H. WATANABE, in: Semi-Insulating III-V Materials, Shiva, Nantwich; Ed. S. MAKRAM-EBEID and B. TUCK, Evian 1982 (p. 54).
[13] M. AKIYAMA, Y. KAWARADA, and K. KANURISHI, J. Crystal Growth **68**, 39 (1984).
[14] D. C. LOOK, in: Semiconductors and Semimetals, Vol. 19, Academic Press, 1983 (p. 87).
[15] G. M. MARTIN, Appl. Phys. Letters **39**, 747 (1981).
[16] A. MITTONEAU and A. MIRCEA, Solid State Commun. **30**, 157 (1979).
[17] G. VINCENT, D. BOIS, and A. CHANTRE, J. appl. Phys. **53** 3643 (1982).
[18] J. LAGOWSKI. H. C. GATOS, J. M. PARSEY, K. WADA, M. KAMINSKA, and W. WALUKIEWICZ, Appl. Phys. Letters **40**, 342 (1982).
[19] YU. E. PERLIN and M. WAGNER (Ed.), The Dynamical Jahn-Teller Effect in Localized Systems, North-Holland Publ. Co.. Amsterdam 1984.
[20] B. CLERJAUD C. NAUD, C. BENJEDDOU, G. GUILLOT, P. LEYRAL, B. DEVAUD, and B. LAMBERT, Semi-Insulating III-V Materials, Shiva, Nantwich; Ed. D. C. LOOK and J. S. BLAKEMORE, Kah-na-tee 1984 (p. 484).
[21] C. PATEL, T. J. PARKER, H. JAMSHIDI, and W. F. SHERMAN, phys. stat. sol. (b) **122**, 461 (1984).
[22] K. CHO, J. Phys. Soc. Japan **25**, 1372 (1968).
[23] K. NASU and T. KOJIMA, Progr. theor. Phys. (Kyoto) **51**, 26 (1974).

(Received May 29, 1985)

phys. stat. sol. (b) 131, 729 (1985)

Subject classification: 13.4 and 20.1; 6; 14.3.4; 22

Sektion Physik der Humboldt-Universität zu Berlin[1])

Electroabsorption for Deep Level to Band Transitions Accompanied by Multiphonon Processes

By

A. SCHENK, R. ENDERLEIN, and D. SUISKY

Dedicated to Prof. Dr. Dr. h. c. Dr. E. h. P. GÖRLICH on the occasion of his 80th birthday

The electroabsorption spectrum for optical transitions from a deep level to a band accompanied by multiphonon processes is calculated within the Lucovsky model. The line shape turns out to be highly sensitive to the Huang-Rhys factor and to the ratio of electrooptical and phonon energy.

Das Elektroabsorptionsspektrum für optische Übergänge aus einem tiefen Niveau in ein Band unter Beteiligung von Multiphonon-Prozessen wird im Rahmen des Lucovsky-Modells berechnet. Die Linienform hängt empfindlich vom Huang-Rhys-Faktor und vom Verhältnis der Phonon-Energie zur optoelektronischen Energie ab.

1. Introduction

Deep impurities and defects in semiconductors have attracted a growing deal of attention during the last years, mainly owing to their great importance for semiconducting devices. Theoretical developments also have contributed to the process of identifying and understanding the properties of several defects. The description of the electronic structure of deep defects and impurities without any external perturbation has been improved by theoretical self-consistent Green's function methods, but if radiative or nonradiative transitions are considered one needs in addition non-perturbative results for the wave functions to get the field- and temperature-dependent transition probabilities. Beside DLTS measurements, EPR and optical methods either with external or internal fields of p–n junctions, Schottky and surface barriers form the most powerful techniques in determining the defect nature, wave functions, binding energies, cross-sections and other properties. Thus, it is useful to study both the essential influence of a strong electric field and of multiphonon excitation and deexcitation processes on optical transitions from deep level to band states.

Compared with the plenty of publications dealing with electric field effects on interband dielectric properties the processes involving deep levels lack improved understanding because of their complexity. But it is possible to cope with the problem by adopting a three-dimensional δ-potential model which allows the exact analytical solubility even in the presence of an electric field (Vinogradov [1]). In [1] the absorption spectrum has been calculated for transitions from the impurity level to the band from which the level splits up. The same model has been applied by Enderlein et al. [2] to the case of transition from valence band to impurity states accompanied by multiphonon processes. In [2] the deep-level field effect on the multiphonon-broadened optical transition has been studied. It was found that the deep level is asymmetrically broadened and its maximum of state density is shifted. In addition density of states

[1]) Invalidenstraße 110, PSF 1297, DDR-1040 Berlin, GDR.

of the initial or final band are changed according to the Franz-Keldysh effect [3]. Which of the two mechanisms is dominating depends on the particular situation, e.g. on the binding energy of the deep level and the band effective mass. In the present paper we will follow the general approach from [2] but concentrate on the second mechanism, sometimes referred to as "half-sided Franz-Keldysh effect".

Starting with the standard multiphonon theory, the general expression for the absorption coefficient $\alpha(\omega)$ will be derived in Section 2. In Section 3 we will calculate the transition rate for impurity to band transitions on the basis of a direct two-band model using the modified Lucovsky model [1] for the deep centre. The result will be discussed for strong and low electric fields in Section 4.

2. General Expression for the Absorption Spectrum

The one-electron potential consists of the periodic lattice part $V_0(\boldsymbol{x})$, the trap potential $V_t(\boldsymbol{x})$, the diagonal part of the electron–phonon interaction $H_{ep}(\boldsymbol{x}, Q)$, the electron–photon interaction, and the potential $-e\boldsymbol{F} \cdot \boldsymbol{x}$ in an external homogeneous electric field $\boldsymbol{F}$. The decoupling procedure of the multiphonon theory is extended to our problem, where the energy spectrum is continuous due to the electric field. This requires the additional assumption that electronic transitions with transition energies comparable with an effective phonon energy can be neglected. However, in the case of deep centres a real gap will remain even for field strengths up to 10^5 V/cm and, consequently we write the total wave function $\psi_{\lambda N}$ of the coupled electron–phonon system in the usual product form $\psi_{\lambda N} = \varphi_\lambda \Phi_{\lambda N}$ with the electron states φ_λ and the lattice states $\Phi_{\lambda N}$ given by the well-known set of coupled equations in its static version,

$$\left\{-\frac{\hbar^2}{2m}\Delta + V_0(\boldsymbol{x}) + V_t(\boldsymbol{x}) - e\boldsymbol{F} \cdot \boldsymbol{x}_{\text{diag}}\right\} \varphi_\lambda = \varepsilon_\lambda^0 \varphi_\lambda\,, \tag{1}$$

$$\{H_p + (\varphi_\lambda|\, H_{ep}\, |\varphi_\lambda) + \varepsilon_\lambda^0\}\, \Phi_{\lambda N} = E_{\lambda N} \Phi_{\lambda N}\,, \tag{2}$$

where H_p denotes the phonon Hamiltonian and $\boldsymbol{x}_{\text{diag}}$ the diagonal part of $\boldsymbol{x}$ with respect to the discrete component of the set of electronic quantum numbers λ, usually being the band or the deep-level index. Optical transitions from deep-level states λ into band states λ' which are assumed to be dipole allowed are calculated by applying the "Golden Rule". The result for the absorption coefficient is (cf. [4])

$$\alpha(\omega, \boldsymbol{F}) = \frac{B}{\omega} \sum_{\lambda\lambda'} \sum_{NN'} p_N\, |(\Phi_{\lambda' N'} \varphi_{\lambda'}|\, \epsilon \cdot \nabla\, |\varphi_\lambda \Phi_{\lambda N})|^2\, \delta(E_{\lambda' N'} - E_{\lambda N} - \hbar\omega) \tag{3}$$

with

$$B = \frac{4\pi^2 e^2 \hbar^2}{m^2 c n V}\,. \tag{4}$$

In (3) ϵ denotes the polarisation vector of the radiation and p_N means the statistical weight of the lattice initial states. The λ, λ' summation extends over occupied electron initial states and empty electron final states. Owing to the static equations (1) the expression for $\alpha(\omega)$ can be written as

$$\alpha(\omega, \boldsymbol{F}) = \frac{B}{\omega} \sum_{\lambda\lambda'} |P_{\lambda'\lambda}(\boldsymbol{F})|^2\, I_{\lambda'\lambda}(\omega) \tag{5}$$

with the electronic transition matrix element

$$P_{\lambda'\lambda}(\boldsymbol{F}) = (\varphi_{\lambda'}|\, \epsilon \cdot \nabla\, |\varphi_\lambda) \tag{6}$$

and $I_{\lambda'\lambda}(\omega)$ the well-known lineshape function of the multiphonon theory [5]. If we define

$$(\varphi_\lambda|\, H_{\text{ep}}\, |\varphi_\lambda) = \sum_q u_q^\lambda Q_q \tag{7}$$

and use the solutions $\Phi_{\lambda N}$ of (2), the explicit result is

$$I_{\lambda'\lambda}(\omega) = \frac{1}{2\pi\hbar} \int_{-\infty}^{\infty} d\tau\, e^{-i\tau\omega}\, e^{\frac{i\tau}{\hbar}(\varepsilon_{\lambda'}-\varepsilon_\lambda)}\, e^{\sum_q S_q g(\omega_q\tau)} . \tag{8}$$

Here

$$S_q = \frac{1}{2}\frac{1}{\hbar\omega_q^3}|u_q^{\lambda'} - u_q^\lambda|^2 \quad \text{and} \quad \varepsilon_\lambda = \varepsilon_\lambda^0 - \frac{1}{2}\sum_q \frac{1}{\omega_q^2}|u_q^\lambda|^2, \tag{9}$$

are, respectively, the differential Huang-Rhys factors and the electronic energies shifted by the lattice relaxation energy, and $g(\omega_q\tau)$ is given by

$$g(\omega_q\tau) = (\bar{N}_q + 1)\, e^{i\omega_q\tau} + \bar{N}_q\, e^{-i\omega_q\tau} - (2\bar{N}_q + 1) \tag{10}$$

($\bar{N}_q$ phonon occupation number, $\hbar\omega_q$ phonon energy).

Restricting to only one effective mode ω_0 the integral in (8) can be easily evaluated yielding [5]

$$I_{\lambda'\lambda}(\omega) = e^{-S(2\bar{N}+1)} \sum_{l=-\infty}^{\infty} \left(\frac{\bar{N}+1}{\bar{N}}\right)^{l/2} I_l[2S\sqrt{\bar{N}(\bar{N}+1)}] \times \\ \times \delta(l\hbar\omega_0 + \varepsilon_{\lambda'} - \varepsilon_\lambda - \hbar\omega) \tag{11}$$

(I_l modified Bessel function of the order l).

Note that according to (7) and (1) the coupling constants u_q^λ become field dependent. In the following we will neglect the electron–phonon coupling for final states. The energy difference is then given by

$$\varepsilon_{\lambda'} - \varepsilon_\lambda = \varepsilon_{\lambda'}^0 - \varepsilon_\lambda^0 + S\hbar\omega_0 \tag{12}$$

with the field-dependent Franck-Condon energy

$$\Delta_{\text{FC}} = S\hbar\omega_0 = \frac{1}{2\omega_0^2}|u_0^\lambda|^2 . \tag{13}$$

The change in $\alpha(\omega)$ is obtained by subtracting from (5) the zero field value,

$$\Delta\alpha(\omega) = \alpha(\omega, \boldsymbol{F}) - \alpha(\omega, \boldsymbol{0}) . \tag{14}$$

3. Wave Functions and Transition Rate

The explicit calculation of electron states is carried out for a two-band model with a parabolic and isotropic valence band $E_v(\boldsymbol{k})$ (effective mass m_v) and a parabolic ellipsoidal conduction band $E_c(\boldsymbol{k})$ (effective masses m_l, m_t). We consider only direct transitions and postulate the possibility to build the deep-level states only from valence states while the continuum states are assumed to be built only from Bloch functions of the conduction band near $\boldsymbol{k} = \boldsymbol{0}$,

$$\varphi_{E\text{c/t}}(\boldsymbol{x}) = \frac{1}{\sqrt{V}} \sum_{\boldsymbol{k}} A_E^{\text{c/t}}(\boldsymbol{k})\, \psi_{\text{c/v}\boldsymbol{k}}(\boldsymbol{x}) . \tag{15}$$

This approach should be reasonable for deep neutral centres attached to the valence band. The set of Schrödinger equations for the two envelope wave functions reads (neglecting interband transitions and the intraband polarisation of the Bloch functions)

$$\{E_c(\boldsymbol{k}) - E + ie\boldsymbol{F} \cdot \nabla_{\boldsymbol{k}}\}\, A_E^c(\boldsymbol{k}) = 0\,, \tag{16}$$

$$\{E_v(\boldsymbol{k}) - E + ie\boldsymbol{F} \cdot \nabla_{\boldsymbol{k}}\}\, A_E^t(\boldsymbol{k}) + \sum_{\boldsymbol{k}'} (v\boldsymbol{k}|\; V_t(\boldsymbol{x})\, |v\boldsymbol{k}')\, A_E^t(\boldsymbol{k}') = 0\,. \tag{17}$$

As has been shown in [6], the exact form of the short-range potential does not affect the energy dependence of the optical cross-section for zero electric field. We take the pseudo-δ-potential

$$V_t(\boldsymbol{x}) = V_0\, \delta(\boldsymbol{x})\, [1 + \boldsymbol{x}\nabla_{\boldsymbol{x}}]\,. \tag{18}$$

It produces one bound state in the gap for zero electric field. The potential strength $V_0 = 4\pi r_0^3 E_B^0$ can be fitted to the observed zero-field binding energy E_B^0, measured from the top of the valence band (r_0 localisation radius). With the potential (18) equation (17) can be solved exactly even in the presence of an electric field [1, 2]. In one-band approximation ($E_v(\boldsymbol{k}) = -E_g - \hbar^2\boldsymbol{k}^2/2m_v$) the normalized envelope wave function $A_E^t(\boldsymbol{x})$ in $\boldsymbol{x}$ space is given by

$$A_E^t(\boldsymbol{x}) = \frac{A}{\sqrt{\mathcal{F}(\eta)}}\, \frac{1}{\sqrt{\mathcal{F}^2(\eta) + \hat{\mathcal{G}}^2(\eta)}} \int\limits_0^\infty \mathrm{d}\varkappa\, \varkappa J_0(\varkappa\sqrt{x^2 + y^2}) \times$$

$$\times \left\{\hat{\mathcal{G}}(\eta)\, \mathrm{Ai}\,[\eta(\varkappa)]\, \mathrm{Ai}\left[\eta(\varkappa) + \frac{eFz}{\hbar\theta_v}\right] - \mathcal{F}(\eta)\, C[\eta(\varkappa); z]\right\} \tag{19}$$

with

$$A = \left[\frac{eF\hbar^2}{\pi(\hbar\theta_v)^3\, 2m_v}\right]^{1/2}, \qquad \eta(\varkappa) = \frac{E - E_v(\varkappa)}{\hbar\theta_v}, \qquad \eta = \frac{E + E_g}{\hbar\theta_v}$$

and

$$\hat{\mathcal{G}}(\eta) = \frac{1}{\pi}\sqrt{\frac{E_B^0}{\hbar\theta_v}} + \mathcal{G}(\eta)\,,$$

$$C[w; z] = \theta(z)\, \mathrm{Ai}\left(w + \frac{eFz}{\hbar\theta_v}\right) \mathrm{Bi}(w) + \theta(-z)\, \mathrm{Ai}(w)\, \mathrm{Bi}\left(w + \frac{eFz}{\hbar\theta_v}\right). \tag{20}$$

The electrooptical functions $\mathcal{F}(\eta)$ and $\mathcal{G}(\eta)$ in (19) and (20) can be expressed by Airy functions in the following way:

$$\mathcal{F}(\eta) = \mathrm{Ai}'^2(\eta) - \eta \mathrm{Ai}^2(\eta)\,, \qquad \mathcal{G}(\eta) = \mathrm{Ai}'(\eta)\, \mathrm{Bi}'(\eta) - \eta\, \mathrm{Ai}(\eta)\, \mathrm{Bi}(\eta)\,, \tag{21}$$

where $\hbar\theta_v$ is the electrooptical energy,

$$\hbar\theta_v = \left[\frac{(e\hbar F)^2}{2m_v}\right]^{1/3}. \tag{22}$$

It is interesting to consider the local density of deep-centre states. Taking (19) at the origin, one obtains after simple algebra

$$\varrho_t(E) = |A_E^t(0)|^2 = (\sqrt{E_B^0 \hbar\theta_v}\, 4\pi^3 r_0^3)^{-1}\, \frac{\mathcal{F}(\eta)}{\mathcal{F}^2(\eta) + \hat{\mathcal{G}}^2(\eta)} \tag{23}$$

($r_0^2 = \hbar^2/2m_v E_B^0$). The spectrum (23) turns out to be an asymmetrically broadened δ-peak close to the zero-field position E_B^0 with increasing broadening for rising field strengths. The zero-field limit of (19) can be easily found using the asymptotic behaviour of $\mathcal{F}(\eta)$ and $\hat{\mathcal{G}}(\eta)$,

$$\lim_{F \to 0} \varrho_t(E) = (2\pi r_0^3)^{-1} \delta(E + E_g - E_B^0) . \tag{24}$$

To determine the position of the maximum of state density one has to solve the equation $\hat{\mathcal{G}}(\eta) = 0$. As long as field strengths are below a certain threshold value the asymptotic form of $\mathcal{G}(\eta)$ for large positive arguments can be applied [7],

$$\mathcal{G}(\eta) \approx -\frac{1}{\pi} \sqrt{\eta}\,(1 - 0.03123/\eta^3) , \tag{25}$$

yielding

$$E(\max\, \varrho_t) \approx E_B^0 \left[1 + 0.0625 \left(\frac{\hbar\theta_v}{E_B^0}\right)^3\right] - E_g , \tag{26}$$

where terms of higher than third order in $\hbar\theta_v/E_B^0$ have been neglected. Equation (26) expresses the quadratic Stark effect and shows that the energy shift will only be remarkable for extremely high field strengths (above 10^6 V/cm). Thus, the field effects on impurity states within the Lucovsky model are much smaller than those on band states.

To calculate the latter for a parabolic ellipsoidal conduction band we define the angle ϑ between the electric field applied in z-direction and the rotational axis of the energy ellipsoid. Then (16) has the solution

$$A^c_{Ek_{\perp 1}k_{\perp 2}}(\boldsymbol{x}) = B \exp\,[ixk_{\perp 1} + iyk_{\perp 2} - iz\beta k_{\perp 2}]\,\mathrm{Ai}\left(\frac{E_\perp - E - eFz}{\hbar\theta_{\|}}\right), \tag{27}$$

where the following abbreviations have been introduced:

$$B = \frac{\sqrt{eF}}{2\pi\hbar\theta_{\|}}, \quad \theta_{\|}^3 = \frac{(eF)^2}{2\hbar m_{\|}}, \quad E_\perp = \frac{\hbar^2 k_{\perp 1}^2}{2m_t} + \frac{\hbar^2 k_{\perp 2}^2}{2m_\perp} \tag{28}$$

and

$$\frac{1}{m_{\|,\perp}} = \frac{1}{m_{t,l}} \sin^2\vartheta + \frac{1}{m_{l,t}} \cos^2\vartheta \tag{29}$$

$$\beta = \frac{m_{\|}(m_l - m_t) \sin\vartheta \cos\vartheta}{m_t m_l} . \tag{30}$$

The local density of states corresponding to (23) now becomes

$$\varrho_c(E) = \int \mathrm{d}k_{\perp 1} \int \mathrm{d}k_{\perp 2}\, |A^c_{Ek_{\perp 1}k_{\perp 2}}(0)|^2 = \frac{1}{4\pi\hbar^3} \sqrt{8m_{\|}m_\perp m_t}\, \sqrt{\hbar\theta_{\|}}\, \mathcal{F}\left(-\frac{E}{\hbar\theta_{\|}}\right). \tag{31}$$

It describes the well-known Franz-Keldysh tails below and the oscillations around the zero-field density of states above the M_0-threshold. Using (19) and (27) for the initial and final states and the relation

$$(\psi_{ck'}|\, -i\hbar\nabla\, |\psi_{vk}) \approx \boldsymbol{p}_{cv}(0)\, \delta_{kk'} \tag{32}$$

for the momentum matrix element built with Bloch functions, the field-dependent absorption coefficient can be written as

$$\alpha(\omega, \boldsymbol{F}) = B \frac{|\epsilon \cdot \boldsymbol{p}_{cv}|^2}{\hbar\omega^2} \int_{-\infty}^{\infty} \mathrm{d}E \int_{-\infty}^{\infty} \mathrm{d}E'\; D(E', E)\, I_{E'E}(\omega) , \tag{33}$$

where $D(E', E)$ denotes the transition rate for a transition $E \to E'$,

$$D(E', E) = \int_{-\infty}^{\infty} \mathrm{d}k_{\perp 1} \int_{-\infty}^{\infty} \mathrm{d}k_{\perp 2} \left| \int \mathrm{d}^3x \, A^{c*}_{E'k_{\perp 1}k_{\perp 2}}(x) \, A^{t}_{E}(x) \right|^2 . \tag{34}$$

By means of (19) and (27) the transition rate can be approximately performed in an analytical expression. Using the assumptions $m_v \gg m_{||}$ and $E + E_g \approx E^0_B$ it follows the simple product form

$$D(E', E) = \frac{V_0^2}{[E + E_g + (m_\perp/m_v) \, E']^2} \varrho_t(E) \, \varrho_c(E') \tag{35}$$

with the potential strength V_0, introduced in (18), and the local densities of states from (23) and (31). The denominator in (35) is known to be typical for the Lucovsky model. The E integration in (33) is due to the field effect on impurity states and includes the electron–phonon coupling constants u_0^E in the lineshape function according to (13). It can be carried out immediately by neglecting the field effect on impurity states completely. The calculation of the absorption coefficient then results in the convolution integral

$$\alpha(\omega, \boldsymbol{F}) = B \frac{|\epsilon \cdot \boldsymbol{p}_{cv}|^2}{\omega \hbar^2} 2 V_0 E^0_B \int_{-\infty}^{\infty} \mathrm{d}E' \frac{\varrho_c(E')}{[E^0_B + (m_\perp/m_v) \, E']^2} I(\hbar\omega - E' - E_B) \tag{36}$$

with

$$I(E) = \frac{1}{2\pi} \int_{-\infty}^{\infty} \mathrm{d}\tau \exp\left[-i\tau E + S g(\hbar\omega_0 \tau)\right] , \tag{37}$$

and the assumption $m_v \gg m_{||}$ is not necessary any longer. In (36) E_B means the effective trap depth measured from the bottom of the conduction band: $E_B = E_g - E^0_B + \Delta_{FC}$.

Equation (36) is useful for a qualitative discussion of the general lineshape of $\Delta\alpha(\omega)$ for various ratios of electrooptical and phonon energy. If we use (14), (11), (36), (38), (31) and the asymptotic form of $\mathcal{F}(\eta)$,

$$\lim_{\boldsymbol{F} \to \boldsymbol{0}} \mathcal{F}(\eta) = \frac{1}{\pi} \theta(-\eta) \sqrt{-\eta} , \tag{38}$$

the field-induced change in $\alpha(\omega)$ finally can be expressed as a sum of thermally weighted electrooptical functions of the first kind,

$$F(\eta) = \pi \mathcal{F}(\eta) - \theta(-\eta) \sqrt{-\eta} , \tag{39}$$

$$\Delta\alpha(\omega) = \alpha_0 \, \mathrm{e}^{-S(2\bar{N}+1)} \sum_{l=-\infty}^{\infty} \left(\frac{\bar{N}+1}{\bar{N}}\right)^{l/2} \mathrm{I}_l[2S \sqrt{\bar{N}(\bar{N}+1)}] \times \\ \times \frac{\sqrt{\hbar\theta_{||}} \, F[(E_B + l\hbar\omega_0 - \hbar\omega) \, (\hbar\theta_{||})^{-1}]}{\omega[E^0_B - (m_\perp/m_v) \, (E_B + l\hbar\omega_0 - \hbar\omega)]^2} . \tag{40}$$

In Fig. 1 we have plotted lineshapes at $T = 10$ K for various field strengths and Huang-Rhys factors using (40). The zero-phonon line exhibits a pure shape of the electrooptical function F, slightly modified by the energy denominator in (40). For high fields the lineshape is determined by F, which gets increasingly broadened with rising coupling strengths and the first zero value of which turns out to be shifted by the Franck-Condon energy Δ_{FC} (Fig. 1a). The discrete nature of the spectrum due to

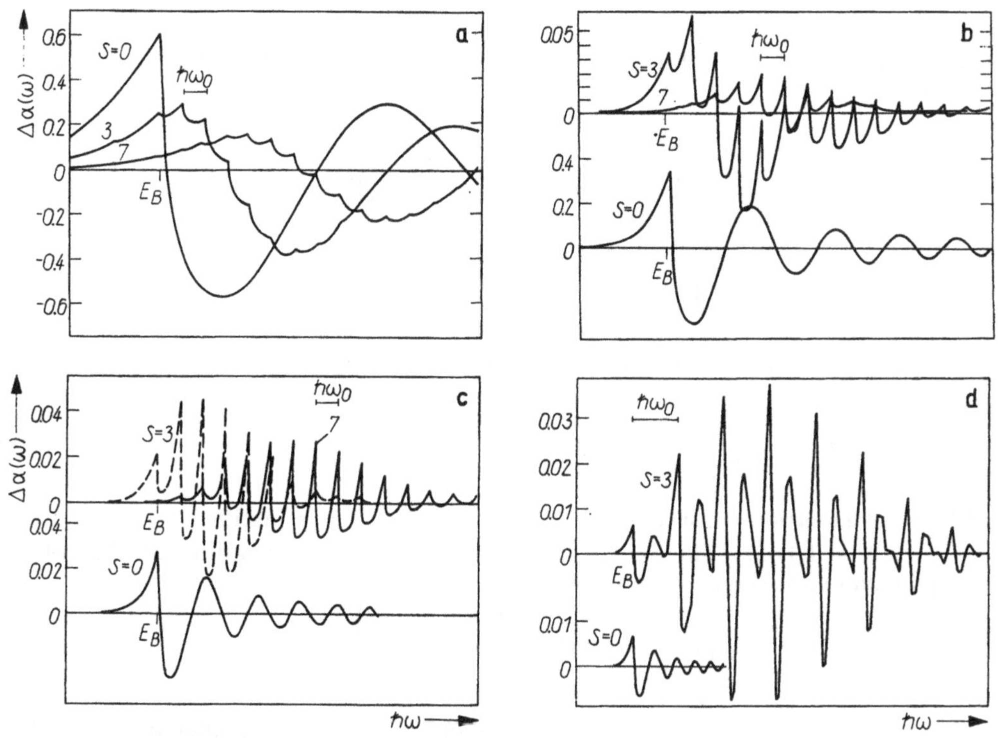

Fig. 1. Electroabsorption spectrum $\Delta\alpha(\omega)$ for different Huang-Rhys factors $S = 0$, $S = 3$, $S = 7$. The following parameters have been used: a) $\hbar\theta_{\parallel} = 113$, b) 39, c) 24, and d) 5 meV. (Here the curve for $S = 7$ has been omitted for clarity.) $\hbar\omega_0 = 20$ meV, $E_B/\hbar\omega_0 = 30$, $T = 10$ K. The energy scale is in units of the effective phonon energy $\hbar\omega_0$. E_B denotes the trap depth

the Einstein model, which is only weak for large field strengths, becomes dominant in the case of lower fields. Below a certain electrooptical energy, when the first oscillation of F has an energetic distance less than the effective phonon energy, additional substructures arise (Fig. 1d). The height of the peaks can grow with decreasing fields now, depending on whether a maximum of $F(\eta)$ (39) coincides with $\hbar\omega_0$ or not. Mironov et al. [8] who have measured the electroabsorption spectrum of a Cr level to conduction band transition report the same effect.

4. Strong- and Low-Field Limits

According to (36) the field-induced change in $\alpha(\omega)$ is given by the convolution integral

$$\Delta\alpha(\omega) = \frac{C\sqrt{\hbar\theta_{\parallel}}}{\omega} \int\limits_{-\infty}^{\infty} \mathrm{d}E' \, \frac{F\left(\dfrac{E'}{\hbar\theta_{\parallel}}\right)}{[E_B^0 - (m_{\perp}/m_v)\, E']^2} \, I(\hbar\omega - E_B + E')$$

$$C \equiv \frac{\alpha_0 \sqrt{8m_{\parallel}m_{\perp}m_t}}{4\pi^2\hbar^3}. \tag{41}$$

4.1 Strong fields

For strong electric fields the general lineshape is determined by the electrooptical function. To see this we write (41) as

$$\Delta\alpha(\omega) = \frac{C\sqrt{\hbar\theta_{||}}}{\omega} \int\limits_{-\infty}^{\infty} dE' \frac{F(E')}{[E_B^0 - (m_\perp/m_v)\,\hbar\theta_{||}E']^2} \times$$

$$\times \frac{1}{2\pi} \int\limits_{-\infty}^{\infty} d\tau\, e^{-i\tau\left[\frac{\omega}{\theta_{||}} - \frac{E_B}{\hbar\theta_{||}} + E'\right]} e^{Sg\left(\frac{\omega_0}{\theta_{||}}\tau\right)}. \tag{42}$$

We expand $g\left(\frac{\omega_0}{\theta_{||}}\tau\right)$ to second order in the phase factors,

$$g\left(\frac{\omega_0}{\theta_{||}}\tau\right) \approx i\tau\frac{\omega_0}{\theta_{||}} - \frac{\tau^2}{2S}\frac{\Gamma^2}{\theta_{||}^2}, \tag{43}$$

with the attenuation constant $\Gamma = \omega_0\sqrt{S(2\overline{N}+1)}$. This is justified if the condition $\omega_0/\theta_{||} \ll 2\overline{N} + 1$ holds. The linear term of (43) yields the Franck-Condon shift of the zero-phonon expression,

$$\Delta\alpha^{(1)}(\omega) = \frac{C\sqrt{\hbar\theta_{||}}}{\omega} \frac{F\left(\frac{E_B + \Delta_{FC} - \hbar\omega}{\hbar\theta_{||}}\right)}{[E_B^0 + (m_\perp/m_v)(\hbar\omega - E_B - \Delta_{FC})]^2}. \tag{44}$$

If the second-order term is taken into account, a Gaussian broadening of $\Delta\alpha^{(1)}(\omega)$ occurs,

$$\Delta\alpha^{(2)}(\omega) = \frac{\hbar}{\sqrt{2\pi}\,\Gamma} \int\limits_{-\infty}^{\infty} d\omega'\, \Delta\alpha^{(1)}(\omega - \omega')\, e^{-\omega'^2/2\Gamma^2}. \tag{45}$$

The resulting lineshape is then given by a shifted and damped electrooptical function F. Expression (45) cannot be evaluated in closed form. Numerical results for 400 K in comparison with those from (40) are given in Fig. 2.

Due to the Gaussian broadening only the general form of the spectrum is preserved, phonon replicas do not occur. This corresponds to the experimental situation of high temperatures.

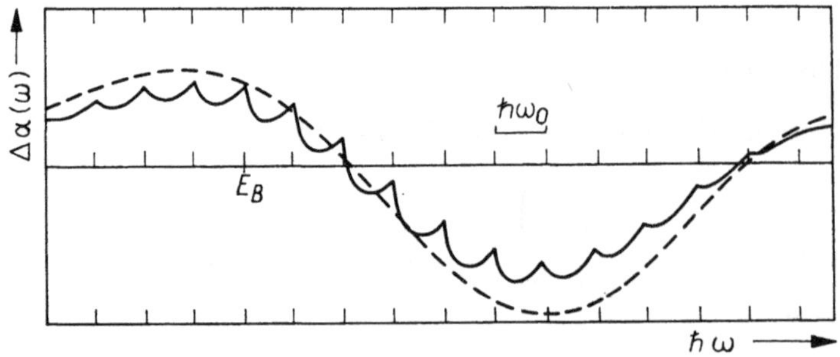

Fig. 2. Comparison of strong-field spectrum (dashed line) with the general form derived from (40). The following parameters have been used: $\hbar\theta_{||} = 113$ meV, $\hbar\omega_0 = 20$ meV, $S = 3$, $T = 400$ K

4.2 Low fields

For sufficiently low electric fields, $\Delta\alpha(\omega)$ can be approximated by the first non-vanishing term of a power series with respect to the field strength. Therefore, we write (41) in the form

$$\Delta\alpha(\omega) = \frac{C}{\omega} \int\limits_{-\infty}^{\infty} d\omega' \left\{ \frac{\left(\mathrm{Ai}\,\frac{\omega' - \omega}{\Omega_{\|}}\right)}{\Omega_{\|}} - \delta(\omega' - \omega) \right\} \times$$

$$\times \int\limits_{-\infty}^{\infty} dE'\, \theta(E') \sqrt{E'} \frac{I(\hbar\omega' - E' - E_B)}{[E_B^0 + (m_\perp/m_v)(E' - [\hbar\omega' - \hbar\omega])]^2} \tag{46}$$

with $\Omega_{\|} = \theta_{\|}/2^{2/3}$ and neglect the $(\omega' - \omega)$-dependence of the energy denominator, since $\mathrm{Ai}\left(\frac{\omega' - \omega}{\Omega_{\|}}\right)$ is localised at $\omega = \omega'$ for low fields. This enables us to express $\alpha(\omega, \boldsymbol{F})$ by convoluting the zero-field value with $\mathrm{Ai}\left(\frac{\omega' - \omega}{\Omega_{\|}}\right)$,

$$\alpha(\omega, \boldsymbol{F}) = \frac{1}{\omega} \int\limits_{-\infty}^{\infty} d\omega'\, \omega' \frac{\mathrm{Ai}\left(\frac{\omega' - \omega}{\Omega_{\|}}\right)}{\Omega_{\|}} \alpha(\omega', \boldsymbol{0})\,. \tag{47}$$

We expand $\mathrm{Ai}\left(\frac{\omega' - \omega}{\beta}\right)\Big/\beta$ to second order in $\boldsymbol{F}$,

$$\frac{\mathrm{Ai}\left(\frac{\omega' - \omega}{\beta}\right)}{\beta} \approx \delta(\omega' - \omega) + \frac{\beta^3}{3} \frac{d^3}{d\omega^3} \delta(\omega' - \omega) \tag{48}$$

to obtain

$$\Delta\alpha(\omega) \approx \frac{\Omega_{\|}^3}{3\omega} \frac{d^3}{d\omega^3} [\omega\alpha(\omega, \boldsymbol{0})]\,. \tag{49}$$

The effective expansion parameter β of (48) can be found from (46) and (45) to be the ratio of $\Omega_{\|}$ and $\sqrt{2}\,\Gamma$. Therefore, (49) is expected to be a good approximation, if the characteristic electrooptical frequency $\Omega_{\|}$ is small compared with the multiphonon broadening frequency Γ, i.e.

$$\frac{\Omega_{\|}}{\omega_0} \ll \sqrt{2S(2\bar{N} + 1)}\,.$$

Expressions (47) and (49) are identical with those derived by Aspnes et al. [4] for the case of optical interband transitions in the low-field limit. There $\hbar\Omega$ must be small compared to the lifetime broadening $\hbar\gamma$ of the transition. In our case the relevant energy of the unperturbed system is the multiphonon broadening energy $\hbar\Gamma$. The use of the third-derivative expression (49) involves the problem of strong divergences at energies $E_B + n\hbar\omega_0$ owing to the Einstein model. However, in reality additional lifetime broadening processes due to phonon–phonon interaction and electronic scattering mechanisms are present giving rise to additional Lorentzian broadening of the spectrum. On the other hand, at high temperatures (large Γ) phonon dispersion and other phonon branches smooth out the multiphonon substructure, which is a consequence of adopt-

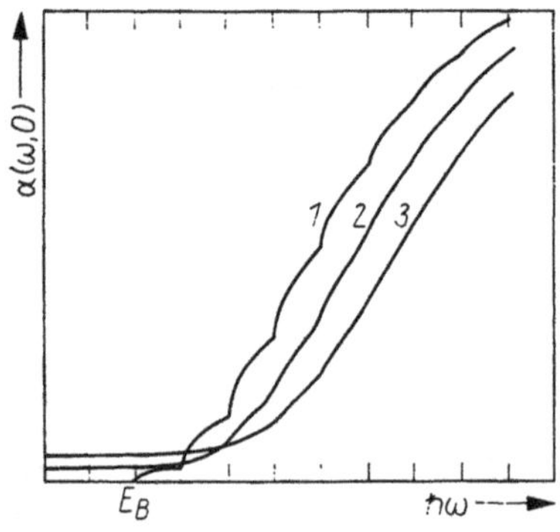

Fig. 3. Zero-field spectrum for various Lorentzian broadening parameters: (1) $\gamma = 0$, (2) 0.005, and (3) 0.01. The curves are shifted by $\hbar\omega_0$ for clarity. Parameters: $S = 3$, $E_B/\hbar\omega_0 = 30$, $T = 10$ K

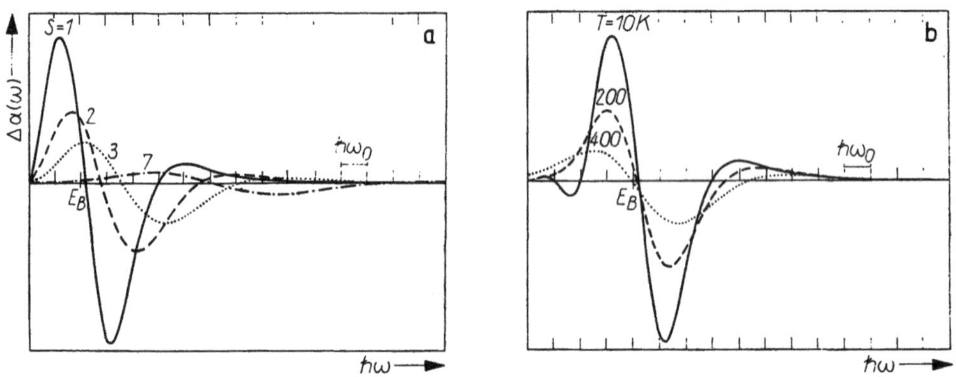

Fig. 4. Low-field spectrum without Lorentzian broadening for a) various electron–phonon coupling strengths ($E_B/\hbar\omega_0 = 30$, $T = 10$ K); b) different temperatures ($E_B/\hbar\omega_0 = 30$, $S = 1$)

ing only one phonon-mode. To illustrate both effects we have plotted the zero-field absorption spectrum for various Lorentzian broadening parameters γ in Fig. 3. Regarding the case when the multiphonon fine structure is smoothed out completely, the envelope curve of $\alpha(\omega,\boldsymbol{0})$ has been utilized to calculate $\Delta\alpha(\omega)$ by means of the third derivative of the former according to (49). The result is the well-known third-derivative lineshape of a damped square-root threshold. Numerical results for various values of S and T are given in Fig. 4. The curves show two characteristic features: The minimum

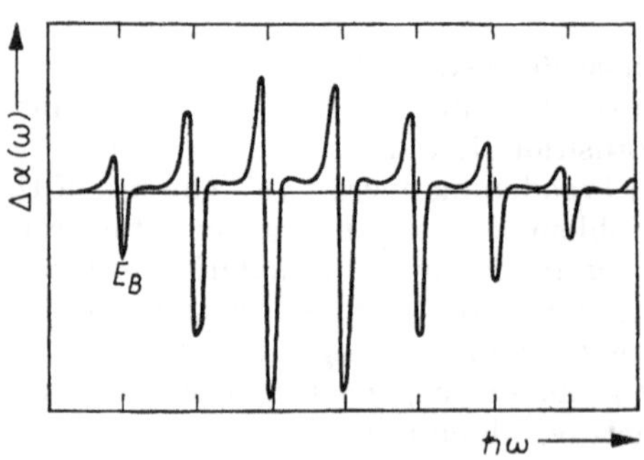

Fig. 5. Low-field spectrum after (49) with Lorentzian broadening, $\gamma = 0.04$. Parameters: $E_B/\hbar\omega_0 = 30$, $T = 10$ K, $S = 1$)

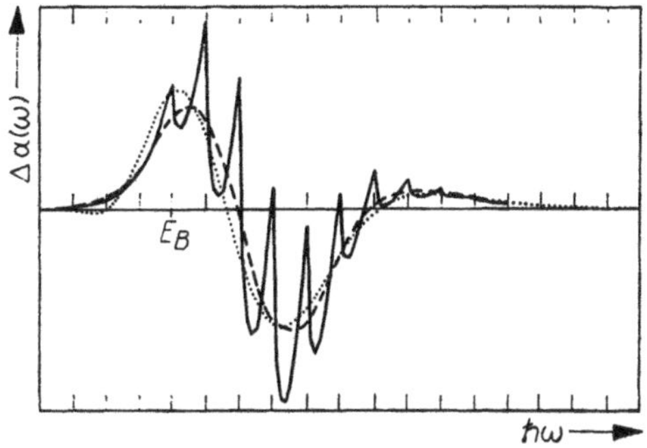

Fig. 6. Comparison of low-field spectrum (dotted line) and averaged spectrum after (40) (dashed line). The following parameters have been used: $\hbar\theta_{||} = 39$ meV, $\hbar\omega_0 = 20$ meV, $T = 10$ K, $S = 3$, $\sqrt{2}\,\Gamma/\theta_{||} = 1.3$

turns out to be localised very close to $E_B + \Delta_{FC}$ and the halfwidth of the negative peak is about $\sqrt{2}\,\hbar\Gamma = \sqrt{2S}\,\hbar\omega_0\sqrt{2\bar{N}+1}$.

If a Lorentzian broadened multiphonon fine structure is taken into account, the low-field spectrum (49) is in general agreement with that calculated from (40) (cf. Fig. 1d). A typical lineshape for this case is shown in Fig. 5. The absence of secondary oscillations is a consequence of the low-field limit.

Fig. 6 exhibits a direct comparison of low-field approximation with smooth zero-field spectrum and lineshape calculated from (40). The dashed line indicates the averaged spectrum of (40) obtained by an average procedure over the energetic distance $\hbar\omega_0$. These lineshape are expected, if phonon dispersion and other phonon branches smooth out the multiphonon substructure due to the Einstein model. A comparison with the low-field lineshape shows the latter to be a good approximation for $\sqrt{2}\,\Gamma/\Omega_{||} \geqq 3$.

References

[1] V. C. Vinogradov, Fiz. tverd. Tela **13**, 3266 (1971).
[2] R. Enderlein, D. Fiddicke, F. Bechstedt, K. Peuker, and R. S. Bauer, Luminescence Conf., Berlin (West) 1981.
[3] W. Franz, Z. Naturf. **13a**, 484 (1958).
L. V. Keldysh, Zh. eksper. teor. Fiz. **34**, 1138 (1958).
[4] D. E. Aspnes, P. Handler, and D. F. Blossey, Phys. Rev. **166**, 921 (1968).
[5] K. Huang and A. Rhys, Proc. Roy. Soc. **A204**, 406 (1950).
[6] G. Lucovsky, Solid State Commun. **3**, 299 (1965).
[7] D. E. Aspnes, Phys. Rev. **153**, 972 (1967).
[8] A. G. Mironov, V. A. Morozova, and V. S. Vavilov, phys. stat. sol. (b) **127**, 359 (1985).

(Received August 7, 1985)

Note added in proof

Experimental electroabsorption spectra have been reported by Sokolov et al. (V. I. Sokolov, A. E. Nikoforov, V. V. Chernyaev, and S. Yu. Shashkin, Zh. eksper. teor. Fiz., Pisma **33**, 189 (1981)) for ZnSe:Ni^{2+} in the energy range between 2.64 and 2.75 eV.

The lineshape can be interpreted in terms of the low field case ($F = 2 \times 10^4$ Vcm^{-1}) where the characteristic Franz-Keldysh oscillations in the energy range between the phonon satellites should appear (see Fig. 1d). In this case the amplitude of the positive and negative scillations is determined by the phononic part of the lineshape function ((see (37) and Fig. 1d).

phys. stat. sol. (b) 131, 741 (1985)

Subject classification: 13.4 and 20.1; 23

Institut für Experimentalphysik der Universität Kiel[1]) *(a)*
and Institut für Atom- und Festkörperphysik der Freien Universität Berlin[2]),
Berlin (West) (b)

Transient Absorption of Xe and Kr Guest Atoms in Ar Crystals

By

T. SCHRÖDER[3]) (a), O. DÖSSEL[4]) (a), H. NAHME (b), and N. SCHWENTNER (b)

Dedicated to Prof. Dr. Dr. h. c. Dr. E. h. P. GÖRLICH on the occasion of his 80th birthday

Time resolved absorption spectra for transitions starting from the lowest excited electronic states of Kr(5s, $^3P_2$) and Xe(6s, $^3P_2$) guest atoms in Ar crystals show broad bands at 1.81 and 1.72 eV, respectively. The bands are assigned to ns → np multiplets of the guest atoms. The blue shift of 0.3 eV with respect to the gas phase and the large band width indicate a strong coupling of the transitions to the matrix. Transitions to higher states are not observed. The initial electronic states are metastable with lifetimes of 10^{-5} s (Kr) and 3×10^{-6} s (Xe).

Zeitaufgelöste Absorptionsspektren aus den niedrigsten elektronisch angeregten Zuständen von Kr(5s, $^3P_2$)- und Xe(6s, $^3P_2$)-Gastatomen in Ar-Kristallen zeigen je eine breite Bande bei 1,81 bzw. 1,72 eV. Sie werden den ns → np-Multiplets der Gastatome zugeordnet. Die Blauverschiebung um 0,3 eV gegenüber den Gasphasenwerten und die große Breite der Banden weisen auf eine starke Kopplung der Übergänge an die Matrix hin. Übergänge zu höheren Zuständen werden nicht beobachtet. Die Ausgangszustände sind metastabil mit Lebensdauern von 10^{-5} s (Kr) und 3×10^{-6} s (Xe).

1. Introduction

Absorption spectra from the electronic ground state of rare gas crystals and of rare gas atoms isolated in rare gas crystals [1, 2] show prominent series of exciton bands. The large linewidth of the exciton bands indicates a strong electron–phonon coupling of these transitions, i.e. a severe local rearrangement around the excited atoms [1, 3]. Emission spectra of Ar, Kr, and Xe crystals are dominated by broad bands due to excimer-like centres [1, 4]. This implies a reduction of the separation to one nearest neighbour atom of about 0.1 nm corresponding 20 to 30% of the ground state equilibrium separation. The strongest emission bands of solid Ne and some weak bands of Ar, Kr, and Xe are attributed to a bubble-like increase of the nearest neighbour separation due to a repulsive interaction of the excited Rydberg electron with the surrounding [1, 5]. Transient absorption spectra gave further evidence for the formation of excimer centres in solid Ne [6, 7], Ar [7, 8], and Kr [7] and for bubble-like centres in solid Ne [6, 7]. In Ne even the growth of the bubble to a real cavity by the agglomeration of thermally created vacancies has been studied [6]. A bubble-like relaxation with an increase of the nearest neighbour separation by about 10% follows also for excitation of rare gas guest atoms in rare gas matrices from a combined discussion of absorption spectra, emission spectra, and lifetimes [9, 10]. With the first transient

[1]) Olshausenstr. 40–60, D-2300 Kiel, FRG.
[2]) Arnimallee 14, 1000 Berlin 33, Berlin (West).
[3]) Present address: Radium GmbH, Wipperfürth, FRG.
[4]) Present address: Philipps GmbH, Hamburg, FRG.

absorption spectra for rare gas guest atoms in rare gas matrices presented here we plan to answer three questions:

(i) Do these experiments support also the notion of bubble formation?

(ii) Are the transitions starting from electronically excited states to higher states strongly or weakly coupled to the lattice, i.e. is a further lattice rearrangement initiated?

(iii) The former transient absorption spectra showed only transitions to the next higher electronic state and to the continuum. Is it possible to identify a series similar to the excitonic series?

2. Experimental

2 cm long optically clear Ar crystals have been doped with Kr and Xe atoms with concentrations between 0.01 to 5 at% in a set-up described earlier [8, 11]. An electron pulse of 3×10^{-9} s duration creates about 10^{18} excited centres per cm^3 within a depth of 0.3 mm. Luminescence light is dispersed by a Seya-Namioka monochromator and recorded by a SIT vidicon system. An 800 W Xe arc is imaged into the excited surface area of the crystal in a total reflection geometry and the transmitted light is dispersed by an Ebert monochromator and recorded by a second SIT vidicon system. Complete transient absorption spectra are taken with a variable delay (5×10^{-8} to 1×10^{-5} s) with respect to the electron pulse using time windows of 10^{-7} to 10^{-6} s by gating the vidicon in combination with a chopper. The continuous time course of the transient absorption at a given wavelength is monitored by a fast photomultiplier.

3. Results and Discussion

The electron energy is deposited primarily in the Ar crystal and converted to Ar excimer centres which cause a strong transient absorption. Some part of the energy

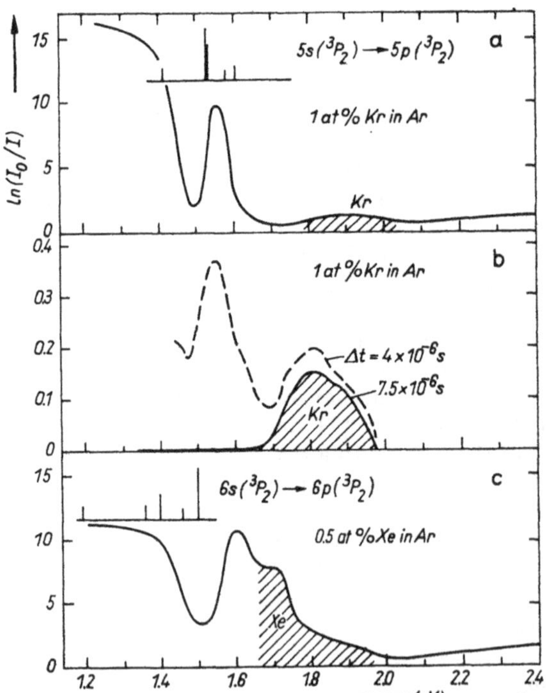

Fig. 1. Transient absorption coefficient, $\ln(I_0/I)$, just after excitation for a) 1 at% Kr in Ar and c) 0.5 at% Xe in Ar with the hatched Kr and Xe absorption bands. b) Transient absorption coefficient of Kr in Ar after delays of 4×10^{-6} s and 7.5×10^{-6} s. Gas phase 5s → 5p and 6s → 6p transition probabilities are given as histograms in inserts

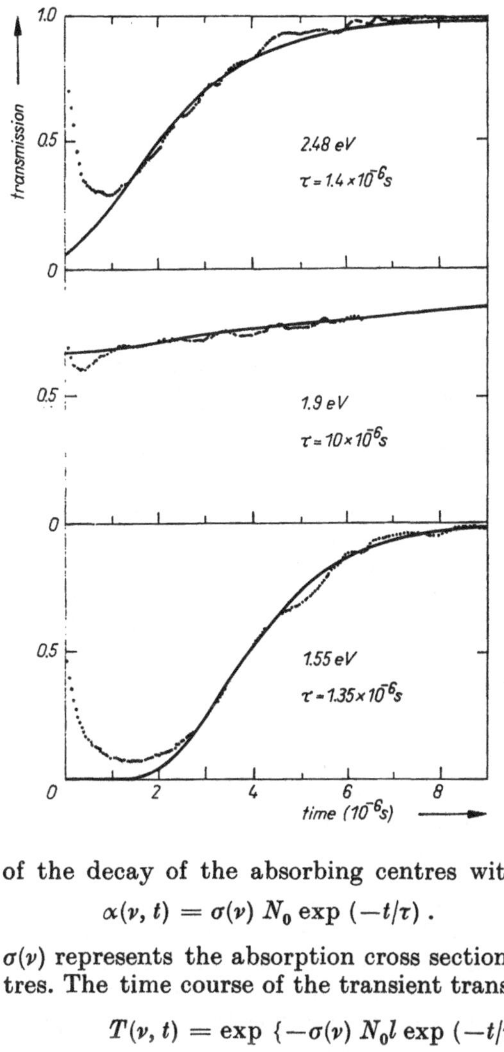

Fig. 2. Points: time course of transmission for 1 at% Kr in Ar in the Kr absorption band (1.9 eV) and in Ar matrix absorption bands (2.48 eV and 1.55 eV). Solid curves: fits with (2) and time constants τ of 1.4×10^{-6}, $10 \times \times 10^{-6}$, and 1.35×10^{-6} s

is transferred to the guest atoms yielding the desired population of excited guest atom states. The transient absorption due to the excited guest atoms has to be extracted from the background of the matrix transient absorption. Fig. 1a shows the transient absorption coefficient for 1 at % Kr in Ar immediately after the electron pulse in the wavelength range of 510 to 1000 nm. Only the weak shaded band marked by Kr ist caused by the Kr guest atoms. All other features up to 400 nm are due to transient absorption by the Ar matrix. Fig. 1b presents the transient absorption coefficient for delay times of 4×10^{-6} and 7.5×10^{-6} s (time window 1×10^{-6} s). The transient absorption by Kr atoms is dominating at long delay times. The transient absorption coefficient $\alpha(\nu, t)$ is time dependent because of the decay of the absorbing centres with their lifetime τ,

$$\alpha(\nu, t) = \sigma(\nu)\, N_0 \exp\,(-t/\tau)\,. \tag{1}$$

$\sigma(\nu)$ represents the absorption cross section and N_0 the initial density of excited centres. The time course of the transient transmission $T(\nu, t)$ follows from

$$T(\nu, t) = \exp\,\{-\sigma(\nu)\, N_0 l \exp\,(-t/\tau)\} \tag{2}$$

with the length l of the light path inside the excited region. A fit of the experimental time dependence in Fig. 2 for photon energies within the Ar transient absorption bands (1.55 and 2.48 eV) yields the well-known lifetime of the Ar excimer triplet state of $(1.4 \pm 0.1) \times 10^{-6}$ s. A lifetime of 10^{-5} s is obtained within the Kr band at 1.9 eV (Fig. 2). The long lifetime of the excited state of the Kr atoms explains the persistence of the transient absorption due to the guest atoms at times when the matrix transient absorption already decayed (Fig. 1b). In this way the energy and shape of the guest absorption band can be derived even for an initially high matrix background.

The transient absorption band of Xe guest atoms in an Ar matrix (shaded area Fig. 1c) lies close to a strong matrix band. This band is clearly identified by its lifetime of 3×10^{-6} s (at $\hbar\omega = 1.77$ eV) in measurements similar to Fig. 2. For the high energy part of the matrix (2.07 eV) again the lifetime of the triplet Ar excimer state of $(1.4 \pm 0.1) \times 10^{-6}$ s is obtained and the region at 1.55 eV yields a mean value of 2.1×10^{-6} s due to a convolution of guest and matrix absorption. In this case the

Table 1

Observed transition energies in Ar matrix E_M with full width at half maximum FWHM lifetime τ, and gas to matrix blue shift ΔE

	Kr	Xe
E_M (eV)	1.81	1.72
FWHM (eV)	0.17	—
transition	$5s(^3P_2) \rightarrow 5p$	$6s(^3P_2) \rightarrow 6p$
$\tau(10^{-6}$ s)	10	3
ΔE	0.3	0.3

bandshape of the guest absorption could not be determined because the lifetimes and transition energies of guest and matrix absorption are too close together. The data derived from the spectra are collected in Table 1.

Our emission spectra of Xe in Ar agree with literature data [10, 12]. Besides other bands they show the prominent decay of the two lowest excited states of Xe — 6 s ($^3P_1$) at $\hbar\omega = 8.92$ eV and 6 s ($^3P_2$) at $\hbar\omega = 8.79$ eV — to the ground state with lifetimes below 10^{-8} s and of 3×10^{-6} s, respectively. The 6 s ($^3P_2$) state is the initial state for the observed transient absorption band because it is the lowest excited state and its lifetime in emission agrees with the lifetime found for the absorbing state (Table 1). Similarly the metastable 5 s ($^3P_2$) state has to be the initial state for the transient absorption band of Kr atoms with a lifetime of 10^{-5} s (Table 1).

The dipole selection rules favour transitions from ns to np or to $(n + 1)$ p with transition energies and transition probabilities well known from the gas phase [13]. The transitions to $(n + 1)$ p are ruled out because they have higher energies (≈ 1 eV) than those of Table 1. The transition energies and transition probabilities for the stronger bands of the $5s(^3P_2) \rightarrow 5p$ and $6s(^3P_2) \rightarrow 6p$ manifolds are shown as histograms in Fig. 1a and Fig. 1c, respectively. A comparison of the measured bands with the envelope of the histograms yields a blue shift for the transitions in the matrix of about 0.3 eV (Table 1).

After identifying the origin of the bands we go back to the three initial questions. The change of the transition energies with the density of the matrix, i.e. the mean nearest neighbour separation has been measured by exposing the guest atoms to the matrix gas with increasing pressure up to the liquid and solid phases [14]. The energies released in the bubble formation are illustrated by these results (Fig. 3). Absorption from the ground state takes place at the nearest neighbour separation.

A structure analysis of Kr and Xe doped Ar samples by EXAFS yielded substitutional places for the guest atoms with a Kr–Ar separation of 0.380 nm and a Xe–Ar separation of 0.399 nm [15]. This separation is shorter than the gas phase potential minima due to the pressure exerted by the next shell of Ar atoms. The increase in energy due to the Kr or Xe repulsive potential is about 0.3 meV per pair and amounts to about 4 meV for the twelve nearest Ar neighbours. The gas to matrix shifts in absorption and emission and the Stokes shift indicate an increase of the nearest neighbour separation by about 5% to about 0.42 nm for Xe in Ar (Fig. 3). Transient absorption starts from the relaxed configuration and terminates at the fine structure components of the 6p states. The $6s(^3P_2)$ and the 6p states do not appear in absorption spectra from the ground state because of dipole selection rules. The splitting of $6s(^3P_1)$ and $(^3P_2)$ by 0.13 eV observed in emission belongs to the relaxed bubble configuration. It is the same as the gas phase splitting showing that at least the $6s^3P_2$ state (dashed line in Fig. 3) has a density dependence very similar to the $6s(^3P_1)$ state. The blue shift of the

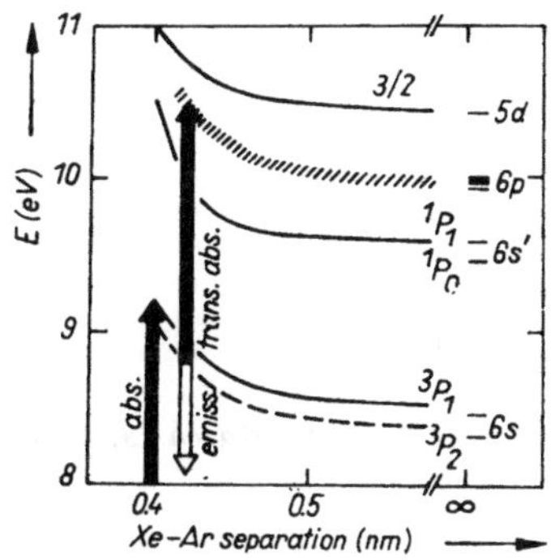

Fig. 3. Xe transition energies versus density (i.e. nearest neighbour separation) of Ar gas for the gaseous, liquid, and solid phases ([14, 15]). The arrows show the transition energies for absorption from the electronic ground state, for emission to the electronic ground state, and for the $6s[^3P_2]$–$6p[^3P_2]$ transient absorption. The latter starts from the relaxed Xe–Ar configuration

transient absorption of 0.3 eV compared to the gas phase can be caused by a steep increase in energy of the 6p states at high density similar to the 6s′ state as suggested by the hatched band in Fig. 3. A similar consideration holds for Kr in Ar. The observed blue shift and the large linewidth indicate a strong coupling to the lattice for the *n*s–*n*p transition of Kr and Xe in Ar. The blue shift represents the energy required to squeeze the *n*p state into the spherical bubble of the *n*s initial state. The large linewidth suggests that the lattice responds with a further rearrangement to a p-like geometry of lower total energy. Such effects have been discussed for emission and transient absorption in liquid He under external pressure [16].

Concerning the number of states, we have to say that only one transition is observed and that there is no indication for further states. We cannot exclude that weak transitions are hidden in the matrix background.

Acknowledgements

We like to thank for the backing by Prof. Haensel. This work has been supported by the Deutsche Forschungsgemeinschaft DFG in the Sonderforschungsbereich für Struktur und Dynamik von Festkörpern.

References

[1] N. Schwentner, E. E. Koch, and J. Jortner, Springer Tracts mod. Phys. Vol. 107 (1985).

[2] V. Saile, Appl. Optics **19**, 4115 (1980).
G. Baldini, Phys. Rev. **137**, A508 (1965).
D. Pudewill, F. J. Himpsel, V. Saile, N. Schwentner, M. Skibowski, and E. E. Koch, phys. stat. sol. (b) **74**, 485 (1976).

[3] Z. Ophir, Master Thesis, Tel-Aviv University (1976).

[4] J. Jortner, L. Meyer, S. A. Rice, and E. G. Wilson, J. chem. Phys. **42**, 4250 (1965).
I. Ya. Fugol, Adv. Phys. **27**, 1 (1978).

[5] P. L. Kunsch and F. Coletti, J. chem. Phys. **70**, 726 (1979).
R. Gaethke, P. Gürtler, R. Kink, E. Roick, and G. Zimmerer, phys. stat. sol. (b) **124**, 335 (1984).
F. Coletti, J. M. Debever, and G. Zimmerer, to be published.

[6] T. Suemoto and H. Kanzaki, J. Phys. Soc. Japan **49**, 1039 (1980); **50**, 3664 (1981).

[7] T. Suemoto and H. Kanzaki, J. Phys. Soc. Japan **46**, 1554 (1979).

[8] O. Dössel, H. Nahme, R. Haensel, and N. Schwentner, J. chem. Phys. **79**, 665 (1983).

[9] U. Hahn and N. Schwentner, Chem. Phys. **48**, 53 (1980).

[10] U. Hahn, R. Haensel, and N. Schwentner, phys. stat. sol. (b) **109**, 233 (1982).

[11] N. Schwentner, O. Dössel, and H. Nahme, in: Laser Techniques for Extreme Ultraviolet Spectroscopy AIP 90, American Institut of Physics, New York 1982 (p. 163).
W. Rudnick, R. Haensel, H. Nahme, and N. Schwentner, phys. stat. sol. (a) **87**, 319 (1985).

[12] A. Gedanken, B. Raz, and J. Jortner, J. chem. Phys. **59**, 5471 (1973).
R. Heumüller, Thesis, University Regensburg (1978).
[13] V. Helbig, Habilitation, University Kiel (1978).
[14] I. Messing, B. Raz, and J. Jortner, Chem. Phys. **23**, 23 (1977).
[15] W. Malzfeldt, Thesis, Kiel (1985).
[16] F. J. Soley, and W. A. Fitzsimmons, Phys. Rev. Letters **32**, 988 (1974).
A. P. Hickman and N. F. Lane, Phys. Rev. Letters **26**, 126 (1971).
A. P. Hickman, W. Steets, and N. F. Lane, Phys. Rev. B **12**, 3705 (1975).

(Received March 19, 1985)

phys. stat. sol. (b) **131**, 747 (1985)

Subject classification: 14.3.4; 22

Institute for Low Temperature and Structure Research, Polish Academy of Sciences, Wrocław[1])

Influence of Real Structure in the Interfacial Region on the Maxwell-Wagner Effect

Spherical Void in a Ionic Crystal

By

J. Poźniak

Dedicated to Prof. Dr. Dr. h. c. Dr. E. h. P. Görlich on the occasion of his 80th birthday

Theoretical calculations of the Maxwell-Wagner polarization are performed by considering thermodynamic equilibrium conditions. The model applied consists of a spherical void inside an ionic matrix of NaCl type containing charge carriers (cation and anion vacancies) subjected to the action of an external dc electric field. Accurate analytical solutions obtained for the stationary state describes i) the distribution of the charge carriers in the vicinity of void and ii) the distribution of potential and electric field both, inside the void and in the crystal matrix.

Unter Berücksichtigung der thermodynamischen Gleichgewichtsbedingungen werden theoretische Berechnungen der Maxwell-Wagner-Polarisation durchgeführt. Das benutzte Modell besteht aus einem kugelförmigen Hohlraum im Inneren einer ionaren Matrix vom NaCl-Typ, der Ladungsträger (Kationen- und Anionenleerstellen) enthält, die der Wirkung eines äußeren elektrischen Gleichfeldes ausgesetzt sind. Exakte analytische Lösungen, die für den stationären Fall erhalten wurden, beschreiben: i) die Verteilung der Ladungsträger in der Nähe des Hohlraumes und ii) die Verteilung des Potentials und des elektrischen Feldes sowohl innerhalb des Hohlraumes als auch in der Kristallmatrix.

1. Introduction

Most of the theoretical works concerned with the description of the Maxwell-Wagner polarization effect apply a model consisting of a system of conducting spheres [1] or rotational ellipsoids (e.g. [2]) immersed in a dielectric matrix. It is assumed that the properties (electric conductivity σ and dielectric susceptibility ε) of the dielectric matrix are the same in the bulk and in the interface regions. The application of such a simple model is good enough for the rough description of the electric properties of composite insulating materials. On the other hand, from a mathematical point of view, considerations involving generation and decay of space charge necessitate the solution of non-linear differential equations. Such analytical solutions are not known for the general case and only some more or less approximate ones have been obtained for some particular cases [3, 4]. Also the void motion under the action of an electric field has been treated theoretically [5 to 8]. In these latter works it has been found that the distribution of charge and potential in the vicinity of the interface is strongly influenced by the real structure of the pertinent regions. For this reason, in the present work an attempt is undertaken to perform theoretical calculations for the Maxwell-Wagner polarization by considering thermodynamic equilibrium conditions in the interface region.

[1]) P.O.B. 937, 50-950 Wrocław, Poland.

2. Distribution of Vacancies and Potential in the Vicinity of a Spherical Void

Let us consider an ionic crystal A^+B^- placed in an external dc electric field and separate a small fragment consisting of N unit cells containing in both sublattices N_1 cations (A^+) and n_1 cation vacancies as well as N_2 anions (B^-) and n_2 anion vacancies. The above fragment is subjected to the action of an external electric field (E_0) as well as to the electric field generated by all remaining fragments of the crystal. Its potential $\varphi = V + V_{\text{ext}}$ where V is the potential created by the charges of all fragments of the system and V_{ext} the potential related to E_0. In that case the Poisson equation can be written in the form

$$\Delta\varphi = \nabla^2\varphi = \frac{4\pi eN}{\varepsilon}\,(\xi'_- - \xi'_+)\,, \tag{2.1}$$

where $\xi'_- = n_1/N$ and $\xi'_+ = n_2/N$, e is the proton charge, ε the dielectric constant.

In a crystal containing a spherical void of radius R the electric field generates directed vacancy fluxes toward and outward the void. In that case the Poisson equation must be completed by the diffusion equations

$$\frac{\partial\xi'_\pm}{\partial t} + \operatorname{div} j_\pm = 0\,, \tag{2.2}$$

$$j_\pm = -\frac{\xi_0}{kT}\,D_\pm\,\nabla\mu_\pm\,, \tag{2.3}$$

or

$$\frac{\partial\xi'_\pm}{\partial t} = D_\pm\left(\Delta\xi'_\pm \pm \frac{e\xi_0}{kT}\,\Delta\varphi\right), \tag{2.4}$$

where $j_\pm$ is the flux of vacancies, $\mu_\pm$ the chemical potential of vacancies, $D_\pm$ the diffusion coefficient of vacancies (the subscripts (+) and (−) refer to anion and cation vacancies, respectively), ξ_0 the equilibrium vacancy concentration in the bulk of the crystal at $E_0 = 0$.

In agreement with [9] for a spherical void in a dielectric we introduce the following boundary conditions:

$$(\varphi_1)_{r\to\infty} = -E_0 r\cos 0\,, \tag{2.5a}$$

$$(\varphi_1)_{r=R} = (\varphi_2)_{r=R}\,, \tag{2.5b}$$

$$\varepsilon_1\left(\frac{\partial\varphi_1}{\partial r}\right)_{r=R} = \varepsilon_2\left(\frac{\partial\varphi_2}{\partial r}\right)_{r=R}. \tag{2.5c}$$

Additionally, we assume that inside the sphere at $r = 0$ the function φ_2 has no singular point. The boundary conditions (2.5) are written for the spherical coordinate system (r, θ) with the origin at the centre of the void.

As a further boundary condition let us assume that the chemical potentials of the vacancies in the vicinity of the interface are the same in both phases $\mu_1(T, P_1) = \mu_2(T, P_2)$ [9]. Then, in agreement with [6] we can write

$$\xi'_+ + \xi'_- = \frac{2\xi_0}{kT}\,P_{nm}^{(R)} \quad \text{for the void–crystal interface}\,,$$

$$\xi'_+ + \xi'_- = \frac{2\xi_0}{kT}\,P_{nm}^{(\infty)} \quad \text{for the crystal–electrode interface}\,,$$

where stresses $P_{nm} = P_1 - P_2$ depend on the surface tension [6].

The thermodynamic potential of our crystal element A^+B^- can be written as

$$\Phi = \Phi_0 + U(N_1 + N_2) + (u_1 - e\varphi)\, n_1 + (u_2 + e\varphi)\, n_2 - $$
$$- kT \prod_{i=1,2} \frac{(N_i + n_i)!}{N_i!\, n_i!}, \tag{2.6}$$

where Φ_0 is a constant, $2U$ the binding energy of A^+B^- molecule, and u_i ($i = 1, 2$) denotes the formation energy of the i-type vacancy.

Minimization of (2.6) by considering equal numbers of lattice sites of both types (we do not introduce the electric neutrality of the whole element) and by introduction of a dimensionless potential

$$y = e\,\frac{V_\infty - \varphi}{kT} \tag{2.6a}$$

results in

$$\xi'_- - \xi'_+ = 2\xi_0 \sinh y\,, \tag{2.6b}$$

where V_∞ denotes the reference potential taken as the potential in the bulk of the crystal under the assumption that the potential at the crystal–electrode interface (at $E_0 = 0$) is also equal to zero.

Introduction of (2.6a) and (2.6b) into (2.1) yields the Poisson equation in the form

$$\Delta\varphi = -r_D^{-2} \sinh y\,, \tag{2.7}$$

where

$$r_D = \omega^{-1} = \left(\frac{8\pi e^2 N}{\varepsilon kT}\,\xi_0\right)^{1/2} \tag{2.7a}$$

is the so-called Debye-Hückel screening length.

In the stationary case ($\partial\xi'_\pm/\partial t = 0$) (2.1) and (2.3) or (2.4) can be written in the form

$$\Delta(\xi'_+ + \xi'_-) = 0 \tag{2.8}$$

and

$$\Delta(\xi'_+ - \xi'_-) = r_D^{-2} \sinh y\,. \tag{2.9}$$

The solution of (2.7) to (2.9) in analytical form is rather difficult (non-linear differential equations). For this reason, like in [5], we consider only the "over-equilibrium" concentration of vacancies ($\xi_\pm$), i.e. the differences between the real values ($\xi'_\pm$) and equilibrium value (ξ_0). This permits to linearize the above equation system (see [5]),

$$\Delta y = -\omega^2\zeta\,, \tag{2.10}$$

$$\Delta\eta = 0\,, \tag{2.11}$$

$$\Delta\zeta = \omega^2\zeta\,, \tag{2.12}$$

where

$$\eta = \frac{\xi_+ + \xi_-}{2\xi_0} \tag{2.12a}$$

and

$$\zeta = \frac{\xi_+ - \xi_-}{2\xi_0}. \tag{2.12b}$$

By taking (2.12a) and (2.12b) we can write the condition of limited concentration values at the interface in the form of a generalized Thomson relation [6],

$$\eta_{r=R} = (NkT)^{-1}\, P_{nm}^{(R)}\,, \tag{2.13a}$$

$$\eta_{r\to\infty} = (NkT)^{-1}\, P_{nm}^{(\infty)}\,. \tag{2.13b}$$

The required equality between the fluxes of anion and cation vacancies to or from the crystal–void interface takes, in agreement with (2.4), the form

$$\left[\frac{\partial \zeta}{\partial n} + f\,\frac{\partial \eta}{\partial n} + \frac{e}{kT}\,\frac{\partial \varphi}{\partial n}\right]_{n=r=R} = 0\,, \tag{2.14}$$

where

$$f = \frac{D_+ - D_-}{D_+ + D_-}\,. \tag{2.14a}$$

3. Analytical Solutions

3.1 Solution of $\Delta\zeta = \omega^2\zeta$

Equation (2.12) in the spherical coordinate system (r, θ) can be written as

$$\left(\frac{\partial^2\zeta}{\partial r^2} + \frac{2}{r}\,\frac{\partial \zeta}{\partial r}\right) + \frac{1}{r^2}\left(\frac{\partial^2\zeta}{\partial \theta^2} + \operatorname{cotg}\theta\,\frac{\partial \zeta}{\partial \theta}\right) = \omega^2\zeta\,. \tag{3.1}$$

Let us now look for the solution of the above equation in the form

$$\zeta(r, \theta) = V(\theta)\; U(r)\,.$$

By substituting the above solutions into (3.1) we obtain the equation system

$$r^2\,\frac{\mathrm{d}^2 U}{\mathrm{d}r^2} + 2r\,\frac{\mathrm{d}U}{\mathrm{d}r} - (\omega^2 r^2 + k_1)\; U = 0 \tag{3.2a}$$

and

$$\frac{\mathrm{d}^2 V}{\mathrm{d}\theta^2} + \operatorname{cotg}\theta\,\frac{\mathrm{d}V}{\mathrm{d}\theta} + k_1 V = 0\,, \tag{3.2b}$$

where k_1 is a constant.

Equation (3.2a) can be compared with the modified Bessel equation [10]

$$t^2\,\frac{\mathrm{d}^2 z}{\mathrm{d}t^2} + (2\alpha + 1)\; t\,\frac{\mathrm{d}z}{\mathrm{d}t} + (\alpha^2 - \beta^2 p^2 + \beta^2\gamma^2 t^{2\beta})\; z = 0\,.$$

The modified Bessel equation will be equivalent to (3.2) in the case of: $U \equiv z$, $t \equiv r$, $\alpha \equiv \frac{1}{2}$, $\beta \equiv 1$, $\gamma \equiv \omega$, and $p \equiv (\frac{1}{4} + k_1)^{1/2}$. Consequently, the general solution of (3.2a) $\big($for $k_1 = \nu(\nu + 1)\big)$ has the form [10]

$$U = r^{1/2} A_1 I_{\nu+1/2}(\omega r) + r^{1/2} B_1 K_{\nu+1/2}(\omega r)\,. \tag{3.3a}$$

Equation (3.2b) can be reduced to the differential Legendre equation by substituting $V(\theta) = y(x)$ and $x = \cos\theta$. For this reason the solutions of (3.2b) are the Legendre polynomials

$$y(x) = A\mathrm{P}_\nu(x) + B\mathrm{Q}_\nu(x)\,,$$

where P_ν and Q_ν are the Legendre functions of the first and second kind. The general solution of $y(x)$ is applied in the case when the z-axis is excluded from the region determined by these solutions. In our case we require continuous solutions also for the z-axis. So, the above solution reduces to

$$y(x) = A P_\nu(x) ,$$

i.e. (3.3b)

$$V(\theta) = A P_\nu(\cos\theta) .$$

As the solution of (3.1) has the form of a product $U(r)\ V(\theta)$, in agreement with (3.3a) and (3.3b) we can write the general solution of (3.1) in the form

$$\zeta(r, \theta) = r^{-1/2} \sum_{\nu=0}^{\infty} \{[C_{1\nu} I_{\nu+1/2}(\omega r) + C_{2\nu} K_{\nu+1/2}(\omega r)]\ P_\nu(\cos\theta)\} . \tag{3.4}$$

3.2 Solution of $\Delta\eta = 0$

Equation (2.11) in spherical coordinate system (r, θ) has the form

$$\left(\frac{\partial^2\eta}{\partial r^2} + \frac{2}{r}\frac{\partial\eta}{\partial r}\right) + \frac{1}{r^2}\left(\frac{\partial^2\eta}{\partial\theta^2} + \operatorname{cotg}\theta\,\frac{\partial\eta}{\partial\theta}\right) = 0 . \tag{3.5}$$

General solutions of this equation are well known [9] and in our case can be written as

$$\eta(r, \theta) = \sum_{l=0}^{\infty} (a_l r^l + b_l' r^{-(l+1)})\ P_l(\cos\theta) . \tag{3.6}$$

3.3 Solution of $\Delta y = -\omega^2\zeta$

The direct calculation of the solutions of (2.10) is difficult. One can do it, however, on comparing (2.10) and (2.12). From the above equations results that $\Delta y + \Delta\zeta = 0$. By introducing the new variable $\alpha = y + \zeta$ the latter equation can be written as $\Delta\alpha = 0$, being equivalent to the Laplace equation in polar coordinates (r, θ),

$$\left(\frac{\partial^2\alpha}{\partial r^2} + \frac{2}{r}\frac{\partial\alpha}{\partial r}\right) + \frac{1}{r^2}\left(\frac{\partial^2\alpha}{\partial\theta^2} + \operatorname{cotg}\theta\,\frac{\partial\alpha}{\partial\theta}\right) = 0 . \tag{3.7}$$

Hence, its solution will take the form $\alpha = V(\theta)\ U(r)$. After substituting the above solution into (3.7) we obtain

$$r^2\frac{\mathrm{d}^2U}{\mathrm{d}r^2} + 2r\frac{\mathrm{d}U}{\mathrm{d}r} - k_1 U = 0$$

and

$$\frac{\mathrm{d}^2V}{\mathrm{d}\theta^2} + \operatorname{cotg}\theta\,\frac{\mathrm{d}V}{\mathrm{d}\theta} + k_1 V = 0 ,$$

where $K_1 = m(m+1)$ is the integration constant. The solutions of the above differential equations are the following:

$$U = A_1 r^m + A_2 r^{-(m+1)} ,$$

$$V = A P_m(\cos\theta) , \tag{3.8}$$

$$\alpha = [C_1 r^m + C_2 r^{-(m-1)}]\ P_m(\cos\theta) .$$

By considering (3.8) and the relation $y = \alpha - \zeta$ we obtain the general solution of $\Delta y = -\omega^2\zeta$ in the form

$$y(r, \theta) = \sum_{m=0}^{\infty} [d_{m1}r^m + d_{m2}r^{-(m+1)}] \, P_m(\cos\theta) - \zeta \, . \tag{3.9}$$

3.4 Particular solutions

In agreement with the boundary conditions (2.13) the solution (3.6) cannot be a function of θ. Because only for $l = 0$ the polynomial $P_l(\cos\theta)$ is independent of θ $[P_0(\cos\theta) = 1]$ so only for $l = 0$ the constants a_1 and b_1 are different from zero. Hence

$$\eta = \eta(r) = a_0 + \frac{b_0}{r}. \tag{3.10}$$

From (3.10) and boundary conditions (2.13) we obtain

$$a_0 = P_{nm}^{(\infty)}/NkT \tag{3.10a}$$

and

$$b_0 = -R\,\frac{P_{nm}^{(\infty)} - P_{nm}^{(R)}}{NkT}, \tag{3.10b}$$

where $P_{nm}^{(R)} = -2\alpha/R$, α being the surface tension [11] of the crystal.

By taking into account the lack of electric charge inside the spherical void ($\zeta = 0$) we can, in agreement with (3.9), write the following expressions for the potential in the crystal $y^{(1)}$ and in the void $y^{(2)}$:

$$y^{(1)}(r, \theta) = \sum_{m=0}^{\infty} [d_{m1}r^m + d_{m2}r^{-(m+1)}] \, P_m(\cos\theta) - \zeta \tag{3.11a}$$

and

$$y^{(2)}(r, \theta) = \sum_{m=0}^{\infty} [f_{m1}r^m + f_{m2}r^{-(m+1)}] \, P_m(\cos\theta) \, . \tag{3.11b}$$

As we require that inside the void the potential $y^{(2)}(r, \theta)$ should not have any singular point, all f_{m2} coefficients should be equal to zero. Hence

$$y^{(2)}(r, \theta) = \sum_{m=0}^{\infty} f_{1m}r^m \, P_m(\cos\theta) \, . \tag{3.12a}$$

Next, in agreement with the boundary condition (2.5a) we must assume that al d_{m1} coefficients except for $d_{11} = d_1$ are equal to zero. From (3.11a) we obtain for $m = 1$

$$y^{(1)}(r, \theta) = \sum_{m=0}^{\infty} \frac{d_{m2}}{r^{m+1}} \, P_m(\cos\theta) - \frac{e}{kT} E_0 r \cos\theta - \zeta \, . \tag{3.12b}$$

From (3.4) we require that it should have a finite value at the point $r = 0$. As $\lim_{x\to 0} K_n(x) \to \infty$, the above condition denotes that all constants $C_{2\nu}$ in (3.4) should be equal to zero (in the range of $0 \leqq r \leqq R$). Hence

$$\zeta^{(2)}(r, \theta) = r^{-1/2} \sum_{\nu=0}^{\infty} C_{1\nu} I_{\nu+1/2}(\omega r) \, P_\nu(\cos\theta) \, . \tag{3.13a}$$

Because of $\lim_{x\to\infty} I_\nu(x) \to \infty$ the solution of (3.4) for $r \geqq R$ accepts the form

$$\zeta^{(1)}(r, \theta) = r^{-1/2} \sum_{\nu=0}^{\infty} C_{2\nu} K_{\nu+1/2}(\omega r) \, P_\nu(\cos\theta) \, . \tag{3.13b}$$

By considering the spherical symmetry of the void ($r = n$) and by substituting into (2.14) the values of the derivatives $\partial\eta/\partial n$ and $\partial y^{(1)}/\partial r$ calculated from (3.10) and (3.12b) we obtain the condition

$$\sum_{m=0}^{\infty} (m+1) \frac{d_{m2}}{R^{m+2}} \mathrm{P}_m(\cos\theta) = -\frac{b_0 f}{R^2} - \frac{eE_0}{kT}\cos\theta \,.$$

Consequently, the values of d_{m2} coefficients will differ from zero only for $m = 0$ and $m = 1$. Hence,

$$d_{02} = -fb_0 \quad \text{and} \quad d_{12} = -\frac{eE_0}{2kT} R^3 \cos\theta \,. \tag{3.14}$$

In agreement with (3.14), (3.12b) can be written as

$$y^{(1)}(r, \theta) = -\frac{fb_0}{r} - \frac{eE_0}{kT} r\cos\theta \left[1 + \frac{1}{2}\left(\frac{R}{r}\right)^3\right] - \zeta \,. \tag{3.15}$$

On the other hand, the condition of equal potentials on both sides of the crystal–void interface (2.5b) by considering the solutions (3.11) and (3.13) takes the form

$$\sum_{m=0}^{\infty} [d_{m1} r^m + d_{m2} r^{-(m+1)}] \mathrm{P}_m(\cos\theta) - r^{-1/2} \sum_{\nu=0}^{\infty} C_{2\nu} K_{\nu+1/2}(\omega r) \mathrm{P}_\nu(\cos\theta) \equiv$$

$$\equiv \sum_{m=0}^{\infty} [f_{1m} r^m + f_{2m} r^{-(m+1)}] \mathrm{P}_m(\cos\theta) \quad \text{for} \quad r = R \,.$$

Tho realization of the above identity requires that only those coefficients at functions r^α (where $\alpha = m$ or $\alpha = -(m+1)$) and $\mathrm{P}_\beta(\cos\theta)$ (where $\beta = m$ or $\beta = \nu$) were different from zero, which secures the appearance of the variables r and $\cos\theta$ in the form of the same functions on both sides of the above identity:

(i) for $\mathrm{P}_\beta(\cos\theta)$ we require that $\mathrm{P}_m(\cos\theta) = \mathrm{P}_\nu(\cos\theta)$, hence $\nu = m$;

(ii) in agreement with (3.14) the values of d_{m1} and d_{m2} coefficients are different from zero in the first sum from the right side of the identity, only for $m = 0$ and $m = 1$.

Hence the boundary condition (2.5b) can be written as

$$-\frac{fb_0}{R} - \frac{3eE_0}{2kT} R\cos\theta - R^{1/2}[C_{20} K_{1/2}(\omega R) + C_{21} K_{3/2}(\omega R)\cos\theta] =$$

$$= f_{10} + f_{11} R\cos\theta \;, \tag{3.16}$$

and the solutions of (3.12a) and (3.12b), by considering the identities $K_{3/2}(\omega r) = K_{1/2}(\omega r)\,[1 + 1/\omega r]$ and $K_{1/2}(z) = (\pi/2z)^{1/2} \exp(-z)$, as

$$\zeta^{(1)}(r, \theta) = \sqrt{\frac{\pi}{2\omega}} \frac{e^{-\omega r}}{r} \left[C_{20} + C_{21}\left(1 + \frac{1}{\omega r}\right)\cos\theta\right] \tag{3.17}$$

and

$$y^{(2)}(r, \theta) = f_{10} + f_{11} r\cos\theta \,. \tag{3.18}$$

From the identity (3.16) we obtain

$$f_{10} = -\frac{fb_0}{R} - \sqrt{\frac{\pi}{2\omega}} C_{20} \frac{e^{-\omega R}}{R} \tag{3.19a}$$

and

$$f_{11} = -\frac{3eE_0}{2kT} - \frac{e^{-\omega R}}{R^2}\sqrt{\frac{\pi}{2\omega}}\, C_{21}\left(1 + \frac{1}{\omega R}\right). \tag{3.19b}$$

The condition of equal field intensities on both sides of the crystal–void interface (3.5c), after substituting

$\partial y^{(i)}/\partial r|_{r=R} \qquad (i = 1, 2)$

calculated from (3.17) and (3.18), takes the form

$$\varepsilon_2 f_{11} \cos\theta = \frac{\varepsilon_1 f b_0}{R^2} + \sqrt{\frac{\pi}{2\omega}}\, \varepsilon_1 \frac{e^{-\omega R}}{R^2}(1 + \omega R)\, C_{20} + \\ + \sqrt{\frac{\pi}{2\omega}}\, \frac{e^{-\omega R}\,[\omega R + 1)^2 + 1]}{\omega R^3}\, \varepsilon_1 C_{21} \cos\theta\,,$$

hence

$$C_{20} = -\sqrt{\frac{2\omega}{\pi}}\, \frac{f b_0}{1 + \omega R}\, e^{\omega R} \tag{3.20}$$

and

$$\varepsilon_2 f_{11} = \sqrt{\frac{\pi}{2\omega}}\, \varepsilon_1 C_{21} \frac{e^{-\omega R}}{\omega R^3}\,[(\omega R + 1)^2 + 1]\,. \tag{3.21}$$

By substituting (3.20) into (3.19a) we obtain

$$f_{10} = -\frac{f b_0 \omega}{1 + \omega R}. \tag{3.22}$$

Further, by substituting the value of f_{11} from (3.21) into (3.19b) we obtain

$$C_{21} = -\sqrt{\frac{2\omega}{\pi}}\, \frac{3eE_0}{2kT}\, \frac{\varepsilon_2 \omega R^3\, e^{\omega R}}{\varepsilon_1[(1 + CR)^2 + 1] + \varepsilon_2(1 + \omega R)} \tag{3.23}$$

and by substituting C_{21} from (3.23) into (3.19b) we obtain

$$f_{11} = -\frac{3eE_0}{2kT}\, \frac{\varepsilon_1[(1 + \omega R)^2 + 1]}{\varepsilon_1[(1 + \omega R)^2 + 1] + \varepsilon_2(1 + \omega R)}. \tag{3.24}$$

The above values of the constants f_{10}, f_{11}, C_{20}, and C_{21} permit us to write the particular solutions sought for in the following way:

(i) The concentration of vacancies as a function of the distance from the centre (r) of the void and of polar angle (θ) is described by

$$\eta(r, \theta) = \eta(r) = a_0 + \frac{b_0}{r} \tag{3.25}$$

and

$$\zeta(r, \theta) = \sqrt{\frac{\pi}{2\omega}}\, \frac{e^{-\omega r}}{r}\left[C_{20} + C_{21}\left(1 + \frac{1}{\omega r}\right)\cos\theta\right], \tag{3.26}$$

where

$$a_0 = P_{nm}^{(\infty)}/kT\,, \tag{3.25a}$$

$$b_0 = -\frac{R\left(P_{nm}^{(\infty)} - P_{nm}^{(R)}\right)}{kT}\,, \tag{3.25b}$$

$$C_{20} = -\sqrt{\frac{2\omega}{\pi}}\,\frac{fb_0}{1+\omega R}\,\mathrm{e}^{\omega R}\,, \tag{3.26a}$$

and

$$C_{21} = -\sqrt{\frac{2\omega}{\pi}}\,\frac{3eE_0}{2kT}\,\frac{\varepsilon_2\omega R^3\,\mathrm{e}^{\omega R}}{\varepsilon_1[(1+\omega R)^2+1]+\varepsilon_2(1+\omega R)}\,. \tag{3.26b}$$

(ii) The potential distribution in the crystal is

$$y^{(1)}(r,\theta) = -\frac{fb_0}{r} - \frac{eE_0}{kT}\,r\cos\theta\left[1+\frac{1}{2}\left(\frac{R}{r}\right)^3\right] - \zeta(r,\theta)\,, \tag{3.27a}$$

and in the spherical void

$$y^{(2)}(r,\theta) = f_{10} + f_{11}r\cos\theta\,, \tag{3.27b}$$

where

$$f_{10} = -\frac{fb_0\omega}{1+\omega R} \tag{3.27c}$$

and

$$f_{11} = -\frac{3eE_0}{2kT}\,\frac{\varepsilon_1[(1+\omega R)^2+1]}{\varepsilon_1[(1+\omega R)^2+1]+\varepsilon_2(1+\omega R)}\,. \tag{3.27d}$$

In addition, basing on the definition of the electric field intensity $E_i(r,\theta) = (kT/e)\left(\partial y^{(i)}(r,\theta)/\partial r\right)$ and (3.27), we can calculate

(iii) the electric field intensity inside the crsystal,

$$E_1(r,\theta) = \frac{kT}{e}\,\frac{fb_0}{r^2}\left[1-\frac{1+\omega r}{1+\omega R}\,\mathrm{e}^{-\omega(r-R)}\right] - \left\{1-\left(\frac{R}{r}\right)^3+\frac{3}{2}\left(\frac{R}{r}\right)^3\frac{\varepsilon_2[(1+\omega r)^2+1]\,\mathrm{e}^{-\omega(r-R)}}{\varepsilon_1[(1+\omega R)^2+1]+\varepsilon_2(1+\omega R)}\right\}\times E_0\cos\theta\,, \tag{3.28a}$$

and the electric field intensity inside the spherical void,

$$E_2(r,\theta) = -\frac{3[(1+\omega R)^2+1]\,\varepsilon_1}{2\{\varepsilon_1[(1+\omega R)^2+1]+\varepsilon_2(1+\omega R)\}}\,E_0\cos\theta\,. \tag{3.28b}$$

4. Some Remarks

The expressions presented in Section 3 can be used in calculations of the magnitude of the charge accumulated in the vicinity of the void. The total charge is $Q = \varrho V$, where ϱ denotes the charge density and V the total volume. As in our case the charge distribution is not uniform ($\varrho = \mathrm{d}Q/\mathrm{d}V$), the above relation must be written as

$$\mathrm{d}Q = \frac{\mathrm{d}Q}{\mathrm{d}V}\,\frac{\mathrm{d}V}{\mathrm{d}r}\,\mathrm{d}r\,.$$

By assuming spherical symmetry of the void and of the crystal we obtain

$$\mathrm{d}Q = 4\pi\varrho r^2\,\mathrm{d}r\,. \tag{4.1}$$

On the other hand, according to our notation see (2.12a),

$$\varrho = Ne(\xi_+ - \xi_-) = 2N\xi_0 e\zeta(r, \theta)\,,$$

where $\zeta(r, \theta)$ is described by (3.26). By substituting ϱ into (4.1), after integration we obtain

$$Q = -\frac{8\pi e R\xi_0}{\omega kT}\left[f\left(P_{nm}^{(\infty)} + \frac{2\alpha}{R}\right) \pm \frac{3eE_0\varepsilon_2(\omega R + 2)\,R^2}{\varepsilon_1[(1 + \omega R)^2 + 1] + \varepsilon_2(1 + \omega R)}\right].$$

The first term in this equation describes the dependence of Q on the thermodynamic parameters, the second term that on the intensity of external electric field E_0 and the size of the void. The signs "+" and "−" refer to two halfes of the void. This means that on opposite sides of the void opposite charges are accumulated. It should be noticed that the magnitude of the space charge is proportional to the intensity of the external field E_0. The value of Q depends non-linearly on the size of the void.

The space charge accumulated in the vicinity of the void induces in the external measuring system a current in the direction opposite to that related to the space charge created at the crystal–electrode interface. Consequently, in measurements of the space charge by the TSD method the shape of experimental current versus time curves will depend on the magnitude of the space charge in the vicinity of the crystal–electrode as well as at the crystal–void interface. In the extreme case, if the charge in the vicinity of the void is higher than in the vicinity of the crystal–electrode interface, one should expect a direction change of the TSD current flow.

The above remarks refer also to a system containing numerous voids. It should be noticed here that according to Lifshits et al. (Section 5 in [4]), the electric field generated by the space charge accumulated in the vicinity of the voids should be equal to zero. This erroneous statement is based on the correct relation

$$\int_v \zeta\,\mathrm{d}V = 0$$

(see equation (46) in [5]), according to which the *total charge* inside the crystal equals zero.

The expressions presented in this paper describe the distribution of charge carriers in the vicinity of a void as well as the distribution of potential and electric field intnsity in a more general way than it has been done so far. In the particular case of $E_0 = 0$ the solutions (3.27) and (3.28) reduce to the simple case valid for a spherical void immersed in an ideal dielectric (see, e.g., [9]).

References

[1] M. Wimmer, H. C. Graham, and N. M. Tallan, in: Electrical Conductivity in Ceramics and Glass, Part B, Ed. N. M. Tallan, M. Dekker, Inc., New York 1974 (p. 619).

[2] H. Fricke, J. phys. Chem. 57, 934 (1953).

[3] G. M. Sessler, Ed., Electrets, in: Topics in Appl. Phys., Vol. 33, Springer-Verlag, Berlin/Heidelberg/New York 1980.

[4] J. van Turnhout, Thermally Stimulated Discharge of Polymer Electrets, Elsevier, Amsterdam 1975.
[5] J. M. Lifshits and Ya. E. Geguzin, Fiz. tverd. Tela 7, 62 (1965).
[6] J. M. Lifshits, A. M. Kosevich, and Ya. E. Geguzin, J. Phys. Chem. Solids 28, 783 (1967).
[7] M. A. Krivoglaz, Fiz. tverd. Tela 7, 431 (1965).
[8] A. M. Kosevich, Fiz. tverd. Tela 10, 3348 (1968).
[9] C. J. F. Böttcher, Theory of Electric Polarization, Elsevier, Amsterdam 1973.
[10] J. S. Gradstein and J. M. Ryzhik, Tablitsy integralov, sum, ryadov i proizvedenii, Izd. FML, Moscow 1963 (in Russian).
[11] P. W. Tasker, Phil. Mag. A39, 119 (1979).

(Received April 23, 1985)

phys. stat. sol. (b) **131**, 759 (1985)

Subject classification: 14.3 and 18.4; 19; 22.8

Université des Sciences et Techniques du Languedoc, G.E.S. associé au CNRS, Montpellier[1]) *(a),*
Concordia University, Department of Electrical Engineering, Montreal[2]) *(b),*
and Université des Sciences et Techniques du Languedoc, Physico-Chimie des Matériaux Solide Inorganique associé au CNRS, Montpellier (c)

$Pb_{1-x}Gd_xTe$ a New Semimagnetic Semiconductor with Rare Earths

By

M. AVEROUS (a), B. A. LOMBOS (b), C. FAU (a), E. ILBNOUELGHAZI (a), J. C. TEDENAC (c), G. BRUN (c), and M. A. BARTKOWSKI (b)

Dedicated to Prof. Dr. Dr. h. c. Dr. E. h. P. GÖRLICH on the occasion of his 80th birthday

$Pb_{1-x}Gd_xTe$, a new semimagnetic semiconductor, containing gadolinium as a rare earth element is synthesized. Its magnetic susceptibility, transport and optical properties are determined in the temperature range of 4.2 to 300 K using magnetic field intensities up to 6 T. The interaction between the Gd^{3+} ions are found to be slightly antiferromagnetic. The measured Hall mobilities of the samples with low gadolinium concentration (1 at% nominal value) are extremely high (4×10^6 cm^2 V^{-1} s^{-1}) at 4.2 K. They are comparable with the mobilities reported in superlattices. The large static dielectric constant of the PbTe host lattice is invoked to serve to shield out the electric fields of the ionized impurities and defects, minimizing impurity scattering. In addition, at low gadolinium concentrations, no Schubnikow-de Haas oscillations are found although all the required conditions for them are satisfied. In this sample the Fermi level is 16.9 meV above the conduction band edge. It is argued consequently, that all the Landau levels passed the Fermi level due to the internal magnetic field induced by the rare earth ions. At higher gadolinium concentrations, 2 and 5 at%, the Fermi level moves up to 338 and 478 meV in the conduction band. Consequently, the internal magnetic field does not pre-empt the effect of the external magnetic field. This results in the appearance of the Shubnikow-de Haas oscillations, as expected.

$Pb_{1-x}Gd_xTe$, ein neuer halbmagnetischer Halbleiter, der Gadolinium als Seltenerdelement enthält wird synthetisiert. Seine magnetische Suszeptibilität und die Transport- und optischen Eigenschaften werden im Temperaturbereich von 4,2 bis 300 K bei Magnetfeldern bis zu 6 T bestimmt. Die Wechselwirkung zwischen den Gd^{3+}-Ionen ist schwach antiferromagnetisch. Die gemessenen Hallbeweglichkeiten der Proben mit niedriger Gadoliniumkonzentration (1 At% Nominalwert) sind extrem hoch (4×10^6 cm^2 V^{-1} s^{-1} bei 4,2 K). Sie sind mit den für Übergitter berichteten Beweglichkeiten vergleichbar. Die große statische Dielektrizitätskonstante des PbTe-Wirtsgitters scheint als Abschirmung für die elektrischen Felder der ionisierten Störstellen und Defekte zu wirken und die Störstellenstreuung zu minimisieren. Darüber hinaus werden bei niedrigen Gadoliniumkonzentrationen keine Shubnikov-de Haas-Oszillationen gefunden, obwohl alle geforderten Bedingungen dafür erfüllt sind. In diesen Proben liegt das Ferminiveau 16,9 meV oberhalb der Leitungsbandkante. Es wird deshalb angenommen, daß alle Landauniveaus das Ferminiveau infolge des durch die Seltenerdionen induzierten inneren Magnetfeldes durchlaufen haben. Bei höheren Gadoliniumkonzentrationen, 2 und 5 At%, steigt das Ferminiveau auf 338 und 478 meV im Leitungsband. Folglicherweise hat das innere Magnetfeld nicht mehr den Vorrang gegenüber dem Effekt des äußeren Magnetfeldes. Das führt wie erwartet zum Auftreten der Shubnikov-de Haas-Oszillationen.

---

[1]) Place Eugene Bataillon, 34060 Montpellier Cédex, France.
[2]) Montreal, Quebec H3G 1M8, Canada.

1. Introduction

Semimagnetic semiconductors (SMSC) are extensively studied lately due to their interesting physical properties. These are the large Faraday rotation and the spin splitting exciton levels and the magnetic field induced non-linear splitting of the Landau levels. All these properties are due to the exchange interaction induced by the magnetic moments localized at the magnetic ions. The electrons of the localized magnetic ions interact with the electrons of the valence and conduction band, respectively, creating the exchange interaction.

One of these magnetic ions is the manganese introduced in the II–VI compounds such as $Cd_{1-x}Mn_xTe$, $Cd_{1-x}Mn_xSe$, $Hg_{1-x}Mn_xTe$, and $Hg_{1-x}Mn_xSe$. These are well treated in the recent literature [1 to 9]. Due to phase diagram restrictions the introduction of the iron in the lattice of the mercury chalcogenide, $Hg_{1-x}Fe_xTe$, is more difficult, however, some investigations were devoted to this material as well. These involved magneto-optical and magneto-transport measurements [10 to 12]. The obtained results in the case of $Hg_{1-x}Fe_xTe$ are rather different from the one of $Hg_{1-x}Mn_xTe$ due to the peculiar magnetism of the iron ions. This could be explained in terms of Van Vleck type paramagnetism [11].

In this paper a new class of semimagnetic semiconductors is presented where the magnetic ions are from the rare earth family. The first of these is gadolinium. One of the reasons to use this approach is the availability of high purity gadolinium, which is not the case for the previously mentioned transition metals. Thus, replacing the manganese and iron with higher purity gadolinium, due to the minimization of scattering by uncontrolled impurities of the low purity transition metals, higher carrier mobilities are to be expected. Indeed, the measured electron mobilities of the prepared $Pb_{1-x}Gd_xTe$ crystals are an order of magnitude higher than the ones obtained in $Pb_{1-x}Mn_xTe$, in the case of equivalent energy gaps. The other reason for the introduction of the rare earth ions is related to their electronic structure. In the case of gadolinium, the electron shell responsible for the magnetic moment is the $4f^7$ one. It has a spin of 7/2 due to the seven electrons of the half filled shell. Furthermore, the 4f shell lies deep inside the rare earth ions, while the 3d shell in the case of the transition metals is an external one. Consequently, the dominant interaction of the transition metals is related to the crystal field. However, in the case of the rare earth ions the spin–orbit interaction is the predominant phenomenon. This introduces new physical properties such as the existence of an internal magnetic field influencing the electrical properties of the crystals. It is proposed that the lack of the Shubnikov-de Haas oscillations in some samples can be interpreted as an effect of this internal magnetic field.

2. Experimental Results

Single crystals of $Pb_{1-x}Gd_xTe$, with $x = 0.01$ were grown from semiconductor grade elements, using the Bridgman technique. A heat pipe controlled zone melter [13] was used for the preparation of crystals having nominal gadolinium concentrations corresponding to $x = 0.02$ and 0.05, respectively. Fig. 1, shows the electron spin resonance, ESR, signals of the samples containing 2 at% nominal gadolinium concentrations. The presence of magnetic ions in the host crystal, can clearly be seen from the depicted curves. The sharp resonance line of the sample is characteristic of lightly interacting Gd (3+) ions. The 5 at% nominal gadolinium concentration sample gave a spectrum similar to the one of Fig. 1. In the case of the sample containing 1 at% Gd, the derivative of the power absorption due to ESR was below the detection limit between 2404 and 4404 G modulated by 10 G at room temperature. The electrical

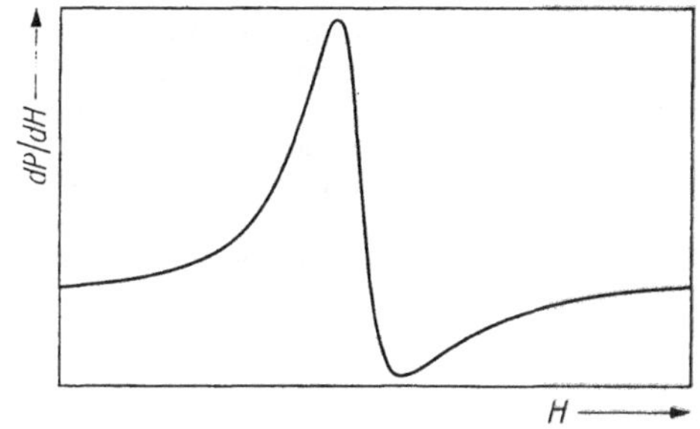

Fig. 1. Electron spin resonance (ESR) absorption spectrum of $Pb_{1-x}Gd_xTe$ with 2 at% nominal gadolinium concentration. The curve was taken between 2406 and 4406 G with a modulation of 10 G at T = 300 K. The experimental g-factor, deduced from the spectrum, is 1.92

transport properties of the crystals were characterized by Hall effect and resistivity measurements up to a magnetic field intensity of 0.8 T in the temperature range of 4.2 to 300 K. The obtained results are collected in Fig. 2 and 3, where the variation of resistivities and Hall mobilities of samples having 1%, 2%, and 5% gadolinium concentrations, respectively, are depicted. The resulting carrier mobilities, are in the range of 6×10^3 to 4×10^6 cm²/Vs at 4.2 K.

One can estimate the position of the Fermi level from the carrier concentrations, determined from the Hall coefficients by the well-known procedure [14]. These calculated values at 4.2 K are: 16.9, 338, and 428 meV above the conduction band edge for the 1, 2, and 5 at% gadolinium concentration samples, respectively. Thus, all three materials are degenerate and with the extremely high mobilities, mentioned above, one would expect very large Shubnikov-de Haas (SdH) oscillations as a function of magnetic field intensities. In Fig. 4 the experimentally determined curves depicting the variation of magnetoresistance as a function of magnetic field intensity is shown measured at 4.2 K. It can be seen that on curve (1) there is no SdH oscillation, while the degeneracy and the quantization conditions are largely satisfied. However, in the case of the other two samples with higher gadolinium concentrations, 2 and 5 at%, curves (2) and (3), the oscillations are clearly identifiable.

In Fig. 5, the magnetic moment M as a function of magnetic field intensity, B, is depicted at different temperatures in the range of 4.2 to 47 K. The behaviour of this curves is classical as it was reported in the case of $Pb_{1-x}Mn_xTe$ [15]. There is an

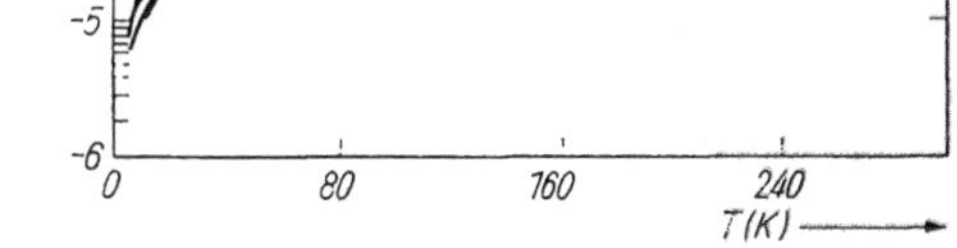

Fig. 2

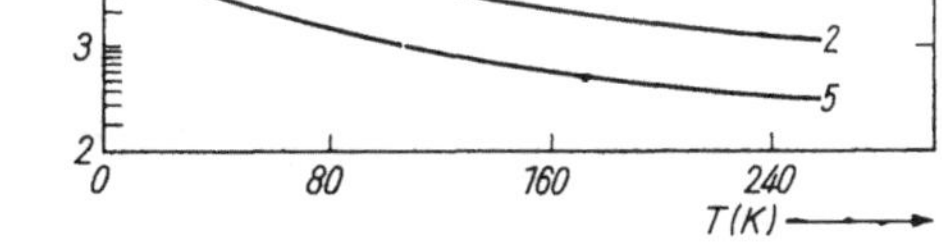

Fig. 3

Fig. 2. Resistivities (Ω cm) as a function of temperature (K) of $Pb_{1-x}Gd_xTe$ with 1 at% (oriented: 1% O (100) and non-oriented: 1% NO), 2, and 5 at.% nominal gadolinium concentrations

Fig. 3. Hall mobilities (cm² V⁻¹ s⁻¹) as a function of temperature (K) of $Pb_{1-x}Gd_x$ Te with 1, 2, and 5 at% gadolinium concentration; the monotonous increase of the mobilities, up to a maximum of 4×10^6 cm² V⁻¹ s⁻¹ characteristic of superlattices, indicates the lack of impurity scattering

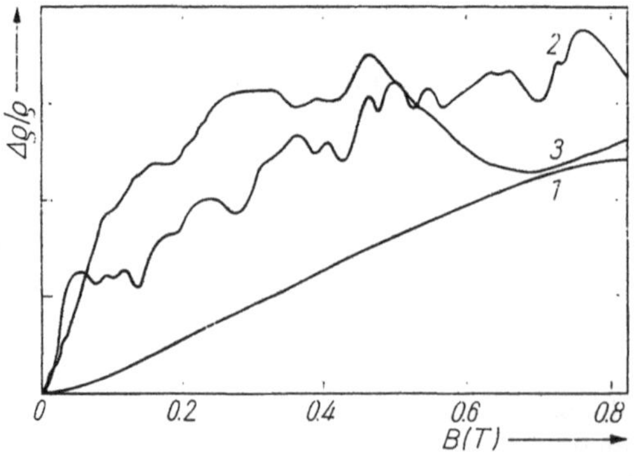

Fig. 4. Magnetoresistances of $Pb_{1-x}Gd_xTe$ as a function magnetic field intensities (T), with (1) 1, (2) 2, and (3) 5 at% nominal gadolinium concentrations at $T = 4.2$ K. The lack of the Shubnikov-de Haas oscillations is noticeable in the case of (1) with 1 at% gadolinium concentration

important difference, however. The curves of the gadolinium containing crystals of this investigation do not converge towards zero. The minimum value at zero field, thus spontaneous magnetization, is about 0.7×10^{-4}. Taking into consideration the weight of the sample and the gadolinium content 1.2 at% deduced from the results of Fig. 6, this corresponds to an internal magnetic field of approximately 0.1 T. In Fig. 6, the curve fitting plots of the Brillouin law [20] on the experimental points of magnetization as a function of applied magnetic field intensity, is shown for the sample with 5 at% nominal gadolinium content. To characterize Gd^{3+}, $J = 7/2$ and $g = 2$ was used in the Brillouin law of the magnetization, for all of the curve fittings. The full lines indicate good agreements with the characteristic temperature of the Curie-Weiss law, $\theta = 1$ K. Other values gave less satisfactory fits. Using $\theta = 0$ K, the curves are depicted, as an example, with the dotted lines for the investigated three temperatures of 4.2, 10.0, and 15.4 K, respectively. In the case of all the curves the

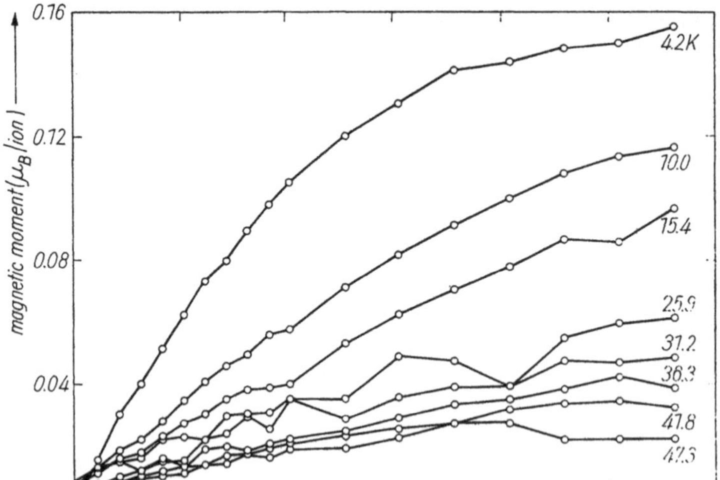

Fig. 5. Magnetic moment (μ_B/ion) as a function of magnetic field intensities (T) of $Pb_{1-x}Gd_xTe$ with 5 at% nominal Gd concentration. The minimum value of the magnetic moment at zero field intensity for temperatures above 4.2 K is $7 \times 10^{-3}\mu_b$/ion. This corresponds to an internal magnetic field intensity of 1.01×10^{-1} T

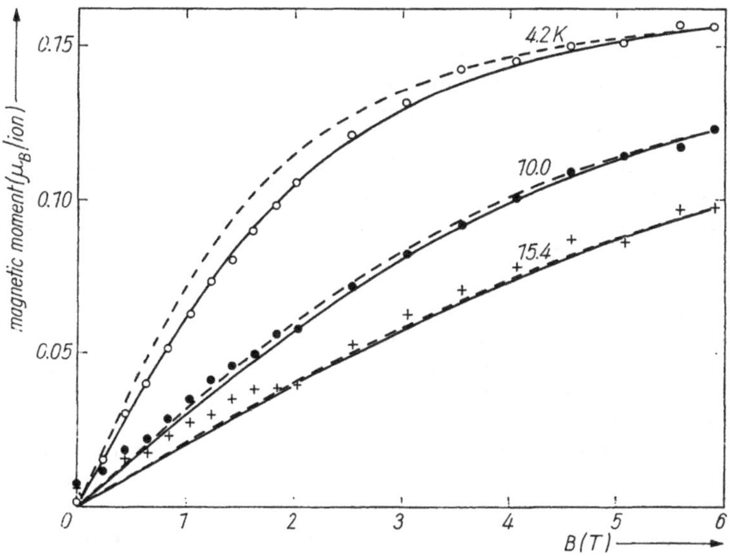

Fig. 6. Curve fittings using the Brillouin function for a total spin of 7/2 and $g = 2$, of the measured magnetic moments as a function of magnetic field intensities (B). Full line $T_c = 1$ K, dashed line $T_c = 0$ K of the Curie-Weiss law. The effective magnetic ion concentration determined from the fit is 2×10^{20} cm^{-3}

best fit was obtained with an effective magnetic ion concentration of 2×10^{20} cm^{-3}. This corresponds to an effective Gd^{3+} content of 1.2 at% in the sample of 5 at% nominal gadolinium concentration.

Fig. 7 shows the reciprocal magnetic susceptibility as the function of temperature in the range of 4.2 to 60 K in the case of the sample with 5 at% gadolinium. It is to be noted that for the other two samples the obtained signal was too small to be measured. The magnetic susceptibility, χ, follows the Curie-Weiss law, $\chi = C/(T + \theta)$, with θ being close to the value obtained in Fig. 6. This positive value clearly indicates some antiferromagnetic interactions between the gadolinium ions.

The temperature dependence of the band gap, deduced from optical absorption measurements in the temperature range of 77 to 300 K, has been determined. The band gap is taken to be between the lowest conduction band edge (L_{3-}^{6-}) and the highest valence band edge (L_{1+}^{6+}). This temperature dependence is found to be $dE_g/dT = 3.6$ meV/K, for the sample containing 1 at% gadolinium. This value is quite close to the one reported for PbTe [16].

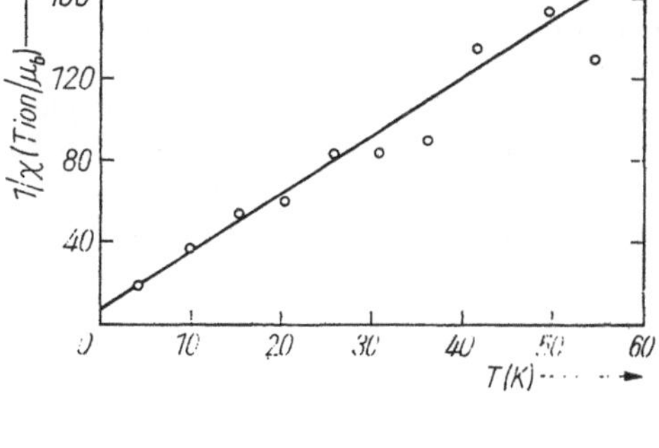

Fig. 7. The reciprocal magnetic susceptibility (T ion/μ_b) as a function of temperature (K) of $Pb_{1-x}Gd_xTe$ with 5 at% Gd, indicating the validity of the Curie-Weiss law

3. Discussion

There are two points to be noted with respect to the first crystal, containing 1% gadolinium. On the one hand, the measured mobilities at 4.2 K are in the range of 4×10^6 cm^2/Vs, as it can be seen from the data collected in Fig. 3. On the other hand, it can clearly be seen in Fig. 4, that the expected SdH oscillations are not found on the magnetoresistance versus magnetic field intensities curve. This is surprising since the material is degenerate, the Fermi level is 16.9 meV above the conduction band, determined from the carrier concentration deduced from Fig. 2 and 3. In addition, due to the extremely high mobility of this crystal the quantization condition for SdH oscillations is largely satisfied.

The temperature dependence of the mobility of this sample exhibits the typical $-3/2$ law down to 10 K (see Fig. 3) indicating acoustic phonon scattering. It is well known that at low temperatures normally the impurity scattering dominates, characterized by the $T^{3/2}$ law. In the intermediate temperature range optical phonon scattering might demonstrate itself. This sample has a carrier concentration of $2.5 \times \times 10^{17}$ (cm^{-3}), thus the effect of the impurity scattering should be seen at low temperatures as it is usually the case in semiconductors having similar band gaps. It is to be noted that PbTe crystals exhibit a large static dielectric constant, as high as 400 in this case. This might serve to shield the electric fields of ionized impurities and defects. Consequently, the impurity scattering appears only at very low temperatures, which might explain the departure from the $T^{-3/2}$ law. The result is an increasing mobility with decreasing temperature, due to the decrease of the effect of lattice scattering. This results in low temperature mobility values in the range of those reported in superlattices.

On the other hand, one can recall the amplitude dependence of the SdH oscillation, which can be given by the following theoretical expression [17]:

$$A(T, B) = \frac{5}{2}\left(\frac{P}{2B}\right)^{1/2} \frac{\beta(T + T_i)m' \exp\left(\frac{-\beta T_D m'}{B}\right)}{\sinh\left(\frac{\beta\,(T + T_i)\,m'}{B}\right)} \times$$
$$\times \cos\frac{\pi m' g^*}{2} \cos 2\pi\left(\frac{1}{PB} - \frac{1}{8} - \frac{1}{\gamma}\right), \qquad (1)$$

where $\beta = 2\pi^2 m_B c m_0/\hbar e$, $p = \hbar e/E_f m^* c$, E_f the Fermi energy, γ the phase factor, g^* the effective spectroscopic splitting factor of the conduction band, T_D the Dingle temperature, T_i the inhomogeneity temperature,

$$m' = \frac{m^*}{m_0}.$$

Here only the fundamental term is taken into consideration, since the higher harmonics are strongly damped. In contrast to normal narrow gap semiconductors in the case of SMSC's the g^* factor is strongly temperature dependent. Consequently the cosine term goes to zero each time its argument equals $\pi/2\ (2n + 1)$. This was discussed in relation to manganese containing mercury chalcogenides [18]. In the case of the gadolinium compound, treated in this investigation, it cannot be envisaged either that the g factor would be temperature independent and fixed to a value were the cosine term in (1) would be zero, since the temperature dependence of the magnetization is known (see Fig. 5).

$$g^* = g^*_{\mathrm{PbTe}} + \frac{4A}{\hbar\,\omega_0}, \qquad (2)$$

where $A = 1/6\, \beta N_0 \langle S_z \rangle$, β is the exchange integral given by

$$\beta = \beta_{\parallel} \sin^2 \theta^- - \beta_{\perp} \cos^2 \theta^- ,$$

where $\beta_{\parallel} = (X_+ \,|J|\, X_+)$ and $\beta_{\perp} = (Z\,|J|\,Z)$, $\theta^{\pm}$ is defined in [9], N_0 is the number of magnetic ions per unit volume, $\langle S_z \rangle$ is the average magnetization.

Consequently, the lack of the SdH oscillation in the sample of 1% gadolinium content, with the Fermi level positioned 16.9 meV above the conduction band edge, can be related to the presence of the gadolinium ions in the lattice. It can be envisaged that a weak ferromagnetism is superimposed on the antiferromagnetism of the gadolinium. Thus, the phenomenon of the lack of SdH oscillation could be accounted for by the existence of a spontaneous magnetic moment introduced by a canted gadolinium sublattice magnetization. It can be clearly seen in Fig. 5 that the magnetization curves are not approaching zero at zero fields, but converge to a value of about 7×10^{-4}. This results in an internal magnetic field of approximately 0.1 T. This perhaps might be related to the formation of rare earth ion clusters. Consequently, due to the extremely high mobility, characterizing this crystal, this relatively small spontaneous magnetic field is sufficient to split the Landau levels. Then, if the Fermi level is low enough in the conduction band, most of the Landau levels would cross it before applying any external magnetic field.

The phenomenon of the superimposed weak ferromagnetism on the basically antiferromagnetic materials, the so-called canted antiferromagnetism, has been discussed by Dzialoshinski [19]. He was taking into consideration some degree of misorientation of the magnetic moments of the antiferromagnetic materials. Then if the angle of this misorientation is α, the resulting canted magnetization can be given by

$$M_s = 2\alpha M_0(T) , \tag{3}$$

where $M_0(T)$ is the magnetization of each sub-lattice species at a tempreature T. The misorientation of the antiferromagnetic arrangement might be induced by an asymmetrical exchange interaction coupling or, on the other hand, by local anisotropy in the lattice.

This could be specifically related to the rare earths, distinguishing them from the effect of the transition metal ions. The exchange and superexchange terms are invariant in the case of the rotation of all the spins. Furthermore, the dipolar interactions, or more generally, the n-polar interactions are resulting in symmetrical potentials as well. However, antisymmetric exchange coupling can be envisaged considering spin–orbital interaction. One can note that in the iron group ions the spin–orbit coupling is weak with respect to the crystal field, while in the case of the rare earths this coupling is predominant compared to the crystal field. Following the treatment of Herpin [20], one can consider two ions, (1) and (2), with their fundamental state $|n_1\rangle$ and $|n_2\rangle$ and their excited states $|n_1'\rangle$ and $|n_2'\rangle$. By spin–orbit interaction ion (1) can be excited from $|n_1\rangle$ to $|n_1'\rangle$, which is a secondary process. This interaction can be expressed by $\lambda \boldsymbol{L}_1 \cdot \boldsymbol{S}_1$. The ion (1) can go back to state $|n_1\rangle$ due to exchange interaction with ion (2). This process can be described by the following expression:

$$2J_{n'_1 n_2 n_1 n_2} \boldsymbol{S}_1 \cdot \boldsymbol{S}_2 , \tag{4}$$

where

$$J_{n'_1,n_2,n_1,n_2} = \int \varphi^*_{n_1}(\boldsymbol{r} - \boldsymbol{R}_1)\, \varphi^*_{n'_2}(\boldsymbol{r}' - \boldsymbol{R}_2) \frac{e^2}{|\boldsymbol{r} - \boldsymbol{r}'|} \varphi'_{n_1}(\boldsymbol{r}' - \boldsymbol{R}_1) \varphi_{n_2}(\boldsymbol{r} - \boldsymbol{r}_2)\, dv_1\, dv_2 \; .$$

There are obviously other processes similar to the above-mentioned one, however, the order of interaction might be reversed. One can find finally that the interaction can be given by

$$\Delta E = \frac{2\langle n_1 | \Lambda \mathbf{L}_1 \cdot \mathbf{S}_1 | n_1' \rangle \, J n_1 n_2 n_1 \, n_2 \, \mathbf{S}_1 \cdot \mathbf{S}_2}{E_{n_1} - E_{n1}} + \\ + \frac{2J n_1, n_2, n_1' n_2' \, \mathbf{S}_1 \cdot \mathbf{S}_2 \, \langle n_1' \, | \Lambda \mathbf{L}_1 \cdot \mathbf{S}_1 \, | n_1 \rangle}{E_{n_1} - E_{n'1}} . \tag{5}$$

Since the matrix elements of the angular momentum are purely imaginary one can write that $\langle n_1 | \, \mathbf{L}_1 \, | n_1' \rangle = -\langle n_1' | \, \mathbf{L}_1 \, | n_1 \rangle$. The exchange interaction does not depend on the order of operation, consequently the two terms of (5) can be regrouped,

$$\Delta E = \frac{2\Lambda J n_1' n_2 n_1 n_2}{E_{n_1} - E_{n'_1}} \langle n_1 | \, \mathbf{L}_1 \, | n_1' \rangle \, \{\mathbf{S}_1(\mathbf{S}_1 \cdot \mathbf{S}_2) - (\mathbf{S}_1 \cdot \mathbf{S}_2) \, \mathbf{S}_1\} = \\ = \frac{2i\Lambda J n_1' n_2 n_1 n_2}{E_{n_1} - E_{n'_1}} \langle n_1 | \, \mathbf{L}_1 \, | n_1' \rangle \, (\mathbf{S}_1 \Lambda \mathbf{S}_2) \; . \tag{6}$$

There exists another term which can be deduced by interchanging ions (1) and (2). This results in the following expression for the antisymmetric interaction potential:

$$V_a = 2i\Lambda \left[\frac{J n_1' n_2 n_1 n_2}{E_{n_1} - E_{n'_1}} \langle n_1 | \, \mathbf{L}_1 \, | n_1' \rangle - \frac{J n_1 n_2' n_1 n_2}{E_{n_2} - E_{n'_2}} \langle n_2 | \, \mathbf{L}_2 \, | n_2' \rangle \right] (S_1 \Lambda S_2) \; . \tag{7}$$

It can be seen that this antisymmetric interaction potential can exist only if the two terms of (7) are different. In the case of a crystal the importance of V_a is enhanced by crystal symmetry. In addition, the peculiar atomic structure of the rare earth gadolinium coupled with the symmetry of the PbTe matrix, makes this phenomenon noticeable.

Considering the other two samples with 2 and 5% gadolinium content one can determine the position of the Fermi level from the data collected in Fig. 3, 4. The calculated values are 338 and 478 meV, respectively. Due to these high positions of the corresponding Fermi levels above the conduction band edge it can be expected that the SdH oscillations will appear. It can be noticed that the measured Hall mobilities are lower in the case of these crystals, compared to the first one, at low temperatures as they can be seen in Fig. 3. The liquid helium temperature values are 33000 and 11000 cm^2/Vs, respectively. This might be related to the higher impurity concentrations. The 5% gadolinium crystal for example has a carrier concentration of 4×10^{19} cm^{-3}. Consequently, the impurity scattering phenomenon cannot be neglected any more as it was the case in the crystal of 1% Gd content.

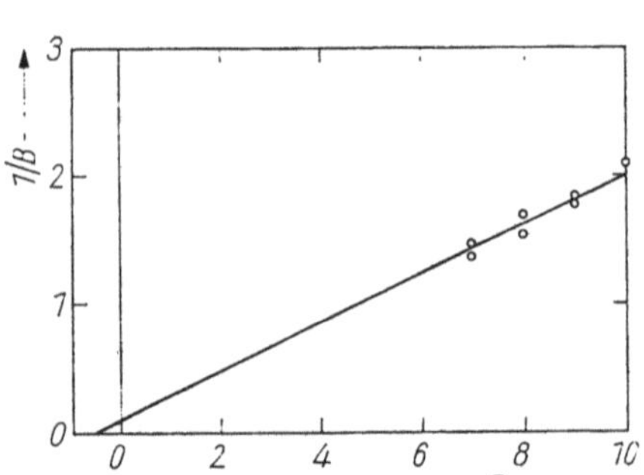

Fig. 8. Typical periodicity of the Shubnikow-de Haas oscillations in $1/B$ as a function of Landau quantum numbers, n, of $Pb_{1-x}Gd_xTe$, with $x = 5$ at% Gd concentrations

As expected in these latter cases the SdH oscillations are observed as they are shown in Fig. 4. The typical periodicity of the oscillations in $1/B$ can be deduced as is depicted in Fig. 8. For the values of quantum numbers of $n = 7$, 8, and 9 the spin splitting of the Landau levels can be distinguished as the function of magnetic field intensity B. It can be noticed that the spin splittings are smaller for $n = 7$ and 9, compared to the one of $n = 8$. The Landau level dependence on the magnetic field intensities could not be determined, since there were not enough samples for each gadolinium concentration with different Fermi levels. Therefore, it is planned that magneto-optic and magneto-transport measurements on several samples will make it possible to determine the exchange interaction.

4. Conclusion

The first experimental results of a new semimagnetic semiconductor $Pb_{1-x}Gd_xTe$ are reported. The measured magnetic susceptibility indicates antiferromagnetic interaction between the gadolinium ions or clusters formed by them. The EPR spectra indicate that the rare earth ions are well incorporated in the host lattice of PbTe. There are two surprising phenomena related to the crystal with a low (1%) gadolinium concentration. On the one hand, it is characterized by a very high mobility at low temperatures (4×10^6 cm²/Vs). On the other hand, the lack of the SdH oscillation is observed, although all the condictions for its appearance are fulfilled.

The results are interpreted in terms of the screening of the impurity potential related to the very large static dielectric constant of the host lattice. On the other hand, the existence of an internal magnetic field due to the canted antiferromagnetism, introduced by the rare earth ions, is invoked to elucidate the lack of the SdH oscillation in the case where the Fermi level lies relatively close to the conduction band edge.

References

[1] R. R. Galazka, in: Proc. Internat. Conf. Phys. Narrow Gap Semicond., Linz (Austria) 1981, Ed. E. Gornik, H. Heinrich, and L. Palmetshofer, Springer-Verlag, 1981 (p. 294).
[2] G. Bastard, J. A. Gaj, R. Planel, and C. Rigaux, J. Physique C5, 247 (1980).
[3] J. Stankiewícz, W. Giriat, and M. V. Bien, phys. stat. sol. (b) 68, 485 (1975).
[4] G. Bastard, Proc. Internat. Conf. Phys. Narrow Gap Semicond., Warsaw (Poland) 1978, Ed. J. Rauluszkiewicz, M. Gorska, and E. Kaczmarek, PWN Polish Sci. Publ., Warsaw 1978 (p. 63).
[5] G. Bastard, C. Rigaux, and A. Mycielski, phys. stat. sol. (b) 79, 585 (1977).
[6] W. Dobrowski, M. Von Ortenberg, A. M. Sandauer, R. R. Galazka, A. Mycielski, and R. Pauthenet, Proc. Internat. Conf. Narrow Gap Semicond., Linz (Austria) 1981, Ed. E. Gornik, Heinrich, and L. Palmetshofer, Springer-Verlag, 1981 (p. 303).
[7] M. Jaczyński, J. Kossut, and R. R. Gałazka, phys. stat. sol. (b) 88, 73 (1978).
[8] P. Byszewski, K. Szlenk, J. Kossut, and R. R. Gałązka, phys. stat. sol. (b) 95, 359 (1979).
[9] J. Niewodniczanska-Zawadska, G. Elsinger, L. Palmetshofer, E. J. Fantner, G. Bauer, and W. Zawadski, Physica 117/118, B/C 458 (1983).
[10] H. Serres, G. Bastard, C. Rigaux, and J. Mycielski, Proc. Internat. Conf. Phys. Narrow Gap Semicond. Linz (Austria) 1981, Ed. E. Gornik, H. Heinrich, and L. Palmetshofer, Springer-Verlag, 1981 (p. 321).
[11] Y. Guldner, C. Rigaux, M. Menant, D. P. Mullin, and J. K. Furdina, Solid State Commun. 33, 133 (1980).
[12] S. Abdel-Maksoud, C. Fau, J. Calas, M. Averous, B. A. Lombos, G. Brun, and J. C. Tedenak, Solid State Commun., in the press.

[13] N. YEMENIDJIAN and B. A. LOMBOS, J. Crystal Growth 56, 163 (1982).
[14] B. A. LOMBOS, M. AVEROUS, C. FAU, and J. CALAS, phys. stat. sol. (b) 112, 237 (1982).
[15] J. NIEWODNICZANSKA-ZAWADZKA, J. KOSSUT, A. SANDERS, and W. DOBROWOLSKI, Proc. Internat. Conf. Phys. Narrow Gap Semicond., Linz (Austria) 1981, Ed. E. GORNIK, H. HEINRICH, and L. PALMETSHOFER, Springer-Verlag, 1981 (p. 326).
[16] YU. I. RAVITCH, B. A. EFIMOVA, and V. I. TAMARCHENKO, phys. stat. sol. (b) 43, 453 (1971).
[17] A. E. STEPFENS, D. G. SEILER, J. R. SYBERT, and H. J. MACKEY, Phys. Rev. B 11, 4999 (1975).
[18] M. JACZYŃSKI and W. DOBROWOLSKI, phys. stat. sol. (b) 102, 195 (1980).
[19] I. DZIALOSHINSKI, Phys. Chem. Solids 4, 241 (1958).
[20] A. HERPIN, Théorie du Magnétisme, Bibl. Sciences et Tech. Nucl., Paris 1968 (p. 538).

(Received June 25, 1985)

phys. stat. sol. (b) 131, 769 (1985)

Subject classification: 13.1 and 14.1; 21.1.1

Wissenschaftsbereich Physik
der Hochschule für Verkehrswesen "Friedrich List", Dresden[1])

The Spontaneous Resistivity Anisotropy of Iron-Based Alloys

By
P. MUTH and V. CHRISTOPH

Dedicated to Prof. Dr. Dr. h. c. Dr. E. h. P. GÖRLICH on the occasion of his 80th birthday

The spontaneous resistivity anisotropy of Fe-based alloys measured at low temperatures and the quotient of the residual sub-band resistivities of the majority and minority spin-band electrons are obviously strongly related. It can be shown that a theoretical expression derived for the spontaneous resistivity anisotropy of Ni-based alloys represents the data of Fe-based alloys, too. Other existing theoretical relations for the spontaneous resistivity anisotropy are analysed.

Zwischen der bei tiefen Temperaturen gemessenen magnetischen Widerstandsanisotropie von Eisenlegierungen und dem Quotienten aus den Rest-Teilwiderständen der Leitungselektronen des Majoritäts- und Minoritäts-Spinbandes besteht offensichtlich ein funktioneller Zusammenhang. Es wird gezeigt, daß eine früher für Nickellegierungen abgeleitete Formel auch die Meßwerte für Eisenlegierungen wiedergibt. Weitere in der Literatur bekanntgewordene theoretische Beziehungen für die magnetische Widerstandsanisotropie werden auf ihre Gültigkeit überprüft.

1. Introduction

It has been demonstrated in a lot of papers that the two-current model describes rather well the electrical conduction in ferromagnetic metals and alloys. At low temperatures there are two independent currents in parallel with the residual sub-band resistivities $\varrho_\uparrow$ and $\varrho_\downarrow$ corresponding to majority and minority carriers, respectively.

In ferromagnetic systems the resistivity depends on the direction of the electrical current with respect to the magnetization. This spontaneous resistivity anisotropy is defined by

$$\frac{\Delta\varrho}{\varrho} = \frac{\varrho_{\parallel} - \varrho_{\perp}}{\varrho}, \tag{1}$$

where $\varrho_{\parallel}$ and $\varrho_{\perp}$ stand for the resistivities measured for currents parallel and perpendicular to the saturation magnetization extrapolated to zero magnetic induction, and $\varrho = \frac{1}{3}\varrho_{\parallel} + \frac{2}{3}\varrho_{\perp}$ is the resistivity of a sample with random domain configuration.

By measurements of the residual resistivity of ternary dilute alloys and analysis of the deviations from Matthiessen's rule the residual sub-band resistivities $\varrho_\uparrow$ and $\varrho_\downarrow$ have been determined for a great number of Ni- and Fe-based alloys [1].

For Ni-based alloys there is a clear correlation between the spontaneous resistivity anisotropy $\Delta\varrho/\varrho$ and the quotient $\alpha = \varrho_\downarrow/\varrho_\uparrow$ [2, 3]. Following a suggestion of Smit [4],

---
[1]) PSF 103, DDR-8072 Dresden, GDR.

Campbell et al. derived a relation between $\Delta\varrho/\varrho$ and α,

$$\frac{\Delta\varrho}{\varrho} = \gamma(\alpha - 1)\,, \tag{2}$$

which approximately represents the data for dilute Ni alloys at low temperatures [2]. The main idea is that in the presence of spin–orbit coupling $d_\uparrow$ and $d_\downarrow$ wave functions are mixed. The mixing is not isotropic because the magnetization direction provides an axis for the spin–orbit interaction, so it provides a mechanism for the resistivity anisotropy. The coefficient γ was chosen as 0.0075 for Ni alloys. It depends on the spin–orbit coefficient and the overlap integrals.

In [5] we derived a relation for the spontaneous resistivity anisotropy,

$$\frac{\Delta\varrho}{\varrho} = \left(\frac{\gamma_1}{\alpha} - \gamma_2\right)\frac{\mu^2 - \alpha^2}{1 + \alpha}\,, \tag{3}$$

where γ_1, γ_2, and μ are constants depending on the spin–orbit coupling and the scattering probabilities for one impurity. Equation (3) represents quite well the data for dilute Ni alloys at low temperatures, too, especially for small quantities of $\Delta\varrho/\varrho$ [5]. Moreover (3) is approximately applicable to concentrated alloys.

2. Relation for the Spontaneous Resistivity Anisotropy

The main ideas in deriving (3) had been discussed in [5] in a very short manner. Therefore, we like to give some comments on this equation.

The spontaneous resistivity anisotropy is given by the general relation [1, 3]

$$\frac{\Delta\varrho}{\varrho} = \frac{\alpha}{1 + \alpha}\left(\frac{\Delta\varrho_\uparrow}{\varrho_\uparrow}\right) + \frac{1}{1 + \alpha}\left(\frac{\Delta\varrho_\downarrow}{\varrho_\downarrow}\right), \tag{4}$$

where $\Delta\varrho_\sigma = \varrho_{\sigma\|} - \varrho_{\sigma\perp}$, $\sigma = \uparrow, \downarrow$. The quantity $\varrho_{\sigma\|}$ is the resistivity contribution of the conduction electrons with spin σ for a current parallel to the saturation magnetization of the sample. Due to Matthiessen's rule the sub-band resistivities $\varrho_\uparrow$, $\varrho_\downarrow$ consist of two terms

$$\varrho_{\sigma\nu} = \varrho_\sigma^{s} + \varrho_{\sigma\nu}^{sd}\,; \qquad \nu = \|, \perp\,,$$

where $\varrho_\uparrow^{s}$, $\varrho_\downarrow^{s}$ describe the conduction electron scattering within the conduction band (intraband scattering) and $\varrho_{\uparrow\nu}^{sd}$, $\varrho_{\downarrow\nu}^{sd}$ are the contributions of the scattering processes of the conduction electrons into states which do not contribute to the electrical current (d-like states, virtual bound states).

For potential scattering these contributions are in the weak-scattering limit proportional to the following transition probabilities (at low temperatures):

$$\varrho_\uparrow^{s} \sim \frac{1}{3}\sum_{k,k'} |\langle k|\, V\, |k'\rangle|^2 \left(\frac{\partial E}{\partial \boldsymbol{k}} - \frac{\partial E}{\partial \boldsymbol{k}'}\right)^2 \delta(E_k - \zeta)\, \delta(E_k - E_{k'})\,,$$

$$\varrho_{\uparrow\nu}^{sd} \sim \sum_{k,q,\lambda} |\langle k\chi_\uparrow|\, V\, |q^{(\lambda)}\rangle|^2 \left(\frac{\partial E}{\partial k_\nu}\right)^2 \delta(E_k - \zeta)\, \delta(E_k - \varepsilon_q^{(\lambda)})\,.$$

The eigenfunctions and the energies in the conduction and in the "non-conducting" bands are given by

$$\begin{aligned} &\text{conduction band: } \varphi_k^{s}\chi_\uparrow,\ \varphi_k^{s}\chi_\downarrow; && E_k \text{ (spin degenerated)}\,, \\ &\text{non-conducting band: } \psi_q^{(1)} = u_q\varphi_q^{(\uparrow)}\chi_\uparrow + v_q\varphi_q^{(\downarrow)}\chi_\downarrow; && \varepsilon_q^{(1)}\,, \\ &\qquad\qquad\qquad\qquad \psi_q^{(2)} = v_q^*\varphi_q^{(\uparrow)}\chi_\uparrow - u_q^*\varphi_q^{(\downarrow)}\chi_\downarrow; && \varepsilon_q^{(2)} \end{aligned}$$

$$(|k\rangle = \varphi_k^{s}; \qquad |q^{(\lambda)}\rangle = \psi_q^{(\lambda)}; \qquad |q^{(\sigma)}\rangle = \varphi_q^{(\sigma)})\,.$$

The coefficients u_q, v_q depend on the spin–orbit interaction. For vanishing spin–orbit interaction we have $v_q \to 0$, $u_q \to 1$ ($|u_q|^2 + |v_q|^2 = 1$). The coefficients v_q sensitively depend on the direction of the magnetization, hence $\varrho^{sd}_{\sigma\parallel} \neq \varrho^{sd}_{\sigma\perp}$, $\sigma = \uparrow, \downarrow$. Using the quite general expressions for the wave functions and considering a spin-conserving scattering potential we easily find $\Delta\varrho_\uparrow = \Delta\varrho_{\uparrow\downarrow} - \Delta\varrho_{\uparrow\uparrow}$, $\Delta\varrho_\downarrow = \Delta\varrho_{\downarrow\uparrow} - \Delta\varrho_{\downarrow\downarrow}$, where

$$\Delta\varrho_{\sigma\sigma'} \sim \sum_{k,q} |\langle k| \; V \,|q^{(\sigma)}\rangle|^2 \left[\left(\frac{\partial E}{\partial k_\parallel}\right)^2 - \left(\frac{\partial E}{\partial k_\perp}\right)^2\right] |v_q|^2 \, \delta(E_k - \zeta)\, \delta(E_k - \varepsilon_q^{\sigma'}) \,.$$

Deriving this expression we have put $\varepsilon_q^{(1)} \approx \varepsilon_q^{(\uparrow)}$ and $\varepsilon_q^{(2)} \approx \varepsilon_q^{(\downarrow)}$, where the $\varepsilon_q^{(\sigma)} (\sigma = \uparrow, \downarrow)$ are the energy eigenvalues for a vanishing spin–orbit interaction corresponding to the eigenfunctions $\varphi_q^{(\sigma)}$, i.e. the spin–orbit interaction has been taken into account only in the eigenfunctions but not in the energy eigenvalues. The isotropic parts of the interband scattering terms are

$$\varrho_\sigma^{sd} \sim \frac{1}{3} \sum_{k,q} |\langle k| \; V \,|q^{(\sigma)}\rangle|^2 \left(\frac{\partial E}{\partial \boldsymbol{k}}\right)^2 \delta(E_k - \zeta)\, \delta(E_k - \varepsilon_q^{(\sigma)}) \,.$$

After averaging over the positions of the scatterers the scattering potential becomes

$$\overline{|\langle k| \; V \,|q\rangle|^2} = c(1 - c)\, |\langle k| \; V^{(1)} \,|q\rangle|^2 \,,$$

where c is the concentration of the scatterers and $\langle k| \; V^{(1)} \,|q\rangle$ is the matrix element of a single scatterer.

In general, the contribution of the intraband s-scattering is negligible compared with the interband scattering. Furthermore, a comparison of $\Delta\varrho_{\sigma\sigma'}$ and ϱ_σ^{sd} suggests the ansatz

$$\Delta\varrho_{\sigma\uparrow} = \gamma_\sigma \varrho_\uparrow^{sd} \,, \qquad \Delta\varrho_{\sigma\downarrow} = \gamma'_\sigma \varrho_\downarrow^{sd} \,,$$

where the coefficients γ_σ, γ'_σ depend on the spin–orbit interaction, but do not depend on the concentration of the scatterers. A further comparison shows

$$\frac{\Delta\varrho_{\uparrow\uparrow}}{\Delta\varrho_{\downarrow\uparrow}} = \frac{\Delta\varrho_{\uparrow\downarrow}}{\Delta\varrho_{\downarrow\downarrow}} \sim \frac{|\langle k| \; V \,|q^{(\uparrow)}\rangle|^2}{|\langle k| \; V \,|q^{(\downarrow)}\rangle|^2} \,,$$

hence

$$\frac{\gamma_\downarrow}{\gamma_\uparrow} = \frac{\gamma'_\downarrow}{\gamma'_\uparrow} \equiv \mu^2 \,, \tag{5}$$

and (4) with the approximation (5) becomes

$$\frac{\Delta\varrho}{\varrho} \approx \frac{\mu^2 - \alpha^2}{1 + \alpha} \,(\alpha^{-1}\gamma_\uparrow - \gamma'_\uparrow) \,.$$

3. Application to Iron-Based Alloys

Because for Fe alloys normally $\varrho_\uparrow > \varrho_\downarrow$ it is more convenient to use $\eta = \varrho_\uparrow/\varrho_\downarrow = 1/\alpha$ instead of α. Then (3) can be rewritten as

$$\frac{\Delta\varrho}{\varrho} = \left(\frac{\gamma_2}{\eta} - \gamma_1\right) \frac{1 - \mu^2\eta^2}{1 + \eta} \,. \tag{6}$$

In Fig. 1 the spontaneous resistivity anisotropy $\Delta\varrho/\varrho$ of dilute Fe alloys is plotted versus η. We find again a clear dependence of $\Delta\varrho/\varrho$ on η. It is easily seen that (2) cannot represent the data of Fe alloys. But the relation used for Ni alloys [5] represents the data for Fe alloys, too. Fitting the published data of 19 Fe alloys measured at $T = 4.2$ K [1] we found a consistent set of values for the parameters $\gamma_1 = 1.78 \times$

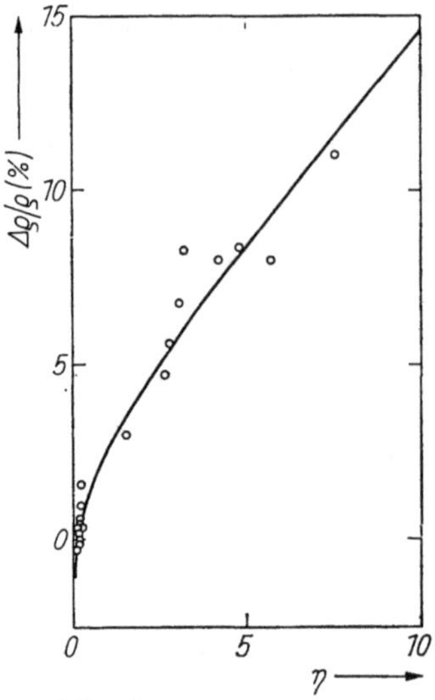

Fig. 1

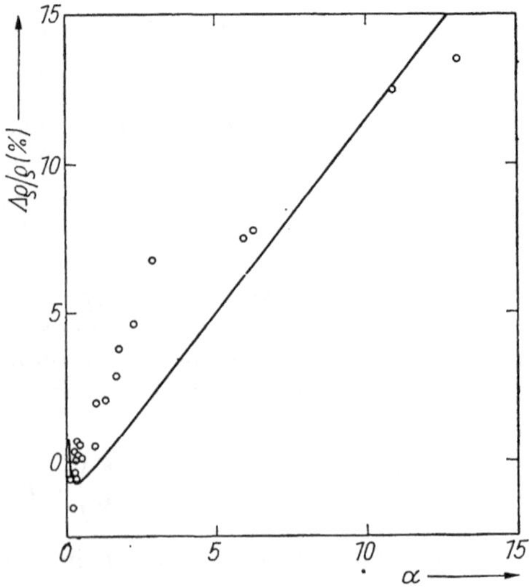

Fig. 2

Fig. 1. Spontaneous resistivity anisotropy of Fe-based alloys at 4.2 K depending on $\eta = \varrho_\uparrow/\varrho_\downarrow$. The full curve represents the theoretical expression of (6) with the fitted parameters γ_1, γ_2, and μ

Fig. 2. Spontaneous resistivity anisotropy of Ni-based alloys at 4.2 K depending on $\alpha = \varrho_\downarrow/\varrho_\uparrow$. The full curve represents the theoretical expression of (7) with the fitted parameters β and γ

$\times\ 10^{-4}$, $\gamma_2 = -6.22 \times 10^{-4}$, and $\mu = 8.18$. The full curve in Fig. 1 exhibits the theoretical expression (6) with the fitted parameters.

Based on the approach of Campbell et al. [2] Ross et al. derived an improved theoretical expression for the spontaneous resistivity anisotropy [6]. For low temperatures they find

$$\frac{\Delta\varrho}{\varrho} = \gamma(\alpha - \beta)\left[(1 - \gamma) - \frac{1 - \beta\gamma}{\alpha}\right], \tag{7}$$

where β and γ are spin–orbit coupling constantes with $\beta \neq 1$ if the coupling strength is different in the two sub-bands. Using 21 data for corresponding values of $\Delta\varrho/\varrho$ and $\varrho_\downarrow/\varrho_\uparrow$ of Ni-based alloys measured at low temperatures [1, 7] and making a computer fit we obtained Fig. 2 with the parameters $\beta = 0.0586$ and $\gamma = 0.0130$.

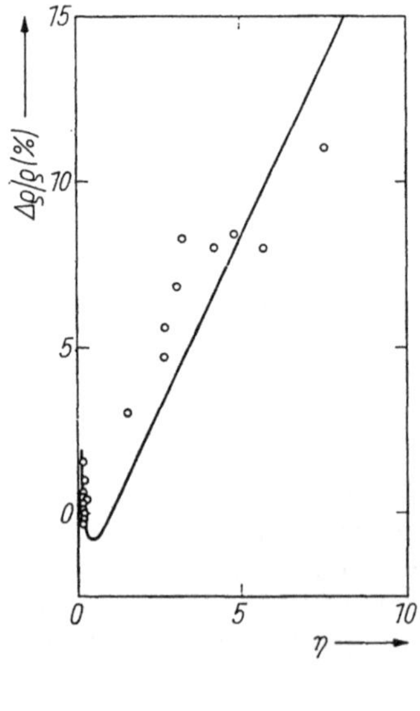

Fig. 3. Spontaneous resistivity anisotropy of Fe-based alloys at 4.2 K depending on $\eta = \varrho_\uparrow/\varrho_\downarrow$. The full curve represents the theoretical expression of (8) with the fitted parameters β and γ

Introducing $\eta = 1/\alpha = \varrho\uparrow/\varrho\downarrow$ and rewriting (7) one gets

$$\frac{\Delta\varrho}{\varrho} = \gamma(1 - \beta\eta)\,\frac{(1 - \gamma) - (1 - \beta\gamma)\,\eta}{\eta}. \tag{8}$$

We used the same 19 data plotted in Fig. 1 and fitted (8) for Fe-based alloys. Fig. 3 shows the resulting curve with the parameters $\beta = 6.300$ and $\gamma = 0.0035$. Relations (7) and (8) do not represent well the behaviour of Ni-based and Fe-based alloys. Obviously the standard deviations from the data are much larger than those for (3) and (6), cf. [5]. Moreover measurements indicate that there is no second change of sign for small α or η [1, 3].

4. Conclusions

The relation we derived for the spontaneous resistivity anisotropy within the framework of the two-current model represents the data of Ni-based alloys as well as the data of Fe-based alloys. The reason for the good agreement is mainly due to the fact that we used very general assumptions only. Actually, we did not specify the type of the spin-orbit interaction as well as the type of the non-conducting states into which the conduction electrons are scattered.

Acknowledgement

The authors are very indebted to Mr. G. Schubert for carrying out the numerical calculations.

References

[1] J. W. F. Dorleijn, Philips Res. Rep. **31**, 287 (1976).
[2] J. A. Campbell, A. Fert, and O. Jaoul, J. Phys. C **3**, S95 (1970).
[3] A. R. Miedema and J. W. F. Dorleijn, Philips tech. Rdsch. **35**, 57 (1975/76).
[4] J. Smit, Physica (Utrecht) **16**, 612 (1951).
[5] P. Muth and V. Christoph, J. Phys. F **11**, 2119 (1981).
[6] R. N. Ross, D. C. Price, and G. Williams, J. Magnetism magn. Mater. **10**, 59 (1979).
[7] O. Jaoul, J. A. Campbell, and A. Fert, J. Magnetism magn. Mater. **5**, 23 (1977).

(Received June 5, 1985)

phys. stat. sol. (b) 131, 775 (1985)

Subject classification: 13 and 14.3.4; 22.2.1

Sektion Physik der Humboldt-Universität zu Berlin[1])

New Aspects in the Theory of Impact Ionization

By

O. ZIEP, M. MOCKER, and R. KEIPER

Dedicated to Prof. Dr. Dr. h. c. Dr. E. h. P. GÖRLICH on the occasion of his 80th birthday

The field induced change of the electron states causes avalanche processes already below the usual ionization threshold. For the calculation of the impact ionization coefficient α_I the field influence of the corresponding microscopical rates has to be considered especially below the characteristic threshold energy due to higher occupation numbers. A comparison of the calculated values α_I with experimental data in GaAs raises essential questions concerning the validity of the used stationary transport models.

Die feldbedingte Zustandsänderung ruft den Ionisationseffekt schon unterhalb der eigentlichen Schwellenergie hervor. Bei der Berechnung des Stoßionisationskoeffizienten α_I ist der Feldeinfluß auf dessen mikroskopische Raten insbesondere unterhalb der charakteristischen Schwellenergie wegen der höheren Besetzungszahlen zu berücksichtigen. Der Vergleich der berechneten Werte α_I mit experimentellen Daten für GaAs wirft prinzipielle Fragen der Gültigkeit der verwendeten stationären Transportmodelle auf.

1. Introduction

Carrier multiplication due to impact ionization as an essential mechanism in the operation of various semiconductor devices has been investigated extensively by experiment and theory [1 to 7]. One of the fundamental parameters of charge multiplication phenomena is the ionization coefficient α_I determined with respect to the experimentally measurable multiplication factor. The ionization coefficient α_I being a characteristic reciprocal distance travelled by the carrier between ionization events is given by [2]

$$\alpha_I = \frac{1}{j} \int_0^\infty \mathrm{d}W \, \nu_i(W) \, f(W) \,, \tag{1}$$

where the current density j, an isotropic scattering rate $\nu_i(W)$ due to impact processes, and the energy distribution function $f(W)$ involving the density of states at a given carrier energy W must be known.

In order to determine the coefficient α_I the microscopical rate $\nu_i(W)$ is approximated in a number of papers using the following models. Supposing the conservation of energy and wave vector of the two interacting carriers, the energy of a carrier in one of the conduction bands must exceed a characteristic threshold value W_t in order to be able to impact ionize [3]. Within a simple model applied in [4 to 7] a high rate ν_i which is assumed above the threshold overwhelms the electron–phonon scattering rate ν_{ep} by orders of magnitude. Carriers reaching the threshold energy W_t cause im-

[1]) Invalidenstr. 110, PSF 1297, DDR-1040 Berlin, GDR.

pact processes and give up all their energy. Often the ionization coefficient α_I is expressed by the threshold energy W_t, the optical phonon energy, and the mean free paths for ionization and phonon scattering as adjustable parameters. In recent work [8, 6, 9] the ionization coefficient α_I is calculated numerically with an empirical expression for the scattering rate ν_i used by Keldysh [10],

$$\nu_i(W) = \nu_{ep}(W_t)\, p\, \frac{(W - W_t)^2}{W_t^2}\, \mathrm{H}(W - W_t)\,, \tag{2}$$

where H denotes the unit step function. By means of a Monte-Carlo simulation for the carrier motion in the whole Brillouin zone experimental values of α_I have been explained choosing appropriate values of the dimensionless parameter p. However, as has been mentioned in [9], a more detailed information about the scattering rate ν_i is needed in order to get a microscopical foundation of the theory.

The present paper aims at calculating numerically the ionization coefficient α_I for GaAs using the definition (1). The electric field causes impact ionization below the usual threshold energy [13, 14]. In difference to similar calculations a field-dependent scattering rate ν_i for the first-order impact process is employed below the theshold W_t whereas a field-independent behaviour of ν_i is used above W_t [11, 12]. Our treatment works without adjustable parameters. In the present work quantitative results for α_I due to the field influence on ν_i are calculated.

The scattering rate $\nu_i(W)$ is discussed in Section 2 for carrier energies W above W_t as well as below W_t where a competitive contribution is obtained. The transport model is considered in Section 3 where also the final results for the ionization coefficient α_I are obtained. The stationary energy distribution function $f(W)$ is calculated for GaAs by means of a Monte-Carlo simulation without inclusion of ionization effects.

The coupling parameters for the electron–phonon interaction are taken from [15]. The simulation procedure for a homogeneous electric field $F = 8 \times 10^5$ V/cm includes intervalley transitions and scattering via acoustic and polar-optical phonons in the lowest conduction band. Reaching higher extrema of the lowest conduction band the carrier is retained and starts drifting at the band edge.

2. The Impact Ionization Rate

For the description of transport we use the field-dependent form of the collision term in the Boltzmann equation discussed in [16, 17]. Due to the presence of a high electric field strong modifications are calculated for the scattering rates near threshold energies of several interaction mechanisms. In the impact ionization process wave vector conservation is only expected for components $\boldsymbol{k}_\perp$ transverse to the electric field $\boldsymbol{F}$ oriented e.g. in z-direction. For an electric field constant in space and time the envelopes $\varphi_{\mu E}(z)$ of the carrier states μ, $\boldsymbol{k}_\perp$, E are Airy functions,

$$\varphi_{\mu E}(z) = (\pi\hbar\theta_\mu l_\mu)^{-1/2}\, \mathrm{Ai}\left(\mp \frac{z}{l_\mu} - \frac{E}{\hbar\theta_\mu}\right), \tag{3}$$

where the upper (lower) sign refers to conduction (valence) band states ($\mu = \mathrm{c}, \mathrm{v}$), respectively, and $\theta_\mu = (e^2F^2/2m_\mu\hbar)^{1/3}$, $l_\mu = (\hbar\theta_\mu/eF)\,\mu$; $eF > 0$. Overlap integrals of the Bloch modulating functions are taken $\boldsymbol{k}$-independent and are approximated by unity. The total energies W of the states (3) are given by $W = E + E_c(\boldsymbol{k}_\perp)$ and $W = -E - E_G - E_v(\boldsymbol{k}_\perp)$, respectively, with the effective masses m_μ, the gap energy E_G, and $E_\mu(k) = \hbar^2k^2/2m_\mu$.

The stationary transport problem is formulated in terms of the frequency-dependent correlation functions [17]

$$G_{\mu R}^{\gtrless}(\boldsymbol{k}_\perp, \Delta z, \omega) = 2\pi \varphi_{\mu E_\mu}\left(R_z + \frac{\Delta z}{2}\right) \varphi_{\mu E_\mu}\left(R_z - \frac{\Delta z}{2}\right) \begin{Bmatrix} 1 - f_\mu(\boldsymbol{k}_\perp, E_\mu, \boldsymbol{R}) \\ f_\mu(\boldsymbol{k}_\perp, E_\mu, \boldsymbol{R}) \end{Bmatrix} \tag{4}$$

which are local in $\boldsymbol{R}$. The distribution functions $f_\mu(\boldsymbol{k}_\perp, E_\mu, \boldsymbol{R})$ depend on the energies $E_c = \hbar\omega - E_c(\boldsymbol{k}_\perp)$ and $E_v = -\hbar\omega - E_G - E_v(\boldsymbol{k}_\perp)$ for conduction and valence band states, respectively. The nonlocal properties of (4) are described by the microscopic variable Δz.

On the line of [16] the generalized collision term for impact processes enters the corresponding transport equation ((10) in [16]) as follows:

$$\begin{aligned} I(\boldsymbol{k}_\perp, \boldsymbol{R}) = & \int_{-\infty}^{\infty} \mathrm{d}\Delta z \int_{-\infty}^{\infty} \frac{\mathrm{d}\omega}{2\pi} \int_{-\infty}^{\infty} \frac{\mathrm{d}\omega'}{2\pi} \frac{1}{\hbar} \times \\ & \times \sum_{\boldsymbol{k}'_\perp} [V_{\mathrm{S}\boldsymbol{R}'}^{<}(\boldsymbol{k}_\perp - \boldsymbol{k}'_\perp, \Delta z, \omega)\, G_{\mathrm{c}\boldsymbol{R}'}^{>}(\boldsymbol{k}_\perp, \Delta z, \omega') \times \\ & \times G_{\mathrm{v}\boldsymbol{R}'}^{<}(\boldsymbol{k}'_\perp, -\Delta z, \omega' - \omega) - V_{\mathrm{S}\boldsymbol{R}'}^{>}(\boldsymbol{k}_\perp - \boldsymbol{k}'_\perp, \Delta z, \omega) \times \\ & \times G_{\mathrm{c}\boldsymbol{R}'}^{<}(\boldsymbol{k}_\perp, \Delta z, \omega')\, G_{\mathrm{v}\boldsymbol{R}'}^{>}(\boldsymbol{k}'_\perp, -\Delta z, \omega' - \omega)] \end{aligned} \tag{5}$$

with $\boldsymbol{R}' = (\boldsymbol{R}_\perp, R_z - \Delta z/2)$. The correlational parts of the screened Coulomb potential $V_{\mathrm{S}\boldsymbol{R}}^{\gtrless}(\boldsymbol{q}_\perp, \Delta z, \omega)$ are used in order to describe the spectrum of intraband excitations [18]. Neglecting the imaginary part and the $\boldsymbol{q}$-dependence of the longitudinal dielectric function $\varepsilon(\boldsymbol{q}, \omega)$ and approximating it by the high-frequency constant ε we get

$$\begin{aligned} V_{\mathrm{S}\boldsymbol{R}}^{\gtrless}(\boldsymbol{q}_\perp, \Delta z, \omega) = & \left(\frac{2\pi e^2}{q_\perp \varepsilon}\right)^2 \iint_{-\infty}^{\infty} \mathrm{d}z\, \mathrm{d}z' \int_{-\infty}^{\infty} \frac{\mathrm{d}\omega'}{2\pi} \mathrm{e}^{-q_\perp|z + (\Delta z/2)| - q_\perp|z' - (\Delta z/2)|} \times \\ & \times \sum_{\boldsymbol{k}_\perp} G_{\mathrm{c}\boldsymbol{R}''}^{\gtrless}(\boldsymbol{k}_\perp, z' - z, \omega')\, G_{\mathrm{c}\boldsymbol{R}''}^{\lessgtr}(\boldsymbol{k}_\perp - \boldsymbol{q}_\perp, z - z', \omega' - \omega)\,, \end{aligned} \tag{6}$$

where $\boldsymbol{R}'' = (\boldsymbol{R}_\perp, R_z - (z + z')/2)$. On the basis of expressions (4) to (6) and in analogy to the derivation of the electron–phonon scattering rate in [16] we find the field-dependent impact ionization rate in the form

$$\begin{aligned} \nu_i(\boldsymbol{k}_\perp, E_2, R_z) = & \frac{2\pi S}{\hbar} \sum_{\alpha\beta\boldsymbol{k}'_\perp\boldsymbol{q}_\perp} \iiint_{-\infty}^{\infty} \mathrm{d}E_1\, \mathrm{d}E_{1'}\, \mathrm{d}E_{2'}\, \delta(W_1^\alpha + W_2^\beta - W_{1'}^\alpha - W_{2'}^\beta) \times \\ & \times \left(\frac{2\pi e^2}{q_\perp \varepsilon}\right)^2 \int_{-\infty}^{\infty} \mathrm{d}z\, \mathrm{e}^{-q_\perp|z - R_z|}\, \varphi_{\mathrm{v}E_1}^\alpha(z)\, \varphi_{\mathrm{c}E_{1'}}^\alpha(z) \iint_{-\infty}^{\infty} \mathrm{d}z'\, \mathrm{d}z''\, \mathrm{e}^{-q_\perp|z' - z''|} \times \\ & \times \varphi_{\mathrm{v}E_1}^\alpha(z')\, \varphi_{\mathrm{c}E_{1'}}^\alpha(z')\, \varphi_{\mathrm{c}E_2}^\beta(z'')\, \varphi_{\mathrm{c}E_{2'}}^\beta(z'')\, \varphi_{\mathrm{c}E_{2'}}^\beta(R_z)\, [\varphi_{\mathrm{c}E_2}^\beta(R_z)]^{-1}\,. \end{aligned} \tag{7}$$

Formula (7) describes the rate of impact processes of the state 2 with participation of the empty states 1′, 2′ and the occupied state 1 (see Fig. 1) where the respective energies

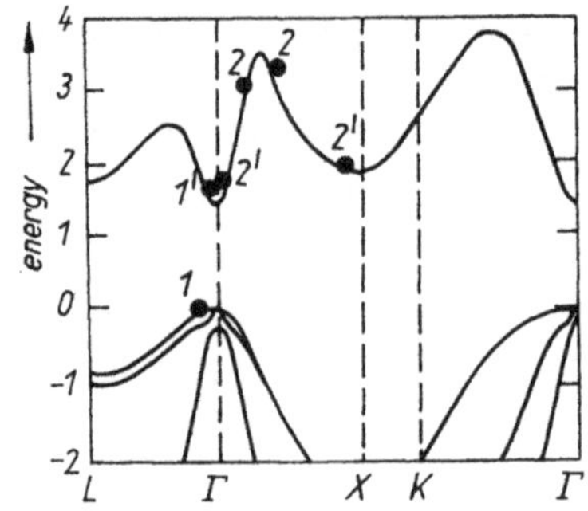

Fig. 1. Band structure for GaAs according to [19]. The carrier states suffering impact ionization in the lowest conduction band are sketched schematically. The interband transition 1–1′ is connected with intraband transitions 2–2′ in the Γ-valley as well as in the X-valleys (energy in eV)

are given by

$$W_1^\alpha = -E_G - \frac{\hbar^2 k_\perp'^2}{2m_v^\alpha} - E_1\,, \qquad W_{1'}^\alpha = \frac{\hbar^2(\boldsymbol{k}'_\perp + \boldsymbol{q}_\perp)^2}{2m_c^\alpha} + E_{1'}\,,$$

$$W_2^\beta = \frac{\hbar^2 k_\perp^2}{2m_c^\beta} + E_2 + \Delta_\beta\,, \qquad W_{2'}^\beta = \frac{\hbar^2(\boldsymbol{k}_\perp - \boldsymbol{q}_\perp)^2}{2m_c^\beta} + E_{2'} + \Delta_\beta$$

and the degeneracy factor is $S = 4$. The carrier states are characterized by valley indices α, β, corresponding band edge energies $\Delta_\alpha = 0$, Δ_β, and effective masses m_μ^α, m_μ^β entering the frequencies θ_μ^α, θ_μ^β in order to consider intra- ($\alpha = \beta$) and intervalley ($\alpha \neq \beta$) processes.

In the following we calculate the electron impact ionization rate for carrier energies below the characteristic threshold energy W_t.

Carrying out the $\boldsymbol{k}'_\perp$ summation in (7) one can use the expression

$$\sum_{\boldsymbol{k}'_\perp} \delta(W_1^\alpha + W_2^\beta - W_{1'}^\alpha - W_{2'}^\beta) = \frac{m_c^\alpha}{4\pi^2\hbar^2(1+r)}\, H\left(\frac{2m_c^\alpha\,\Delta E}{\hbar^2} - \frac{q_\perp^2}{\gamma} + 2r'\boldsymbol{k}_\perp\boldsymbol{q}_\perp\right), \tag{8}$$

where

$$r = \frac{m_c^\alpha}{m_v^\beta}, \qquad r' = \frac{m_c^\alpha}{m_c^\beta}, \qquad \gamma = \left[r' + \frac{r}{1+r}\right]^{1}, \qquad \Delta E = E_2 - E_1 - E_{2'} - E_{1'} - E_G\,.$$

For carriers propagating mainly in field direction ($k_\perp = 0$), the values of $q_\perp$ are limited by $q_{\perp 0} = \sqrt{2m_c^\alpha\,\Delta E\gamma}/\hbar$.

If field-induced tail states 1, 1′, and 2′ participate the usual conservation law for the wave vector components in field direction is violated and $\varphi_{\mu E}(z)$ gets the asymptotic form valid for $(-E \mp eFz)/\hbar\theta_\mu \gg 1$,

$$\varphi_{\mu E}^\alpha(z) = \frac{(m_\mu^\alpha)^{1/4} \exp\left[-\frac{2}{3}\left(\frac{-E \mp eFz}{\hbar\theta_\mu^\alpha}\right)^{3/2}\right]}{2^{3/4}\sqrt{\pi\hbar}\,(eFz \pm E)^{1/4}}\,. \tag{9}$$

For tail states 1 and 1′ the product $\varphi_{vE_1}^\alpha(z)\,\varphi_{cE_1'}^\alpha(z)$ in (7) is localized around the point z^* at a length Δz^* defined by

$$z^* = \frac{E_1 - rE_{1'}}{(1+r)\,eF}\,; \qquad -\frac{E_1}{eF} < z^* < \frac{E_{1'}}{eF}, \qquad \Delta z^* = \left(\frac{|E_1 + E_{1'}|}{\hbar\theta_c^\alpha(1+r)}\right)^{1/4} \sqrt{\frac{r}{1+r}}\, l_c^\alpha \tag{10}$$

and decreases exponentially outside this region.

For tail states the quantities ΔE in $q_{\perp 0}$ and $E_1 + E_{1'}$ in Δz^* are of the order of $\hbar\theta_\mu^\alpha$. Correspondingly, for a flat valence band ($r \ll 1$) the values in the exponent in (7) are small because of $q_{\perp 0}\,\Delta z^* \ll 1$. Compared with the product $\varphi_{\mathrm{v}E_1}^\alpha(z)\,\varphi_{\mathrm{c}E_1}^\alpha(z)$ the z-, z'-dependence of $\mathrm{e}^{-q_\perp|R_z - z|}$ and $\mathrm{e}^{-q_\perp|z' - z''|}$ in (7) is weak and can be taken out of the integrals at $z = z' = z^*$, respectively. With (8) to (10) expression (7) changes over into

$$\nu_\mathrm{i}(0, E_2, R_z) = \frac{8m_\mathrm{c}^\alpha e^4}{\hbar^3\varepsilon^2(1 + r)} \sum_{\alpha\beta} \iiint_{-\infty}^{\infty} \mathrm{d}E_1\,\mathrm{d}E_{1'}\,\mathrm{d}E_{2'} \int_0^{q_{\perp 0}} \frac{\mathrm{d}q_\perp}{q_\perp} H(\Delta E)\,\mathrm{e}^{-q_\perp|R_z - z^*|} \times$$
$$\times \left[\int_{-\infty}^{\infty} \mathrm{d}z\,\varphi_{\mathrm{v}E_1}^\alpha(z)\,\varphi_{\mathrm{c}E_1'}^\alpha(z)\right]^2 \int_{-\infty}^{\infty} \mathrm{d}z'\,\varphi_{\mathrm{c}E_2}^\beta(z' + z^*)\,\varphi_{\mathrm{c}E_2'}^\beta(z' + z^*) \times$$
$$\times\ \mathrm{e}^{-q_\perp|z'|}\,\varphi_{\mathrm{c}E_2'}^\beta(R_z)\,[\varphi_{\mathrm{c}E_2}^\beta(R_z)]^{-1}\,. \tag{11}$$

The integrals in (11) with Airy functions take the standard form

$$\int_{-\infty}^{\infty} \mathrm{d}z\,\varphi_{\mathrm{v}E_1}^\alpha(z)\,\varphi_{\mathrm{v}E_1'}^\alpha(z) = \frac{1}{\sqrt{\pi}\,\hbar\theta_\mathrm{c}^\alpha(1 + r)^{1/3}}\,\mathrm{Ai}\left(-\frac{E_1 + E_{1'}}{\hbar\theta_\mathrm{c}^\alpha(1 + r)^{1/3}}\right) \tag{12}$$

and

$$\int_{-\infty}^{\infty} \mathrm{d}z'\,\varphi_{\mathrm{c}E_2}^\beta(z' + z^*)\,\varphi_{\mathrm{c}E_{2'}}^\beta(z' + z^*)\,\mathrm{e}^{-q_\perp|z'|} = \frac{q_\perp}{2\pi^{3/2}\hbar\theta_\mathrm{c}^\beta\sqrt{l_\mathrm{c}^\beta}} \int_{-\infty}^{\infty} \frac{\mathrm{d}q_z}{q^2\sqrt{|q_z|}} \times$$
$$\times \exp\left\{i\left[\frac{(q_z l_\mathrm{c}^\beta)^3}{12} - \frac{(E_2 + E_{2'} + 2eFz^*)\,q_z}{2eF} - \frac{m_\mathrm{c}^\beta(E_2 - E_{2'})^2}{2\hbar^2 eFq_z}\right] + \frac{i\pi}{4}\,\mathrm{sign}\,q_z\right\}. \tag{13}$$

In the characteristic energy range $E_2 \approx E_\mathrm{G} \gg \hbar\theta_\mathrm{c}^\beta$ the expression for the scattering rate ν_i resulting from (11) to (13) shows strong oscillations in the kernel of the integral. Regarding its $E_{2'}$ dependence the corresponding exponent has a stationary point at $E_{2'0} = E_2 - (\hbar^2 q_z^2)/2m_\mathrm{c}^\beta$. With respect to the q_z dependence at $E_2 = E_{2'0}$ one finds stationary points at $q_{z0} = \pm(1/\hbar)\,[2m_\mathrm{c}^\beta(E_2 - eFz^*)]^{1/2}$. For the characteristic values $E_{2'0}$ and q_{z0} the quantities ΔE and q^2 are given by $\Delta E = E_2 + eFz^* - E_1 - E_{1'} - E_\mathrm{G}$ and $q^2 = q_\perp^2 + (2m_\mathrm{c}^\beta/\hbar^2)\,(E_2 + eFz^*)$. In good approximation we can neglect the $q_\perp^2$ dependence of (13) due to the relations $q_{\perp 0}^2 = 2m_\mathrm{c}^\beta\,\Delta E\gamma/\hbar^2 \ll q_{z0}^2$, $|\Delta E| \ll E_\mathrm{G}$, and $\gamma \approx 1$ in applying the saddle-point method.

Considering the asymptotic form of the wave function

$$\varphi_{\mathrm{c}E}^\beta(R_z) = \frac{(m_\mathrm{c}^\beta)^{1/4}\exp\left\{\pm i\left[\frac{2}{3}\left(\frac{E + eFR_z}{\hbar\theta_\mathrm{c}^\beta}\right)^{3/2} + \frac{\pi}{4}\right]\right\}}{2^{3/4}\sqrt{\pi\hbar}\,(E + eFR_z)^{1/4}} \tag{14}$$

for the impact-ionizing carrier state 2 where E_2 is comparable with E_G, we find the scattering rate

$$\nu_\mathrm{i}^{\alpha\beta}(W) = \frac{8m_\mathrm{c}^\alpha e^4 v_{\alpha\beta} w^{1/4}}{\pi^{3/2}\varepsilon^2\hbar^3(1 + r)^{4/3}} \int_0^1 \frac{\mathrm{d}t}{t} \int_0^\infty \mathrm{d}s\,H(s + \Delta w)\,(w + st)^{-5/4}\,\mathrm{Ai}\,(st) \times$$
$$\times\ \mathrm{Ai}^2(\gamma' s(1 - t))\cos\left[\frac{2}{3}\,(w + st)^{3/2} - \frac{2}{3}\,w^{3/2}\right]\left[1 - \exp\left(-st\sqrt{\gamma(s + \Delta w)}\right)\right], \tag{15}$$

where the locally defined energy $W = E_2 + eFR_z$ and the following denotations have been used:

$$w = \frac{W}{\hbar\theta_c^\beta} \gg 1\,, \qquad \Delta w = \frac{W - E_G}{\hbar\theta_c^\beta} \gg 1\,, \qquad \gamma' = \left(\frac{r'}{1+r}\right)^{1/3},$$
$$s = \frac{eF(z^* - R_z) - E_1 - E_1'}{\hbar\theta_c^\beta}\,; \qquad t = \frac{eF(z^* - R_z)}{eF(z^* - R_z) - E_1 - E_1'}\,. \tag{16}$$

Depending on whether the interband transitions 1, 1′ and the intraband transitions 2, 2′ take place near a definite symmetric point in $\boldsymbol{k}$-space ($\alpha = \beta$) or in non-equivalent valleys ($\alpha \neq \beta$), one has to choice a characteristic multiplicity $v_{\alpha\beta}$ of the process, i.e. the parameter gets $\gamma = (1 + r)/(1 + 2r)$ or $\gamma = [r' + r/(1 + r)]^{-1}$, respectively. As it can be seen from (15) the main contribution to the field-induced tail of the scattering rate results from local impact processes near the point R_z. Above the threshold energy W_t in accordance to [16] a weak field dependence of the rate (7) is expected as far as $(W - W_t)/\hbar\theta_c^\beta \gg 1$. This can be seen by the following discussion. With the representation $(1/2q_\perp)\,\mathrm{e}^{-q_\perp|z|} = \int_{-\infty}^{\infty} (\mathrm{d}q_z/2\pi)\,\mathrm{e}^{-iq_z z}\,q^{-2}$ the overlap integrals

$$B_{\mu\mu'}^\alpha(q_z, E, E') = \int_{-\infty}^{\infty} \mathrm{d}z\,\mathrm{e}^{-iq_z z}\,\varphi_{\mu E}^\alpha(z)\,\varphi_{\mu' E'}^\alpha(z) \tag{17}$$

appear in (7). For extended states 1, 1′, and 2′ with high energies $E \gg \hbar\theta_c^\beta$ the respective z integrations can be carried out by means of the saddle-point method. As a result the wave functions $\varphi_{\mu E}(z)$ enter the expression for the rate ν_i at points of the stationary phase $z_{11'}$ and $z_{22'} = R_z$ in the integrals (17). Substituting the energies E_1, E_2, $E_{2'}$ and $E_{1'}$ in (7) by locally defined wave vectors

$$k_{z1} = \frac{1}{\hbar}\,[2m_v^\alpha(E_1 - eFz_{11'})]^{1/2}\,, \qquad k_{z1'} = \frac{1}{\hbar}\,[2m_c^\alpha(E_{1'} + eFz_{11'})]^{1/2}\,,$$

$$k_{z2} = \frac{1}{\hbar}\,[2m_c^\beta(E_z + eFR_z)]^{1/2} \quad \text{and} \quad k_{z2'} = \frac{1}{\hbar}\,[2m_c^\beta(E_{2'} + eFR_z)]^{1/2}$$

one obtains the field-independent standard expression

$$\nu_i^{\alpha\beta}(\boldsymbol{k}_2) = \sum_{\boldsymbol{k}_1\boldsymbol{k}_2\boldsymbol{k}_{1'}\boldsymbol{k}_{2'}} \frac{2\pi}{\hbar}\left[\frac{4\pi e^2}{(\boldsymbol{k}_1 - \boldsymbol{k}_{1'})^2\,\varepsilon}\right]^2 \delta_{\boldsymbol{k}_1+\boldsymbol{k}_2,\boldsymbol{k}_{1'}+\boldsymbol{k}_{2'}} \times$$
$$\times\,\delta\big(E_c^\beta(\boldsymbol{k}_2) - E_c^\alpha(\boldsymbol{k}_{1'}) - E_v^\alpha(\boldsymbol{k}_1) - E_c^\beta(\boldsymbol{k}_{2'}) - E_G\big) \tag{18}$$

investigated in a number of papers [10, 11, 12, 20]. For energies above the threshold and small values of $[E_c(\boldsymbol{k}_2) - W_t]/W_t \ll 1$ the approximate expression

$$\nu_i^{\alpha\alpha}(\boldsymbol{k}_2) = \frac{6m_c^\alpha e^4\,(1 + 2r)^{5/2}\,E_G^2}{\pi^2\varepsilon^2\hbar^3(1 + r)^4\,W_t^2}\left(1 - \frac{W_t}{E_c^\alpha(\boldsymbol{k}_2)}\right)^2 \tag{19}$$

holds where the high-energetic state 2 is characterized by $E_c^\alpha(\boldsymbol{k}_2) = \hbar^2k_2^2/(2m_t^\alpha)$ with the parameter m_t^α and the threshold energy is given by

$$W_t = \frac{1 + 2r}{1 + r - (1 + 2r)\,(1 - r'')}\,E_G \quad \text{with} \quad r'' = \frac{m_c^\alpha}{m_t^\alpha}\,.$$

For a parabolic band structure the common value $W_t = (1 + 2r)/(1 + r)\,E_G$ is valid.

The approximation (19) corresponds qualitatively to the results of [12, 13] for this case and differs e.g. from the expression obtained in [13] only by the factor

$$\frac{6}{\pi^2}\left(\frac{1+2r}{1+r}\right)^2\left(\frac{W_t}{W}\right)^2 \approx 1 .$$

3. Calculation of the Ionization Coefficient and Discussion

Employing conservation of energy and momentum the electron-initiated threshold state in the lowest conduction band of GaAs leading to $W_t = 1.7$ eV is attributed to the ⟨110⟩ direction [21]. In GaAs only the central valley and the X-valleys extend to energies beyond the direct band gap what rules out impact ionization of carriers in the L-valleys. The consideration of tail states enables a variety of impact processes in the Γ- and X-valleys due to the breakdown of $\boldsymbol{k}$-conservation in field direction. According to (15) the respective scattering rates $\nu_i^{\Gamma\Gamma}$ and $\nu_i^{\Gamma X}$ are calculated with the parameters $m_c^\Gamma = 0.066 m_0$, $m_v^\Gamma = 0.46 m_0$, $m_c^X = 0.43 m_0$, $E_G = 1.4$ eV, $\varepsilon = 10.9$, $v_{\Gamma\Gamma} = 1$ and $v_{\Gamma X} = 3$ [22]. For energies $W > W_t$ the expression (19) is used with r'' adjusted in order to obtain $W_t = 1.7$ eV. The corresponding results are represented in Fig. 2 for three values of electric field strength. At energies W below E_G (i.e. if $-\Delta w \gg 1$) the essential contribution to the integral (15) arises from $t \ll 1$ $(s \approx -\Delta w)$ and its field dependence is mainly given by

$$\exp\left[-\frac{4}{3}\left(\frac{2m_v^\alpha}{m_c^\beta(m_c^\alpha + m_v^\alpha)}\right)^{1/2}\frac{m_c^\alpha(E_G - W)^{3/2}}{\hbar eF}\right].$$

The calculated field-dependent values $\nu_i^{\alpha\beta}(W)$ are of the order of 10^{10} to 10^{11} s^{-1} in a wide energy range below and above $W \approx E_G$ in distinction to the behaviour of the electron–phonon scattering rate near the phonon emission threshold [16]. This behaviour results from the participation of the three field-induced tail states 1, 1′, and 2′ with characteristic energies $\hbar\theta_c^\Gamma = 0.154$ eV, $\hbar\theta_c^X = 0.083$ eV, $\hbar\theta_v^\Gamma$ (heavy holes) $= 0.081$ eV and $\hbar\theta_v^\Gamma$ (light holes) $= 0.14$ eV, e.g. at $F = 8 \times 10^5$ V/cm in qualitative agreement with earlier calculations [13]. It is interesting to note that values of the impact ionization rate $\nu_i(W)$ at $W > W_t$ employed in [6, 9] according to (2) are more than two orders of magnitude higher than our result. The difference arises from the highfrequency dielectric constant omitted by some authors in estimating ν_i. Our calculated values ν_i are even smaller than the electron–phonon scattering rates having typical values of 10^{14} s^{-1}.

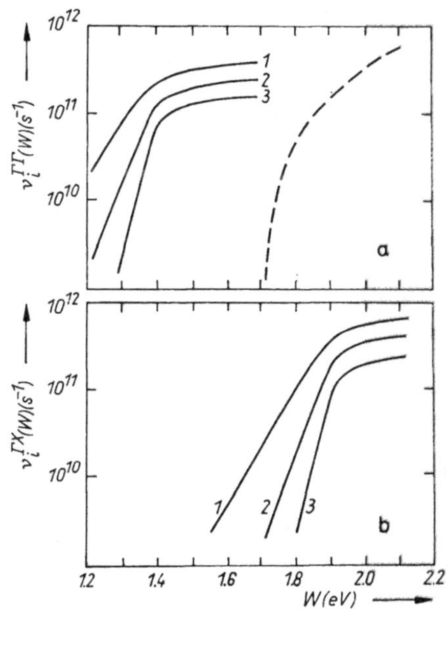

Fig. 2. Scattering rates a) $\nu_i^{\Gamma\Gamma}$ and b) $\nu_i^{\Gamma X}$ in GaAs according to (15) in dependence on the energy W measured from the bottom of the conduction band for various electric field strength (1) $F = 8 \times 10^5$, (2) 4×10^5, and (3) 2×10^5 V/cm. The dashed line is calculated from the field-independent expression (19) assuming a threshold energy $W_t = 1.7$ eV

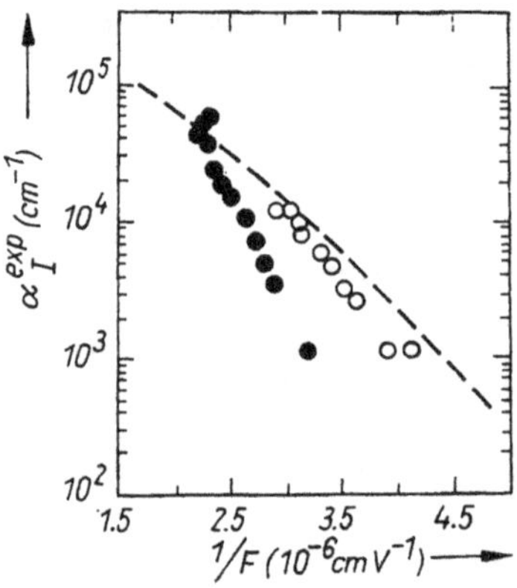

Fig. 3. Experimental values of the electron impact-ionization coefficient α_I^{exp} in GaAs in dependence on the reciprocal field strength: ○ [23] and ● [24], – – – [25] for fields oriented in ⟨100⟩ direction

In calculating α_I we regard an electric field $F = 8 \times 10^5$ V/cm in ⟨100⟩ direction being in the upper range for available data near breakdown. Some of the experimental data on the electron-initiated ionization coefficient α_I^{exp} for GaAs summarizes Fig. 3. The experimental verification of α_I depending on the ratio of electron and hole ionization coefficients is rather sensitive to the device structure which explains different values of α_I^{exp} in different papers. The regarded high fields are attainable in space charge layers of p–n junctions. The spatial variation of the field is neglected in the present calculations. We assume that a steady state has been reached in which the distribution function depends on the position only through the electric field. It is mentioned that the field dependence of the scattering rate ν_i is similar to that of the tunnel generation rate $g_T(F)$ [26],

$$g_T(F) = \frac{\sqrt{m_c^\Gamma}\, e^2F^2 \exp\left(-\dfrac{\pi}{\sqrt{1+r}}\, \dfrac{\sqrt{m_c^\Gamma}\, E_G^{3/2}}{eF\hbar}\right)}{2\pi^3\hbar^2 E_G^{1/2}\sqrt{1+r}}, \tag{20}$$

having the value 1.2×10^{22} cm^{-3} s^{-1} in our case. In order to neglect interband tunneling we are confronted with not too small structures and a sufficiently high potential drop across the junction.

For the calculation of the ionization coefficient α_I the energy distribution function $f(W)$ has to be determined including a more realistic band structure computed by adjustable functions. In the central valley of the lowest conduction band the expression

$$E(\boldsymbol{k}) = p_1 - \{p_2 + p_3[\cos(p_0k_x)\cos(p_0k_y) + \cos(p_0k_x)\cos(p_0k_z) + \cos(p_0k_y)\cos(p_0k_z)]\}^{1/2}$$

resulting from a tight-binding calculation [27] is used. The adjustable parameters $p_0, \ldots, p_3$ are determined by energies and position of the two nearest maxima in ⟨111⟩ and ⟨100⟩ direction. The X- and L-valleys are modeled by anisotropic parabolic bands with effective masses $0.9m_0$, $0.19m_0$ for longitudinal and transverse components with respect to the ⟨100⟩ and ⟨111⟩ directions, respectively, and by the valley separation energies 0.52 and 0.36 eV with respect to the central valley. At energies 0.2 eV above the band edges the corresponding effective mass parameters $1.32m_0$, $0.66m_0$ are chosen in order to correspond approximately to the tops of the conduction band near the X- and L-valleys.

In spite of the simplifications connected with the regarded band structure model the results which we obtain are useful to discuss peculiarities of the energy distribution function. The advantage of this model is the convenient choice of the energy for the final state after the scattering event in difference to the computation of the band structure by a set of meshpoints where it is only possible to choose a characteristic energy interval [6]. For the applied Monte-Carlo procedure simulating the carrier motion in small time steps we must know the electron–phonon scattering rates far from symmetry points in $\boldsymbol{k}$-space. Due to a lack of additional information about coupling parameters for scattering processes at higher energies we have chosen values used in simulation at low fields [15]. The scattering rate is assumed to be isotropic and only energy-dependent. Whereas in the central valley scattering by acoustic and polar optical phonons as well as intervalley transitions are simulated, in the satellite valleys only acoustic phonon and intervalley transitions are considered. Due to relatively small values of ν_i the impact ionization as an additional scattering process is neglected what supresses slightly the high-energy tail of the distribution function $f(W)$. The maximum values of the scattering rates are of the order of 10^{14} s^{-1}. This treatment allows to draw principal conclusions on the ionization coefficient. The simulation keeps track of the carrier until it attains the energy maxima of the lowest conduction band and retains it to the band edges. The stationary energy distribution function $f^\alpha(W)$ for the valley α is determined as the ratio of the time spent by the carrier near the valley α in a definite energy interval, to the total simulation time. Thus, $\sum_\alpha f^\alpha(W)$ is normalized to unity. The calculated values of $f^\Gamma(W)$ and $f^X(W)$ are represented in Fig. 4 for an electric field $F = 8 \times 10^5$ V/cm applied in $\langle 100 \rangle$ direction for GaAs at $T = 300$ K. Whereas 58% of the carriers are in L-valleys and 39% remain in X′-valleys, only 3% are observed in the central valley. The mean energies with respect to the band edges $\langle W \rangle_\Gamma = 0.9$ eV, $\langle W \rangle_X = 0.7$ eV calculated according to $\langle W \rangle_\alpha = \int f^\alpha(W) \times W \, dW / \int f^\alpha(W) \, dW$ for the valley α as well as the obtained drift velocity $v_D = 7.3 \times 10^6$ cm/s are in good agreement with results of [6, 9]. With the calculated values of the distribution function $f^\alpha(W)$ and the scattering rates $\nu_i(W)$ we obtain from relation (1) for the ionization coefficient of GaAs the value $\alpha_I^{theor} = 3.5 \times 10^3$ cm^{-1}. The relative amount of impact processes occurring in the central valley is only 10% of α_I^{theor} and that with both energy and momentum conserving processes is 40% of α_I^{theor} or 1.5×10^3 cm^{-1} according to (19).

Thus, the field-induced contribution to the impact ionization rate $\nu_i(W)$ dominates in calculating α_I^{theor} due to higher occupation numbers at smaller energies. However, compared with the experimental value $\alpha_I^{exp} \gtrapprox 10^5$ cm^{-1} our theoretical result is remarkable smaller. Whereas experimental values α_I^{exp} were reproduced using a similar transport model but several adjustable parameters p in (2) [6, 9], a confirmation is not possible working with the correct expressions (15) and (19). Analyzing the reason for the observed discrepancy we note that processes of higher order can increase the scattering rate ν_i. Furthermore, a more accurate model of band structure and scattering rates with enlarged densities of states would lead to a weak suppression of $f(W)$ at higher energies which strengthens the obtained discrepancies. Otherwise, uncertainties of the description of scattering processes at higher energies can influence the distribu-

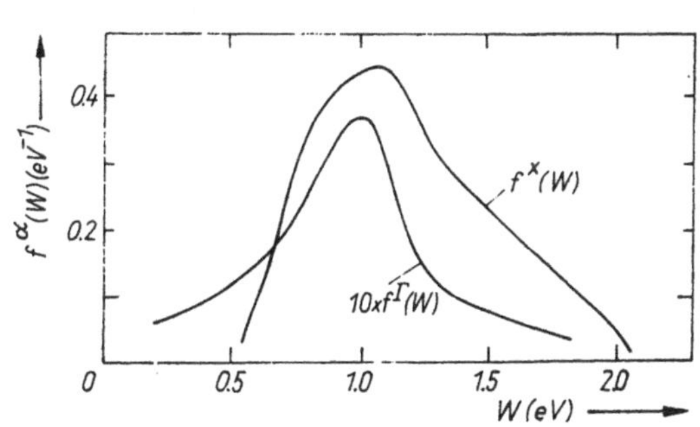

Fig. 4. Calculated energy distribution functions f^X and f^Γ for the lowest conduction band of GaAs at $F = 8 \times 10^5$ V/cm and $T = 300$ K in dependence on the energy W

tion function in several directions. Furthermore, essential changes can be expected due to instabilities of the distribution function, in analogy to the situation in SiO_2 [28] and the inclusion of higher bands and inhomogeneous fields. In the present calculation the carrier is retained after reaching the upper band extrema what simulates a steady state. Future work has especially to include several field profiles in order to investigate the possibility of an inflation of $f(W)$ at higher energies.

References

[1] A. G. Chynoweth, Semiconductors and Semimetals, Vol. 4, Ed. R. K. Willardson and A. C. Beer, Academic Press, New York 1968 (p. 263).
[2] G. A. Baraff, Phys. Rev. **133**, A26 (1964).
[3] C. L. Anderson and C. R. Crowell, Phys. Rev. B **5**, 2267 (1972).
[4] P. A. Wolff, Phys. Rev. **95**, 1415 (1954).
[5] W. Shockley, Solid State Electronics **2**, 35 (1961).
[6] H. Shichijo and K. Hess, Phys. Rev. B **23**, 4197 (1981).
[7] B. K. Ridley, J. Phys. C **16**, 3373, 4733 (1983).
[8] R. Chwang, C. W. Kao, and C. R. Crowell, Solid State Electronics **22**, 599 (1979).
[9] K. Brennan, and K. Hess, Phys. Rev. B **29**, 5581 (1984).
[10] L. V. Keldysh, Zh. eksper. teor. Fiz. **37**, 713 (1959); **48**, 1692 (1965).
[11] H. G. Holzhütter, Dissertation, Humboldt-Universität, Berlin 1976.
[12] D. J. Robbins, phys. stat. sol. (b) **97**, 9, 387 (1980); **98**, 11 (1980).
[13] M. Takeskhim, Phys. Rev. B **12**, 575 (1975).
[14] D. Hill, J. Phys. C **9**, 3527 (1976).
[15] M. A. Littlejohn, J. R. Hauser, and T. H. Glisson, J. appl. Phys. **48**, 4587 (1977).
[16] O. Ziep and R. Keiper, phys. stat. sol. (b) **128**, 779 (1985).
[17] R. Keiper and O. Ziep, phys. stat. sol. (b) **131**, K91 (1985).
[18] L. P. Kadanoff and G. Baym, Quantum Statistical Mechanics, W. A. Benjamin, Inc., New York 1962.
[19] M. L. Cohen and T. K. Bergstresser, Phys. Rev. **141**, 789 (1966).
[20] J. Geist and W. K. Gladden, Phys. Rev. B **27**, 4833 (1983).
[21] C. L. Anderson and C. R. Crowell, Phys. Rev. B **5**, 2267 (1972).
[22] T. S. Moss, G. J. Burrell, and B. Ellis, Semiconductor Opto-Electronics, Izd. Mir, Moscow 1976.
[23] G. E. Stillman and C. M. Wolfe, Semiconductors and Semimetals, Vol. 12, Ed. R. K. Willardson and A. C. Beer, Academic Press New York 1977 (p. 291).
[24] T. P. Pearsall, F. Capasso, R. E. Nahory, M. A. Pollack, and J. R. Chelikowsky, Solid State Electronics **21**, 297 (1978).
[25] G. E. Bulman, V. M. Robbins, K. F. Brennan, K. Hess, and G. E. Stillman, IEEE Electron Devices, Letters **EDL-4**, 181 (1983).
[26] C. B. Duke, Tunneling Phenomena in Solids, Ed. E. Burstein and S. Lundquist, New York 1969.
[27] D. J. Chadi and M. L. Cohen, phys. stat. sol. (b) **68**, 405 (1975).
[28] D. K. Ferry, J. appl. Phys. **50**, 1422 (1979).

(Received August 6, 1985)

Short Notes

phys. stat. sol. (b) 131, K97 (1985)
Subject classification: 1 and 7

Laboratorium voor Kristallografie en Studie van de Vaste Stof, Gent[1)]

X-Ray Diffraction by Vibrating Crystals

By

L. T. WILLE and J. VENNIK

Dedicated to Prof. Dr. Dr. h. c. Dr. E.h. P. GÖRLICH
on the occasion of his 80th birthday

In a recent note in this journal Gabrielyan and Aslanyan /1/ studied the diffraction of X-rays by an ultrasonically vibrating crystal. These authors calculated the time-dependent intensity of the diffracted beam by a numerical integration of Takagi's equations /2/. In the present work we point out how the problem can be recast in terms of known differential equations. This has the advantage of allowing an analytical study of approximate solutions, although the exact solution still has to be obtained by numerical means, upon which we also comment.

Following Gabrielyan and Aslanyan /1/ we consider the case of X-ray diffraction in the Laue geometry from a crystal vibrating in a thickness-shear mode. Since the ultrasound frequency Ω is negligible compared to that of the electromagnetic radiation we may consider scattering from static distorted planes and consequently Takagi's general analysis /2/ applies. In the case under study the equations for the electric field components D_o and D_h in the o and h direction, respectively, reduce to

$$\frac{d D_o(z)}{dz} = i \frac{\pi}{\Lambda} D_h(z) , \qquad D_o(0) = 1 , \tag{1a}$$

$$\frac{d D_h(z)}{dz} = i \frac{\pi}{\Lambda} D_o(z) + \frac{2\pi i}{\Lambda}\left(\eta - \Lambda \frac{d\vec{h} \cdot \vec{u}}{dz}\right) D_h(z), \qquad D_h(0) = 0 . \tag{1b}$$

Here Λ is the extinction length which can be written as

$$\Lambda = \lambda \cos \Theta_B / C |\chi_h| , \tag{2}$$

where λ is the wavelength of the incoming radiation, Θ_B is the Bragg angle,

1) Krijgslaan 281/S1, B-9000 Gent, Belgium.

C is the polarisation factor, and χ_h is the Fourier component of the polarisability. In general the incident beam can have a dispersion $\Delta\Theta$ around the Bragg angle and then η is given by

$$\eta = \sin\Theta_B \Delta\Theta / C|\chi_h| \ . \tag{3}$$

However, in what follows we will assume that the radiation is incident precisely under the Bragg angle so that $\Delta\Theta$ and hence η are zero. When this is not the case angular averages have to be taken. Finally $\vec{h}$ is the reciprocal lattice vector which in the present case can be written in terms of the interplanar distance d as

$$\vec{h} = \left(\frac{1}{d}, 0, 0\right) \tag{4}$$

and the displacement $\vec{u}$, as a function of time τ, is given by

$$\vec{u} = \left(A_o \sin\frac{2\pi}{\lambda_s} z \sin\Omega\tau, 0, 0\right) , \tag{5}$$

where A_o is the vibration amplitude and λ_s is the ultrasound wavelength, which is related to the crystal thickness t by

$$\lambda_s = 2t \ . \tag{6}$$

The coupled set of differential equations (1) was solved numerically by Gabrielyan and Aslanyan /1/ using parameter values taken from the experiment. It was found that the diffracted beam intensity was modulated in phase with the absolute value of the ultrasound vibration, in perfect agreement with the experimental results.

It would be very pleasing from a theoretical point of view if more information of an analytic nature could be extracted from (1) and we will now show how this can be done. It is a straightforward matter to deduce a second order differential equation for $D_h(z)$:

$$D_h'' - \frac{2\pi i}{\Lambda}\left(\eta - \Lambda\frac{d\vec{h}\cdot\vec{u}}{dz}\right)D_h' + \left(\frac{\pi^2}{\Lambda^2} + 2\pi i\frac{d^2\vec{h}\cdot\vec{u}}{dz^2}\right)D_h = 0 \tag{7a}$$

with the initial conditions

$$D_h(0) = 0 , \qquad D_h'(0) = i\frac{\pi}{\Lambda} \ . \tag{7b}$$

Next, using (4) and (5) substituting

$$Y\left(\frac{\pi}{\lambda_s} z\right) \longrightarrow D_h(z) \tag{8}$$

we find

$$Y'' - 2i\frac{\lambda_s}{\Lambda}\left(\eta - \Lambda\frac{2\pi}{\lambda_s}\frac{A_o}{d}\cos 2z \sin\Omega\tau\right)Y' +$$

$$+\left(\left(\frac{\lambda_s}{\Lambda}\right)^2 - 8\pi i\frac{A_o}{d}\sin 2z \sin\Omega\tau\right)Y = 0 , \tag{9a}$$

$$Y(0) = 0 , \quad Y'(0) = i\frac{\lambda_s}{\Lambda} . \tag{9b}$$

This is a homogeneous second order differential equation with periodic coefficients, i.e. a Hill equation /3/. Using one further substitution

$$y(z) \longrightarrow Y(z) \exp\left[-i\frac{\lambda_s}{\Lambda}\left(\eta z - \Lambda\frac{\pi}{\lambda_s}\frac{A_o}{d}\sin 2z \sin\Omega\tau\right)\right] \tag{10}$$

we can reduce this equation to its standard form

$$y'' + \left[\left(\frac{\lambda_s}{\Lambda}\right)^2 - 4\pi i\frac{A_o}{d}\sin 2z \sin\Omega\tau + 4\left(\frac{\pi A_o}{d}\cos 2z \sin\Omega\tau\right)^2\right] y = 0 , \tag{11a}$$

$$y(0) = 0 , \quad y'(0) = i\frac{\lambda_s}{\Lambda} , \tag{11b}$$

where now we have put $\eta = 0$. The electric field intensity along the h direction is related to the solutions of (11) by

$$D_h(z) = y\left(\frac{\pi}{\lambda_s}z\right)\exp\left(-i\pi\frac{A_o}{d}\sin\left(\frac{2\pi}{\lambda_s}z\right)\sin\Omega\tau\right). \tag{12}$$

The intensity of the transmitted beam is

$$I_h = \frac{|D_h(t)|^2}{|D_o(0)|^2} = \left|y\left(\frac{\pi}{2}\right)\right|^2 . \tag{13}$$

Thus the problem is completely solved once we can determine the value of the solution of (11) in $z = \pi/2$.

A full investigation of (11) will be given elsewhere and we now merely outline how this will proceed. We rewrite (11) more concisely as

$$y'' + \left[\left(\frac{\lambda_s}{\Lambda}\right)^2 + \frac{1}{2}q^2 - 2iq\cos 2z - \frac{1}{2}q^2\cos 4z\right] y = 0 , \tag{14}$$

where

$$q = 2\pi\frac{A_o}{d}\sin\Omega\tau \tag{15}$$

and the initial conditions are

$$y(\pi/4) = 0 , \quad y'(\pi/4) = -i\frac{\lambda_s}{\Lambda} . \tag{16}$$

This is a Whittaker-Hill equation /3/ with complex coefficients and we need the solution $y(-\pi/4)$. In general this equation has to be solved numerically. Extensive investigations of several integration methods for finite Hill equations have been conducted by Wagenführer and Lang /4/. These authors conclude that the best results are obtained by a method based on a high order Taylor expansion. Moreover, they have also performed a full error analysis, so that a reliable numerical solution of (14) is feasible. On the other hand, several approximations to (14) can be made in order to keep the problem analytically tractable. When q is sufficiently small we can neglect terms in q^2 in (14) and we obtain a Mathieu equation with a complex parameter, about which a wealth of information is available /5/. Alternatively, when either $(\lambda_s/\Lambda)^2 \ll q^2$ or $(\lambda_s/\Lambda)^2 \gg q^2$ we can use a WKB approximation /6/. The method, as applied to the Mathieu equation /5/, has recently been improved in several ways /7 to 9/ and is now capable of providing accurate information on the eigenvalues and eigenfunctions. Using some additional approximations the WKB method is also applicable to the Whittaker-Hill equation (14).

In summary, we have shown that the diffraction of X-rays from an ultrasonically vibrating crystal is governed by the solutions of a Whittaker-Hill equation with complex parameters. The same is true of neutron or electron diffraction, the main difference being that the electric field intensities in (1) are replaced by wave functions /2/. Analytical and numerical work along the lines indicated in the text is in progress.

References

/1/ R.G. GABRIELYAN and H.A. ASLANYAN, phys. stat. sol. (b) 123, K97 (1984).
/2/ S. TAKAGI, J. Phys. Soc. Japan 26, 1239 (1969).
/3/ W. MAGNUS and S. WINKLER, Hill's Equation, Dover, New York 1979.
/4/ E. WAGENFÜHRER and H. LANG, Numer. Math. 32, 31 (1979).
/5/ N.W. MC LACHLAN, Theory and Application of Mathieu Functions, Dover, New York 1964.
/6/ N. FRÖMAN and P.O. FRÖMAN, JWKB Approximation: Contributions to the Theory, North-Holland, Amsterdam 1965.
/7/ N. FRÖMAN, J. Phys. A 12, 2355 (1979).
/8/ T.P. GROZDANOV and S. YU. SLAVYANOV, J. Phys. A 14, 1513 (1981).
/9/ J.N.L. CONNOR, T. UZER, R.A. MARCUS, and A.D. SMITH, J. chem. Phys. 80, 5095 (1984).

(Received August 8, 1985)

phys. stat. sol. (b) 131, K101 (1985)
Subject classification: 6 and 20.1; 14.4.1; 22.8.1

Institute of Solid State Physics, Bulgarian Academy of Sciences, Sofia[1)]

Isotopic Shift and Mode Interaction in $K(D_xH_{1-x})_2PO_4$ Crystals

By

E. ANACHKOVA and I. SAVATINOVA

Dedicated to Prof. Dr. Dr. h.c. Dr. E.h. P. GÖRLICH
on the occasion of his 80th birthday

Introduction It is well known that the stretching vibration ν_1 of PO_4^{3-} ions in crystals of the KH_2PO_4 (KDP) family changes its frequency with deuteration /1 to 3/. The isotope shift being about 35 cm^{-1} for DKDP /1/ and RbD_2PO_4 (DRDP) crystals /2, 3/ and only 10 cm^{-1} for CsD_2PO_4 (DCDP) /4/. This is one of the reasons for some authors to consider the spectrum of crystal internal vibrations starting not from PO_4^{3-} ions but from $H_2PO_4^-$ complexes. Such point of view can explain an isotopic shift of about 10 cm^{-1}, but it cannot answer the question why by deuteration one and the same vibration of the $H_2PO_4^-$ ion changes strongly its frequency in some crystals (KDP, RDP), while in other crystals the change is very small (see Table 1).

Savatinova et al. /5/ have pointed out that a resonant interaction can take place in DKDP crystals between the ν_1 (PO_4) mode and the neighbouring in-plane bending mode δ(OD). Ratajczak and Baran /6/ consider the spectrum in this region as mixed ν_1 (PO_4) and δ(OD) modes. Recently, Tanaka et al. /7/ and Martinez et al. /8/ who have observed not only a large shift of the $\nu_1(PO_4)$ vibration to lower frequencies but also a shift of the δ(OD) vibration in opposite direction, have suggested that a Fermi resonance between the weak δ(OD) and the strong ν_1 (PO_4) modes occurs.

Table 1
$\nu_1(PO_4)$ frequencies in some crystals of the KDP family (cm^{-1})

crystal	KDP	DKDP	RDP	DRDP	CDP	DCDP
ν_1	914	879	916	881	921	911
$\Delta\nu_1$	35		35		10	
ref.	this work,/1, 5/		/2,3/		/4/	

1) blvd. Lenin 72, Sofia 1184, Bulgaria.

The aim of the present note is a quantitative analysis of the spectral features of $K(D_xH_{1-x})_2PO_4$ crystals in the region of $\nu_1(PO_4)$ and $\delta(OD)$ vibrations by means of the model of two coupled oscillators using the Green's function formalism.

Experiment The Raman spectra were excited by the 488 nm radiation of an Ar^+ ion laser ILA 120 (Carl-Zeiss Jena) and analysed with a DFS-12 double monochromator equipped with gratings of 1200 grooves/mm. Light detection was done by the photon counting system PARC 1105 using an EMI 9658 B photomultiplier with magnetic lens and EMI-ZD-MK-III cooler. 90°-scattering geometry was used to register the A_1-type vibrations in the paraelectric phase.

Analysis of the spectra The polarized Raman spectra in the region of 800 to 1000 cm^{-1} of $K(D_xH_{1-x})_2PO_4$ crystals with x = 0, 0.80, and 0.96 are shown in Fig. 1. At higher degree of deuteration the ν_1 (PO_4) frequency strongly decreases. At the same time the broad band at 960 cm^{-1} increases in intensity and both band profiles have an asymmetric form. The observed spectra are analysed in terms of a system with two coupled oscillators. It is known (see e.g. /9/) that the phonon density of states is related to the imaginary part of the dielectric susceptibility $\chi(\omega)$ by the fluctuation-dissipation theorem, which for high frequencies $\omega_i \gg \delta_i$ is given by

$$I(\omega) \sim \mathrm{Im}\,\chi(\omega) \sim \mathrm{Im} \sum_{i,k}^{2} G_{ik} M_i M_k \;, \qquad (1)$$

where M_i is the mode strength,

$$G_{ii} = \tilde{G}_i = \frac{1}{G_i^{-1} + \gamma^2 G_k}\,, \qquad G_{ik} = -\frac{\gamma G_i G_k}{1 - \gamma^2 G_i G_k} \;; \qquad (2)$$

G_i is the one-particle Green function for uncoupled oscillators of the type

$$G_i(\omega) = \frac{1}{\omega_i - \omega - i\delta_i}\;, \qquad (3)$$

where ω_i is the frequency of the uncoupled oscillator, δ_i is half of the bandwidth, and γ is the mode coupling constant. Thus, the system of two coupled oscillators is described with seven parameters: ω_1, ω_2, δ_1, δ_2, M_1, M_2, and γ.

For ω_1 we have used the frequency of 914 cm^{-1} measured in the spectrum of the KH_2PO_4 crystal. The values of δ_1 and δ_2 have been roughly determined out of the coupling region. The other parameters are found by fitting.

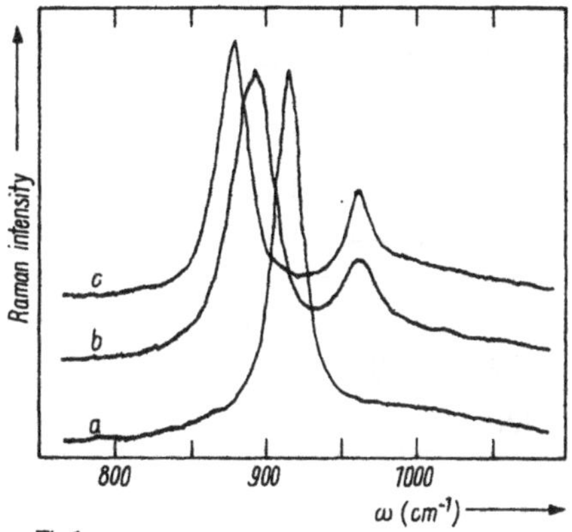

Fig.1

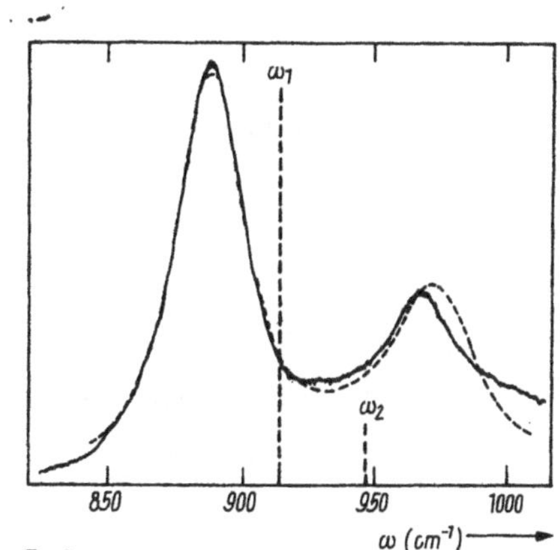

Fig.2

Fig. 1. Raman spectra of $K(D_xH_{1-x})_2PO_4$ crystals at 300 K: (a) x = 0, (b) 0.80, (c) 0.96. Scattering geometry x(zz)y

Fig. 2. Experimental and calculated (using (1)) Raman spectra of $K(D_xH_{1-x})_2PO_4$ crystals with x = 0.96. The dashed line shows the results of calculation

The results of calculation for the highest deuterated crystal are illustrated in Fig. 2, the obtained agreement between the experimental and calculated curves for other crystals is also so good. The used parameters are summarized in Table 2.

Discussion and comments Applying the model of two coupled oscillators allows us to explain qualitatively as well as quantitatively the main features observed in the spectra: With increasing D concentration the coupling constant γ also increases and the frequency ν_1 shifts to lower frequencies up to 879 cm^{-1} for highly deuterated crystals (x = 0.96). The calculated curves describe well the asymmetrical band wings in the region between the two coupled modes. One can see from Table 2 that a redistribution of intensity between the two bands takes place causing an increase of the δ(OD) band intensity at higher deuteration.

It should be noted that the model of only two coupled oscillators used by us is rather approximated: The spectral picture in this region is complicated by

Table 2

Parameters of the oscillators used in the calculation of the spectra in Fig. 1 and 2

x (D)	ω_1 (cm^{-1})	ω_2	δ_1	δ_2	γ	M_1 (arb. units)	M_2
0.0	914	-	12.5	-	-	17	-
0.80	914	950	16	22	35	22	0.5
0.96	914	947	14.5	23	41	26	1.5

the presence of some weak bands with A_1 symmetry probably due to ν_3 (PO_4) vibrations. These bands appear in spite of the group-theoretical considerations /7, 8/. The usual isotopic shift due to different masses of H and D should also take place, but we suppose it cannot cause such a large shift of ν_1 as observed in DKDP and DRDP crystal spectra. There the interaction between ν_1 (PO_4) and δ(OD) plays the major role.

The effect of ν_1 shift reported here allowed us to relate successfully spectroscopic data to D concentrations. The $\Delta\nu_1$ (D) dependence is not linear in the whole interval D = 0 to 100%, $\Delta\nu_1$ increasing more rapidly at higher degrees of deuteration. However, the crystals purchased from suppliers are usually with D = 90 to 98% and we turned our attention to this concentration region. The calibration line obtained (1.05 cm^{-1}/%) was proved to be very useful for practice.

This work is a part of a study of interactions between OH(OD) and PO_4 vibrations which, as shown by Davydov and Chisler /10/, are very important for understanding the complicated Raman spectra of KDP crystals.

The authors thank Dr. E.V. Chisler and Dr. V.Yu. Davydov for the acquaintance with some unpublished results and for the helpful discussions.

References

/1/ D.K. AGRAWAL and C.H. PERRY, Proc. 2nd Internat. Conf. Light Scattering in Solids, Flammarion Sciences, Paris 1971 (p. 429).

/2/ E.V. CHISLER, I.T. SAVATINOVA, and V.YU. DAVYDOV, Soviet Phys.- Optics and Spectroscopy 32, 544 (1972).

/3/ B.N. MAVRIN, KH.E. STERIN, A.V. BOBROV, L.N. RASHKOVICH, and A.V. MISHCHENKO, Fiz. tverd. Tela 15, 1682 (1973).

/4/ B. MARCHON and A. NOVAK, J. chem. Phys. 78, 2105 (1983).

/5/ I.T. SAVATINOVA, E.V. CHISLER, and M.L. SUBEVA, C.R. Bulg. Acad. Sci. 25, 1475 (1972).

/6/ H. RATAJCZAK and J. BARAN, J. Raman Spectroscopy 7, 54 (1978).

/7/ H. TANAKA, M. TOKUNAGA, and I. TATSUZAKI, Solid State Commun. 49, 153 (1984).

/8/ J.L. MARTINEZ, J.M. CALLEJA, and J.A. GONZALO, Solid State Commun. 52, 499 (1984).

/9/ A.S. DAVYDOV, Theory of Solid State, Izd. Nauka, Moscow 1976 (Chap. IX) (in Russian).

/10/ V.YU. DAVYDOV and E.V. CHISLER, Fiz. tverd. Tela 23, 3407 (1981).

(Received June 26, 1985)

phys. stat. sol. (b) 131, K107 (1985)
Subject classification: 4 and 20.1; 22.5.2

Institute of Physics, Technical University of Łódź[1)]

On the Nature of the Quadratic Electrooptic Effect in LiF[2)]

By
W. KUCHARCZYK

Dedicated to Prof. Dr. Dr. h.c. Dr. E.h. P. GÖRLICH
on the occasion of his 80th birthday

Nonlinear susceptibilities responsible for the primary electrooptic effect originate from two contributions. The first is purely an electronic term in which the electronic polarizability is directly modulated by the applied low-frequency field. The second, the ionic contribution, is a hybrid term originating from interaction of the modulating field with the lattice, which, in turn, changes the electronic polarizability. The latter contribution to the cubic susceptibility seems to be of special interest in connection with second-order Raman scattering and electric-field-induced first-order Raman scattering recently observed in alkali halide crystals and calculated on the basis of the shell model /1/.

In this note the ratio of ionic and purely electronic contributions to the primary quadratic electrooptic effect is discussed. The analogous ratio in the case of the linear electrooptic effect is usually called the Faust-Henry coefficient C_1.

It is known that nonlinear susceptibilities can be described to arise from the anharmonic motion of charge of individual chemical bonds /2, 3/. In crystals with inversion symmetry the second-order susceptibility exists on the bond level and vanishes on the macroscopic level. Taking into account the results based on the bond charge model /3/, the C_1 originating on the bond level from the variation of bond length in NaCl structure crystal can be expressed by

$$C_1 = \frac{-3qn^2(\varepsilon - n^2)\left[f_i(1 + k_s a/8) + 2.48\, f_c - 1.5\right]}{e_s(n^2+2)(n^2-1)(1.33\, f_i + 23.24\varrho\, f_c)}\ , \tag{1}$$

1) Wólczańska 219, 93-005 Łódź, Poland.

2) This work was carried out under the Research Project M.R.I.5.

where n is the refractive index, f_i and f_c are the ionicity and covalency parameters of the bond in the Phillips-Van Vechten theory /4/, ε denotes the relative permittivity, k_s is the Thomas-Fermi screening wave number, a is the lattice constant, $\rho = (r_\alpha - r_\beta)/(r_\alpha + r_\beta)$ and r_α is the distance between the α atom and the centre of gravity of the bond charge q. The Szigeti ionic charge e_s can be related to the ratio of the Phillips electronegativity difference C /4/ and the plasma energy $\hbar\omega_p$ due to the valence electrons by the relation /5/

$$e_s = Ce/\hbar\omega_p \quad , \tag{2}$$

where e is the electron charge. According to Levine /2/, the total bond charge can be calculated by

$$q = n_v e(1/n^2 + f_c/3) \quad , \tag{3}$$

where n_v is the number of valence electrons per bond. Treating the ionic charge and bond charge as the charges of two oscillators it is possible to predict, on the basis of the model of two anharmonic coupled oscillators /6/, the Faust-Henry coefficient generalized for the quadratic effect, as

$$C_q = C_1(C_1 + 2) \quad . \tag{4}$$

The experimental value of C_q can be obtained by

$$C_q = (R^p - R^e)/R^e \quad , \tag{5}$$

where R^p is the value of the primary quadratic electrooptic coefficient and R^e is the purely electronic term obtained from the cubic susceptibility $\chi^{(3)}$ measured in third harmonic generation or from the frequency-mixing experiments. The purely electronic term R^e is related to $\chi^{(3)}$ by

$$R^e = -12\,\chi^{(3)}/n^4 \quad . \tag{6}$$

For the coefficient $C_{q_{1111}}$ in LiF crystal the experimental data concerning the primary quadratic coefficient R^p_{1111} and cubic susceptibility $\chi^{(3)}_{1111}$ /7/ are known. The primary coefficient R^p_{1111} was obtained in /8/ from "absolute" measurements /9/ by takinginto account the constituents of the total effect, namely the piezooptic term, the change of optical path length induced by attraction of the electrodes and that introduced by electrostriction. The values of

r_α and r_β used in calculations of $C_{q_{1111}}$ were taken, as in the calculation of $\chi^{(3)}_{1111}$ for LiF performed in /10/ by the bond charge model, from X-ray data.

For LiF the calculated value of $C_{q_{1111}}$ from (1) and (4) and the experimental value obtained by (5) are +9.5 and +13.3, respectively. The agreement is fairly good. From the present results it can be stated that the quadratic electrooptic coefficient in LiF is essentially ionic. For the other rock-salt structure alkali halides, particularly for LiCl, LiBr, NaF, KF, and RbF, (1) and (4) also predict a predominantly ionic nature of the primary quadratic electrooptic effect.

The author is grateful to Dr. K. Sangwal for going through the manuscript.

References

/1/ B.G. DICK, phys. stat. sol. (b) 127, 131 (1985).
/2/ B.F. LEVINE, Phys. Rev. B 7, 2600 (1973).
/3/ C. SHIH and A. YARIV, J. Phys. C 15, 825 (1982).
/4/ J.A. VAN VECHTEN, Phys. Rev. 182, 891 (1969).
/5/ P. LAWAETZ, Phys. Rev. Letters 26, 697 (1971).
/6/ A.S. BARKER and R. LOUDON, Rev. mod. Phys. 44, 18 (1972).
/7/ M.D. LEVENSON, IEEE J. Quantum Electronics 10, 110 (1974).
/8/ L. BOHATÝ and S. HAUSSÜHL, Acta cryst. A33, 144 (1977).
/9/ S. HAUSSÜHL and G. WALDA, phys. stat. sol. (a) 5, K163 (1971).
/10/ V.G. TSIRELSON, O.V. KOROLKOVA, I.S. REZ, and R.P. OZEROV, phys. stat. sol. (b) 122, 599 (1984).

(Received April 3, 1985)

phys. stat. sol. (b) 131, K111 (1985)

Subject classification: 10.2 and 13.4; 22.5.2

Istituto di Fisica del Politecnico di Milano[1)] and
Gruppo Nazionale di Struttura della Materia del C.N.R., Milano

An Analysis of the Classical Arguments Concerning Forecasts for the Z_1 Centre Formation in Alkali Halides

By

L. BOSI and M. NIMIS

Dedicated to Prof. Dr. Dr. h.c. Dr. E.h. P. GÖRLICH
on the occasion of his 80th birthday

In explaining the reduction of Mg^{2+} to a monovalent ion in alkali halides, various authors /1, 2/ put forward the argument of its high second ionization potential (see Table 1), and it is indeed well-known that Z_1 centres are not always formed in alkali halides doped by divalent metals. These authors essentially note that alkaline-earth ions (Ca^{2+}, Sr^{2+}, Ba^{2+}) and rare earth ions (Eu^{2+}, Yb^{2+}, Sm^{2+}) do not trap electrons directly, and consequently form Z_1 centres due to their relatively low second ionization potential (< 15 eV). On the contrary, for a second class of ions (Mg^{2+}, Mn^{2+}, Cd^{2+}, Zn^{2+}, Pb^{2+}, Hg^{2+}), whose ionization potential is higher than 15 eV, formation of Z_1 centres is impossible.

To the best of our knowledge - as well as from their partial experimental check - this classification is to be attributed to Paus et al. /3, 4/. Furthermore, these authors attempted to give a criterion for the forecasting of Z_1 centre formation by adopting an oversimplified scheme based on the use of the Born-Haber cycle. On the basis of his scheme, Paus /5/ suggested the possibility of Z_1 centre formation in NaF:Mg attributing to this centre a band peaked at ≈ 3.32 eV obtained by F bleaching of X-ray coloured samples. On the other hand, the same band was recently observed also by Egranov and Nepomnyachikh /6/ who attributed it to the Mg_c^+ centre.

The aim of this note is to clarify the theoretical aspect of this intriguing problem, and to establish a criterion for the reduction of divalent to monovalent ions in "ideal" alkali halides, i.e. without presence of impurities - especially OH^- in the fluorides - which might drastically change the behaviour

1) piazza Leonardo da Vinci 32, 20133 Milano, Italy.

Table 1

Second ionization potentials (eV) of some divalent metals

Ca	11.87	Mg	15.04
Sr	11.03	Zn	17.96
Ba	10.00	Cd	16.91
Eu	11.25	Hg	18.76
Yb	12.17	Mn	15.64
Sm	11.07	Pb	15.03

of "real" samples. Our object is to distinguish between the "a priori" possibility of reduction of divalent to monovalent ions and what really happens during an experiment on such samples.

The energy of the electron at the divalent cation can be estimated by making suitable changes in the Born-Haber cycle as used in the past by many authors /7/ for the calculation of the exciton energy. It is worthwhile to summarize their calculations /7/.

The transition of an electron from the halogen ion to the alkali ion of the lattice immediately next to it takes place upon the absorption of light. Hilsch and Pohl found that the position of the first absorption band is remarkably well represented by the relation

$$h\nu = E - I + A_M \frac{e^2}{d} , \tag{1}$$

where E is the electron affinity of halogen ion, I is the work of ionization, A_M is the Madelung constant, d is the shortest distance between two ions in the lattice, and e is the charge of one ion.

Wolf and Herzfeld stressed that the above relationship is not correct and proposed the following scheme of calculation /7/. If one ion (the halogen ion) is neutralized within the lattice, the amount of energy necessary is $A_M e^2/d$; the neutralization of another ion (the alkali metal ion) in the lattice would require the same amount again, if this ion were also surrounded on all sides by charges. If, however, an alkali metal ion is neutralized, and this ion lies immediately next to the halogen ion which loses its electron, the quantity of energy necessary for the neutralization of this alkali atom is only $(A_M - 1)e^2/d$, since one of the neighbouring ions (one e^2/d term) was no longer present (the halogen ion being already neutralized). They therefore gave the following relationship:

$$h\nu = E - I + (2\,A_M - 1)\,\frac{e^2}{d} \qquad (2)$$

which, however, yields results about 30% too high.

Klemm /7/ called attention to the fact that, because of the formation of two electrically neutral atoms, the ions are no longer completely symmetrically surrounded by charges, these ions being polarized in one direction, which polarization is the source of an important gain in energy: Ψ_{pol}. He also considered the adsorption energies, Ω_{hal} and Ω_{alk}, of the halogen atom and of the alkali atom formed in the lattice. Moreover, the value of $h\nu$ for the first absorption band may be calculated from the following formula:

$$h\nu = E - I + (2\,A_M - 1)\,\frac{e^2}{d} - \Psi_{pol} - \Omega_{hal} - \Omega_{alk} \quad . \qquad (3)$$

We would add that agreement with the experiment is good.

In conclusion, the soundness of (1) is fortuitous, whereas a correct treatment of the subject necessitates the use of (3) above.

We can therefore adopt the following analogy: to estimate the energy, W_f, of the electron at the divalent cation, the energy necessary to transfer an electron from the nearest halogen ion to the foreign cation, corresponding to a transition from the top of the valence band to the desired ground state Me^+ of the metal, must be sought. To do this, while also taking into account the requirement of electroneutrality in the crystal - the divalent cation must be accompanied by a cation vacancy - we use the same relationship (3) but substitute $(2\,A_M - 1)\,e^2/d$ by $2\,A_M\,e^2/d$, and the ionization energy, I, of the host alkali by I_f, which is the second ionization potential of the foreign cation. In addition, we make the drastic assumption of disregarding changes in Ψ_{pol}, Ω_{hal} and Ω_{alk} also to avoid the difficulty of their calculation. Finally, encouraged by the soundness of relationship (3), we apply to it the experimental /8/ value $(h\nu)_{exp}$. In conclusion, we calculate W_f as

$$W_f = (h\nu)_{exp} + \frac{e^2}{d} - I_f + I \quad . \qquad (4)$$

The value W_f must be compared with the energy value W_v which represents the distance between the F centre ground state and the valence band. We can ap-

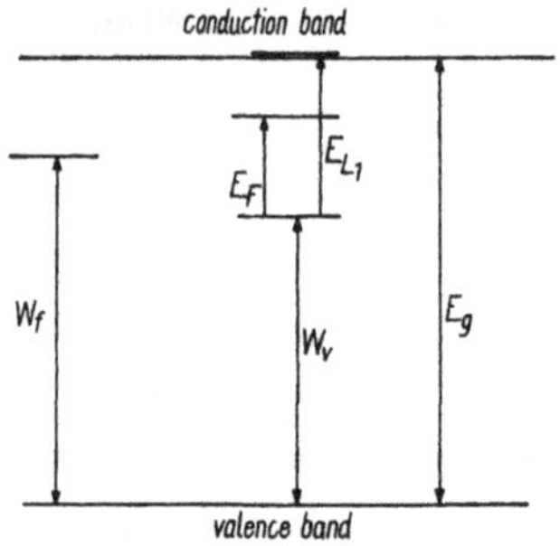

Fig. 1. Scheme for energies involved in the Z_1 centre formation (if $W_f > W_v$), and in the reduction of the foreign cation (if $W_f < W_v$)

proximate the above by $W_v = E_g - E_{L_1}$, where E_g is the band gap and E_{L_1} is the absorption energy of the L_1 band. According to the photoconductivity measurements of Crandall and Mikkor /9/, we are dealing with the first important transition from the F centre ground state to the conduction band. If W_f is higher than W_v (see Fig. 1), the ground state of the foreign ion lies above the F centre ground state and we can then expect that Z_1 centre formation is possible; on the contrary, if W_f is lower than W_v a reduction of the divalent cation into its monovalent state should be favoured. Thus, if we turn our attention to a particular host crystal, the essential physical parameter seems to be the ionization potential, I_f, of the impurity cation.

We would stress, however, that, following this suggestion, we can also determine an upper threshold value, $(I_f)_{thr}$, of the ionization potential for which Z_1 centre formation can be observed, i.e. the value for which $(W_f - W_v)$ is near to zero. $(I_f)_{thr}$ can of course be very different depending on the host crystal. Consequently, the above mentioned classification of the divalent ions in two classes is <u>not</u> correct.

If, according to our criterion, we use E_g and $(h\nu)_{exp}$ values from /10/ and /11/, and E_{L_1} values from /12/, we obtain the following results for the most important crystals investigated in the literature: Z_1 centre formation is favoured in KCl for $I_f \leqq 11.6$ eV, in KBr for $I_f \leqq 11.3$ eV, in KI for $I_f \leqq 10.7$ eV. It may be added that in spite of the drastic assumption we made, these forecasts are in agreement with experimental results /3, 4/, while good Z_1 formation is observed in KCl, difficulties in KBr, and no Z_1 formation in KI.

As regards NaF and LiF, i.e. the crystals which, as already stated, are at the core of this discussion /1, 2/, we are forced to calculate W_v by using, in NaF, the high energy tail of the K band (≈ 5 eV), and in LiF the high energy tail of the F band (≈ 6 eV), because of the uncertainty on the spectral position of the L_1 band and of the K band in LiF. We therefore expect that Z_1 centre formation is favoured in NaF for $I_f < 15.5$ eV, and in LiF for $I_f < 17.2$ eV. Taking into account the assumption we made on W_v calculation such values are of course only a lower limit for $(I_f)_{thr}$.

To conclude, it is evident that Z_1 centre formation is favoured in LiF and NaF "ideal" crystals doped with Mg^{2+}: the argument /1, 2/ of a high second ionization potential which might explain the reduction of Mg^{2+} to a monovalent ion is in fact not valid. Such a reduction, if it exists, may be due, in our opinion, to the presence of impurities in "real" samples, or, possibly, to local effects connected with X-ray coloration /1, 2, 6/.

We think, however, that one possible experimental approach to the present problem concerning "real" crystals is to analyse the properties of the relaxed excited state of the F centre when perturbed by the presence of magnesium before the bleaching of the F band: such measurements are now in progress in our laboratories.

References

/1/ A. WATTERICH and I. FÖLDVÁRI, Internat. Conf. Defects in Insulating Crystals, Riga (USSR) 1981 (p. 460).

/2/ A. WATTERICH, I. FÖLDVÁRI, G. CORRADI, and H. OPYRCHAŁ, phys. stat. sol. (b) 121, 117 (1984).

/3/ G. KENNTNER and H.J. PAUS, Internat. Conf. Colour Centres in Ionic Crystals, Reading (U.K.) 1971 (Abstr. No. 55).

/4/ K.H. UMBACH and H.J. PAUS, ibid. (Abstr. No. 77).

/5/ H.J. PAUS, private communication.

/6/ A.V. EGRANOV and A.I. NEPOMNYACHIKH, phys. stat. sol. (b) 122, 249 (1984).

/7/ J.H. DE BOER, Electron Emission and Adsorption Phenomena, Cambridge 1935 (Chap. X).

/8/ R.S. KNOX and K.J. TEEGARDEN, in: Physics of Color Centers, Ed. W.B. FOWLER, Academic Press, New York/London 1968 (Appendix A).

/9/ R.S. CRANDALL and M. MIKKOR, Phys. Rev. 138, 1247 (1965).

/10/ K. TEEGARDEN and G. BALDINI, Phys. Rev. 155, 896 (1967).

/11/ G. BALDINI, in: Elementary Excitations in Solids, Ed. A.A. MARADUDIN and G.F. NARDELLI, Plenum Press, New York/London 1969 (p. 437 to 454).

/12/ F. LÜTY, Z. Phys. 160, 1 (1960).

(Received July 17, 1985)

phys. stat. sol. (b) 131, K117 (1985)
Subject classification: 18.1; 22.5

Magnetism Theory Division, Institute of Physics,
A. Mickiewicz University, Poznań[1)]

The Theory of Hydrostatic Pressure Dependence of the Tricritical Temperature of the $FeCl_2$ Metamagnet[2)]

By
A. LIPOWSKI and Z. ONYSZKIEWICZ

Dedicated to Prof. Dr. Dr. h.c. Dr. E.h. P. GÖRLICH on the occasion of his 80th birthday

Introduction The anhydrous dihalide $FeCl_2$ is a typical metamagnet. Hexagonal sheets of Fe^{2+} spins are ferromagnetically aligned by a strong anisotropy along the hexagonal c-axis. Successive sheets are weakly antiferromagnetically coupled /1 to 3/. When the internal magnetic field H (the external field H_{appl} minus the demagnetizing field) is directed along the c-axis, then, at a threshold field H_{AP}, a transition occurs to a paramagnetic (P) phase of high magnetic moment. At a given pressure p this transition is of first order (or metamagnetic) from temperatures T = 0 to T_t, where T_t is the tricritical temperature, and of second order (or continuous) between T_t and T_N, where T_N is the Néel temperature /1 to 3/. At a pressure p of about 2 kbar a structural phase transition occurs from the rhombohedral r-$FeCl_2$ phase to a closer packed hexagonal phase, h-$FeCl_2$, which has been shown to have the CdI_2 structure /2, 4 to 6/ and is therefore isomorphous to another ferrous halide of considerable interest, $FeBr_2$ /2, 6 to 10/. In the report we apply simple spin models to the theoretical investigation of the influence of hydrostatic pressure on the tricritical temperature of the metamagnetic $FeCl_2$. The basic assumption we take is that the parameters of the Hamiltonian are linear functions of pressure p. This assumption is justified by the fact that both Néel temperature T_N and threshold field H_{AP} (T = 0) are linear functions of pressure in wide ranges of their variation in r-$FeCl_2$ and h-$FeCl_2$ phases /6/.

---

1) ul. Matejki 48/49, 60-769 Poznań, Poland.

2) Supported by the Institute for Low Temperatures and Structural Research, Polish Academy of Sciences, Wrocław.

First model The two-sublattice collinear metamagnet $FeCl_2$ to be examined is described by the following Hamiltonian /11 to 14/:

$$\mathcal{H} = -\frac{1}{2}\sum_{\vec{f} \neq \vec{f}'}\sum_{\alpha\gamma} I_{11}^{\alpha\gamma}(\vec{f} - \vec{f}', p)\, S_{\vec{f}}^{\alpha} S_{\vec{f}'}^{\gamma} - \frac{1}{2}\sum_{\vec{g} \neq \vec{g}'}\sum_{\alpha\gamma} I_{22}^{\alpha\gamma}(\vec{g} - \vec{g}', p)\, S_{\vec{g}}^{\alpha} S_{\vec{g}'}^{\gamma} +$$
$$+ \sum_{\vec{f} \neq \vec{g}}\sum_{\alpha\gamma} I_{12}^{\alpha\gamma}(\vec{f} - \vec{g}, p) S_{\vec{f}}^{\alpha} S_{\vec{g}}^{\gamma} + D(p)\left(\sum_{\vec{f}} (S_{\vec{f}}^{z})^2 + \sum_{\vec{g}} (S_{\vec{g}}^{z})^2\right) -$$
$$- g\mu_B H \left(\sum_{\vec{f}} S_{\vec{f}}^{z} + \sum_{\vec{g}} S_{\vec{g}}^{z}\right) , \qquad (1)$$

where fixed spins (S = 1) in two different sublattices 1 and 2 refer to spin-down and spin-up positions, respectively. The indices $\vec{f}$, $\vec{f}'$ and $\vec{g}$, $\vec{g}'$ refer to spin sites in the sublattice 1 and 2, respectively. The symbol g denotes the Landé factor, μ_B the Bohr magneton, H the magnetic field aligned in the z direction. The symbol $\alpha\gamma$ means

$$\alpha\gamma = +- \quad \text{or} \quad -+ \quad \text{or} \quad zz ,$$

and

$$I_{11,12}^{\alpha\gamma} = \begin{cases} \frac{1}{2} I_{1,2} & \text{for} \quad \alpha\gamma = -+ \text{ or } +- , \\ \frac{1}{\eta} I_{1,2} & \text{for} \quad \alpha\gamma = zz . \end{cases}$$

Now we assume that $I_{1,2}$ and D are linear functions of hydrostatic pressure p:

$$I(p)/I(0) = I_o + I_1 p + O(p^2) ,$$
$$K(p)/I(0) = K_o + K_1 p + O(p^2) ,$$
$$D(p)/I(0) = D_o + D_1 p + O(p^2) , \qquad (2)$$

where $I = \sum_{\vec{f}'} I_1$ and $K = \sum_{\vec{g}} I_2$.

In order to obtain the magnetical properties of the model (1) we resort to the Haley and Erdös method /15, 16/ in which we obtain the spectrum of collective excitations in RPA and thermodynamical properties in MFA. Then, the expressions obtained in this way as well as experimental data obtained in the previous papers /6, 17/ for $T \ll T_N$ served to determine the parameters of Hamiltonian (1). This allowed a numerical calculation of the hydrostatic pressure dependence of the tricritical and Néel temperatures. The results obtained are presented in

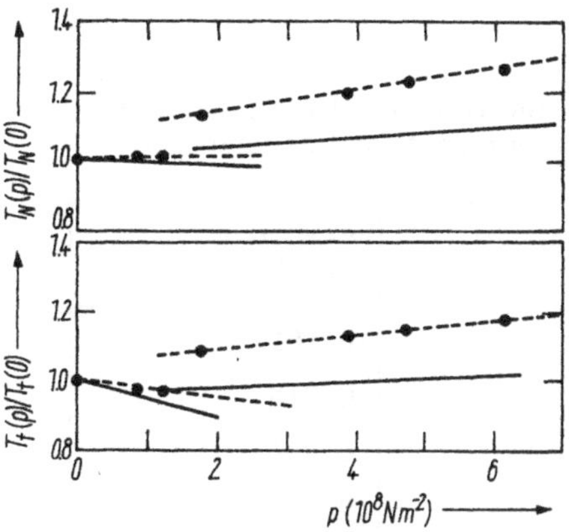

Fig. 1. The experimental Néel and tricritical temperature (·●·) versus hydrostatic pressure p obtained by Vettier et al. /6/ compared with the ones calculated in our communication for the first model (1) (——) and for the second one (3) (- - -)

Fig. 1. The agreement between theoretical results and experimental data is only qualitative (despite the choice of the optimum parameters of the Hamiltonian) and this cannot be treated only as the result of the approximation. Therefore we shall analyse the tricritical behaviour of $FeCl_2$ on the grounds of another model.

Second model Let us consider the properties of the following Hamiltonian (s = 1/2) /18, 19/:

$$\mathcal{H} = -\frac{1}{2}\sum_{\vec{f} \neq \vec{f}'} I_{1\vec{f}\vec{f}'}(p)\, S^z_{\vec{f}} S^z_{\vec{f}'} - \frac{1}{2}\sum_{\vec{g} \neq \vec{g}'} I_{1\vec{g}\vec{g}'}(p)\, S^z_{\vec{g}} S^z_{\vec{g}'} +$$

$$+\sum_{\vec{f} \neq \vec{g}} I_{2\vec{f}\vec{g}}(p)\, S^z_{\vec{f}} S^z_{\vec{g}} - g\,\mu_B\, H\Big(\sum_{\vec{f}} S^z_{\vec{f}} + \sum_{\vec{g}} S^z_{\vec{g}}\Big) +$$

$$+\frac{1}{2}\sum_{\vec{f}\vec{g}\,\vec{f}'\vec{f}''}{}' D_{\vec{f}\vec{g}\,\vec{f}'\vec{f}''}(p)\, S^z_{\vec{f}} S^z_{\vec{g}} S^z_{\vec{f}'} S^z_{\vec{f}''} + \frac{1}{2}\sum_{\vec{f}\vec{g}\,\vec{g}'\vec{g}''}{}' D_{\vec{f}\vec{g}\,\vec{g}'\vec{g}''}(p)\, S^z_{\vec{f}} S^z_{\vec{g}} S^z_{\vec{g}'} S^z_{\vec{g}''}\,, \quad (3)$$

where Σ' stands for a summation over different lattice points, but $\sum'_{\vec{f}\vec{g}\,\vec{f}'\vec{f}''}$ denotes a summation over the lattice $\vec{f}$ points and their nearest neighbours and $\sum'_{\vec{f}\vec{g}\,\vec{g}'\vec{g}''}$ denotes a summation over the lattice sites $\vec{g}$ and their nearest neighbours, respectively.

In order to obtain the tricritical properties of the model (3) we resort to the MFA. Now we assume that the parameters of the Hamiltonian (3) are linear functions of hydrostatic pressure p. Using the following reduced magnitudes:

$$h = 2\,g\,\mu_B\,H\,B^{-1}(p=0),\quad t = 4\,k_B\,T\,B^{-1}(p=0)\,,$$

$$a(p) = C(p)\,B^{-1}(p=0),\quad b(p) = B(p)B^{-1}(p=0),\quad d(p) = (4B(p=0))^{-1}\,D(p)\,,$$

where

$$NB(p) = \sum_{\vec{f} \neq \vec{f}'} I_{1\vec{f}\vec{f}'}(p), \qquad NC(p) = \sum_{\vec{f} \neq \vec{g}} I_{2\vec{f}\vec{g}}(p)\,,$$

Table 1

The parameters used in calculations of the tricritical temperature of $FeCl_2$ metamagnet for first model (upper part) and second model (lower part)

parameter	rhombohedral r-$FeCl_2$ phase	hexagonal h-$FeCl_2$ phase
D_o	-0.171	-0.181
D_1	-0.026	-0.018
I_o	1.000	0.981
I_1	-0.024	0.004
K_o	0.059	0.070
K_1	0.004	0.004
η	0.78	0.78
a_o	0.1840	0.2500
a_1	0.0560	0.0162
b_o	1.0000	1.0300
b_1	-0.0489	0.0217
d_o	-0.0300	-0.0779
d_1	-0.0455	-0.0038

$$ND(p) = \sum_{\vec{f}\vec{g}\vec{f}'\vec{f}''}' D_{\vec{f}\vec{g}\vec{f}'\vec{f}''}(p) = \sum_{\vec{f}\vec{g}\vec{g}'\vec{g}''}' D_{\vec{f}\vec{g}\vec{g}'\vec{g}''}(p)$$

we obtain

$$a(p) = a_o + a_1 p + O(p^2) ,$$
$$b(p) = b_o + b_1 p + O(p^2) ,$$
$$d(p) = d_o + d_1 p + O(p^2) .$$

Moreover we can obtain the following expression /19/:

$$t_t(p) = t_N(p)\left(1 - \frac{1}{3}a(p)\right) + 2\,d(p)\,t_t(p)\left[\frac{a(p)\,t_t(p)}{t_N(p)} + \right.$$
$$\left. + \frac{3}{2}\left(1 - \frac{t_t(p)}{t_N(p)}\right)\left(1 - \frac{4}{3}\,\frac{t_N(p)}{t_t(p)} + 2\,\frac{d(p)\,t_t(p)}{t_N^2(p)}\right)\right] , \tag{4}$$

where

$$t_N(p) = a(p) + b(p)$$

is the formula for the Néel temperature $t_N(p)$. From (4) the tricritical temperature $t_t(p)$ can be determined as a function hydrostatic pressure p. The result of our calculation of $t_t(p)$ for the $FeCl_2$ metamagnet is illustrated in Fig. 1, which presents the experimental results obtained by Vettier et al. /6/ compared with the theoretical one calculated in our work for Hamiltonians (1) and (3). The theoretical results are obtained for the parameters given in Table 1.

Conclusion The results obtained allow us to state that the simple two-sublattice model (3) proposed in /18/ with the four-spin terms added describes the dependence of the tricritical temperatures of $FeCl_2$ on hydrostatic pressure much better than the model (1) commonly used up to now to describe $FeCl_2$ metamagnets /11 to 14, 17/. We see that the agreement between the theoretical (for model (3)) and experimental results is evident.

References

/1/ H. COFTA, Metamagnets, PWN, Warszawa 1971 (in Polish).
/2/ E. STRYJEWSKI and N. GIORDANO, Adv. Phys. 26, 487 (1977).
/3/ Z. ONYSZKIEWICZ and H. COFTA, Acta Phys. Polon. A57, 525 (1980).
/4/ A. NARATH and I.E. SCHIRBER, J. appl. Phys. 37, 1124 (1966).
/5/ K.C. JOHNSON and A.J. SIEVERS, Phys. Rev. B 7, 1081 (1973).
/6/ C. VETTIER, H.L. ALBERTS, and D. BLOH, Phys. Rev. Letters 31, 1414 (1973).
/7/ Z. ONYSZKIEWICZ, phys. stat. sol. (b) 99, K151 (1980).
/8/ Z. ONYSZKIEWICZ, Physica 103A, 274 (1980).
/9/ Z. ONYSZKIEWICZ, phys. stat. sol. (b) 122, 543 (1984).
/10/ Z. ONYSZKIEWICZ, Acta Magnetica 1, 23 (1984).
/11/ K. ONO, A. ITO, and T. FUJITA, J. Phys. Soc. Japan 19, 2119 (1964).
/12/ R. ALBEN, J. Phys. Soc. Japan 26, 261 (1969).
/13/ R.J. BIRGENEAU, W.B. YELON, E. COHEN, and J. MAKOVSKY, Phys. Rev. B 5, 2607 (1972).
/14/ M.G. PINI, E. RASTELLI, A. TASSI, and V. TOGNETTI, J. Phys. C 14, 3041 (1981).
/15/ S.B. HALEY and P. ERDÖS, Phys. Rev. B 5, 1106 (1972).
/16/ B. WESTWAŃSKI and K. SKROBIŚ, Physica 87A, 515 (1977) and references cited therein.
/17/ C. VETTIER and W.B. YELON, Phys. Rev. B 11, 4700 (1975).

/18/ Z. ONYSZKIEWICZ, Phys. Letters A 68, 113 (1978).
/19/ Z. ONYSZKIEWICZ, Physica 103A, 226 (1980).

(Received July 8, 1985)

phys. stat. sol. (b) 131, K123 (1985)
Subject classification: 13 and 18.2; 21

Department of Chemistry, University of California, San Diego[1)]

On the Spin Reorientation in $R_2Fe_{14}B$ Compounds[2)]

By
H. OESTERREICHER

Dedicated to Prof. Dr. Dr. h.c. Dr. E.h. P. GÖRLICH
on the occasion of his 80th birthday

$Nd_2Fe_{14}B$ has a ferromagnetic structure with moment alignment along the c-axis near room temperature but a spin reorientation to an easy cone with an angle of order 25° from the axis occurs around $T_S = 150$ K (first observed /1/ on a material slightly off stoichiometry and subsequently amply corroborated /2 to 7/). This phenomenon stems from the Nd lattice only and poses interesting questions concerning the origin of anisotropy in this technologically important compound. We shall here use a crystal field point charge (PC) approach to explain this feature. We will also deal with the tendencies in T_S both observed /6/ and predicted in partly rare earth substituted compound series.

In the first PC analysis /2/ explaining the axial anisotropy field for $Nd_2Fe_{14}B$ (of order $H_A = 80$ kOe) it was pointed out that this compound is an anisotropy antitype to $SmCo_5$ because R^{3+} coordination occurs to first approximation only in planes (σ_s planes) in this layer type compound. By contrast $SmCo_5$ is anisotropic because of the close axial approach of R^{3+} connected with deviation from ideal axial ratios (c/a). Concerning T_S it was assumed that this feature may be connected with different signs of V_2^o of the two rare earth sites (4g, 4f in the notation of /8, 9/) or with the importance /3, 7/ of terms V_2^2. Experiments concerning the response of T_S to substitution, however, now lead us to consider V_4^o to be an even more important contribution. We collect pertinent data (as presented earlier /7/) in Table 1. One notes that T_S has so far been observed only for $R_2Fe_{14}B$ with R = Nd. One also notes that T_S is strongly depressed on partial substitution with R = Tb for Nd but less so for R = Ho which is not easily explained on the earlier approaches /6/.

---

1) La Jolla, California 92093, USA.

2) Work supported by CMRR (Center for Magnetic Recording Research).

Table 1

Spin reorientation temperatures, T_S, for selected $R_2Fe_{14}B$ type compounds

compound	$T_S(K)$ a)
$Pr_2Fe_{14}B$	(4.2)
$Nd_2Fe_{14}B$	150
$Tb_2Fe_{14}B$	(4.2)
$Ho_2Fe_{14}B$	(77)
$Nd_2Fe_{13}CoB$	142
$Nd_2Fe_{10}Co_4B$	132
$Nd_2Fe_7Co_7B$	120
$Nd_2Co_{14}B$	37
$NdYFe_{14}B$	132
$NdTbFe_{14}B$	96
$NdHoFe_{14}B$	120

a) Values in parentheses stand for temperatures down to which no T_S was observed.

In the PC approach /10/ one considers crystal structure to establish a potential (given in terms of A_n^m, corresponding to structure factor terms) at the site of magnetic ions. These ions interact with this potential in terms of that part of the relevant 4f wave functions which is, by its own symmetry, capable of such interaction (equivalent to $S_n O_n^m$, where S_n is the Stevens factor and O_n^m represents a polynomial in the angular momentum). Accordingly, the crystal field Hamiltonian is written perhaps most didactically as

$$H_c = \sum_{n=0}^{n'} \sum_{m=0}^{n} F_n (A_n^m)(S_n O_n^m) ,$$

where F_n represents a factor varying smoothly through the rare earths. One also usually defines a crystal field intensity parameter $B_n^m = F_n A_n^m S_n$ or $B_n^m = S_n V_n^m$. We note proportionality /11/ between the anisotropy constant K_1 and terms $B_2^0 O_2^0(J) = E_2$ and $B_4^0 O_4^0(J) = E_4$. In the following we shall show that trends in the competition of these terms can explain the tendencies in T_S

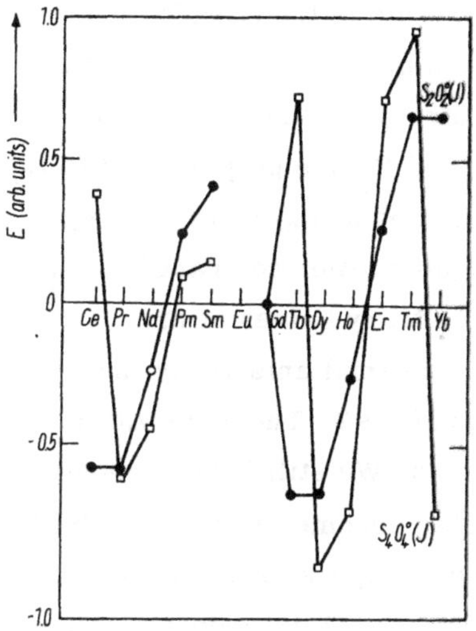

Fig.1

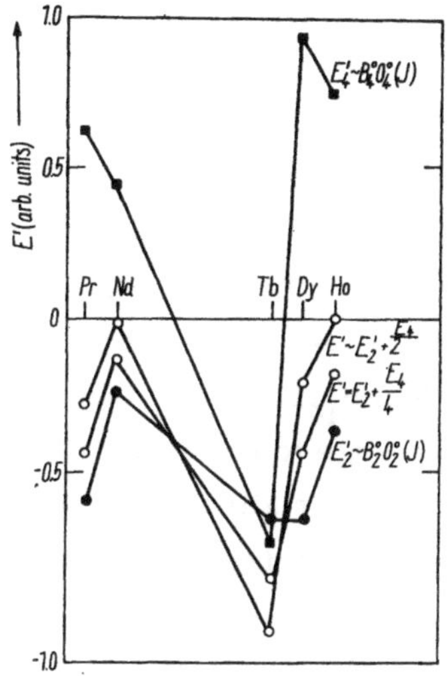

Fig.2

Fig. 1. Crystal field terms $S_2O_2^o(J)$ and $S_4O_4^o(J)$ for rare earths (see text)

Fig. 2. Energies E' in arbitrary units of crystal field terms $E_2' = S_2O_2^o(J)G_2^o$, $E_4' = S_4O_4^o(J)G_4^o/3$ (see text). For $E_2' + E_4'/4$ both Nd and Ho indicate susceptibility to a spin orientation at cryogenic temperatures

of Table 1. For this purpose we first show values of $S_nO_n^m(J)$ for various rare earths in Fig. 1. The usually dominant second-order term determines the sign of the anisotropy (e.g., axial for R = Nd and planar for R = Sm in $R_2Fe_{14}B$). We now determine for $R_2Fe_{14}B$ compounds with axial anisotropy the simplest part of the geometric factors A_n^m referring to them as G_n^m. They are

$$A_2^o \sim \sum \frac{3z^2 - r^2}{r^5} = G_2^o \quad \text{and} \quad G_4^o = \sum \frac{35z^4 - 30z^2r^2 + 3r^4}{r^7} .$$

As we shall only consider five nearest R neighbors which are in the σ_s layer (z = 0), $G_2^o = -1/r^3$ and $G_4^o = +3/r^3$. One notes the different sign of second- and fourth-order terms, respectively. Therefore, in cases where $S_2O_2^o(J)$ is of the same sign as $S_4O_4^o(J)$ (Fig. 1), the respective terms in the Hamiltonian (E_2, E_4) will oppose each other and their relative size will determine the magnetic structure for R with axial anisotropy in $R_2Fe_{14}B$. This is the case for R = Pr, Nd, Dy, and Ho but not for Tb. We can now reach a qualitative understanding of the relative influence of E_2 and E_4 on the anisotropy by plotting (Fig. 2) $S_2O_2^0(J)G_2^0 = E_2'$ and $S_4O_4^0(J)G_4^0/3 = E_4'$ where E_2' and E_4' are proportional to E_2 and E_4, respectively. We do not know the actual ratios of E_2/E_4 (shielding, etc.) but E_4 and therefore E_4' will have to be weighted lower compared to E_2. We therefore take sums of $E' = E_2' + E_4'/n$ with arbitrary values of n.

For $Nd_2Fe_{14}B$ we note that with pure E_2' we have axial anisotropy (negative E') but a change to non-uniaxial occurs for values of $n < 2$. The case with $n = 4$ appears representative of the experimental situation. We still have predominance of axial anisotropy but the fourth-order term has become rather competitive. This forces moment alignment into a cone at lower temperatures. However, as the fourth-order component has a steeper temperature (T) dependence, a spin orientation occurs on increasing T. We see that for R = Ho, E_2 and E_4 are also rather competitive while Pr and Dy remain of relatively strong axial anisotropy. For Tb both terms reinforce each other leading to increased axial anisotropy beyond the contribution from E_2. The general trends in Table 1 are now understandable. We note, accordingly, that partial Ho substitution does not change T_S much but partial Tb substitution does. In fact, we predict that $Ho_2Fe_{14}B$ may be a likely candidate to show a spin reorientation below T = 77 K. We also understand that $Nd_2(Co_xFe_{1-x})_{14}B$ shows decreases in T_S with increasing x as R-transition metal exchange coupling generally decreases from Fe to Ni.

References

/1/ F. SPADA, C. ABACHE, and H. OESTERREICHER, J. less-common Metals 99, L21 (1984).

/2/ H. OESTERREICHER, F. SPADA, and C. ABACHE, Mater. Res. Bull. 19, 1069 (1984).

/3/ D. GIVORD, H.S. LI, and R. PERRIER DE LA BATHIE, Solid State Commun. 51, 857 (1984).

/4/ S. SINNEMA, R.J. RADWANSKI, J.J.M. FRANSE, D.B. DE MOOIJ, and K.H.J. BUSCHOW, J. Magnetism magnetic Mater. 45, 333 (1984).

/5/ R. GROSSINGER, P. ORBITSCH, X.K. SUN, R. EIBLER, H.R. KIRCHMAY, and H. SASSIK, Mater. Letters 2, 539 (1984).

/6/ N.C. KOON, B.N. DAS, M. RUBINSTEIN, and J. TYSON, J. appl. Phys. 57, 4091 (1985).

/7/ C. ABACHE and H. OESTERREICHER, J. appl. Phys. 57, 4112 (1985).

/8/ F. HERBST, J.J. CROAT, F.E. PINKERTON, and W.B. YELON, Phys. Rev. B 29, 4176 (1984).

/9/ H. BOLLER and H. OESTERREICHER, J. less-common Metals 103, L5 (1984).

/10/ T. HUTCHINGS, Solid State Phys. 16, 227 (1966).

/11/ B. SZPUNAR and P.A. LINDGARD, J. Phys. F 9, L55 (1979).

(Received June 3, 1985)

phys. stat. sol. (b) 131, K129 (1985)
Subject classification: 18.2

Departamento de Física, Universidade Federal de Minas Gerais, Belo Horizonte[1]

A Cooperative System in a Random Field: A Look on Its Thermodynamics[2]

By
P.R. SILVA

Dedicated to Prof. Dr. Dr. h.c. Dr. E.h. P. GÖRLICH
on the occasion of his 80th birthday

There is a controversy in the current literature about the lower critical dimension (LCD) of a cooperative system in a field, when this field conjugated to the order parameter is a random variable. A recent review on this subject has been done by Imry /1/.

For the Ising model in a random field (RFIM), according to Krey /2/, there are two different predictions:

(i) considerations of stability of domains by Imry and Ma /3/ or of the roughening of domain walls at low temperatures by Grinstein and Ma /4/ and Villain /5/ suggest LCD = 2, whereas

(ii) diagramatic renormalization by Aharony et al. /6/, Grinstein /7/, and Young /8/ and a supersymmetric-field theory by Parisi and Sourlas /9/ predict LCD = 3. The latter prediction (LCD = 3) has also been obtained by Krey who modified the calculations of domain wall roughening by the inclusion of certain "mode couplings" between fluctuations with different wave vector $\vec{q}$.

In this note, we intend to study the effect of a random field in a cooperative system within the framework of a free energy expansion in the order parameter. However, this expansion will not be the traditional Landau's free energy expansion, but a modification of that expansion, proposed recently by the author /10/, which contemplated various models of cooperative systems with near exact critical exponents in the absence of a random field.

Let us consider a random field variable, h_R, which couples linearly to the order parameter M. Then, the contribution of this field for the expansion of

---

1) Caixa Postal 702, 30.000 Belo Horizonte, M.G., Brazil.

2) Work partially supported by CNPq, FINEP (Brazilian Agenciés).

the free energy density g, will be given by a function of the product $h_R M$. So we will have

$$\Delta g = \Delta g(h_R M) . \tag{1}$$

In this note, we are going to consider only distributions of random fields with equal probability to assume positive and negative values. So, expanding Δg in power series of the product $h_R M$, and taking the configurational average, we have

$$\overline{\Delta g} = A\, h^2\, M^2 + B\, h^4\, M^4 + \ldots \quad , \tag{2}$$

where the bar stands for the configurational average, $h^2 \equiv \overline{h_R^2}$, $h^4 \equiv \overline{h_R^4}$, and A and B are the coefficients of the second- and fourth-order terms of the expansion, odd terms averaging out to zero. The coefficient A is positive because the random field contribution is due to a disorder-type field, whereas a uniform field gives a contribution which favours the ordering.

Now we are going to add the contribution of the random field, given by (2) to the free energy expansion in the absence of a random field as given by /10/. Doing that, we have for $T \geqq T_c$

$$g_{new} = g_{old} + \overline{\Delta g} \quad , \tag{3a}$$

where g_{old} is given by (11) of /10/, and g_{new} includes the effect of the random field. More explicitly, we can write

$$g_{new} \approx \left[b(T - T_c)^{\gamma} + A\, h^2 \right] M^2 + UM^p + Bh^4 M^4 - HM. \tag{3b}$$

The term $C(T - T_c)^{2\beta + \gamma}$ (see /10/) has been detached from g_{old}, once it has been included in the non-random version due to its homogeneity with the RM_o^2 term of the ordered phase.

Now, the free energy expansion (3b) can describe only a phase transition in zero uniform field (H = 0), if we have a temperature which makes the coefficient of the M^2 term equal to zero. This leads to

$$b(T_o - T_c)^{\gamma} = -A\, h^2 . \tag{4}$$

Equation (4) can be satisfied only in the case $\gamma = 1$, once A is a positive constant. But in the case $\gamma = 1$ and $p = 4$, (3b) represents the first terms of a Landau's free energy expansion, which takes into account the perturbation effect due to the random field. In this case, we have

$$T_o = T_c - \frac{A}{b} h^2 \quad , \tag{5}$$

where T_o is the new critical temperature which decreases with increasing random field strength. In fact, this result is consistent with the mean field treatment of the Ising model in a random field (RFIM), which has been done by Aharony /11/.

Thus according to our free energy expansion, and the values of γ given by /10/, two- and three-dimensional Ising models will not undergo phase transition, in the presence of random fields. The same thing can be said about three-dimensional spherical and Heisenberg models.

If we allow for a slowly varying space-dependent order parameter, the free energy, G, for $T \geqq T_c$, will be a kind of Landau-Ginzburg expansion

$$G \approx \int d^3x \left\{ \left[\vec{\nabla} M(x) \right]^2 + \left[b(T - T_c)^{\gamma} + Ah^2 \right] M^2 - H(x)\, M(x) \right\}, \tag{6}$$

where we have neglected powers of M higher than the second order. The minimization of relation (6) leads to the following differential equation:

$$-\nabla^2 M(x) + \left[b(T - T_c)^{\gamma} + A\, h^2 \right] M(x) = H \delta^3(x) \quad , \tag{7}$$

where we suppose H(x) is a δ-function localized at x = 0.

The asymptotic behavior of the solution of (7) will be given by

$$M(x) \sim \frac{H}{|\vec{x}|^{(d-1)/2}} \exp\left\{ - \left[b(T - T_c)^{\gamma} + Ah^2 \right]^{1/2} |\vec{x}| \right\}. \tag{8}$$

Equation (8) implies that the correlation length, ξ, will be given by

$$\xi \sim \left[b(T - T_c)^{\gamma} + A\, h^2 \right]^{-1/2} \quad . \tag{9}$$

Then, just at $T = T_c$, the inverse of the correlation length ξ^{-1}, will be given by

$$(\xi^{-1})_{T_c} \sim \sqrt{A}\, h \; . \tag{10a}$$

Relation (10a) must be compared with the experimental result for $Rb_2Co_{0.7}Mg_{0.3}F_4$ in a field (an experimental realization of a two-dimensional RFIM), obtained by Birgeneau et al. /12/,

$$(\xi^{-1})_{T_c,\, exp} \sim h^{0.7 \pm 0.2} \quad , \tag{10b}$$

and with a result obtained by Aharony and Pytte /13/, for the two-dimensional RFIM,

$$(\xi^{-1})_{T_c,\ \text{Aharony-Pytte}} \sim h^{(1-n/2)^{-1}} = h^{1.14} . \qquad (10c)$$

The relaxation of the system towards equilibrium can be obtained, applying the Landau-Khalatnikov equation /14/ to relation (3b). We have

$$\frac{dM}{dt} = -\Gamma \frac{\partial g}{\partial M} . \qquad (11)$$

The solution of the linearized equation of motion, for small fluctuations of the order parameter ($M(t) = M_o + \delta M(t)$) around equilibrium, can be obtained in an analogous way as we have done in /10/. The solution is

$$\delta M(t) = \delta M(0) e^{-t/t_2} , \qquad (12)$$

where

$$t_2 \equiv \frac{1}{\Gamma \chi^{-1}} , \quad \chi^{-1}\Big|_{\substack{H = 0 \\ T \geqq T_c}} = 2 \left[b(T - T_c)^{\gamma} + A h^2 \right] . \qquad (12a, b)$$

Then, just at $T = T_c$, the relaxation time, t_2, is proportional to h^{-2}.

We are grateful to Dr. A.S.T. Pires for a critical reading.

References

/1/ Y. IMRY, J. statist. Phys. 34, 849 (1984).

/2/ U. KREY, J. Phys. C 18, 1455 (1985).

/3/ Y. IMRY and S.K. MA, Phys. Rev. Letters 35, 1399 (1975).

/4/ G. GRINSTEIN and S.K. MA, Phys. Rev. B 28, 2588 (1983).

/5/ J. VILLAIN, J. Physique, Lettres 43, L551 (1982).

/6/ A. AHARONY et al., Phys. Rev. Letters 37, 1367 (1976).

/7/ G. GRINSTEIN, Phys. Rev. Letters 37, 944 (1976).

/8/ A.P. YOUNG, J. Phys. C 10, L257 (1977).

/9/ G. PARISI and N. SOURLAS, Phys. Rev. Letters 43, 744 (1979).

/10/ P.R. SILVA, submitted to phys. stat. sol. (b).

/11/ A. AHARONY, Phys. Rev. B 18, 3318 (1978).

/12/ R.J. BIRGENEAU et al., Phys. Rev. B 28, 1438 (1983).

/13/ A. AHARONY and E. PYTTE, Phys. Rev. B 27, 5872 (1983).

/14/ L.D. LANDAU and I.M. KHALATNIKOV, Dokl. Akad. Nauk SSSR 96, 469 (1954).

(Received August 5, 1985)

phys. stat. sol. (b) 131, K133 (1985)

Subject classification: 13.1

Physics Department (a) and Mathematics Department (b), North Staffordshire Polytechnic, Stoke-on-Trent[1] and Department of Pure and Applied Physics, University of Salford[2] (c)

Contributions to the Muffin-Tin Potential in Band Structure Calculations

By

G.R. EVANS (a), B.C. JONES (b), D.J. SHAUGHNESSY (a), and M.I. DARBY (c)

Dedicated to Prof. Dr. Dr. h.c. Dr. E.h. F. GÖRLICH on the occasion of his 80th birthday

At each stage of a self-consistent band structure calculation involving a muffin-tin potential V(r), the potential has to be recalculated using the electron density of the previous iteration. The usual construction of the potential /1/ is based on the fact that, since each unit cell is neutral, it produces no Coulomb potential in any other cell. Thus, if higher multipole moments are neglected, it is only necessary to consider a single unit cell. Corrections can then be made for the non-spherical symmetry in the interstitial region between the muffin-tin sphere and the Wigner-Seitz cell by recalculating the mean value of 1/r in that region.

The contribution to V(r) from the charge in the interstitial region can be evaluated from $e^2 Z_{int} \langle 1/r \rangle_{int}$, where Z_{int} is the number of electrons in this region and $\langle 1/r \rangle_{int}$ is defined by

$$\left\langle \frac{1}{r} \right\rangle_{int} = \frac{1}{\Omega_{WS} - \Omega_{sphere}} \int_{\text{sphere surface}}^{\text{WS cell boundary}} \frac{1}{\vec{r}} \, d\vec{r} \; .$$

Koelling /2/ quotes values of $\left\langle \frac{1}{r} \right\rangle_{int}$ of 2.04276/a and 2.41603/a for b.c.c. and f.c.c. lattices, respectively, for the cases of touching muffin-tin spheres, but both these values appear to be in error. The mean value $\left\langle \frac{1}{r} \right\rangle_{int}$ has been determined analytically for arbitrary values of the muffin-tin sphere radius s, for both b.c.c. and f.c.c. lattices. The results are

1) College Road, Stoke-on-Trent ST4 2DE, Great Britain.

2. Salford M5 4WT, Great Britain.

b.c.c. lattice

$$\left\langle\frac{1}{r}\right\rangle_{int} = \frac{a^2\left\{\frac{6}{\sqrt{2}}\ln\left(\frac{\sqrt{10}+1}{\sqrt{10}-1}\right) + 6\tan^{-1}\sqrt{0.8} + 9\tan^{-1}\sqrt{0.2} - 3\pi\right\} - 2\pi s^2}{0.5\,a^3 - 4\pi s^3/3}$$

$$= (1.516904\,a^2 - 2\pi s^2)/(0.5\,a^3 - 4\pi s^3/3) \;;$$

f.c.c. lattice

$$\left\langle\frac{1}{r}\right\rangle_{int} = \frac{a^2\left\{\frac{\sqrt{3}}{2}\ln\left(\frac{2\sqrt{3}+2}{\sqrt{3}-1}\right) - \pi/4\right\} - 2\pi s^2}{0.25\,a^3 - 4\pi s^3/3}$$

$$= (0.9554039\,a^2 - 2\pi s^2)/(0.25\,a^3 - 4\pi s^3/3) \;.$$

For the case of touching spheres, the above give

b.c.c. lattice: $s = \sqrt{3}\,a/4$; $\left\langle\frac{1}{r}\right\rangle_{int} = 2.118697/a$,

f.c.c. lattice: $s = a/2\sqrt{2}$; $\left\langle\frac{1}{r}\right\rangle_{int} = 2.620315/a$.

It can be seen that both results differ significantly from those quoted by Koelling.

References

/1/ D.A. LIBERMANN, Phys. Rev. 153, 704 (1967).

/2/ D.D. KOELLING, Rep. Progr. Phys. 44, 139 (1981).

(Received July 25, 1985)

phys. stat. sol. (b) 131, K135 (1985)

Subject classification: 13.5.1 and 20.1; 22.6

Institute of Physics, Academy of Sciences of the Ukrainian SSR, Kiev[1] (a)
and State University, Chernovtsy (b)

Temperature Anomalies of Exciton Absorption in GeO_2

By

B.M. NITSOVICH (a), G.M. PESTRYAKOV (a),
and V.D. FALENCHUK (b)

Dedicated to Prof. Dr. Dr. h.c. Dr. E.h. P. GÖRLICH
on the occasion of his 80th birthday

Recently, in their studies of GeO_2 crystals Fröhlich et al. /1/ have observed a number of specific peculiarities of the temperature behavior of an exciton spectral line which lack an adequate interpretation in the framework of the classical theory. The most remarkable among these anomalies are the following ones. a) The absorption band halfwidth grows with temperature increasing up to $T_{cr} = 50$ K; then, within a narrow temperature interval this growth practically vanishes and with the further increase of temperature the halfwidth augments again. b) The shift of the absorption band maximum behaves anomalously, too, i.e. its temperature dependence is represented by a curve which consists of two almost linear sections with different slopes separated by a plateau where the shift does not change with temperature. c) The peculiar temperature-independent behavior of both spectral characteristics is observed in the same temperature region from 50 to 70 K. Below these observations are tentatively interpreted in terms of the anharmonic crystal model.

The Green function calculational procedure applied to exciton absorption of light with frequency ω results in a quasi-Lorentzian shape of the absorption band /2/ such as

$$K(\omega) = 4\pi T_0^2 \gamma(\omega) \left\{ \left[\omega - E_0 - \Delta(\omega)\right]^2 + \gamma^2(\omega) \right\}^{-1} . \quad (1)$$

Here T_0 is the exciton-photon coupling, E_0 is the bottom of the exciton energy band; Δ and γ denote the real and imaginary parts of the exciton self-energy operator which describe the absorption band shift and halfwidth, respectively. Quantitative investigations of exciton absorption based on (1) were carried by

---

1) prospekt Nauki 144, 252028 Kiev, USSR.

quite a few authors. The main result of their studies was that in the lowest approximation the self-energy operator might be presented as a linear function of phonon occupation numbers. This fact allows one to use the following expressions in evaluating the position of the exciton band maximum ω_M and the halfwidth γ:

$$\omega_M(T) = E_o - \Delta_o - \sum_q \Delta(q)\, n_q(T) \ , \tag{2}$$

$$\gamma(T) = \gamma_o + \sum_q \gamma(q)\, n_q(T) \ . \tag{3}$$

Here Δ_o and γ_o are the temperature-independent shift and radiation width of the exciton level, respectively; $\Delta(q)$ and $\gamma(q)$ are theoretical parameters proportional to the exciton-phonon coupling, n_q denotes the occupation number of phonons with energy ν_q and quasi-momentum q. It should be noted that (2), (3) follow from the usual model of relaxation processes where exciton-phonon interaction is supposed to be a linear function of phonon operators. Thus, in this model the exciton-phonon scattering leads to a monotonous broadening of the absorption band with temperature and a gradual shifting of the maximum to lower energies. In order to obtain some other temperature dependence of Δ and γ one must consider additional mechanisms of exciton relaxation.

Earlier it has been shown in /3/ that deviations from the linear temperature dependence of the spectral characteristics may be attributed to phonon-phonon interaction. The latter results from anharmonic terms in the expansion of the lattice potential energy and reduces phonon lifetimes. In other words, the phonon-phonon interaction gives rise to quantum transitions between phonon eigenstates and redistributes energy between phonon branches. And, so far as different phonon branches are of importance in various interactions, the possibility of energy trade-off between the branches suggests that, in fact, there exists an additional channel of exciton energy relaxation. In our case the account for phonon-phonon interaction leads to the following expressions /3/:

$$\omega_M(T) = E_o - \Delta_o - \sum_q \Delta(q)\, n_q \left[1 - \sum_{q_1} V_a(q, q_1)(1 + 2\, n_{q_1})\right] , \tag{4}$$

$$\gamma(T) = \gamma_o + \sum_q \gamma(q)\, n_q \left[1 - \sum_{q_1} V_a(q, q_1)(1 + 2\, n_{q_1})\right] , \tag{5}$$

where V_a is the force constant of the phonon anharmonicity. The anharmonicity effects manifest themselves in a specific temperature dependence of the ab-

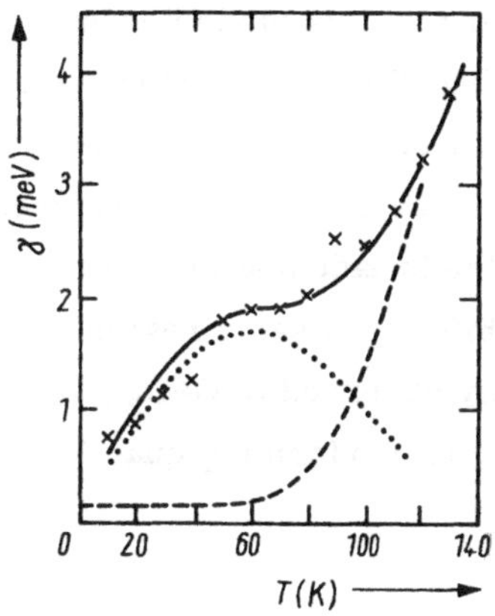

Fig. 1. Absorption band halfwidth versus temperature: theoretical results (solid curve) and data given in /1/ (crosses); the dotted and dashed curves correspond to the halfwidth due to either acoustic or optical modes, respectively

sorption band halfwidth. Now the latter increases with temperature growing only up to a certain T_{cr} above which the broadening becomes constant and, with the further increase of temperature the band may even become narrower. The same is true for the shift of the band maximum, at T_{cr} its temperature dependence should change sign.

Supposing that in exciton scattering two phonon branches (optical and acoustic) are involved, rewrite (5) in the form appropriate for calculations:

$$\gamma(T) = \gamma_o + aT(1 - bT) + c\, n_o \ . \tag{6}$$

In deriving (6) the anharmonicity of the optical branch has been neglected. since in GeO_2 crystals there are only high-frequency phonons with energies greater than 300 cm^{-1}, and the high-temperature expansion of the occupation numbers for the acoustic branch $n_q \sim k_B T/\nu_q$ has been employed; the parameters a, b, c can be readily figured out from (5). In line with /1/ we have set the energy of the optical phonon equal to ν_o = 42.8 meV and calculated γ as a function of temperature (Fig. 1) with the following numerical values: a = 0.057 meV/K, b = 0.0083 K^{-1}, c = 0.24 eV, γ_o = 0.15 meV. Comparison of the calculated curve with the data reveals their good fit which enables one to arrive at the following conclusions. In the low-temperature region scattering of excitons by the acoustic mode is dominant, a small discrepancy between the data and theoretical results is likely to be due to the high-temperature expansion employed in the calculations. With temperature increasing, the number of acoustic phonons grows and the anharmonicity effects become more pronounced, it is these effects that, first, slow down the growth of the band halfwidth and result, afterwards, in the saturation at ≈ 50 K. Further increase of temperature should lead to band narrowing (Fig. 1, dotted curve). Yet, in this temperature region the optic phonons are being excited as well and at $T > 70$ K exciton scattering by this mode becomes dominant and governs the high-temperature broadening (Fig. 1, dashed curve). According to (4) and (5), the value of T_{cr} depends only on the energy of the phonon, active in scattering, and the anharmonicity force constant. Therefore, in complete analogy to the temperature behavior of the broadening there is a plateau in the temperature dependence of the band shift in the same region from 50 to 70 K. It is this picture that was reportedly observed in GeO_2 /1/. The lack of experimental information prevented us from performing quantitative investigations of the function $\omega_M(T)$.

Thus, it may be stated that in GeO_2 crystals

- the relaxation mechanism corresponds to the exciton-phonon scattering which is linear with respect to phonon operators, i.e. displacements of atoms;
- in the low-temperature region ($T < 50$ K) excitons scatter mainly on acoustic modes, whereas in the high-temperature region ($T > 70$ K) the scattering on optical phonons becomes more effective;
- the anomalous temperature behavior of the exciton band within the interval from 50 to 70 K is due to anharmonic interaction of the low-frequency acoustic mode with the continuum of phonon states. The latter conclusion suggests that in GeO_2 the non-parabolic dependence of the lattice potential energy may be significant in spectroscopic effects. This invites, of course, more detailed studies of heat properties of GeO_2 crystals below 70 K.

The authors are indebted to Dr. I.V. Blonskii for helpful and stimulating discussions.

References

/1/ D. FRÖHLICH, H.-J. WEBER, R. WILLE, and R. FRANCINI, phys. stat. sol. (b) 126, 305 (1984).

/2/ A.S. DAVYDOV, Teoriya tverdogo Tela, Izd. Nauka, Moscow 1976.

/3/ I.L. KARELSKICH and B.M. NITSOVICH, Fiz. tverd. Tela 23, 308 (1981).

(Received June 14, 1985)

phys. stat. sol. (b) 131, K139 (1985)
Subject classification: 13.1 and 20; 4; 22.8

Institute of Physics, Rostov State University[1)]

X-Ray Spectroscopic Investigation of the Electronic Structure and Chemical Bond of the Semiconductors $Sn_2P_2S_6$ and Tl_3PS_4

By

A.N. GUSATINSKII, A.A. LAVRENTEV, M.A. BLOKHIN, and V.YU. SLIVKA

Dedicated to Prof. Dr. Dr. h.c. Dr. E.h. P. GÖRLICH on the occasion of his 80th birthday

The semiconducting compounds $Sn_2P_2S_6$ and Tl_3PS_4 are perspective for use in optoelectronic devices due to their nonlinear acousto-optical and electro-optical characteristics /1/. $Sn_2P_2S_6$ has ferroelectric properties, too. However, there are no works dedicated to a systematic investigation of the electronic energy band structure of these compounds.

Because of the complicated crystal structures of the compounds investigated /1/ it is advisable to single out: in $Sn_2P_2S_6$ the coordination octahedron $[P_2S_6]$, and in Tl_3PS_4 the coordination tetrahedron $[PS_4]$. The distances between P and S atoms in these compounds, and also between P atoms themselves in $Sn_2P_2S_6$, do not exceed the sum of atomic covalent tetrahedron radii of P (0.110 nm) and S (0.104 nm). Hence it is possible to consider the compounds investigated (as well as $InPS_4$ /2/) to have strong covalent interactions between P and S atoms.

To investigate the electronic interactions of the P and S atoms in $Sn_2P_2S_6$ and Tl_3PS_4, X-ray fluoresence K-bands and the absorption K-edge of phosphorus and sulphur were measured on the spectrograph DRS-2 with a resolution of 0.1 eV, and emission $L_{2,3}$ bands of phosphorus and sulphur on the spectrometer RSM-500 with a resolution of 0.4 eV. The energetic superposition of the K and L spectra of P and S was realized using the energy of the $K\alpha_1$ line of P and S, respectively. To match the P and S spectra an X-ray photoelectron spectrometer was used to obtain energies of the inner 2p levels of P and S in the same compounds. The K spectra were corrected with respect to their broadening by the inner K level and apparatus distortions. The high energy

---

1) prospekt Stachki 194, 344090 Rostov-on-Don, USSR.

drop of the corrected SK band was approximated by a straight line, the intersection of which with the energy axis was chosen for the valence band top and for energy scale zero.

The X-ray spectra of P and S in $Sn_2P_2S_6$ and Tl_3PS_4 (Fig. 1) show the upper part of the valence band to contain p-states of sulphur (the A_3 maxima of SK bands) and phosphorus (the B_1 maxima of PK bands). The valence band top is formed mainly by the S p-states, and the P p-states are shifted deep into the valence band approximately by 3.5 eV.

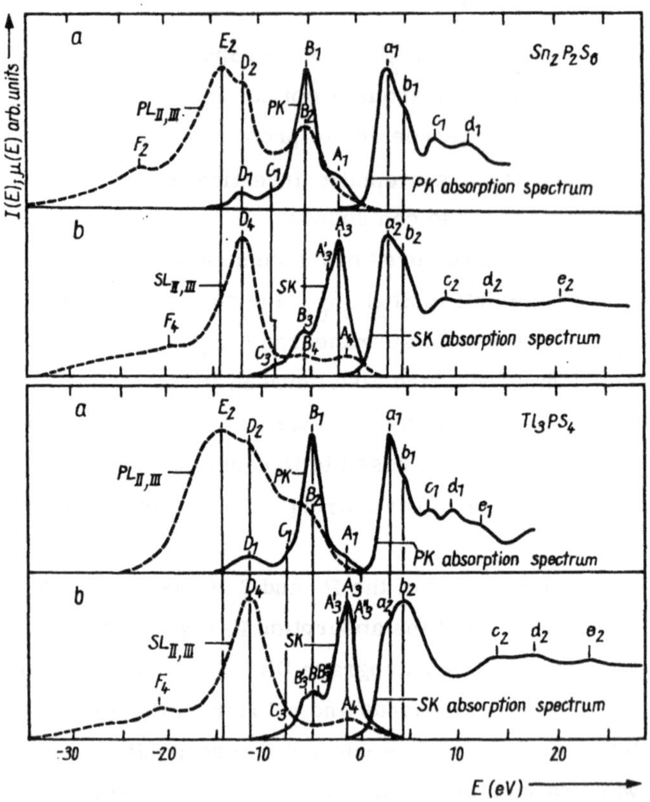

Fig. 1. Experimental results for $Sn_2P_2S_6$ and Tl_3PS_4: a) X-ray primary $L_{2,3}$ emission band, fluorescence K band, and K absorption spectrum of P; b) the same for S. As zero energy the valence band top is chosen

T a b l e 1

Energies (in eV) of the main characteristic peaks of the $Sn_2P_2S_6$ and Tl_3PS_4 X-ray emission spectra

compound \ peak	$A(A_3)$	$B(B_1)$	$C(C_1, C_3)$	$D(D_4)$	$E(E_2)$
$Sn_2P_2S_6$	1.7	5.1	8.4	11.6	14.1
Tl_3PS_4	1.5	5.0	7.0	11.5	14.0

The lower part of the valence band in the compounds investigated is formed by the s-states of sulphur (the D_4 maxima of $SL_{2,3}$ bands) and phosphorus (the E_2 maxima of $PL_{2,3}$ bands). As in the case of p-states, the s-states of P are located deeper in comparison to the s-states of S approximately by 2.5 eV and they form the valence band bottom.

In Table 1 are given energies of the main characteristic peaks of the $Sn_2P_2S_6$ and Tl_3PS_4 X-ray emission spectra.

The above mentioned relative locations of separate P and S energy subbands in $Sn_2P_2S_6$ and Tl_3PS_4 can be explained by the influence of effective charges on the atoms in solids. For example the greater electronegativity (EN) of S in comparison to P atoms ($EN^S = 2.44$, $EN^P = 2.06$ /3/) leads to lower energy bands of the s- and p-states of S compared to P if these atoms are considered as neutral. Obviously the relative locations of the P and S subbands are observed reversely to this expectation. As a consequence the P-S chemical bond in the compounds considered has an essential ionic contribution (e.g. S^+, P^-).

The resemblance in main features of the shapes of corresponding P and S X-ray emission subbands and of their relative locations seems to be connected with the similarity of the interaction type of the P and S atoms with the atoms of their respective neighbouring environment in the compounds investigated.

The P and S X-ray emission spectra of these compounds, as it is seen from Fig. 1, contain intensive additional maxima energetically coinciding with the main maxima of the S and P subbands, respectively. It points out a considerable admixture of p- and s-states of P and S to each other and is indicative of a substantial covalent contribution to the chemical interaction of P and S atoms in the compounds investigated.

Thus, in these phosphorus containing sulphides a strong interaction of P and S atoms is revealed with a considerable amount of both ionic and covalent contributions to the chemical bond.

The K bands and K absorption edges of P and S for both compounds investigated were corrected with respect to their broadening by the K level width and by apparatus distortions. The corrected spectra were used to obtain the energy gap ΔE between the high energy drop of the emission spectrum (linearly extrapolated from the inflection point to the energy axis) and the beginning of the main absorption edge spectrum. The values of ΔE^P and ΔE^S are included in Table 2 together with the energy gaps E_g from literature.

The agreement of the ΔE values with E_g values taken from literature (within errors of the X-ray spectra determination of ΔE) demonstrates the S and P p-states to give appreciable contributions to the density of states near the valence band top and the conduction band bottom of the compounds investigated.

The analysis of P and S K-absorption edges shows the conduction band bottom of Tl_3PS_4 and $Sn_2P_2S_6$ to be formed by highly hybridized p-states of P and S.

Table 2

Comparison of values of ΔE (in eV) with E_g

compound	ΔE^P	ΔE^S	E_g
$Sn_2P_2S_6$	2.1 ± 0.3	2.3 ± 0.2	2.3 /4/
Tl_3PS_4	2.2 ± 0.3	2.0 ± 0.2	2.3*)

*) The value of $E_g \approx 2.3$ eV for Tl_3PS_4 was estimated by the transmission limit of the visible light from /5/.

References

/1/ D.M. BERCHA, YU.V. VOROSHILOV, et al., Complex Chalcogenides and Chalcohalogenides, Vishcha Shkola, Lvov 1983 (in Russian).

/2/ A.N. GUSATINSKII, M.A. BLOKHIN, A.A. LAVRENTEV, et al., Abstr. Internat. Conf. X-Ray and Inner-Shell Processes in Atoms, Molecules, and Solids, Vol. 1, Leipzig 1984 (p. 170).

/3/ M.C. DAY and J. SELBIN, Theoretical Inorganic Chemistry, Reinhold Book Co., New York/Amsterdam/London 1976.

/4/ M.I. GURSAN, A.P. BUTURLAKIN, et al., Fiz. tverd. Tela 19, 3068 (1977).

/5/ V.A. STEFANOVICH, YU.V. VOROSHILOV, et al., Zh. prikl. Spektroskopii 31, 80 (1979).

(Received May 3, 1985)

phys. stat. sol. (b) 131, K145 (1985)

Subject classification: 13.4 and 18; 22.2.3

Sektion Physik der Humboldt-Universität zu Berlin[1)] (a) and
Institute of Metal Physics, Academy of Sciences of the USSR,
Ural Scientific Centre, Sverdlovsk[2)] (b)

Magneto-Optical Investigations in the Submillimeter Wavelength Range on p-InSb Bicrystals

By

B. MENGEL (a), H.-U. MÜLLER (a), F. LUDWIG (a),
V.L. KONSTANTINOV (b), and R. HERRMANN (a)

Dedicated to Prof. Dr. Dr. h.c. Dr. E.h. P. GÖRLICH
on the occasion of his 80th birthday

Magneto-optical measurements in the submillimeter wavelength range from $\lambda = 57\ \mu m$ to $\lambda = 164\ \mu m$ (21.78 to 7.57 meV) were performed on p-InSb bicrystals (d = 1.2 mm) with a carrier concentration $n_A - n_D = 5\times10^{15}\ cm^{-3}$ and an n-type grain boundary (tilt angle of the grain boundary $\Theta = 13^o$ along the rotation axis [112] of two p-InSb crystals) at a temperature T = 1.8 K and magnetic fields up to B = 6.5 T.

Shubnikov-de Haas investigations by Herrmann et al. /1/ on the same samples showed that due to the existence of potential barriers on both sides of the grain boundary the electron motion in the grain boundary is quantized and therefore discrete subbands are formed at low temperatures and magnetic fields $B > 0.45$ T. From the oscillation period three occupied subbands in the inversion layer could be identified /2/.

First theoretical calculations by Schurig and Herrmann /3/ of the band structure of an n-inversion layer for a symmetric grain boundary indicate that the transition energy between the last occupied and the next unoccupied subband lies in the range of the SMM radiation and that the effective masses of the carriers in the subbands could be determined by absorption measurements.

During the transmission measurements performed by means of the SMM spectrometer /4/ the propagation vector of the SMM wave $\vec{q}$ was parallel to the applied magnetic field $\vec{B}$ ($\vec{q} \parallel \vec{B}$, Faraday configuration). The transmitted signal was detected by a polished Allen-Bradley carbon resistor.

---

1) Invalidenstr. 110, DDR-1040 Berlin, GDR.

2) Kovalevskii ul. 18, 620219 Sverdlovsk, USSR.

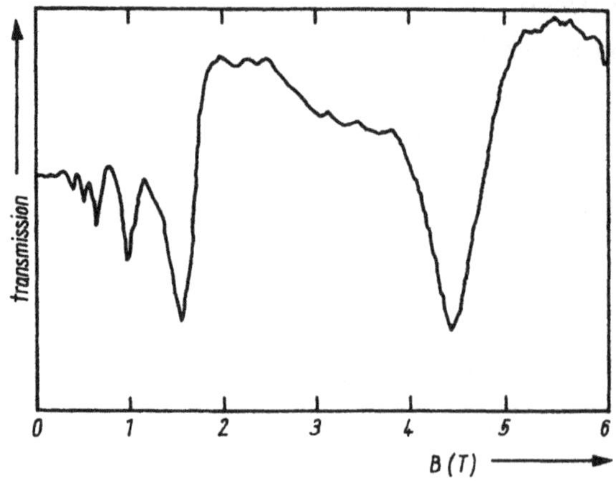

Fig. 1. Transmission spectrum at λ = 70.6 μm and T = 1.8 K

Fig. 1 shows the transmission spectrum at λ = 70.6 μm (E_{SMM} = 17.55 meV) for a temperature T = 1.8 K. Six distinct absorption maxima at different values of the magnetic field can be recognized.

Comparative measurements on samples of the same crystal with the same thickness containing no grain boundaries showed the same absorption spectrum. Therefore the absorption resonances detected at T = 1.8 K are caused by transitions between the s-like acceptor ground state and bound p-like light-hole states /5/. In agreement with measurements by Little et al. /6/ who investigated bound holes in p-InSb samples with concentrations $n_A - n_D \approx \approx 10^{14}\ cm^{-3}$ using higher excitation energies (60 to 140 meV) the binding energy of the ground state of the acceptors was determined from our experimental data, i.e. from the dependence of the transition energies of bound holes on the magnetic field (Fig. 2) as E_A = (9.0 ± 0.6)meV at B = 0 T. This result shows that a very high amount of shallow acceptors (Cd or Zn) exists in the p-InSb bicrystals used and that the strong absorption effects in the bulk material connected with these acceptors mask the expected, relatively small absorption effects in the n-inversion layer of the grain boundary (in n-inversion layers of InSb MIS structures the change of the absorption signal in

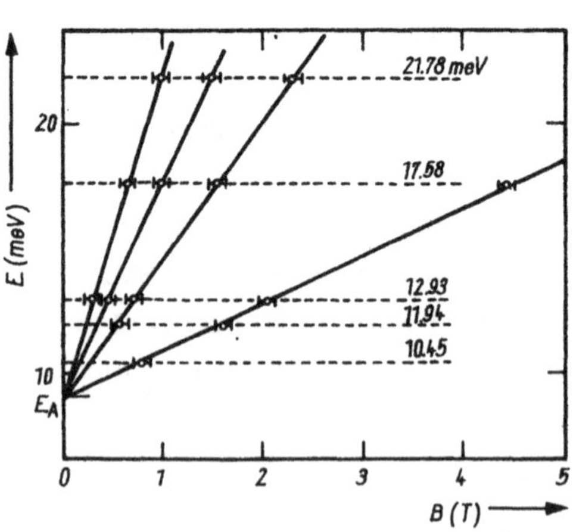

Fig. 2. Transition energies of bound holes versus magnetic field

measurements of the cyclotron resonance is of the order of 1 to 3%, cf. /7/. Investigations of the absorption behaviour of Ge-doped p-InSb crystals with carrier concentrations $n_A - n_D = 5.7 \times 10^{14}$ cm^{-3} showed that due to the higher binding energy of the ground state of Ge acceptors (≈ 100 meV) no resonance effects can be observed in the absorption signal for excitation energies E = = 7.56 meV to E = 21.74 meV. Therefore this bulk material should be used when investigating the n-inversion layers in p-InSb bicrystals.

References

/1/ R. HERRMANN, W. KRAAK, G. NACHTWEI, and G. WORM, Solid State Commun. 52, 843 (1984).

/2/ W. BRAUNE, U. PREPPERNAU, N. KUBICKI, N. PRUSS, and R. HERRMANN, 24. Internat. Conf. Low Temp. Phys. and Techn., Berlin 1985, in the press.

/3/ TH. SCHURIG and R. HERRMANN, phys. stat. sol. (b) 127, K159 (1985).

/4/ R. HERRMANN, H.-U. MÜLLER, TH. SCHURIG, U. PREPPERNAU, R. KOCH, and B. MENGEL, Exper. Tech. Phys. 32, 449 (1984).

/5/ P.J. LIN-CHUNG and B.W. HENVIS, Phys. Rev. B 12, 630 (1975).

/6/ C.L. LITTLE, D.G. SEILER, R. KAPLAN, and R.J. WAGNER, Phys. Rev. B 27, 7473 (1983).

/7/ J.H. CRASEMANN, U. MERKT, and J.P. KOTTHAUS, Phys. Rev. B 28, 2271 (1983).

(Received August 15, 1985)

phys. stat. sol. (b) 131, K149 (1985)
Subject classification: 13.4 and 18; 22.2.3

Sektion Physik der Humboldt-Universität zu Berlin[1)]

Magneto-Optical Investigations on n-InSb in the Far Infrared

By

F. LUDWIG, H.-U. MÜLLER, B. MENGEL, and R. HERRMANN

Dedicated to Prof. Dr. Dr. h.c. Dr. E.h. P. GÖRLICH
on the occasion of his 80th birthday

For many years magneto-optical measurements in the far infrared play an important role in the investigation of the electronic and donor states in n-InSb quantized by high magnetic fields. Besides experimental spectra obtained by SMM spectrometers, theoretical calculations were performed of the energy spectrum of shallow donors in high magnetic fields ($\gamma = \hbar\omega_c/2R_y^* \gg 1$) and of the selection rules for optical dipole transitions.

Since the difference between the transition energies of different allowed transitions is essentially smaller than the linewidth of the transitions an unambiguous identification of different transitions is a problem. Whereas the magneto-optical transitions allowed by the selection rules /1/ from the donor ground state $(000)^+$ to the excited states $(0\bar{1}0)^+$, $(001)^+$, and $(110)^+$ have been detected in good agreement with theoretical calculations, there are still some unsolved questions concerning the identification of spin-flip transitions and transitions from excited donor levels. Thus, it follows from theoretical calculations of the energy levels of shallow donors in a high magnetic field /2/ that the transition energies of the cyclotron resonance of free electrons $0^+ \rightarrow 1^+$ and the donor transition $(0\bar{1}0)^+ \xrightarrow{+} (100)^+$ allowed at equal polarisation of radiation /1/ differ by less than 50 μeV in magnetic fields from 1 to 3 T, whereas the experimentally observed halfwidth of this resonance line is about 200 μeV. This transition is generally denoted as cyclotron resonance of free electrons (e.g. /3/).

In the following it will be shown that at temperatures $T \lessapprox 5$ K this line is mainly caused by electronic transitions from the first excited state of the donor levels.

---

1) Invalidenstr. 110, DDR-1040 Berlin, GDR.

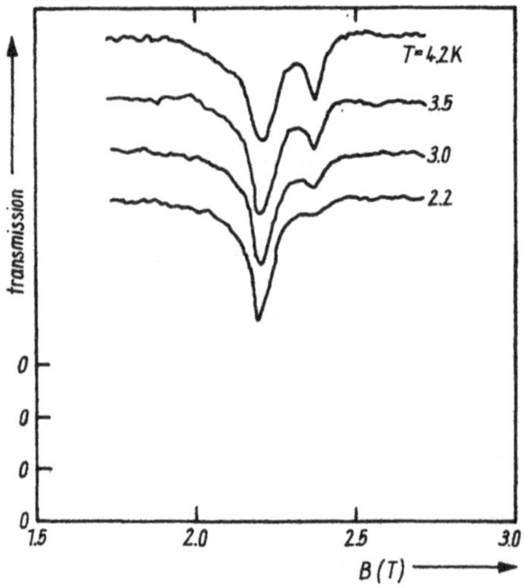

Fig. 1. The transmission spectrum at $\lambda = 70.6\ \mu m$ for different sample temperatures. The intensity of the SMM radiation was chosen so that the equilibrium distribution of the electron system was only negligibly disturbed

Magneto-optical measurements were performed at $\lambda = 70.6\ \mu m$ (17.58 meV) in transmission for different sample temperatures. The radiation source of the SMM spectrometer used /4/ was an optically pumped molecular gas laser containing CH_3OH vapour. The SMM wavelength chosen allows a sufficiently high resolution. The intensity was chosen so that the equilibrium distribution of the electron system was only negligibly disturbed. The transmitted laser beam was detected by a polished Allen-Bradley carbon resistor.

By means of a superconducting solenoid magnetic fields up to 6.5 T could be realized. During the transmission measurements performed the propagation vector of the SMM wave $\vec{q}$ was parallel to the applied magnetic field $\vec{B}$ (Faraday configuration).

The sample used was polished to a thickness of d = 50 μm, etched with CP4, and has a net carrier concentration $n_D - n_A = 3.2\times10^{13}\ cm^{-3}$ and electron mobilities $\mu_{77K} = 3.8\times10^5\ cm^2/Vs$.

At $\lambda = 70.6\ \mu m$ transmission spectra were measured in the temperature range from 1.8 to 5.0 K (Fig. 1). From the experimental spectra the temperature dependence of the absorption coefficient was determined for both transitions.

It follows from the known equations for the absorption coefficient of optical dipole transitions that the maximum of the absorption coefficient of a magneto-optical transition depends on the carrier concentration n of the initial state of the transition investigated, its oscillator strength f, and its halfwidth Γ in the following way:

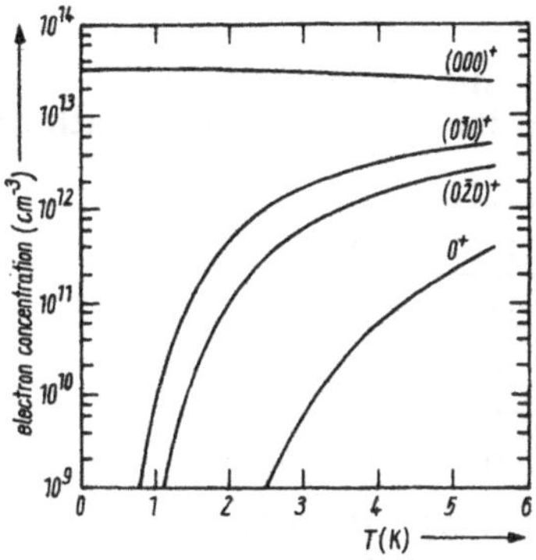

Fig. 2. The theoretically calculated electron concentrations of the energetically lowest donor states $(000)^+$, $(0\bar{1}0)^+$, $(0\bar{2}0)^+$ and the lowest Landau level 0^+ as a function of temperature; B = 2.3 T, $n_D - n_A = 3.2 \times 10^{13}$ cm^{-3}, K = 0.97

$$\alpha_{max} \sim \frac{nf}{\Gamma} .$$

It has been shown experimentally as well as theoretically that the line broadening of cyclotron resonance /3, 5/ and of transitions between discrete energy levels of shallow donors /6, 7/ in pure n-InsB in strong, quantizing magnetic fields at low temperatures is determined by scattering on ionized impurities and does not show an explicit temperature dependence. Since therefore under the experimental conditions described the oscillator strength as well as the half-width are temperature independent, the temperature dependence of the maximum of the absorption coefficient of a magneto-optical transition is determined only by the electron concentration of the initial state.

Fig. 2 shows the electron concentrations of the energetically lowest donor levels and of the lowest Landau level as a function of temperature self-consistently calculated for the case of thermal equilibrium.

It comes out that the temperature dependence of the absorption coefficient α_{max} of the impurity cyclotron resonance $(000)^+ \rightarrow (110)^+$ coincides with that of the electron concentration calculated for the ground state $(000)^+$.

Table 1 presents the experimentally determined temperature dependence of the absorption coefficient for the line observed at B = 2.33 T in comparison with the calculated temperature dependence of the levels $(0\bar{1}0)^+$ and 0^+. It is clearly seen that the temperature dependence of the absorption coefficient α_{max} of the investigated transition agrees with that of the electron concentration of the first excited donor level $(0\bar{1}0)^+$.

Therefore it can be concluded from these investigations that the line observed in the spectra originates not from the cyclotron resonance of free electrons but from the excited donor transition $(0\bar{1}0)^+ \rightarrow (100)^+$. This is caused by the essentially stronger occupation of the $(0\bar{1}0)^+$ level than of the Landau level 0^+ at temperatures $T \lesssim 5$ K for highly compensated samples.

Table 1

Comparison of the temperature dependence of the experimentally determined absorption coefficient with that of the electron concentrations of the levels $(0\bar{1}0)^+$ and 0^+

T(K)	5.0	4.2	3.5	3.0	1.8
α_{max} (cm^{-1})	54.0	41.6	30.5	21.5	4.8
norm.	1.29	1	0.73	0.51	0.11
$n(0^+)$ (10^{10} cm^{-3})	22.8	8.27	2.29	0.63	0.00148
norm.	2.76	1	0.28	0.08	0.0002
$n(0\bar{1}0)^+$ (10^{12} cm^{-3})	4.48	3.51	2.50	1.74	0.28
norm.	1.27	1	0.71	0.50	0.08

References

/1/ W. ZAWADZKI and J. WLASAK, Proc. Advanced Study Inst. Theoretical Aspects and New Developments in Magneto-Optics, Antwerpen 1979.

/2/ P.J. LIN CHUNG and B.W. HENVIS, Phys. Rev. B 12, 630 (1975).

/3/ O. MATSUDA and E. OTSUKA, J. Phys. Chem. Solids 40, 809 (1979).

/4/ R. HERRMANN, H.-U. MÜLLER, TH. SCHURIG, U. PREPPERNAU, R. KOCH, and B. MENGEL, Exper. Tech. Phys. 32, 449 (1984).

/5/ B.D. MC COMBE, R. KAPLAN, R.J. WAGNER, E. GORNIK, and W. MÜLLER, Phys. Rev. B 13, 2536 (1976).

/6/ D.M. LARSEN, Phys. Rev. B 8, 535 (1973).

/7/ F. LUDWIG, Diplomarbeit, Humboldt-Universität zu Berlin, 1985.

(Received August 15, 1985)

phys. stat. sol. (b) 131, K153 (1985)

Subject classification: 13.4 and 20.1; 19; 22.5.3

Department of Physics, Tbilisi State University[1)] (a) and
Wihuri Physical Laboratory and Department of Physical Sciences,
University of Turku[2)] (b)

Discrete Saturation of the Optical Absorption Line of Pr^{3+} in CaF_2 under the Condition of Pumping

By

T.SH. ABESADZE (a) and M. PUNKKINEN (b)

Dedicated to Prof. Dr. Dr. h.c. Dr. E.h. P. GÖRLICH
on the occasion of his 80th birthday

Recently experiments have been performed to investigate a) hole burning in inhomogeneously broadened lines of the optical absorption of Pr^{3+} ions and b) the optical NMR detection of ^{19}F nuclei, both in CaF_2 crystals /1, 2/. Hole burning was due to the pumping of superhyperfine sublevels of Pr^{3+} ions, split by the superhyperfine interaction (SHFI) with F^- nuclei, by a polarized laser beam. It is established that the origin of non-equilibrium population of the superhyperfine sublevels of paramagnetic centers is connected with spin flip-flop processes between the nuclei surrounding Pr^{3+}, while the ion is in the excited state. This mechanism of hole formation is predominant, since the probability of optical transitions with ^{19}F nuclear spin flip (which we further call forbidden transitions) is rather small. But is should be noticed that if superhyperfine interaction is strong enough in an excited state of a paramagnetic ion, the above mentioned mechanism will not work. On the other hand, it has been shown in /3/ that in the case of the fulfilment of the so-called discrete saturation (DS) condition, the relative probability of the allowed optical transitions (i.e. those which are not accompanied by a change of the fluorine nucleus spin projection) and that of the forbidden ones may become of the same order of magnitude. In the case of transition between the lowest crystal field components of $^3H_4$ and $^1D_2$ of a Pr^{3+} ion at 16 830 cm^{-1}, the above-mentioned condition is

$$A_0 \approx \hbar \omega_I , \qquad (1)$$

1) 380028 Tbilisi, USSR.

2) SF-20500 Turku, Finland.

where A_0 is the constant of the diagonal part of the SHFI, and ω_I is the Larmor frequency of the fluorine nuclear spin in the permanent external magnetic field. The excited state of the term D_2 is a singlet, the respective SHFI being ≈ 0, which corresponds to the case (b) in /3/.

Since the constant of the diagonal part of SHFI is different for different nuclei, the condition (1) may be selectively satisfied only for magnetically equivalent nuclei. Thus, under DS conditions, side by side with the hole-burning mechanism described in /1, 2/, another mechanism caused by forbidden optical transitions is possible. The hole-burning rate should increase at external magnetic field values satisfying the condition (1).

We should note that such mechanism of hole burning takes place at optical pumping of hyperfine sublevels of the Pr^{3+} nucleus in the absence of the external magnetic field /4, 5/.

After hole burning in the absorption line of Pr^{3+}, the scanning of the adjacent regions by the weak detecting laser beam made it possible to detect regions of increased absorption, or antiholes /1/. These antiholes are separated by ≈ 10 MHz from the burnt hole. Antiholes were not revealed at larger distances. This points to the fact that due to the optical pumping of the superhyperfine sublevels of the ion ground state, the increased population is created only at sublevels with corresponding z-projection of the total nuclear spin of the nearest nuclei differing by unity from the projection of the sublevel emptied by the laser beam. The pumping has a much smaller effect on the sublevels, which have a projection of the total spin differing by more than unity from the projection of the emptied sublevel, because of several nuclear flip-flops required at the optically excited state. Therefore, spectral diffusion and spin-lattice relaxation have sufficient time to destroy the non-equilibrium populations of these sublevels.

We are proposing a different situation which should appear when the DS condition is fulfilled. For instance, when the external magnetic field is directed along the symmetry axis C_4, and the above condition is satisfied for the equivalent nearest-neighbour nuclei, instead of two antiholes, a series of antiholes should appear with the following interval between them:

$$\varepsilon_0 = \sqrt{(\hbar\omega_0 - A_0)^2 + |B_0|^4/4} \quad , \tag{2}$$

where B_0 is the constant of non-diagonal SHFI of $S^zI^{\pm}$ type in the ground state of Pr^{3+} /3/.

In fact, according to the mechanism described in the present note, the increased population is simultaneously created at all the sublevels involving the nearest-neighbour fluorine nuclei indicated above, which leads to the appearance of an antihole series.

Here NMR saturation of these nuclei at the frequency $\varepsilon_0/\hbar$ leads to a decrease in the hole depth and in the heights of all antiholes.

In conclusion, we should note that according to /3/, the DS phenomenon, when the condition (1) satisfied, can be observed especially clearly in the absorption spectra of polarized light, if hole burning and scanning of the line in its vicinity is performed during a time shorter than the excited state lifetime. This means a time less than 0.5 ms for the $^3H_4 \leftrightarrow {}^1D_2$ transition of the Pr^{3+} ion in CaF_2.

One of us (T.Sh.A.) would like to express his gratitude to L.L. Buishvili and B.G. Berulava for fruitful discussiots. He would also like to thank the Ministry of High Education of the USSR and the Ministry of Education of Finland for sponsoring his stay in Finland.

References

/1/ R.M. MAC FARLANE, R.M. SHELBY, and D.P. BURUM, Optics Letters 6, 593 (1981).

/2/ D.P. BURUM, R.M. SHELBY, and R.M. MAC FARLANE, Phys. Rev. B 25, 3009 (1982).

/3/ T.SH. ABESADZE, Fiz. tverd. Tela 21, 3114 (1979).

/4/ L.E. ERIKSON, Phys. Rev. B 16, 4731 (1977).

/5/ R.M. SHELBY, R.M. MAC FARLANE, and C.S. YANNONI, Phys. Rev. B 21, 5004 (1980).

(Received December 27, 1984)

phys. stat. sol. (b) 131, K157 (1985)

Subject classification: 13.4 and 20.3; 22.8

Laboratoire de Physique du Solide, Université de Perpignan[1)]

Donor-Acceptor Pair Transitions in $AgInS_2$

By

E. REDJAI and G. MASSÉ

Dedicated to Prof. Dr. Dr. h.c. Dr. E.h. P. GÖRLICH
on the occasion of his 80th birthday

The I-III-VI_2 compounds, which $AgInS_2$ belongs to, are direct energy gap semiconductors (chalcopyrite structure) which show very interesting electrooptical properties /1, 2/. $AgInS_2$ is unique among these compounds as it can exist in two crystal forms: chalcopyrite and orthorhombic /3, 4/. The orthorhombic form is stable at high temperature ($T > 620$ °C) and the quadratic form at low temperature ($T < 620$ °C). The band structure at the Γ point has been given by Shay et al. /5/ according to electroreflectance measurements. In this note we are only interested in the chalcopyrite form, whose energy gap is 1.90 eV at low temperature and 1.87 eV at room temperature /5/. $AgInS_2$, annealed in sulphur at 600 °C for several days, shows a characteristic luminescence spectrum which is investigated in this note.

The crystals used are parts of an ingot obtained by slow cooling (2 K/h) of the liquid formed by the stoichiometric mixture of the elements (99.999% purity) sealed in a quartz tube and heated at 1000 °C during 48 h. We have verified by X-ray analysis that the ingot obtained is formed of chalcopyrite $AgInS_2$. The crystals were then annealed at 600 °C in the sulphur during 10 days.

The luminescence was excited by an electron gun with an acceleration voltage of 20 kV. The light was analyzed by means of an Oriel spectrometer followed by a photomultiplier and a PAR box-car integrator. Fig. 1 shows the luminescence spectrum at liquid He at various times t after the excitation pulse end. A shift of the emission peak towards low energies is observed when t increases. The weaker emission bands lying on the low energy side of the spectrum and spaced rather regularly also shift towards low energies. They could be phonon replicas of the main emission. Fig. 2 shows the luminescence spectrum for two different excitation intensities. A shift of the emission peak

---

1) av. de Villeneuve, 66025 Perpignan, France.

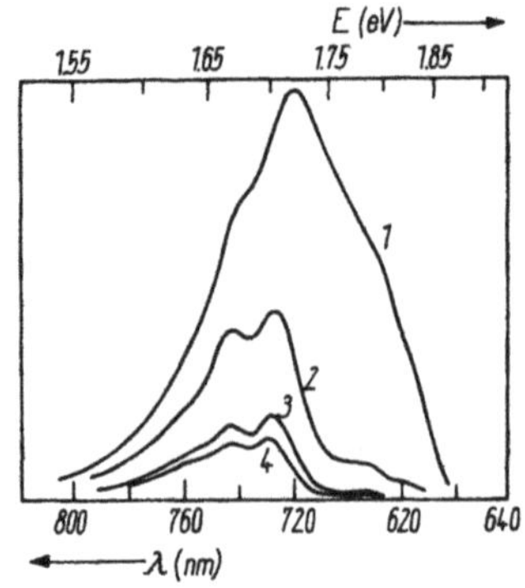

Fig.1

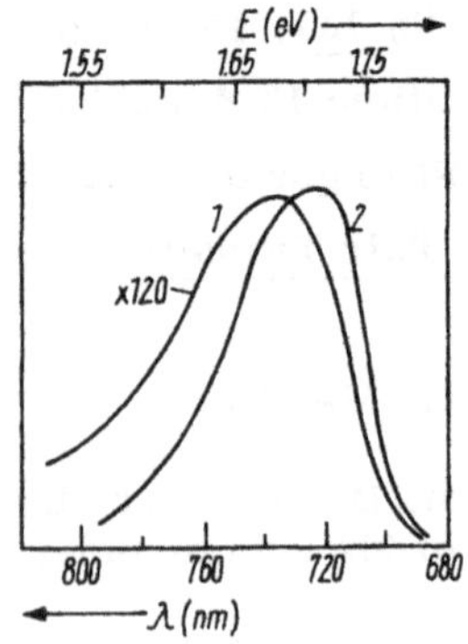

Fig.2

Fig. 1. Time resolved spectra at different times after the pulse end, at liquid He: (1) t = 0, (2) 0.8, (3) 1.6, (4) 5 μs

Fig. 2. Emission at 77 K for two excitation intensities: (1) low excitation (i = 5 mA/cm^2, U = 20 kV), (2) high excitation (i = 150 mA/cm^2, U = 20 kV)

towards low energies can be seen when the excitation intensity decreases. Whence, we interpret the luminescence of S annealed $AgInS_2$ by donor-acceptor pair transitions for which the energy of the photon emitted is given by /6/

$$h\nu = E_g - (E_A + E_D) + \frac{e^2}{\varepsilon r} + f(r) \quad , \tag{1}$$

where r is the intrapair separation and f(r) is only important when r is of the order of the lattice constant. From the TRS spectra, we may evaluate the ionization energies of the impurities involved, using the equation /7/

$$y = \ln t = \ln\left(1 - \frac{h\nu_{MX} - h\nu_{\infty}}{E_B}\right) + \frac{4\,E_B}{h\nu_{MX} - h\nu_{\infty}} - \ln W_{MX} \tag{2}$$

Here E_B is the ionization energy of the shallowest level, t is the time after the pulse end, W_{MX} is the time constant, and $h\nu_{MX}$ is the peak energy:

t (μs)	0	0.3	0.8	1.6	2.5	5
$h\nu_{MX}$ (eV)	1.720	1.710	1.706	1.703	1.701	1.699

$h\nu_{\infty} = E_g - (E_A + E_D)$ can be evaluated from the last TRS spectrum, as the low energy base of this spectrum:

$$h\nu_{\infty} \approx 1.675 \text{ eV}.$$

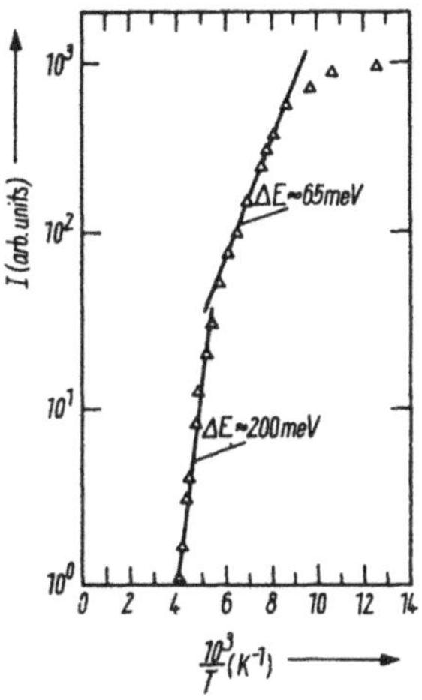

Fig. 3. Thermal quenching of the 1.720 eV emission

There is a fair amount of uncertainty about this energy, as the spectrum is generally broadened by the phonons.

Fitting (2) gives: $E_B \approx 50$ meV, $W_{MX} \approx 3 \times 10^8\ s^{-1}$. The depth of the other level is: $E_{A,D} \approx 180$ meV. We consider the error on the energies to be < 10 meV.

We can compare these values with the activation energies found by thermal quenching (Fig. 3). The first slope corresponds to the thermal emptying of the shallowest level /8, 9/. The value found for the activation energy ($\approx$65 meV) is of the same order of magnitude as the fitting result. The second slope is difficult to interpret and does not generally represent a significant result.

References

/1/ J.L. SHAY and J.H. WERNICK, Ternary Chalcopyrite Semiconductors, Pergamon Press, New York 1975.

/2/ G.D. HOLAH (Ed.), Ternary Compounds, Third Internat. Conf. Ternary Compounds, Edinburgh 1977.

/3/ R.S. ROTH, H.S. PARKER, and W.S. BROWER, Mater. Res. Bull. 8, 333 (1973).

/4/ N.V. JOSHI, L. MARTINEZ, and R. ECHEVERRIA, J. Phys. Chem. Solids 42, 281 (1981).

/5/ J.L. SHAY, B. TELL, L.M. SCHIAVONE, H.M. KASPER, and F. THIEL, Phys. Rev. B 9, 1719 (1974).

/6/ D.G. THOMAS, J.J. HOPFIELD, and W.M. AUGUSTINYAK, Phys. Rev. 140, 202 (1965).

/7/ N. LAHLOU and G. MASSE, J. appl. Phys. 52, 978 (1981).

/8/ K. MAEDA, J. Phys. Chem. Solids 26, 595 (1965).

/9/ A. HALPERIN and E. ZACKS, Phys. Rev. B 11, 2237 (1974).

(Received June 18, 1985)

phys. stat. sol. (b) 131, K161 (1985)
Subject classification: 20.3; 22.5.3

Physics Department, Memphis State University[1)]

Changes in the Spectrum of Thermally Stimulated Emission from CaF_2:Mn at Low Temperatures

By
V.K. JAIN[2)] and M. SHAH JAHAN

Dedicated to Prof. Dr. Dr. h.c. Dr. E.h. P. GÖRLICH
on the occasion of his 80th birthday

Calcium fluoride doped with Mn is a highly thermoluminescent (TL) material. The high sensitivity of this material has led to its use in environmental and personnel monitoring /1/. This widespread use led to detailed investigations of its TL characteristics. The glow curve of CaF_2:Mn above room temperature (RT) is relatively simple, with the main peak at about 260 ^{o}C. The concomitant emission spectrum is a broad band with maximum at 500 nm /1, 2/. Schulman attributed this band to holes trapped at Mn^{2+} sites, and mentioned that Mn^{2+} centers act as hole traps /2/. Puite and Arends later, in their combined TL and thermally stimulated exo-electron (TSEE) studies, confirmed Schulman's suggestion /3/. In recent years, TL of CaF_2:Mn has been studied at liquid nitrogen temperature (LNT). The glow curve has variously been reported to have three peaks /4/ or just one peak /5/, with wide variations in peak temperatures. The main peak temperature is reported as 200 to 210 K by some /4, 9/ and 175 K by others /5/. Part of the explanation may lie in the different amounts of the impurity Mn present in the host lattice. The emission spectrum, on the other hand, has consistently been observed to be a broad band with maximum around 500 nm /5 to 8/, as is the case for room temperature irradiation. CaF_2:Mn also exhibits an absorption band at 450 nm upon X-irradiation. This absorption has been assigned to the $^7S(3d^5 4s) \longrightarrow {}^7P(3d^5 4p)$ transition of Mn^+ /9/. The electron spin resonance (ESR) measurements on CaF_2:Mn, X-irradiated at 77 K, show the presence of a new signal which has been identified as due to Mn^+ centers /10, 11/. Baranov reached this conclusion after combined ESR and optical absorption (OA) studies /10/. Jahan

1) Memphis, Tennessee 38152, USA.

2) Permanent address: Health Physics Division, Bhabha Atomic Research Center, Bombay, India.

and Cooke observed the thermal decay of the ESR signal, calculated its activation energy, correlated it with the 200 K TL peak, and concluded the signal to be from Mn^+ /11/. Alcala et al., after a more detailed study, concluded that the 500 nm emission at low temperatures is due to the de-excitation of Mn^{2+} as described below /9/.

During low temperature irradiation of $CaF_2:Mn^{2+}$, Mn^+ ions and modified V_K centers (V_{KA} centers) are created. Bleaching in the Mn^+ band (450 nm) ionizes these ions producing free electrons and Mn^{2+} in an excited state. Inversely, holes released by bleaching V_{KA} centers, optically or thermally, recombine with Mn^+, transforming it to Mn^{2+} in an excited state. The de-excitation of Mn^{2+} from its excited state gives the 500 nm emission. TL peaks are, therefore, said to be due to the recombination of holes released from several hole traps with electrons trapped in the impurity ion.

Ce is known to act as a sensitizer for the Mn^{2+} emission in CaF_2 and $CaSO_4$ /12 to 14/. In fact, Mn alone is said to be poor as a luminescent impurity dopant. A co-dopant such as Ce acts as sensitizer and enhances its luminescence efficiency /7, 12 to 14/. But CaF_2:Mn, without deliberate Ce doping, has been reported to be highly thermoluminescent /1, 2, 15/, giving a strong 500 nm emission. No other emissions have been reported to have been observed in CaF_2:Mn. In our recent investigations we found another band with maximum around 345 nm in the emission spectrum of 0.1% Mn-doped CaF_2, and tentatively attributed it to Ce emission /16/. This note reports on further changes observed in the TL spectrum of CaF_2:Mn.

The crystal used was obtained from Optovac, Inc., and had an impurity content of 0.1% Mn which was found to be 0.11% by neutron activation analysis /17/. Neutron activation analysis also revealed the presence of several other impurities, e.g., Na (96), Fe (80), Sr (240), Ba (55), La (15), Yb (0.18), and Ce (0.75 ppm). The crystal was used for ESR and TL experiments from LNT to RT and exposed to a total dose of more than 1 Mrad. It was then annealed at 450 °C for 1 h and used for further TL, photobleaching and phototransferred TL (PTTL) measurements from LNT to RT and never heated above RT. Thus, the crystal was once again exposed to a total dose of about 1 Mrad. The irradiation always took place at about 83 K. Experimental apparatus and methods used in this work are described elsewhere /16/. During the course of these studies, the glow curve and emission spectrum were repeatedly recorded. A

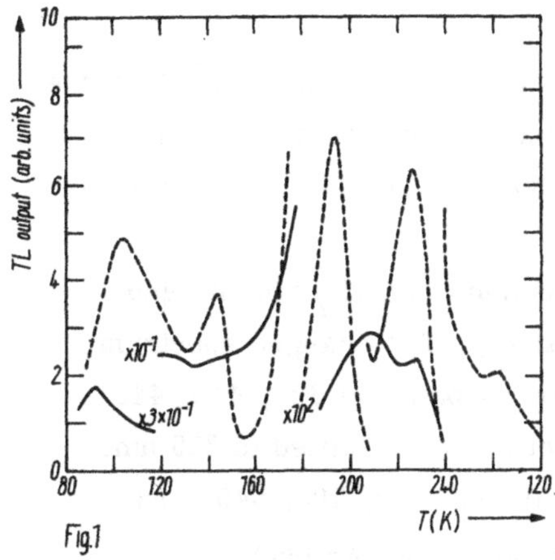

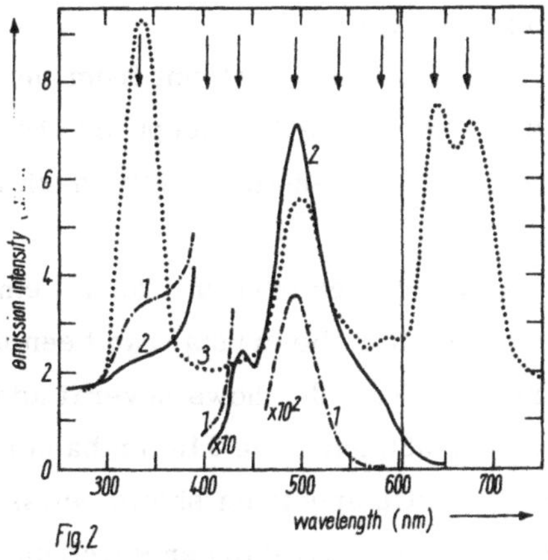

Fig. 1. Typical glow curve of CaF_2:Mn (----) and CaF_2:Ce (——)

Fig. 2. Emission spectra of thermoluminescence: (1) typical in CaF_2:Mn (—·—), (2) changed in CaF_2:Mn (———), (3) typical in CaF_2:Ce (· · · ·)

typical glow curve is shown in Fig. 1 by the dashed line curve. Contrary to most reports seen in literature, five peaks were resolved at 97, 145, 194, 230, and 265 K, and one peak was indicated at 250 K. The emission spectrum always showed a weaker band with maximum at about 345 nm in addition to the well-known intense emission at 495 nm. This is represented by the dot-dashed line, curve 1 of Fig. 2. The 345 nm emission intensity was maximum for the 97 K peak and diminished with increasing temperature although some increase corresponding to the 194 K peak temperature was noted. This simple spectrum of two bands changed dramatically after the crystal had been exposed repeatedly to X-rays for a total dose of over 1 Mrad. The changed spectrum taken at 163 K is shown in Fig. 2 by the solid line, curve 2. However, as the temperature was increased, the spectrum reverted gradually to that of curve 1. At temperatures of 200 K and higher, the spectrum was the same as depicted by curve 1.

The glow curve of a 0.1% Ce-doped sample (also obtained from Optovac, Inc.) is shown in Fig. 1 by the continuous line curve. The first and the third peaks of the CaF_2:Ce sample appear to correspond to the 97 and 145 K peaks, respectively, of the CaF_2:Mn crystal. Initially, instead of two peaks at 97 and

145 K, CaF_2:Mn showed a peak at 124 K, which corresponds to the second peak from CaF_2:Ce.

The emission spectrum from the CaF_2:Ce sample is shown in Fig. 2 by the dotted line, curve 3. Except for changes in relative intensities, the emission from CaF_2:Ce remains pretty much the same over all the glow peak temperatures.

As mentioned earlier, the TL emission spectrum of CaF_2:Mn changed radically after the crystal had been used repeatedly. The changed spectrum, curve 2 in Fig. 2, shows several additional narrow bands at 325, 410, 440, 545, and 580 nm. The 345 nm band is either not there or shifted to 325 nm. The CaF_2:Ce spectrum shows emission bands at 337, 420, 500, 545, and 580 nm. The second order spectrum from this sample shows peaks at 625 and 675 nm corresponding to the real ones at 312 (which is not otherwise resolved) and 337 nm. These two (312 and 337 nm) are well-known emission bands of Ce in CaF_2 /18/. The 410 and the 440 nm bands of CaF_2:Mn seem to merge into one at 420 nm in CaF_2:Ce. But for these differences, the two emission spectra resemble each other and it may be said that the new changed emission spectrum from CaF_2:Mn matches with that from CaF_2:Ce.

To summarize, the emission spectrum of the CaF_2:Mn (0.1%) crystal after repeated X-ray and ultraviolet light (UV) exposures, changes radically; however, there was no serious loss in the TL output. It is very difficult to say whether the extra lines in the emission spectra of the two samples, besides the ones reasonably known, are due to one or the other of the two impurities, viz., Mn and Ce, or due entirely to some other impurities, which are unintentional but are always present. The presence of a 500 nm band in the emission from CaF_2:Ce shows the inclusion of Mn in this sample. The fact that in the CaF_2:Mn sample, the changed spectrum has been brought about after long use, which matches with that present in a fresh sample of CaF_2:Ce, leads one to speculate. In the first place, if there is some transfer of energy from Ce to Mn, some sort of spatial association/nearness is required. In samples having very low concentrations of impurities, this may be present only to a limited extent. Even in quite heavily doped crystals, the association is not complete /19/. This, perhaps, is the reason for observing the 345 nm emission which may be attributed to Ce. Further, its is known that several energy levels of

Mn^{2+} are possible, but normally do not lead to emission transitions /10, 20/. However, after use, including heavy irradiation and annealing, there may be configurational change which makes some of these transitions possible. Additionally, low levels of doping (0.1%), as is the case in this study, may be quite important in observing emissions normally not seen at higher concentrations of the dopant Mn. In the CaF_2:Ce sample, with very low impurity concentrations, these transitions are seen to take place anyway. The five band complex, viz., 410, 440, 495, 545, and 580 nm, is perhaps due to Mn. In any case, these observations point to a dramatic change in the luminescent transitions brought about by irradiation (X-rays and UV) and warming, which result in several new emission bands normally not observed.

This work was supported in part by the Naval Surface Weapons Center.

References

/1/ Y.S. HOROWITZ, in: Thermoluminescence and Thermoluminescent Dosimetry, Vol. 1, Chap. 4, Ed. Y.S. HOROWITZ, CRC Press, Inc., 1984.

/2/ J.H. SCHULMAN, Proc. Symp. Solid State and Chemical Radiation Dosimetry in Medicine and Biology, IAEA, Wien 1967 (p. 3).

/3/ K.J. PUITE and J. ARENDS, Proc. 3rd Internat. Conf. Luminescence Dosimetry, Ed. V. MEJDAHL, IAEA/AEC,Risø 1971 (p. 680).

/4/ P.J. ALONSO and R. ALCALA, J. Lum. 21, 147 (1980).

/5/ D.W. COOKE, E.P. GAVATHAS, and M.D. BROWN, J. appl. Phys. 54, 1165 (1983).

/6/ R.J. ABBUNDI, D.W. COOKE, V.K. MATHUR, G.A. ROYCE, and M.D. BROWN, Radiat. Prot. Dosimetry 6, 329 (1984).

/7/ R.J. GINTHER, J. Electrochem. Soc. 101, 248 (1954).

/8/ V.E. BOZHEVOLNOV, L.N. IVANOV, V.K. KOZLOV, YU.V. VORONOV, YU.P. TINOFSEV, and V.V. KARELIN, phys. stat. sol. (b) 78, 483 (1976).

/9/ R. ALCALA, P.J. ALONSO, G. LALINDE, and A. CARRETERO, phys. stat. sol. (b) 98, 315 (1980).

/10/ P.G. BARANOV, Soviet Phys. - Solid State 22, 133 (1980).

/11/ M.S. JAHAN and D.W. COOKE, phys. stat. sol. (b) 126, 687 (1984).

/12/ S.C. SEN and H.N. BOSE, Z. Phys. 201, 368 (1967).

/13/ R. LEACH, J. Electrochem. Soc. 105, 27 (1958).

/14/ V.N. BAPAT, Proc. 5th Internat. Conf. Luminescence Dosimetry, São Paulo 1977 (p. 61).

/15/ J.H. SCHULMAN, F.H. ATTIX, E.J. WEST, and R.J. GINTHER, Rev. sci. Instrum. 31, 1263 (1960).

/16/ V.K. JAIN and M.S. JAHAN, to be published.

/17/ JOANNE F. RHODES, R.J. ABBUNDI, D.W. COOKE, and M.D. BROWN, Phys. Rev. B 31, 5393 (1985).

/18/ C.M. SUNTA, Radiat. Prot. Dosimetry 8, 25 (1984).

/19/ J. RUBIO O., H. MURIETA S., R.C. POWELL and W.A. SIBLEY, Phys. Rev. B 31, 59 (1985).

/20/ W.L. MEDLIN, in: Thermoluminescence of Geological Materials, Ed. D.J. MCDOUGALL, Academic Press, 1968 (p. 193 to 223).

(Received June 21, 1985)

phys. stat. sol. (b) 131, K167 (1985)
Subject classification: 13.5.4 and 19; 22.8.1

J. Stefan Institute, E. Kardelj University of Ljubljana[1)]

EPR of Tl^{2+} Doped Incommensurate K_2SeO_4

By

R. BLINC, P. CEVC, J. SELIGER, and M. KOREN

Dedicated to Prof. Dr. Dr. h.c. Dr. E.h. P. GÖRLICH
on the occasion of his 80th birthday

Introduction K_2SeO_4 is hexagonal and belongs to the space group D_{6h}^4 from the melting point down to T = 745 K /1/. An orthorhombic distortion sets in below this temperature and the crystal becomes isomorphous with Rb_2ZnCl_4. The space group is now D_{2h}^{16} with four formula units per unit cell (z = 4). An incommensurate modulation /2/ along the pseudo-hexagonal c-axis - first discovered by Iizumi et al. /3/ - sets in below T_I = 129.5 K. The "lock-in" transition connected with a tripling of the unit cell along the c-axis and the appearance of the spontaneous polarization along the b-axis occurs at T_c = 93 K. The space group is again orthorhombic, C_{2v}^9, and z = 12 /4/.

The eigenvectors of the incommensurate modulation wave in K_2SeO_4 have been recently studied in great detail by X-rays /2/. The paraelectric unit cell contains two chemically non-equivalent potassiums, K(I) and K(II), which are distributed among eight K sites. All K nuclei have site symmetry m. The incommensurate modulation is characterized by a rotation of the SeO_4 groups and an associated displacement of the K ions /2/. The transition to the ferroelectric commensurate phase below T_c is associated with a "locking" of the modulation wave to the basic lattice as well as with a homogeneous shift of the K(I) ions /4/. It should be noted, however, that the above analysis was performed in the plane-wave approximation and that very little is known on the transition from the plane-wave to the multi-soliton lattice modulation regime.

The incommensurate-commensurate (I-C) transition in Rb_2ZnCl_4 and K_2ZnCl_4 has been shown to be greatly influenced by defects /5, 6/ which are pinning the incommensurate modulation wave. In order to contribute to the further understanding of nature of the modulation wave and the role of defects in I-C transitions we decided to study the electron paramagnetic resonance

---

1) P.O.B. 53, Jamova 39, 61111 Ljubljana, Yugoslavia.

(EPR) spectra of Tl^{2+}-doped K_2SeO_4. The EPR spectra of the SeO_4^{4-} and VO_2^{+} centres in K_2SeO_4 have been already investigated /7, 8/.

Results and discussion The temperature and the angular dependences of the EPR spectra of Tl^{2+}-doped K_2SeO_4 single crystals have been studied between 300 and 3 K. The Tl^{2+} ions were found to substitute for K^+ ions.

The Tl^{2+} spectra can be described by a spin Hamiltonian

$$\mathcal{H} = \beta \vec{H} g \vec{S} + \vec{S} A \vec{I} \quad , \qquad (1)$$

where S = 1/2, I = 1/2, β is the Bohr magneton and where the g and A tensors are functions of the order parameter and thus vary in space in the phase I in a way which reflects the spatial variation of the incommensurate modulation. The ^{203}Tl and ^{205}Tl isotopes - both of which have I = 1/2 - could not be resolved in view of the fact that their magnetic moments are nearly identical. In view of the huge hyperfine coupling the energy levels are characterised by the quantum numbers $F = I \pm S = 0, 1$ and magnetic quantum numbers $m_F = 0, \pm 1$. The room temperature g and A tensors for the two non-equivalent defect sites - Tl(I) and Tl(II) - are presented in Table 1. It should be noted that the second term in (1) is much larger than the first one so that only the low field transitions were studied.

The angular dependence of the low field Tl^{2+} EPR transitions is shown in Fig. 1 for T = 250, 135, and 100 K.

The Tl(I) spectra vary only little with T and differ significantly from the Tl(II) ones which undergo a significant shift as well as a huge broadening on going into the phase I. The temperature dependence of the spectra between 150 and 10 K is shown in Fig. 2 for $\vec{H} \parallel \vec{a}$.

The Tl(I) spectra are practically unchanged down to 90 K where they start to split into three components in agreement with the triplication of the unit cell.

Table 1

Principal values of the g and A tensors (in GHz) for the Tl^{2+} centres in K_2SeO_4

	g_{xx}	g_{yy}	g_{zz}	A_{xx}	A_{yy}	A_{zz}
Tl(I)	1.9944	1.9954	1.9949	115.05	114.85	115.13
Tl(II)	1.9913	1.9929	1.9940	107.03	107.15	107.32

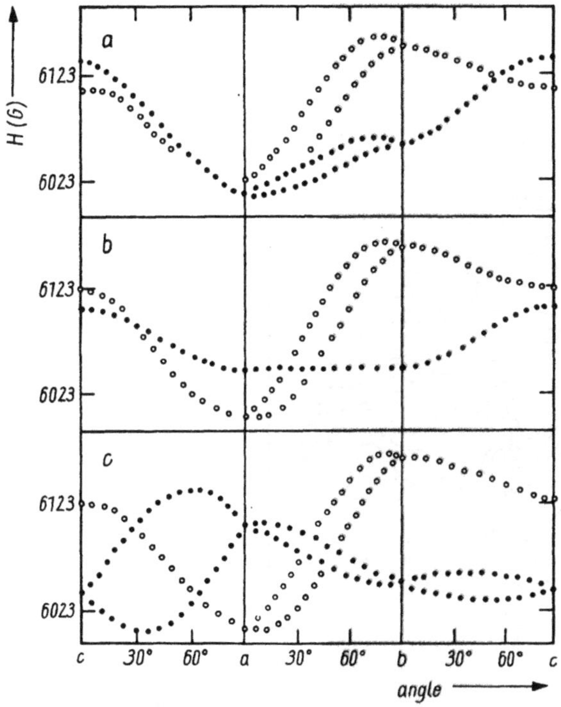

Fig. 1. Angular dependence of the low field Tl^{2+} EPR transitions at T = a) 250, b) 135, and c) 100 K; o Tl(I), ● Tl(II)

The Tl(II) spectra, on the other hand, undergo a significant shift of the centre of gravity between 140 and 90 K and in addition exhibit a typical inhomogeneous incommensurate line broadening, limited by two edge singularities. Commensurate lines characteristic of a multi-soliton lattice appear in the low temperature part of the I phase.

The Tl(II) spectra cannot be described in the usually used "local" approximation /8, 9/ where the resonance field H_i at a given site depends only on the real displacement at this site, $H_i = H\left[u_i(x_i)\right]$, and where $\vec{u}_i = \vec{u}_o \cos\varphi(x_i)$. If, however, non-local terms are taken into account, or the real displacement of the paramagnetic centre is given by an admixture of two components

$$\vec{u}_j = \vec{u}_o^c \cos\varphi(x_i) + \vec{u}_o^s \sin\varphi(x_i) \ , \qquad (2)$$

where $\vec{u}_o^c$ and $\vec{u}_o^s$ are not parallel, the theory /10/ fits the experimental data

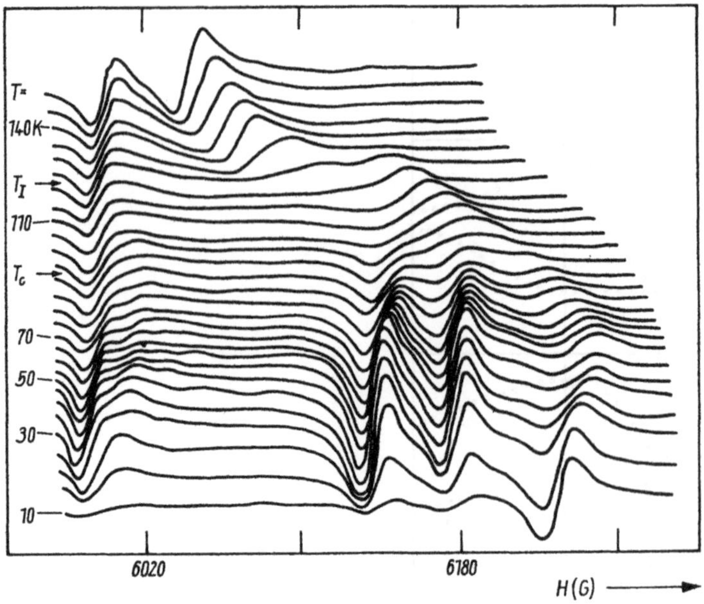

Fig. 2. Temperature dependence of the Tl^{2+} EPR lineshapes for $\vec{H} \parallel \vec{a}$

rather well. The resonant field H(x) now becomes

$$H(x) = H_o + H_1 \cos(\Psi(x) + \varphi) + H_2' + H_2 \cos^2\Psi(x) \quad , \tag{3}$$

where $\Psi(x)$ is a renormalized phase, $\cos \Psi(x)$ takes on nearly continuously all values between -1 and +1 and where $H_o \neq f(T)$, $H_1 \sim (T - T_I)^\beta$ and H_2' as well as $H_2 \sim (T - T_I)^{2\beta}$.

Since the Tl sites lie on the mirror plane, $H_1 = 0$ by symmetry if the external magnetic field lies in this plane or is perpendicular to m. This is the case for the spectra shown in Fig. 2. In the plane-wave limit one thus finds the lineshape $f(\nu) = (d\nu/d\Psi)(d\Psi/dx)$ as

$$f(\nu) = \frac{\text{const}}{\sqrt{(H - H_2' - H_o)(H_2 + H_2' + H_o - H)}} \quad ; \tag{4}$$

$f(\nu)$ now exhibits edge singularities at

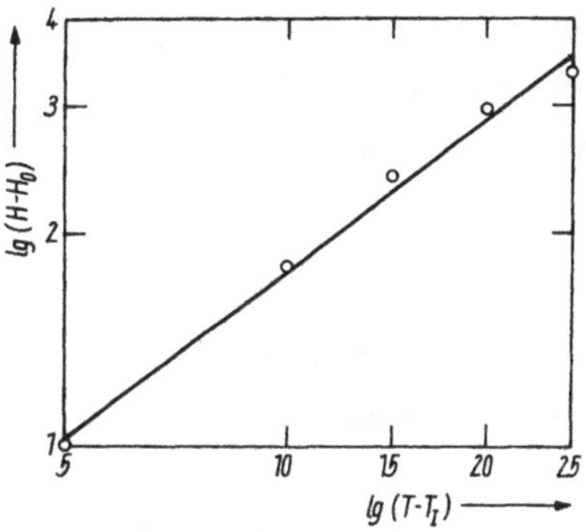

Fig. 3. Double-logarithmic plot of the position of one of the "quadratic" edge singularities $H - H_o = H_2'$ versus $T - T_I$ yielding a critical exponent $2\beta = 0.77$

$$H - H_o = H_2', \quad H - H_o = H_2 + H_2' \tag{5}$$

in contrast to the "local" case /8/ where edge singularities appear at $H - H_o = 0$ and $H - H_o = H_2$. The positions of the two edge singularities as well as the splitting between them is now proportional to $(T_I - T)^{2\beta}$. At lower temperature the plane-wave approximation breaks down and $d\Psi(x)/dx$ is obtained from the sine-Gordon equation as

$$\frac{d\Psi}{dx} = \text{const}\sqrt{\Delta^2 + \cos^2\left[n(\Psi - \Psi_o)/2\right]} \quad , \tag{6}$$

where n = 6 for K_2SeO_4 and Ψ_o is an initial phase of the modulation wave. Δ is an integration constant related to the soliton density n_s by

$$n_s = \frac{\pi}{2K\left[1/(1+\Delta^2)^{1/2}\right]} \quad , \tag{7}$$

where K is the complete elliptic integral of the first kind.

The log-log plot of the position $H - H_o = H_2'$ of one of the two Tl(II) edge singularities versus $T - T_I$ is shown in Fig. 3. The plot is a straight line yielding a critical exponent $2\beta = 0.77 \pm 0.1$ for the square of the amplitude of the modulation wave.

The soliton density n_s does not become zero in the C phase below T_c in contrast to the predictions of the equilibrium Landau theory. The value of n_s is non-zero down to 20 K. It is only below this temperature that the Tl(II) spectra become of the form expected for a C phase where $n_s = 0$. This demonstrates that the Tl^{2+} centres pin the discommensurations (i.e. phase solitons) and hinder the modulation pattern from reaching thermal equilibrium. Thus a defect induced phase with a stochastic variation of the inter-soliton distance is created intermediate between the I and the C phases. The pinned soliton density is here much larger than in nominally pure Rb_2ZnCl_4 and Rb_2ZnBr_4 /11/.

References

/1/ G. GATTOW, Acta cryst. 15, 419 (1962).

/2/ N. YAMADA and T. IKEDA, J. Phys. Soc. Japan 53, 2555 (1984).

/3/ M. IIZUMI, J.D. AXE, G. SHIRANE, and K. SHIMAOKA, Phys. Rev. B 15, 4392 (1977).

/4/ N. YAMADA, Y. ONO, and T. IKEDA, J. Phys. Soc. Japan 53, 2565 (1984).

/5/ K. HAMANO, K. EMA, and S. HIROTSU, Ferroelectrics 36, 343 (1981).

/6/ K. DEGUCHI, SHIN-ICHI SATO, K. HIRANO, and E. NAKAMURA, J. Phys. Soc. Japan 53, 2790 (1984).

/7/ A.S. CHAVES, R. GAZZINELLI, and R. BLINC, Solid State Commun. 37, 123 (1981).

/8/ M. FUKUI and R. ABE, J. Phys. Soc. Japan 51, 3942 (1982); Japan. J. appl. Phys. 20, 1533 (1981).

/9/ See, for instance, R. BLINC, Phys. Rep. 79, 331 (1981) and references therein.

/10/ R. BLINC, J. SELIGER, and S. ŽUMER, J. Phys. C, to be published.

/11/ P. PRELOVŠEK and R. BLINC, J. Phys. C 17, 577 (1984).

(Received August 2, 1985)

phys. stat. sol. (b) 131, K1 73 (1985)

Subject classification: 14.2

Institute for Low Temperature and Structure Research, Polish Academy of Sciences, Wrocław[1]

Sum Rules for Tunneling Density of States

By

J.J. WNUK

Dedicated to Prof. Dr. Dr. h.c. Dr. E.h. F. GÖRLICH on the occasion of his 80th birthday

The difference between the tunneling density of states

$$N_T(E) = \mathrm{Re}\,\frac{E}{\sqrt{E^2 - \Delta^2(E)}} \tag{1}$$

for the strong-coupling superconductor /1/ and the BCS density

$$N_{BCS}(E) = \mathrm{Re}\,\frac{E}{\sqrt{E^2 - \Delta_o^2}} \tag{2}$$

arises from the energy dependence of the complex gap parameter $\Delta(E)$. One may expect the average deviations of $N_T(E)$ from the BCS behaviour to vanish, as soon as the electron-phonon interaction strength in the 'strong-coupling' superconductor is not so great in average over energies as strongly energy dependent.

From the well known analytic properties of $\Delta(E)$ and the superconvergence theorem it may be shown that /2/

$$\int_{\Delta_o}^{\infty} \left[\frac{N_T(E)}{N_{BCS}(E)} - 1\right] dE = 0 \ , \tag{3}$$

what confirms that suggested above, but only under the assumption

$$\Delta(E \lessgtr \Delta_o) = \Delta_o \ . \tag{4}$$

However, it is possible to get rid of (4) defining a function

$$f(z) = f_1(z) + i\,f_2(z) = \frac{z}{\sqrt{z^2 - \Delta^2(z)}} - \frac{z}{\sqrt{z^2 - \Delta_o^2}} \ , \tag{5}$$

---

1) Próchnika 95, 53-529 Wrocław, Poland.

for which

$$f(-z^*) = -f^*(z) ,$$

$$f(0) = f(\infty) = 0 ,$$

$$f_1(x<\Delta_o) = 0 , \qquad f_1(x)/x \xrightarrow{x\to\infty} 0 .$$

The dispersion relation in this case has the form

$$\int_{\Delta_o}^{\infty} \frac{x\, f_1(x)}{x^2 - E^2} dx = -\frac{\pi}{2} f_2(E) , \tag{6}$$

and, for E = 0, we get

$$\int_{\Delta_o}^{\infty} f_1(E) \frac{dE}{E} = 0 \tag{7}$$

or

$$\int_{\Delta_o}^{\infty} \left[N_T(E) - N_{BCS}(E)\right] \frac{dE}{E} = 0 . \tag{8}$$

This is an exact result suggesting that the difference between $N_T(E)$ and $N_{BCS}(E)$ caused by the energy dependence of $\Delta(E)$ may be removed by taking an average over energies.

Both (3) and (8) are of importance for the tunneling spectroscopy calculations based on the method of Galkin et al. /3, 4/ as they may help to choose the proper extrapolation of $N_T(E)$ out of the experimental range (in practice, extrapolated are the functions $N_T(E)/N_{BCS}(E) - 1$ or $N_T(E) - N_{BCS}(E)$). This will make available the elimination of the Coulomb pseudopotential μ^* from calculations of the Eliashberg function $\alpha^2F(E)$ and allow to derive μ^* separately.

For practical purposes, the integral in (7) may be separated into two parts at sufficiently high cutoff energy E_{co} to obtain, expanding f(E) in powers of Δ^2/E^2,

$$\int_{\Delta_o}^{E_{co}} f_1(E) \frac{dE}{E} = \left[\Delta_o^2 - \Delta_1^2(E_{co}) + \Delta_2^2(E_{co})\right] / 4E_{co}^2 . \tag{9}$$

Next, using the dispersion relation from /5/:

$$\int_{\Delta_o}^{\infty} \frac{f_1(x)}{E^2 - x^2}\, dx = \frac{\pi}{2}\, \mathrm{Im}\, \frac{1}{\sqrt{E^2 - \Delta^2(E)}} \quad , \quad E > \Delta_o \tag{10}$$

it may be similarly derived the following approximate formula:

$$\int_{\Delta_o}^{E_{co}} f_1(x)\, dx = \left[\Delta_o^2 - \Delta_1^2(E_{co}) + \Delta_2^2(E_{co})\right] / 2E_{co} \ . \tag{11}$$

The last two rules ((9) and (11)) give now a reliable method of extrapolation of $N_T(E)$ out of the experimental energy region. This may be done as follows: the rule (9) serves as a criterion in selecting the lower energy extrapolation because the integral in (9) in converges rapidly, and then, (11) helps to choose the higher energy extrapolation. The procedure may be repeated to obtain convergence.

We have made first steps in applying of the sum rules presented here to extrapolations of the tunneling density of states. Rule (3) applied to the $N_T(E)$ data of Pb given up to 11 meV (with straight line extrapolation in the lower energy region) showed that there should be additional structure in $N_T(E)$ above 11 meV with a positive 'peak' in the $N_T(E)/N_{BCS}(E) - 1$ curve, what coincides with measurements made up to 27 meV contained in /6/. Using rules (9) and (11) with the right sides approximated by $\Delta_o^2/4E_{co}^2$ and $\Delta_o^2/2E_{co}$, respectively, we have also obtained in the high energy region a qualitative agreement with results of /6/. The $N_T(E) - N_{BCS}(E)$ data has been extrapolated by appropriately defined polynomials and the convergence between the low and high energy extrapolation has been obtained after six cycles of calculations.

It should be underlined that it is not necessary for the extrapolation to reconstruct the experimental density of states, but only to ensure that the high and low energy parts of the integral in (10) will given an appropriate contribution into the whole result. However, extrapolation which well approximates the behaviour of the density of states out of the experimental range allows one to estimate the values of the energy gap function calculated from (10) in this energy region.

References

/1/ D.J. SCALAPINO, J.R. SCHRIEFFER, and J.W. WILKINS, Phys. Rev. 148, 263 (1966).

/2/ J.J. WNUK, phys. stat. sol. (b) 128, K187 (1985).

/3/ A.A. GALKIN, A.I. DYACHENKO, and V.M. SVISTUNOV, Zh. eksper. teor. Fiz. 66, 2262 (1974).

/4/ V.M. SVISTUNOV, A.I. DYACHENKO, and O. CHERNYAK, Fiz. tverd. Tela 19, 1994 (1977).

/5/ V.M. SVISTUNOV, A.I. DYACHENKO, and M.A. BELOGOLOVSKII, J. low-Temp. Phys. 31, 339 (1978).

/6/ W.L. MC MILLAN and J.M. ROWELL, in: Superconductivity, Ed. R.D. PARKS, Marcel Dekker, New York 1969.

(Received June 25, 1985)

phys. stat. sol. (b) 131, K177 (1985)
Subject classification: 14.4; 6; 22.8.1

Institute of Physics, Academy of Sciences of the Ukrainian SSR, Kiev[1)]

The Anharmonicity Influence on the Proton Conductivity of an Hydrogen-Bonded Chain

By
V. V. KRASNOGOLOVETS and P. M. TOMCHUK

Dedicated to Prof. Dr. Dr. h.c. Dr. E.h. P. GÖRLICH
on the occasion of his 80th birthday

In our previous paper /1/ a polaron conductivity mechanism of protons (or proton holes) injected into a quasi-linear molecular H-bonded chain has been suggested. In complex compounds ($Li(N_2H_5)SO_4$ and others, proton conducting proteins) such chain is formed by means of AH groups (atom A is O, N or others) the positions of which being rather rigidly fixed by the side chemical bonds. Therefore the distances between the sites (AH groups) and, thus, the chain rigidity appear to be functions of external parameters, i.e. thermostat functions. In other words, if we neglect a comparatively weak influence of the light H atoms upon the vibrations of A atoms, then the A atom participation in acoustic vibrations of the molecular chain will be entirely environmental, A atoms being rigidly connected to the environment by the side chemical bonds. Therefore, the above-mentioned acoustic vibrations appear to be given, external with respect to the chain. Owing to this fact, a bilinear connection of the acoustic vibrations with the longitudinal optical polarizational phonons (the chain proton vibrations on the H bonds), responsible for the proton formation appears. On calculating the proton conductivity in /1/, all these peculiarities of an H-bonded chain have been taken into account.

Apart from the bilinear connection between acoustic and optical vibrations the usual phonon anharmonicity takes place in the chain. The acoustic vibrations modulate the distances between A atom and thus influence the frequency of proton vibrations on H bonds. The proton vibrational frequency can be presented as

$$\omega(q) \rightarrow \omega_q + \frac{\partial \omega(q)}{\partial U_q(r)} \; \delta U_q(r) \; , \tag{1}$$

1) prospekt Nauki 144, 252650 Kiev 28, USSR.

where $U_q(r)$ is the coordinate of the normal acoustic vibrations of the chain. The calculation of the second member in (1) results in an addendum of the type

$$H' = \sum_q (\gamma_q B_q^+ + \gamma_q^* B_q) b_q^+ b_q \ , \tag{2}$$

appearing in the Hamiltonian that describes the movement of an excessproton (or a proton hole) along an H-bonded chain; here γ_q is the anharmonicity constant, $B_q^+(B_q)$ are the acoustic phonon creation(annihilation)operators, $b_q^+(b_q)$ are the analogous optical phonon operators.

In this note we are going to estimate the influence of anharmonicity (2) on the proton conductivity, neglecting the phonon-phonon interaction considered in /1/. The proton current density is obtained from formula (20) in /1/ (in this formula we should formally believe that $\nu = 1$, $|\Delta^{(\nu)}(q)| \to |\Delta_q|$, $S_{T,1} \to S_T$, $\Omega_1(q) \to \omega_q \left[1 + (\gamma_q B_q^+ + \gamma_q^* B_q)\right]$), where the thermal averaging after Gibbs should be additionally performed:

$$\exp(-\beta \sum_q \hbar \Omega_q (B_q^+ B_q + \tfrac{1}{2}))/\mathrm{Tr}(\exp(-\beta \sum_q \hbar \Omega_q (B_q^+ B_q + \tfrac{1}{2}))) \ , \tag{3}$$

where $\beta \equiv 1/k_B T$, T is the temperature; $\hbar \Omega_q$ is the energy of the thermal vibrational quantum, i.e. the lattice acoustic phonon energy. As a result, we obtain

$$I = 2^{3/2} \pi^{1/2} \, e g n \, J^2 \hbar^{-2} \left\{ \frac{\exp(-\beta E_a)}{\left[\sum_q |\Delta_q|^2 \omega_q^2 \operatorname{cosech}(\beta \hbar \omega_q/2)\right]^{1/2}} \right\} \times$$
$$\times \sinh \frac{\beta e E g}{2} \exp\left\{ \sum_q |\gamma_q|^2 |\Delta_q|^2 \left(\frac{\beta \hbar \omega_q}{2}\right)^2 \operatorname{cosech}^4 \left(\frac{\beta \hbar \omega_q}{2}\right) \times \right.$$
$$\left. \times \left(1 - \operatorname{cosech} \frac{\beta \hbar \omega_q}{2}\right)^2 \coth \frac{\beta \hbar \Omega_q}{2} \right\} \tag{4}$$

Here g is an H-bonded chain constant, n is the charge carrier concentration in the chain, J is the overlap integral of the charge carrier wave functions between the nearest-neighbour sites, $|\Delta_q|$ is a function of the carrier coupling with optical phonons, E_a is the activation energy for carrier hopping (see /1/), E is the electric field strength. The expression (4) for I is correct for $E < 10^8$ V/m (see /1/) and $|\gamma_q| \ll 1$. For $\gamma_q = 0$ the current density (4) coincides with that in the small polaron theory /2/.

We assume that $|\gamma_q| = 0.1$, $|\Delta_q|^2 = 20$ to 30, $\omega_q = 2\pi \cdot 6\times10^{12}\ s^{-1}$., $T = (300$ to $400)$K. Then the last exponential factor in (4) can be estimated by

$$\exp\left\{\alpha \coth \frac{\beta\hbar\Omega}{2}\right\}, \tag{5}$$

$\alpha = 10^{-2}$ to 10^{-3}; Ω is the characteristic frequency of the acoustic phonons conditioning the anharmonicity. Since the heavy charge carriers (protons or proton holes) in an H-bonded chain should be described by strongly localized wave functions (see also /3/), then they should also interact with the shortest-wave vibrations. Let us take as the Debye value $\Omega_D = (1$ to $5)\times10^{12}\ s^{-1}$, then we can assume that $\Omega = (0.01$ to $0.1)\Omega_D$. Substituting this value into (5) we obtain a rough estimation: $e^{0.2}$ to e^1.

Thus, the adduced estimations show that under the indicated values of parameters the anharmonicity can appreciably influence the proton current in compounds with H-bonded chains.

References

/1/ V.V. KRASNOGOLOVETS and F.M. TOMCHUK, phys. stat. sol. (b) 123, 365 (1984).

/2/ YU.A. FIRSOV (Ed.), Polarony, Izd. Nauka, Moscow 1975.

/3/ C. KITTEL, Introduction to Solid State Physics, Izd. Nauka, Moscow 1978 (in Russian).

(Received July 18, 1985)

phys. stat. sol. (b) 131, K181 (1985)

Subject classification: 14.3.2; 22.2.3

Sektion Physik der Humboldt-Universität zu Berlin,
Bereich Tieftemperatur-Festkörperphysik[1])

Carrier Mobility Determination in Inversion Layers of Anodized InSb MOS Structures

By

W. BRAUNE, M. SCHNÜRER, N. KUBICKI, and R. HERRMANN

Dedicated to Prof. Dr. Dr. h.c. Dr. E.h. P. GÖRLICH
on the occasion of his 80th birthday

The measurement of the dynamic conductivity in the microwave region in the presence of a perpendicular magnetic field is a simple contactless method for carrier mobility determination in a two-dimensional system /1/. For common microwave frequencies ω and typical relaxation times τ at semiconductor-isolator interfaces $\omega\tau \ll 1$ results. Under this condition the conductivity can be described with classical quasi-static methods. For an isotropic thin conducting surface layer the conductivity is given by

$$\sigma_S(B) = \frac{N_S e^2 \tau}{m_\parallel^*} \frac{1}{1 + \omega_c^2 \tau^2} \tag{1}$$

with N_S surface carrier concentration, $m_\parallel^*$ effective mass of the surface carriers, ω_c cyclotron frequency.

Fig. 1 shows the microwave transmission arrangement for surface conductivity measurements in a conventional laboratory magnet with $B \leqq 2.5$ T and with microwave frequencies in the range 80 to 110 GHz. The high microwave frequencies allow the application of comparatively small gate areas of about 0.75 mm^2. The 20 nm thick Cr gate enables sufficient microwave transmission. The change of the transmission through the surface charge layer is only a few percent, therefore a nearly linear proportionality of variation between transmission and conductivity of the layer can be assumed /2/.

The measurements are carried out on anodized p-InSb MOS structures with bulk carrier concentrations $p = 1 \times 10^{13}$ cm^{-3} (T = 77 K). $(\bar{1}\bar{1}\bar{1})$ B_5 surfaces are anodized in the dark in a constant current regime at 0.3 mA cm^{-2} using

1) Invalidenstr. 110, DDR-1040 Berlin, GDR.

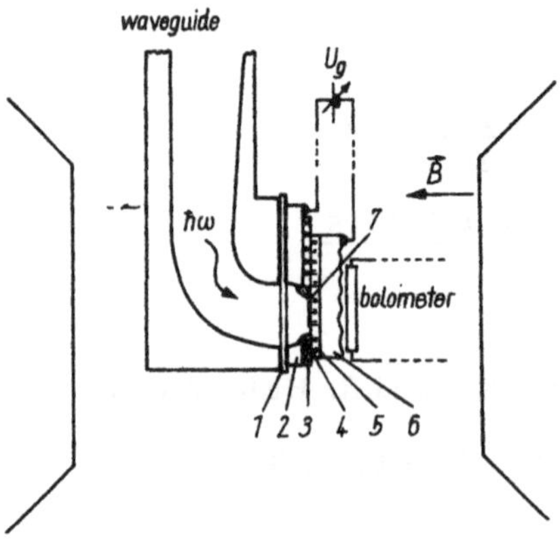

Fig. 1. Experimental arrangement for microwave conductivity measurements on inversion layers of MOS structures in a perpendicular magnetic field:(1) mylar foil, (2) copper diaphragm, (3) isolation, (4) conductive silver paste, (5) anodic oxide, (6) InSb, (7) Cr gate

0.1 n KOH. The thickness of the grown oxide films was 400 to 500 nm /3/.

Fig. 2 demonstrates the typical correlation between the C-U characteristic and the microwave transmission. At T = 1.4 K the inversion branch of the C-U characteristic can be measured /3/. Fig. 2 shows a distinct decrease of the microwave transmission (increase of the surface conductivity σ_S) at the beginning of inversion. This is also obvious in the corresponding hysteresis behaviour. Further, Fig. 2 shows the existence of a positive effective interface charge inverting the semiconductor surface already at the gate voltage U_g = = 0 V. The positive interface charge is expected for the used oxide thickness /4/.

In the region of inversion $\sigma_S(B)/\sigma_S(0)$ was measured at various gate voltages. By means of a normalized representation according to (1) the relaxation time respectively the mobility μ was found. The dependence of the inversion carrier density on the gate voltage $n_S(U_g)$ was determined from the C-U characteristic. Fig. 3 shows the obtained dependence of the inversion carrier mobility on the density. The appearance of a mobility maximum corresponds qualitatively with results found on SiO_2-InSb MIS structures by means of Shubnikov-de Haas measurements, but the maximum occurs there at higher n_S and μ values.

The absolute values of mobility should be realistic considering that the mean interface state density $\bar{N}_{SS} \approx$ $\approx 5\mathrm{x}10^{12}$ cm^{-2}. Further, the results

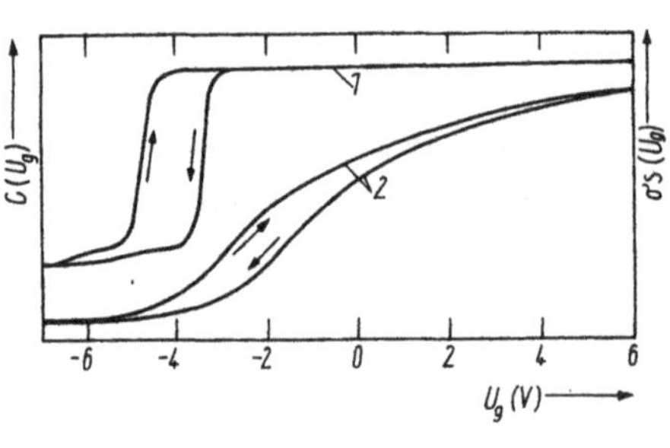

Fig. 2. Capacitance (1) and microwave conductivity (2) versus gate voltage for an anodized p-InSb MOS structure observed at T = 1.4 K

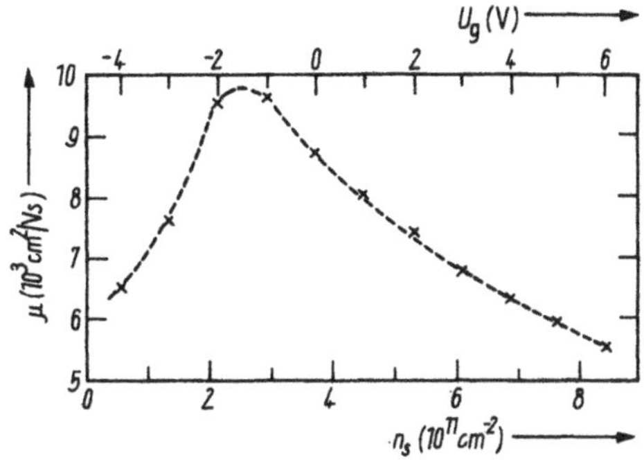

Fig. 3. Mobility of inversion electrons versus their density n_S respectively gate voltage U_g determined for an anodized p-InSb MOS structure

show that also in the interface system anodized native oxide-InSb high mobility values (analogous to SiO_2-InSb structures) can be expected, if the oxidation process wil be optimized and values of $\bar{N}_{SS} \lessapprox 10^{11}$ cm^{-2} will be achieved /4, 5/.

References

/1/ U. MACKENS and U. MERKT, Thin Solid Films 97, 53 (1982).

/2/ J. BINDER, K. GERMANOVA, A. HUBER, and F. KOCH, Phys. Rev. B 20, 2382 (1979).

/3/ W. BRAUNE, M. SCHNÜRER, N. KUBICKI, and R. HERRMANN, phys. stat. sol. (a) 86, 427 (1984).

/4/ O.V. ROMANOV, S.G. SASONOV, M.A. SOKOLOV, V.V. BALAKIREV, A.V. POPOV, and N.V. MOTALEVA, Poverkhnost 2, 95 (1983).

/5/ V.N. DAVYDOV and T.D. LEZINA, Mikroelektronika 12, 117 (1983).

(Received July 17, 1985)

Pre-Printed Titles

Papers to be published
in "physica status solidi (b) - basic research"
Vol. 132, No. 1

Review Article

S. A. SHABALOVSKAYA, Tomsk
Phase Transitions in the Intermetallic Compound TiNi with Charge-Density Wave Formation

Original Papers

A. V. MITIN and E. A. POPOV, Kazan
GMR Effect on Conversion Electron Polarization

R. R. ATABEKYAN, V. L. VINETSKII, V. A. GEVORKYAN, L. P. GODENKO, and R. Y. EZOYAN, Erevan and Kiev
Photoconductivity of Corundum Irradiated with High-Energy Electrons at Excitation of F and F^+ Centers

E. ERBARUT and M. TOMAK, Ankara
A Localized Orbital Description of Ge-Si Alloy

B. DUTTA, S.K. SARKAR, and D. ROY, Calcutta
Three-Body Interaction in H.C.P. Metals

I.R. ENTIN, Chernogolovka
On Suppression of X-Ray Anomalous Transmission by Acoustic Oscillations

H. A. KASSIM, Basrah
Energy Levels of Z_2^+-Centres in KCl:Ca^{2+}

F. M. GASHIMZADE and E.V. TAGIROV, Baku
Selection Rules for Matrix Elements of Electron-Phonon Scattering Processes in Crystals with TlSe Structure

TH. NATTERMANN, Berlin
Some Remarks on the Random Field Ising Model on a Lattice

MARIA MATOS and B. MAFFEO, Rio de Janeiro
Non-Equivalence Effects in Cluster Self-Consistent Calculations of Ionic Solids

F.F. CHARFI and M. ZOUAGHI, Tunis
Hot-Exciton Orientation in Indirect-Gap $GaAs_{1-x}P_x$ Alloys

B. ČABRIĆ, LJ. RISTOVSKI, and G. DAVIDOVIĆ, Kragujevac and Belgrade
Impurity Induced Energy Transfer in a Linear Polymer Chain

A.B. HENRIQUES and J.P. VON DER WEID, Rio de Janeiro
Decay Kinetics of Triplet Molecular Excitons in Alkali Cyanides

K. YOSHIDA, Uji, Kyoto
Displacement of the Fermi Level Due to Impurities in a Random Distribution

M. CZAPELSKI and M. SUSZYŃSKA, Wrocław
Effect of Plastic Deformation upon Optical Absorption of $KCl:Eu^{2+}$ Crystals

S.W. BIERNACKI, Warsaw
Parametrization of Crystal Field Spectra of Mn^{2+} in ZnSe and ZnS

V.P. SETH, A. YADAV, and R.S. BANSAL, Rohtak
Solution of the Spin-Hamiltonian with Orthorhombic HF and g-Tensors
(S = 1/2) (I)

A.K. HEAD, Clayton
The Sextic Polynomial of Crystal Elasticity

P. ENDERS, Berlin
Acoustic and Optical Deformation Potentials in Cubic IV-VI Compounds

L. VAN ROBBROECK, A. BOUWEN, and D. SCHOEMAKER, Wilrijk-Antwerp
Electron Spin Resonance Study of Ni^+ Centers in AgCl (Cu, Ni)

L.L. BUISHVILI, N.P. FOKINA, and L.T. CHELIDZE, Tbilisi
Cross-Relaxation of Nuclear Spins at Low Temperatures

S. BANDYOPADHYAY and S.K. DEB, Kalyani
Solution Enthalpies of Aliovalent Defects in Alkali Halides

M. KORKMAZ and B. AKTAŞ, Ankara
A Fast Computational Method to Obtain Spin-Hamiltonian Parameters of S and
I = 5/2 Ions

K. BÄRNER and E. GMELIN, Göttingen and Stuttgart
Coupling of a Structural and a High-Spin-Low-Spin Transition in $MnAs_{1-x}P_x$ Compounds

V.D. ISKRA, Kiev
On the Theory of Scattering by Short-Range Potentials

L.A. BUGAEV and R.V. VEDRINSKII, Rostov
Hartree-Fock Approximation in the Scattering Wave Method

M. INOUE, M. KOYANO, H. NEGISHI, Y. UEDA, and H. SATO, Hiroshima and Osaka-ju
Localized Impurity Level and Carrier Concentration in Self-Intercalated TiS_2 Crystals

N. DUTT, G.G. AGRAWAL, and J. SHANKER, Agra
Evaluation of the First- and Higher-Order Volume and Pressure Derivatives of Dielectric Constants of Alkali Halide Crystals

M.L. DALAL, S. SIVARAMAN, and Y.V.G.S. MURTI, Khandwa, Sagar, and Madras
Optical Absorption Studies on Colored KCl:Na:Pb Crystals

P.N. RAM, R.D. BIJALWAN, P.D. SEMALTY, and M. ISRAEL, Uttar-Pradesh and Roorkee
Resonant Scattering of Phonons in $ZnS:Fe^{2+}$ and $CdTe:Fe^{2+}$

S.N. LUKIN and O.P. TESLYA, Donetsk
The Structure of the Magnetic Centre in Cs_3CoCl_5

F. BOLDUAN, W. HÖNLE, H.D. HOCHHEIMER, and W. HENKEL, Stuttgart
Temperature and High Pressure Raman Study of M_3P_7 (M = Li, Na, K, Rb, Cs) Compounds

E. KOLLEY and W. KOLLEY, Leipzig
Dynamic and Static Aspects in the Theory of Itinerant Spin Glasses

E.A. POPOPV, M.M. KOTLYARSKII, and I.S. EDELMAN, Krasnoyarsk
Temperature and Field Dependence of Exciton-Magnon Absorption of $NaMnCl_3$

T. KATTERLE and W. LORENZ, Leipzig
Quantumchemical Study of Imperfect Semiconductor Surfaces Including Step and Kink Sites

V.B. KOKSHENEV, Kharkov

Spin-Phonon Excitations in Paramagnetic Crystals

F. CAMPABADAL and X. AYMERICH-HUMET, Bellaterra, Barcelona

On the Effects of Localized States and Tunnel Charge in Thin Insulating Films

K.T. STOYCHEV, Sofia

Non-Linear Optical Susceptibilities of Molecular Crystals in the Region of Vibronic Excitations (III)

M.S. ISLAM, Rajshabi

Chemical Shifts and EXAFS of the L_{III} Absorption Discontinuity of Dysprosium in Some of Its Compounds

A.I. MELKER, S.N. ROMANOV, and N.L. TARASENKO, Leningrad

Defect Clusters and Subcascades (I)

H.-J. MÜLLER, Rossendorf

Influence of Local-Environment on Bound Excitons and Bound Excitonic Molecules in III-V Alloys (I)

Papers to be published
in "physica status solidi (a) - applied research"
Vol. 91, No. 2

Review Articles

B. L. GELMONT, B. T. KOLOMIETS, and K. D. TSENDIN, Leningrad
Impurity Conduction of Chalcogenide Vitreous Semiconductors

E. P. WOHLFARTH, London
Magnetic Materials for Recording

Original Papers

T. C. LEE, S. SHEKHAR, and K. N. SUBRAMANIAN, East Lansing, Michigan
In-Situ Straining of Cu-10Ni-6Sn Spinodal Alloy in the High Voltage Electron Microscope (II)

HJ. BERNHARDT, Jena
Investigations of the Orange Luminescence of $PbMoO_4$ Crystals

D. COLAITIS, D. VAN DYCK, and S. AMELINCKX, Meudon-Bellevue and Antwerpen
Electron Microscopic Study of the Domains and Interfaces in MV_4S_8 (M = Fe, Co, Ni)

J. ČERMÁK, M. ČERŇANSKÝ, P. DUHAJ, and J. HORVÁTH, Praha and Bratislava
Crystallization of the Metallic Glass $Fe_{40}Ni_{40}B_{20}$ during Isochronal Annealing in Argon and Hydrogen

G. VAN TENDELOO, S. AMELINCKX, C. MANOLIKAS, and WEN SHULIN, Antwerpen, Mol, Thessaloniki, and Shangai
The Direct Observation of "Discommensurations" in Barium Sodium Niobate (BSN) and Its Homologues

PAN XIAO-QING, HU MEI-SHEN, YAO MING-HUI, and FENG DUAN, Nanjing
A TEM Study of the Incommensurate Phase and Related Phase Transitions in Barium Sodium Niobate

K. URBAN, N. MOSER, and H. KRONMÜLLER, Stuttgart
Phase Transitions between the Quasicrystalline, Crystalline, and Amorphous Phases in Al-14 at% Mn

L. URAY, Budapest
Investigation of Dilute Tungsten Alloys by Electrical Transport Measurements

I. GAAL and J.M. NEUGEBAUER, Budapest and Miskolc
Effect of Grain Boundaries on the Out-Diffusion Rate of Volatile Solutes from Thin Wires

G. GRAISS and G. SAAD, Cairo
Recovery of Internal Friction in Cold-Worked Aluminium Containing 0.35 wt% Fe

A. NOHARA and T. IMURA, Nagoya
Dislocation Behaviour and the Anomalous Yield-Stress Peak in a β-Brass Single Crystal

W.M. STOBBS, Cambridge
The Short Range Order Exhibited by Au-Fe Alloys

MINH HOANG, J.-C. NIEPCE et J.-C. MUTIN, Dijon
Mise en évidence du rôle chimique joué par l'oxyde de niobium Nb_2O_5 lors du frittage d'un mélange des perovskites $BaTiO_3$ et $CaZrO_3$

M. NOGUES, M. HAMEDOUN, J.L. DORMANN, and G. VILLERS, Meudon
Magnetic Domains in Antiferromagnetic $ZnCr_2X_4$ (X = S, Se) as Seen by Magnetization Measurements

H. LEMKE, Berlin
Dotierungseigenschaften von Kobalt in Silizium

S. ÅSBRINK, L. GERWARD, and J. STAUN OLSEN, Stockholm and Lyngby
Investigations of the Phase Transition in V_3O_5 Using Energy Dispersive X-Ray Diffraction and Synchrotron Radiation White Beam X-Ray Topography

D.K. FOWLER, D.C. CREAGH, R.W.N. KINNEAR, and G.V.H. WILSON, Duntroon, New South Wales
Two-Pulsed Stimulated Echo in Ferromagnetic Metals

J. PIQUERAS and E. KUBALEK, Duisburg
Cathodoluminescence from Electron-Irradiated ZnO

G. WENDROCK, H. LÖFFLER, O. KABISCH, and CHU THIEN TRUONG, Halle
On the Position of the Critical Temperature for the Dissolution of Guinier-Preston Zones in an Al-15 at% Zn Alloy

YE MING, E. KRÄTZIG, and R. ORLOWSKI, Osnabrück and Hamburg
Photorefractive Effects in $LiNbO_3$:Cr Induced by Two-Step Excitation

Y. NISHIO, Y. OOTUKA, K. KAJITA, T. IWATA, and W. SASAKI, Funabashi, Tokyo, and Ibaragi
Reverse Annealing Effect in the Neutron-Irradiated Si:P System

R. SCHÜLBE and U. SCHMIDT, Halle
On the Influence of the Pre-Ageing Conditions on the Reversion Behaviour of an Al-1.87 at% Cu Alloy

P. GAWORZEWSKI and E. HILD, Berlin and Budapest
The Effect of Pre-Heat-Treatments on the Formation Kinetics of Thermal Donors in Silicon

J. KULDA and P. MIKULA, Řež
The Three-Beam Case of Neutron Diffraction in a Distorted Crystal

L. JAHN and R. SCHUMANN, Dresden
The Mechanism of Thermal Remagnetization of Permanent Magnets

E.Y. GUTMANAS and Y. ESTRIN, Haifa and Hamburg-Harburg
Inertial Mechanism of Discontinuous Plastic Flow of Lead in the Superconducting State

A. WITZMANN, A. DITTMAR, K. GÄRTNER, and G. GÖTZ, Jena
Formation of Ni Silicides on As-Doped Silicon

S.A. DYOMIN, A.I. USTINOV, and K.V. CHUISTOV, Kiev
Structural Aspects of Reversible Plastic Deformation of Martensitic Mn-Cu Alloys

S.E. ABDULGAFAROV, V.L. AVERYANOV, B.T. KOLOMIETS, and V.M. LYUBIN, Leningrad
Photoconductivity of Vitreous As_2Se_3 Films Modified with Nickel

W. FRENTRUP, M. GRIEPENTROG, and U. MÜLLER-JAHREIS, Berlin
Negative Secondary Ion Emission Influenced by Alkali Atoms

K. MAASS, K. IRMSCHER, and H. KOSE, Berlin
On the Determination of Deep Level Concentration Profiles by DLTS Measurements

F. APPEL, H. BETHGE, and M. KROHN, Halle
Direct Evidence of Post-Nucleation Decoration

K. SCHEERSCHMIDT and R. HILLEBRAND, Halle
Calculated Defocus Diffraction Contrast and Lattice Fringe Imaging of Small Crystal Defects

E. GRILLI, M. GUZZI, E. CAMERLENGHI, and F. PIO, Milano
Photoluminescence of $Zn_xCd_{1-x}In_2S_4$ Layered Alloys ($0.7 < x < 1$)

R.W. CHANTRELL and E.P. WOHLFARTH, Preston and London
Rate Dependence of the Field-Cooled Magnetisation of a Fine Particle-System

S. KEMMLER-SACK, A. EHMANN, and D. WOLF, Tübingen
Sensitized IR-Emission of Ho^{3+}, Er^{3+}, and Tm^{3+} in 12R-$Ba_3LaBW_2O_{12}$ (B = Er, Tm, Yb) and 12R-$Sr_3LaYbW_2O_{12}$

L.N. ALEKSANDROV, A.N. KOGAN, and N.P. TIKHONOVA, Novosibirsk and Saransk
Study of the Peculiarities of Film Growth from Molecular Beams by the Method of Simulation

M. FÄHNLE and A. SEEGER, Stuttgart
On the Magnetic Contribution to the Free Enthalpy of Vacancy Formation in Ferromagnetic Crystals

D. BRAUNIG, K.H. YANG, T.Y. TAN, and C.P. SCHNEIDER, Hopewell Junction, New York, and Yorktown Heights, New York
In-Depth Generation Lifetime Profiling of Heat-Treated Czochralski Silicon

19. Magnetic Resonance

20. Optical Properties
    20.1 Spectra. Optical Constants (X-Ray Spectra See 20)
    20.2 Lasers
    20.3 Luminescence (X-Ray Spectra See 20)

21. Metals. Alloys
    21.1 Transition Metals (Group Sc to Zn) and Their Alloys
        21.1.1 Fe and Fe Alloys
    21.2 Alkali Metals
    21.3 Alkali-Earth Metals
    21.4 Rare-Earth Metals
    21.5 Actinides
    21.6 Noble Metals
    21.7 Semi-Metals (Semiconducting Alloys 22.7)

22. Semiconductors and Ionic Crystals
    22.1 Elements
        22.1.1 Germanium
        22.1.2 Silicon
        22.1.3 Group VI Elements
    22.2 III–V Compounds
        22.2.1 Arsenides
        22.2.2 Phosphides
        22.2.3 Antimonides
        22.2.4 Mixed Crystals
    22.3 IV–IV, II–V, and III–VI Compounds
        22.3.1 Mixed Crystals
    22.4 II–VI Compounds
        22.4.1 Sulphides
        22.4.2 Selenides
        22.4.3 Tellurides
        22.4.4 Mixed Crystals
    22.5 Halides
        22.5.1 Silver Halides
        22.5.2 Alkali Halides
        22.5.3 Alkali-Earth Halides
        22.5.4 Mixed Crystals
    22.6 Simple Oxides (CdO, ZnO See 22.4)
        22.6.1 Mixed Crystals
    22.7 Semiconducting Intermetallic Compounds
        22.7.1 Mixed Crystals
    22.8 Tertiary and Higher Compounds
        22.8.1 Salts of Oxy-Acids (Carbonates, Phosphates, Silicates Including Mica etc.)
        22.8.2 Spinels, Garnets, Ferrites
    22.9 Organic Semiconductors

23. Solidified Gases

Attention

The preceding classification scheme is closely akin to the scheme used by Physics Abstracts to classify the various fields of solid state physics. A paper will either be classified under one or more of the subsections or, if none of these are relevant, it will just appear under the general section heading (i. e. sections and subsections are equivalent in this sense).

The *subject classification* given at the head of each paper begins with the main subject matter of the paper, and this is followed, if necessary, by figures referring to subsidiary subject matter and the type of solids investigated.

The papers published in this issue are listed in the order of their subject classification at the end of the table of contents.

82918

Manuscripts and letters for physica status solidi (b) — basic research and physica status solidi (a) — applied research

should be sent to

DDR-1086 Berlin, Leipziger Straße 3–4, P. B. 1233

or

Internationaler Buchversand GmbH, 1000 Berlin 19, Eschenallee 21.

The object of "physica status solidi (b)" and "physica status solidi (a)" is to publish rapidly important and topical results concerning fundamental research in the field of solid state physics and aspects of application and experimental techniques of fundamental solid state research, respectively. The issues of "physica status solidi (b)" and "physica status solidi (a)" will be published on the 1st and the 16th day of each month, respectively. Either series will contain the following categories:

1. Review articles written by arrangement with the editors. They will be remunerated.
2. Original papers selected from those which have not been submitted for publication elsewhere, after being reviewed. (The latest date for acceptance of contributions is the 8th and the 23rd day of each month for "physica status solidi (b)" and "physica status solidi (a)", respectively, 50 days before the delivery of the journal.)
3. Short notes published in the off-set part of the journal. They are, if necessary after being reviewed, selected by editorial staff. (The latest date for acceptance of manuscripts for this part is the 12th and the 27th day of each month for "physica status solidi (b)" and "physica status solidi (a)", respectively, 16 days before the delivery of the journal.)
4. Pre-printed titles of papers to be published in the next issue of both series.

Editorial Information

Manuscripts should be submitted preferentially in English. Besides, papers in German, French, and Russian are admitted. The final decision on the incorporation of the papers in the series (b) or (a) is taken by the editorial staff after consultation of the referee. Authors are asked to observe the following rules:

1. Manuscripts have to be written as concisely as possible. At most, 20 and 4 pages (28 lines per page; abstracts, tables, references, figures, and captions of figures included, where 3 figures are counted as about one page) are admitted for original papers and short notes, respectively. All manuscripts should be type-written on one side only. Two copies are required. Badly arranged manuscripts are returned.
2. The author's name, initials, and institution should be given.
3. One abstract of about ten lines in English and one in German, French, or Russian are required for original papers.
4. Because of lack of space conclusions (if any) must not have the character of a summary (i. e. substantially a repetition of the contents of the abstract).
5. Formulae should not be written too small and not with pencil. Separate lines for formulae are desirable. Indistinct writing will lead to a return of the manuscript.
6. Drawings should be enclosed on separate sheets of paper, not exceeding the maximum and minimum sizes of 20×30 and 10×15 cm^2, respectively. Photographs should be submitted, on white paper, only for review articles and original papers. Duplicates are required.
7. Captions for the figures should be submitted on separate sheets of paper, and special instructions with regard to the size of illustrations should be given on these.
8. References should be numbered in order of appearance and should be given, on a separate sheet, in the form used in physica status solidi.
9. SI-units should be used.
10. Each paper should contain a proposal for the subject classification according to the classification scheme appearing on the second and third cover of each issue.

The authors will be sent galley proofs of review articles and original papers. Proof reading of short notes and pre-printed titles will be done by the editorial staff.

For each original paper and short note a total of 75 reprints is available free of charge. The number of reprints free of charge per review article is 25. Requests for additional reprints (against payment) should be given together with the submitted manuscripts. **The deadline for orders of additional reprints is the 15th and the 30th of each month** for "physica status solidi (b)" and "physica status solidi (a)", respectively, two weeks before the delivery of the journal.

Bei Fragen zur Produktsicherheit wenden Sie sich bitte an:
If you have any questions regarding product safety,
please contact:

Walter de Gruyter GmbH
Genthiner Straße 13
10785 Berlin
productsafety@degruyterbrill.com